Old and New Aspects in Spectral Geometry

Mathematics and Its Applications

Volume 534

Old and New Aspects in Spectral Geometry

by

Mircea Craioveanu

Mircea Puta

Facultatea de Matematică,
Universitatea de Vest din Timişoara,
Timişoara, Romania

and

Themistocles M. Rassias

Department of Mathematics,
National Technical University of Athens,
Athens, Greece

KLUWER ACADEMIC PUBLISHERS
DORDRECHT / BOSTON / LONDON

A C.I.P. Catalogue record for this book is available from the Library of Congress.

ISBN 1-4020-0052-9

Published by Kluwer Academic Publishers,
P.O. Box 17, 3300 AA Dordrecht, The Netherlands.

Sold and distributed in North, Central and South America
by Kluwer Academic Publishers,
101 Philip Drive, Norwell, MA 02061, U.S.A.

In all other countries, sold and distributed
by Kluwer Academic Publishers,
P.O. Box 322, 3300 AH Dordrecht, The Netherlands.

Printed on acid-free paper

Printed in the Netherlands

CONTENTS

Preface

It is known that to any Riemannian manifold (M,g), with or without boundary, one can associate certain fundamental objects. Among them are the Laplace-Beltrami operator and the Hodge-de Rham operators, which are natural [that is, they commute with the isometries of (M,g)], elliptic, self-adjoint second order differential operators acting on the space of real valued smooth functions on *M* and the spaces of smooth differential forms on *M*, respectively. If *M* is closed, the spectrum of each such operator is an infinite divergent sequence of real numbers, each eigenvalue being repeated according to its finite multiplicity. Spectral Geometry is concerned with the spectra of these operators, also the extent to which these spectra determine the geometry of (M,g) and the topology of *M*. This problem has been translated by several authors (most notably M. Kac), into the colloquial question "Can one hear the shape of a manifold?" because of its analogy with the wave equation. This terminology was inspired from earlier results of H. Weyl. It is known that the above spectra cannot completely determine either the geometry of (M,g) or the topology of *M*. For instance, there are examples of pairs of closed Riemannian manifolds with the same spectra corresponding to the Laplace-Beltrami operators, but which differ substantially in their geometry and which are even not homotopically equivalent. Thus, one of the main tasks in this subject is to find methods allowing to extract geometric and topological information from spectral data. On the other hand, the only way to identify specific geometric invariants, which are not spectrally determined, is through explicit constructions of isospectral closed Riemannian manifolds.

This text is designed to introduce spectral geometry to graduate students and interested mathematicians and physicists. It is readily accessible to anyone whose background includes introductory Riemannian geometry and functional analysis. However, the book is largely self-contained, the prerequisite in mathematics being contained in the first two chapters. Many significant applications are developed, making obvious the ways in which the spectrum of a closed Riemannian manifold (M,g) influences the geometry of (M,g) and the topology of *M*, as well as pointing out the fascinating interplay between analysis, topology and geometry. This work includes an extensive bibliography of books and papers of both current and historical interest. The references appear at the end of each chapter. At the end of the book a symbol index is also included.

After reviewing some basic facts on smooth differential manifolds, Chapter 1 provides some fundamental concepts and results of Riemannian geometry. In particular, we explain the notions of curvature, Singer-Thorpe irreducible decomposition of the algebraic curvature tensor space, and geodesic, with many examples. The second chapter is devoted to canonical differential operators (gradient, divergence, the Laplace-Beltrami operator and the Hodge-de Rham operators on forms) in the context of Riemannian manifolds. The third chapter covers spectral properties of the Laplace-Beltrami operator associated to a compact Riemannian manifold. Among them included are Courant's principle, the mini-max characterization of the eigenvalues, the continuous dependence of the eigenvalues with respect to the Riemannian metric in the C^∞-topology and some of their consequences such as Uhlenbeck's theorem on the genericity of the Riemannian metrics with the property that the eigenvalues of the corresponding Laplace-Beltrami operators have multiplicity one. We also treat Ebin's theorem on the genericity of the Riemannian metrics with finite isometry groups and estimation of various geometric features of a compact Riemannian manifold in terms of spectral data, usually the dimension, the volume and the eigenvalues of the Laplace-Beltrami operator as well as of geometric data such as the diameter and curvature bounds. Some basic examples concerning the explicit computation of spectra for some standard closed Riemannian manifolds are presented. In Chapter 4 the method of heat asymptotics is discussed and geometric information is derived (which may be inferred from the heat coefficients and thus from the spectrum on functions). For example, it is proved that the round sphere and the real projective space with the standard Riemannian metric are spectrally determined in all dimensions less than or equal to six. Examples are constructed of non-isometric isospectral flat tori in all dimensions greater than or equal to four. The classical technique of Gordon as well as other ones for constructing isospectral closed Riemannian manifolds are discussed in this chapter. Among them Sunada's technique and Pesce's approach to isospectrality are presented in detail. The example of Gordon and Webb of non-isometric convex domains in $\mathbb{R}^n$ $(n \geq 4)$ that are isospectral for both Dirichlet and Neumann boundary conditions is described. The subject matter of Chapter 3 and Chapter 4, §1 is extended in Chapter 5, in which we study spectral properties of the Hodge-de Rham operators (Laplacians on forms) associated to a compact Riemannian manifold. Thus, we present various characterizations of the eigenvalues and the continuity property of the eigenvalues with respect to the Riemannian metric in the C^∞-topology. The corresponding heat flow as well as the corresponding heat asymptotics are described in a detailed way. In particular, one derives Patodi's classical result showing that from the spectra of the Laplace-Beltrami operator acting on functions and of the Hodge-de Rham operators acting on 1-forms and 2-forms one can tell whether the closed oriented Riemannian manifold has constant scalar curvature (resp. is Einsteinian) in all dimensions greater than or equal to four. Using an argument of Chanillo and Trèves, one can show that the first positive eigenvalue of the Hodge-de Rham operator on the regular level sets of a real analytic function f on S^n $(n \geq 2)$ is bounded below by a constant times a power of the distance between the value of f and the set of critical values of f. Some aspects of the interplay between analysis, topology and geometry are also discussed in Chapter 6. The main tool in the subject is the heat flow associated to a Hodge-de Rham operator of a closed Riemannian manifold. Accordingly, the Hodge-de Rham decomposition theorem, vanishing theorems for the real cohomology of closed Riemannian manifolds, the Lefschetz fixed point theorem and the Chern-Gauss-Bonnet theorem are treated with complete proofs. Chapter 7 provides an introduction to the Witten-Helffer-Sjöstrand theory, which allows one to relate some of the spectral properties of the Hodge-de Rham operators associated to a Riemannian metric g on a closed smooth manifold M to the combinatorial Laplacians associated to a certain smooth triangulation defined by a pair (f, g), where $f : M \to \mathbb{R}$ is a Morse function. In particular, this can be used in order to obtain a short proof of the weak and strong Morse inequalities – a very useful result in topology. In writing this chapter we have greatly profited by Burghelea's lecture notes on Witten-Helffer-Sjöstrand theory delivered at Ohio State University as well as at the Third International Workshop of Differential Geometry and its Applications (Sibiu, September 18-23, 1997). Finally, the last chapter contains open problems as well as related comments. The Appendix presents spectral properties of square matrices with entries in $\mathbb{R}$ or $\mathbb{C}$. These include the existence and uniqueness of the Jordan canonical form, density theorems of the set of matrices, whose eigenvalues have certain properties, in the space of all square matrices, the existence of Hermitian (symmetric) matrices with prescribed eigenvalues as well as the theorem of Rayleigh-Ritz.

One should mention that some of the topics presented in this text-book are still in rapid progress and are in the front line of current research.

The writing of this book itself has been under way for at least five years, and has been the subject of extensive correspondence and discussions with several mathematicians, who provided us with valuable information, new ideas, and the necessary encouragement during the preparation of this manuscript. We are sincerely grateful to the following colleagues: D. Alekseevski, A. Bacopoulos, M. Berger, P. Bérard, G. Besson, M. Bordoni, D. Burghelea, J. Dodziuk, H. Donnelly, J. Eichhorn, Th. Friedrich, P.S. Gilkey, L. Gligor, St. Ianuş, R. Iordănescu, E. Macias-Virgós, St. Marchiafava, P. Michor, P.T. Nagy, D.I. Papuc, A. Savo, R. Schimming, D. Schüth, Z.I. Szabó, A. Török, Gr. Tsagas, L. Vanhecke, C. Vizman, S. Zelditch, and others. Special thanks go to C.S. Gordon for her many useful comments and remarks on different topics of the book. She and Z.I. Szabó have also provided us with their recent unpublished results for Section 3 in Chapter 4.

Some of the above topics have been presented in courses given by M. Craioveanu in 1995/1996 and 1997-2000 at the West University of Timişoara, the University of Craiova, the University of Dortmund and the University of Udine, as well as by Th.M. Rassias in the period 1995-1999 at the National Technical University of Athens.

We wish to thank several former students of ours who have read and commented on earlier versions of various chapters of the manuscript: S. Benczik, P. Birtea, O. Bodrogean, D. Boros, M. Buligă, L. David, V.-D. Lalescu, N. Laos, F. Sakellaridis, V. Slesar and R. Tudoran.

Work on the book has been much aided by visits of M. Craioveanu and M. Puta to the Erwin Schrödinger Institute of Mathematical Physics in Vienna, to the Mathematical Institutes of the Humboldt University of Berlin, to the University of Debrecen, the University of Dortmund, the Technical University of Lausanne, the Catholic University of Leuven, the Technical University of Munich, the University of Rome "La Sapienza" and to the University of Santiago de Compostela. Also, Th.M. Rassias expresses his thanks to the Departments of Mathematics of the following Universities: Harvard University, M.I.T., Oxford University, The Technion (Israel Institute of Technology) and the University of Rome "La Sapienza", which provided facilities during the preparation of part of the manuscript. We thank our colleagues in these Institutes for making these visits possible, productive, memorable and enjoyable.

The preparation of this book has been partially supported by travel and other expenses awarded to M. Craioveanu by the West University of Timisoara and the companies S.C. Argirom International SA, Bucharest, and Petrom S.A., Timişoara Branch. Th.M. Rassias also thanks the Department of Mathematics of the National Technical University of Athens, for granting him partial financial support during this project.

We would also like to express our thanks to X. Hadjiliadis for the time and energy devoted to the preparation of the camera ready manuscript. Last, but not least, we want to thank Kluwer Academic Publishers for their generous co-operation.

Timişoara/Athens
June 2001

Mircea Craioveanu
Mircea Puta
Themistocles M. Rassias

Chapter 1
INTRODUCTION TO RIEMANNIAN MANIFOLDS

1. Tensor Fields on Smooth Differential Manifolds

Let M (resp. N) be a connected, smooth $(=C^\infty)$ n-dimensional manifold without boundary. We denote by $C^\infty(M)$ the ring of smooth real valued functions on M and by $\mathfrak{X}(M)$ the Lie-algebra of all smooth vector fields on M. Recall that $X \in \mathfrak{X}(M)$ is a smooth map

$$X : M \to TM = \bigcup_{x \in M} T_x M$$

such that $X(x) = X_x \in T_x M$ (= the tangent space of M at x) for each $x \in M$. $T_x M$ may be characterized as the space of all derivations of the algebra of smooth real valued functions defined on neighborhoods of x. Note that a vector field X and a function $f \in C^\infty(M)$ give rise to a new function $X(f) \in C^\infty(M)$ defined by

$$X(f)(x) = X_x(f), \ x \in M.$$

Therefore a smooth vector field $X \in \mathfrak{X}(M)$ leads to an $\mathbb{R}$-linear map, denoted also by X, namely

$$X : f \in C^\infty(M) \mapsto X(f) \in C^\infty(M)$$

which is obviously a derivation, i.e.

$$X(f_1 f_2) = f_1 X(f_2) + f_2 X(f_1), \quad \text{for all } f_1, f_2 \in C^\infty(M).$$

Conversely, a mapping $X : C^\infty(M) \to C^\infty(M)$ which is $\mathbb{R}$-linear and a derivation defines an element of $\mathfrak{X}(M)$. Indeed, one defines $X_x \in T_x M, x \in M$, as follows. First note that if $f_1 \in C^\infty(M)$ has the property $f_1|_U = 0$, where U is a neighborhood of x, then $(X(f_1))(x) = 0$. In fact, choose a $f_2 \in C^\infty(M)$ such that $f_2(x) = 0$ and $f_2|_{M \setminus U} = 1$. For the existence of such a f_2 assume that M satisfies the second countability axiom, so that in particular M is paracompact and admits partitions of unity. Then one gets $f_1 = f_2 f_1$, and consequently

$$X(f_1) = f_2 X(f_1) + f_1 X(f_2),$$

which implies that $(X(f_1))(x) = 0$. Now for a smooth real valued function $\tilde{f}$ defined on a neighborhood of x, let us define

$$X_x(\tilde{f}) = (X(f))(x),$$

where $f \in C^\infty(M)$ is an extension of $\tilde{f}$. This definition does not depend on the choice of f and note that $X_x \in T_x M$. Since in a local coordinate system $(U, x^1, \ldots, x^n)$ one may write $X = \sum_{j=1}^{n} X(x^j) \partial / \partial x^j$, the mapping $x \mapsto X_x$ defines an element of $\mathfrak{X}(M)$.

For $X, Y \in \mathfrak{X}(M)$, the Lie bracket of X and Y is defined by

$$[X, Y] = X \circ Y - Y \circ X .$$

Then $[X, Y] \in \mathfrak{X}(M)$ and

$$[X + Y, Z] = [X, Z] + [Y, Z],$$
$$[X, Y] = -[Y, X]$$

and

$$[[X, Y], Z] + [[Y, Z], X] + [[Z, X], Y] = 0 \text{ (the Jacobi identity)}$$

for all $X, Y, Z \in \mathfrak{X}(M)$, that is $\mathfrak{X}(M)$ carries the structure of a real Lie algebra with respect to the Lie bracket. Note that

$$[fX, Y] = f[X, Y] - Y(f)X$$

for all $f \in C^\infty(M)$ and $X, Y \in \mathfrak{X}(M)$.

Any smooth diffeomorphism $\Phi : M \to N$ induces linear real vector space isomorphisms

$$\Phi^* : C^\infty(N) \to C^\infty(M), \; \Phi_* : \mathfrak{X}(M) \to \mathfrak{X}(N)$$

defined respectively by

$$\Phi^*(\tilde{f}) = \tilde{f} \circ \Phi, \; [\Phi_*(X)](\tilde{f}) = [X(\tilde{f} \circ \Phi)] \circ \Phi^{-1}$$

for all $\tilde{f} \in C^\infty(N)$ and $X \in \mathfrak{X}(M)$. Note that

$$[\Phi_*(fX)](\tilde{f}) = (f \circ \Phi^{-1}) \cdot [\Phi_*(X)](\tilde{f})$$

for all $f \in C^\infty(M), \tilde{f} \in C^\infty(N)$ and $X \in \mathfrak{X}(M)$, that is $\Phi_*(fX) = (\Phi^{-1})^*(f)\Phi_*(X)$.

Let $\mathfrak{X}_0(M) = C^\infty(M)$ and for each $k \in \mathbb{N}^*$, let $\mathfrak{X}_k(M) = \mathfrak{X}(M) \times \ldots \times \mathfrak{X}(M)$ (k factors). Let $r \in \mathbb{N}$ and $s \in \{0, 1\}$. A smooth tensor field of type (s,r) on M is a map from $\mathfrak{X}_r(M)$ to $\mathfrak{X}_s(M)$ which is r-linear with respect to the ring $C^\infty(M)$. In particular, each element from $C^\infty(M)$ [resp. $\mathfrak{X}(M)$] can be conceived as a smooth tensor field of type (0,0) [resp. (1,0)] on M. For each $f \in C^\infty(M)$, the map $df : \mathfrak{X}(M) \to C^\infty(M)$ given by

$$df(X) = X(f)$$

defines a smooth tensor field of type (0,1) on M called the **differential** of f.

Let $TM_x^0 = \mathbb{R}, TM_x^k = T_x M \times \cdots \times T_x M$ (k-factors) and $\mathcal{T}_r^s(M)$ be the space of all smooth tensor fields of type (s,r) on M. If $L \in \mathcal{T}_r^s(M)$, $x \in M$ and $X_i, \tilde{X}_i \in \mathfrak{X}(M)$ have the property

$$X_i(x) = \tilde{X}_i(x),$$

for each $i = 1,2,\ldots,r$, then

$$\left[L(X_1,\ldots,X_r)\right](x) = \left[L(\tilde{X}_1,\ldots,\tilde{X}_r)\right](x),$$

i.e. the value of $L(X_1,\ldots,X_r)$ at x depends only of the values of its arguments at x. Therefore, for each $x \in M$ we can define an r-linear map $L_x : TM_x^r \to TM_x^s$, such that

$$L_x(v_1,\ldots,v_r) = \left[L(X_1,\ldots,X_r)\right](x),$$

where $X_1,\ldots,X_r \in \mathfrak{X}(M)$ satisfy

$$X_i(x) = v_i,$$

for each $i = 1,2,\ldots,r$.

Conversely, if at each point $x \in M$ we can associate an r-linear map $L_x : TM_x^r \to TM_x^s$ such that for each $X_1,\ldots,X_r \in \mathfrak{X}(M)$, $L(X_1,\ldots,X_r) \in C^\infty(M)$ [or $\mathfrak{X}(M)$], where

$$\left[L(X_1,\ldots,X_r)\right](x) = L_x\left(X_1(x),\ldots,X_r(x)\right),$$

then $L \in \mathcal{T}_r^s(M)$. Also, let us note that for $L \in \mathcal{T}_1^1(M)$, one can define the map $Tr(L) \in C^\infty(M)$ given by

$$\left[Tr(L)\right](x) = Tr(L_x),$$

for each $x \in M$.

If $L \in \mathcal{T}_r^1(M)$, $r \geq 1$, then L defines canonically r smooth tensor fields on M of type $(0,r\text{-}1)$, denoted $c_1L,\ldots,c_rL$ and given by

$$c_jL : \mathfrak{X}_{r-1}(M) \to C^\infty(M),$$
$$\left[(c_jL)(X_1,\ldots,\hat{X}_j,\ldots,X_r)\right](x) = Tr\left(v \mapsto L_x\left(X_1(x),\ldots,X_{j-1}(x),v,X_{j+1}(x),\ldots,X_r(x)\right)\right),$$

where $x \in M$, $v \in T_xM$ and $X_1,\ldots,X_r \in \mathfrak{X}(M)$, for each $j = 1,2,\ldots,r$. The tensor field c_jL is called the **contraction of** L with respect to the index j.

A smooth tensor field ω on M of type $(0,p)$ with the property that the map $\omega : \mathfrak{X}_p(M) \to C^\infty(M)$ is skew-symmetric is called **a smooth differential p-form** on M. The $\mathbb{R}$-linear space of all smooth differential p-forms on M is denoted by $A^p(M)$. Note that $A^0(M)$ can be identified with $C^\infty(M)$ and $df \in A^1(M)$ for each $f \in C^\infty(M)$.

Let $A(M) = \sum_{p=0}^{n} A^p(M)$ be the exterior algebra of smooth differential forms on M. On $A(M)$ are defined the exterior differential operator d and, for each $X \in \mathfrak{X}(M)$, the interior multiplication operator $i(X)$.

The exterior differential d is a derivation of $A(M)$ of degree +1, which is defined on p-forms by

$$d\omega(X_0,\dots,X_p)=\sum_{j=0}^{p}(-1)^j X_j\left(\omega\left(X_0,\dots,\hat{X}_j,\dots,X_p\right)\right)+$$
$$+\sum_{j<k}(-1)^{j+k}\omega\left([X_j,X_k],X_0,\dots,\hat{X}_j,\dots,\hat{X}_k,\dots,X_p\right),$$

where $\omega\in A^p(M)$, $X_0,\dots,X_p\in\mathfrak{X}(M)$ and $\hat{X}_j$ means that X_j is deleted. The operator d is $\mathbb{R}$-linear and satisfies the properties:

(i) $d\circ d=0$;

(ii) $d(\omega_1\wedge\omega_2)=d\omega_1\wedge\omega_2+(-1)^p\omega_1\wedge d\omega_2$ for all $\omega_1\in A^p(M)$ and $\omega_2\in A^q(M)$.

The operator $i(X)$ is a derivation of $A(M)$ of degree −1, which is defined on smooth functions by

$$i(X)f=0,\ f\in C^\infty(M),$$

for 1-forms by

$$i(X)\omega=\omega(X),\quad \omega\in A^1(M),$$

and for p-forms $(p\geq 2)$, by

$$[i(X)\omega](Y_2,\dots,Y_p)=\omega(X,Y_2,\dots,Y_p)$$

for $Y_2,\dots,Y_p\in\mathfrak{X}(M)$.

This operator satisfies the following properties, which follow immediately from the definitions:

(i) $i(X)\circ i(X)=0$ for every $X\in\mathfrak{X}(M)$;

(ii) The mapping of $\mathfrak{X}(M)\times A^p(M)$ into $A^{p-1}(M),(X,\omega)\mapsto i(X)\omega$, is bilinear with respect to $C^\infty(M)$;

(iii) If $\omega_1\in A^p(M),\omega_2\in A^q(M)$, then
$i(X)(\omega_1\wedge\omega_2)=(i(X)\omega_1)\wedge\omega_2+(-1)^p\omega_1\wedge(i(X)\omega_2)$.

For $X\in\mathfrak{X}(M)$ the Lie derivative operator L_X in the direction X is a derivation on $A(M)$, which acts by X on functions, that is $L_X(f)=X(f)$, is defined on $\mathfrak{X}(M)$ by

$$L_X(Y)=[X,Y], \tag{1.1}$$

$Y\in\mathfrak{X}(M)$, and on $A^p(M)$, $p\geq 1$, by

$$(L_X\omega)(Y_1,\dots,Y_p)=X\left(\omega(Y_1,\dots,Y_p)\right)-\sum_{k=1}^{p}\omega\left(Y_1,\dots,[X,Y_k],\dots,Y_p\right), \tag{1.2}$$

$Y_1,\dots,Y_p\in\mathfrak{X}(M)$. So for each p, $0\leq p\leq n$, the Lie derivative gives an $\mathbb{R}$-linear map

$$L_X : A^p(M) \to A^p(M),$$

that satisfies the following properties on $A(M)$:

$$L_X = d \circ i(X) + i(X) \circ d, \tag{1.3}$$

$$L_X \circ d = d \circ L_X,\ L_X \circ i(X) = i(X) \circ L_X, \tag{1.4}$$

$$L_X(\omega_1 \wedge \omega_2) = L_X(\omega_1) \wedge \omega_2 + \omega_1 \wedge L_X(\omega_2), \tag{1.5}$$

for every $X \in \mathfrak{X}(M)$.

Any smooth diffeomorphism $\Phi : M \to N$ induces linear real vector space isomorphisms

$$\Phi^* : A^p(N) \to A^p(M),\ 0 \le p \le n.$$

For $p = 0$ this has been previously defined and for $p = 1,\ldots,n$, Φ^* is given by

$$\left[\Phi^*(\tilde{\omega})\right](X_1,\ldots,X_p) = \tilde{\omega}\left(\Phi_*(X_1),\ldots,\Phi_*(X_p)\right) \circ \Phi$$

for any $\tilde{\omega} \in A^p(N)$ and $X_1,\ldots,X_p \in \mathfrak{X}(M)$, that is $\Phi^*(\tilde{\omega}) = \Phi^* \circ \tilde{\omega} \circ (\Phi_* \times \cdots \times \Phi_*)$ (p-times). Note that

$$\Phi^*(\tilde{f}\tilde{\omega}) = \Phi^*(\tilde{f})\Phi^*(\tilde{\omega})$$

for all $\tilde{f} \in C^\infty(N)$ and $\tilde{\omega} \in A^p(N)$. More generally, Φ induces isomorphisms $\Phi^* : \mathcal{T}_r^0(N) \to \mathcal{T}_r^0(M)$ for each $r \in \mathbb{N}$, where Φ^* is formally defined as above.

2. RIEMANNIAN STRUCTURES. EXAMPLES

Let M be a connected, smooth $(= C^\infty)$ n-dimensional manifold without boundary.

Definition 2.1 A Riemannian structure on M is a smooth $(= C^\infty)$ tensor field g on M of type (0,2), i.e. $g : \mathfrak{X}(M) \times \mathfrak{X}(M) \to C^\infty(M)$ is a bilinear map with respect to $C^\infty(M)$ such that the following conditions are satisfied:

(i) $g(X,Y) = g(Y,X)$ for each $X,Y \in \mathfrak{X}(M)$, i.e. g is symmetric;

(ii) $\left[g(X,X)\right](x) > 0$ for each $x \in M$ and each $X \in \mathfrak{X}(M)$ with $X_x \neq 0$, i.e. for each $x \in M$ the real quadratic form on the tangent space T_xM induced by g is positive definite.

g is called the **metric tensor field** on M and the pair (M,g) is called a **Riemannian manifold**.

Note that if g is a metric tensor field on M and $f \in C^\infty(M)$ is strictly positive, i.e. $f > 0$, then fg is also a metric tensor field on M.

Let $\left(\dfrac{\partial}{\partial x^i}\right)_{1 \le i \le n}$ be the coordinate vector fields in a local chart (U,ϕ) around $x \in M$

and $u, v \in T_x M$ with

$$u = \sum_i u^i \left(\frac{\partial}{\partial x^i} \right)_x,$$

$$v = \sum_i v^i \left(\frac{\partial}{\partial x^i} \right)_x.$$

Then

$$g_x(u,v) = \sum_{i,j} g_{ij}(x) u^i v^j,$$

where

$$g_{ij}(x) = g_x\left(\left(\frac{\partial}{\partial x^i} \right)_x, \left(\frac{\partial}{\partial x^j} \right)_x \right),$$

$1 \leq i, j \leq n$. By the definition of the tensor product of two tensor fields, it follows that

$$g|_U = \sum_{i,j} g_{ij} dx^i \otimes dx^j,$$

or abusively

$$g = \sum_{i,j} g_{ij} dx^i dx^j.$$

Theorem 2.1 If *M* is paracompact, then there exists at least one metric tensor field defined on it.

Proof. We first built up metric tensor fields on the domains of the local charts of an atlas of *M* satisfying an additional property and then one globalizes with the help of a partition of unity. Indeed, let $\{(U_k, \phi_k) \mid k \in I\}$ be an atlas on *M* such that $\{U_k\}_{k \in I}$ is a locally finite cover of *M*. Let $\{\alpha_k\}_{k \in I}$ be a partition of unity subordinate to this cover. For a scalar product q on $\mathbb{R}^n$ and $k \in I$, let $q_k = \phi_k^*(q)$. Then q_k is obviously a Riemannian structure on U_k and

$$g = \sum_{k \in I} \alpha_k q_k$$

is a Riemannian structure on *M*.

Q.E.D.

Remark 2.1 If (U, ϕ) and (U', ϕ') are local charts on *M* and *g* is given by $[g_{kl}]$ [resp. $[g'_{ij}]$] with respect to the related coordinate systems (x^k) [resp. (x'^j)] then

$$g'_{ij} = \sum_{k,l} \frac{\partial x^k}{\partial x'^i} \frac{\partial x^l}{\partial x'^j} g_{kl},$$

or equivalently in matrix form:

$$[g'_{ij}] = J[g_{kl}]J^T,$$

where *J* is the Jacobian matrix of the transition function $\phi \circ \phi'^{-1}$ and J^T the transpose of *J*.

Example 2.1 The simplest example of a Riemannian manifold is $(\mathbb{R}^n, g_{can})$, g_{can} being the canonical metric tensor field on $\mathbb{R}^n$, i.e.

$$g_{can} : \mathfrak{X}(\mathbb{R}^n)\times\mathfrak{X}(\mathbb{R}^n) \to C^\infty(\mathbb{R}^n),$$
$$[g_{can}(X,Y)](x) = \langle X_x, Y_x\rangle,$$

where $X, Y \in \mathfrak{X}(\mathbb{R}^n)$, $x \in \mathbb{R}^n$ and $\langle\cdot,\cdot\rangle$ denotes the usual inner product on $\mathbb{R}^n$. With respect to this metric tensor field, the tangent vectors $\left(\frac{\partial}{\partial x^i}\right)_x$, $i = 1, 2, \ldots, n$, form an orthonormal basis for the tangent vector space $T_x\mathbb{R}^n \cong \mathbb{R}^n$ at each point $x \in \mathbb{R}^n$, i.e.

$$\begin{aligned} g_{ij}(x) &= \delta_{ij} \\ &= \begin{cases} 1 & \text{if } i = j \\ 0 & \text{if } i \neq j \end{cases} \end{aligned}$$

for each $x \in \mathbb{R}^n$.

Note that if we identify $T_x\mathbb{R}^n$ with $\mathbb{R}^n$, any smooth vector field $X \in \mathfrak{X}(\mathbb{R}^n)$ may be considered as a C^∞-mapping $X : \mathbb{R}^n \to \mathbb{R}^n$.

Example 2.2 Let $\mathbb{R}^2$ be equipped with the canonical Riemannian structure g_{can} (see Example 2.1). In polar coordinates this metric has the following form:

$$g_{can} = dr^2 + r^2 d\theta^2.$$

Indeed, if $S^1 = \left\{(x^1, x^2) \in \mathbb{R}^2 \,\middle|\, (x^1)^2 + (x^2)^2 = 1\right\}$, the polar coordinate system $P : [0, +\infty)\times S^1 \to \mathbb{R}^2$ is given by $(x^1, x^2) = P(r, (\cos\theta, \sin\theta)) = (r\cos\theta, r\sin\theta)$, so that we have successively

$$\begin{aligned} \left[g'_{can}\right] = J\begin{bmatrix} 1 & 0 \\ 0 & 1 \end{bmatrix} J^T &= \begin{bmatrix} \frac{\partial x^1}{\partial r} & \frac{\partial x^1}{\partial \theta} \\ \frac{\partial x^2}{\partial r} & \frac{\partial x^2}{\partial \theta} \end{bmatrix} \cdot \begin{bmatrix} 1 & 0 \\ 0 & 1 \end{bmatrix} \cdot \begin{bmatrix} \frac{\partial x^1}{\partial r} & \frac{\partial x^2}{\partial r} \\ \frac{\partial x^1}{\partial \theta} & \frac{\partial x^2}{\partial \theta} \end{bmatrix} \\ &= \begin{bmatrix} \cos\theta & \sin\theta \\ -r\sin\theta & r\cos\theta \end{bmatrix} \cdot \begin{bmatrix} \cos\theta & -r\sin\theta \\ \sin\theta & r\cos\theta \end{bmatrix} = \begin{bmatrix} 1 & 0 \\ 0 & r^2 \end{bmatrix}, \end{aligned}$$

as required.

Example 2.3 Let B^n be the open unit ball in $\mathbb{R}^n$ with the canonical smooth differential structure induced by that of $\mathbb{R}^n$, and consider the mapping $h : \mathfrak{X}(B^n)\times\mathfrak{X}(B^n) \to C^\infty(B^n)$ defined by

$$[h(X,Y)](x)=\frac{4\langle X_x,Y_x\rangle}{\left[1-\sum_{i=1}^{n}(x^i)^2\right]^2},$$

where $X,Y\in\mathfrak{X}(B^n)$, $x=(x^1,\ldots,x^n)\in B^n$ and $\langle\cdot,\cdot\rangle$ denotes the usual inner product on $\mathbb{R}^n$ (identified with T_xB^n). This gives a Riemannian structure on B^n called the **Poincaré metric** or **hyperbolic metric**. Note that with respect to this metric tensor field, one has

$$h_x\left(\left(\frac{\partial}{\partial x^i}\right)_x,\left(\frac{\partial}{\partial x^i}\right)_x\right)=\frac{4}{\left[1-\sum_{i=1}^{n}(x^i)^2\right]^2},\ i=1,2,\ldots,n,$$

and

$$h_x\left(\left(\frac{\partial}{\partial x^i}\right)_x,\left(\frac{\partial}{\partial x^j}\right)_x\right)=0,\ i\neq j$$

for each $x=(x^1,\ldots,x^n)\in B^n$.

Example 2.4 Let H^n be the open half-space $\{(x^1,\ldots,x^n)\in\mathbb{R}^n\mid x^n>0\}$ of $\mathbb{R}^n$ with the canonical smooth differential structure induced by that of $\mathbb{R}^n$ and consider the mapping $g:\mathfrak{X}(H^n)\times\mathfrak{X}(H^n)\to C^\infty(H^n)$ defined by:

$$[g(X,Y)](x)=\frac{\langle X_x,Y_x\rangle}{(x^n)^2},$$

$X,Y\in\mathfrak{X}(H^n)$, $x=(x^1,\ldots,x^n)\in H^n$, where $\langle\cdot,\cdot\rangle$ denotes the usual inner product on $\mathbb{R}^n$ (identified with T_xH^n). This defines a Riemannian structure called also the **Poincaré** or **hyperbolic metric tensor field** on H^n.

Example 2.5 If (M,g) is a Riemannian manifold, then each submanifold N of M has a canonical Riemannian structure $\bar{g}:\mathfrak{X}(N)\times\mathfrak{X}(N)\to C^\infty(N)$ given by

$$[\bar{g}(\bar{X},\bar{Y})](x)=g_x(\bar{X}_x,\bar{Y}_x),$$

$\bar{X},\bar{Y}\in\mathfrak{X}(N)$, $x\in N$, which will be also denoted with g. Thus, in view of this example, every smooth submanifold of $\mathbb{R}^n$ has a canonical Riemannian structure. In particular, the n-dimensional unit sphere S^n, considered as a smooth submanifold of $\mathbb{R}^{n+1}$, inherits a canonical Riemannian structure, denoted also by g_{can}.

Example 2.6 Let G be a finite dimensional Lie group and $\mathfrak{g}$ its Lie algebra. The bilinear form $K:\mathfrak{g}\times\mathfrak{g}\to\mathbb{R}$ given by

$$K(X,Y)=-trace(adX\circ adY),$$

$X,Y\in\mathfrak{g}$, where $ad:\mathfrak{g}\to End(\mathfrak{g})$ denotes the **adjoint representation** of $\mathfrak{g}$ defined by

$$(adX)(Z)=[X,Z],$$

$Z \in \mathfrak{g}$, is called the **Killing form** on $\mathfrak{g}$. This symmetric bilinear form is invariant under the adjoint representation $Ad : G \to Aut(\mathfrak{g})$ of G given by

$$Ad(g) = (dI_g)_e,$$

where $I_g : G \to G$ is defined by

$$I_g(x) = gxg^{-1},$$

$x \in G,\ e \in G$ being the identity element of G. This means that

$$K(X,Y) = K(Ad(g)(X), Ad(g)(Y)),$$

for all $g \in G$ and $X, Y \in \mathfrak{g}$. Indeed, first note that

$$ad(\alpha(X)) = \alpha \circ adX \circ \alpha^{-1}$$

for any Lie algebra automorphism $\alpha : \mathfrak{g} \to \mathfrak{g}$:

$$\begin{aligned} ad(\alpha(X))Y &= [\alpha(X), Y] \\ &= [\alpha(X), \alpha\alpha^{-1}(Y)] \\ &= \alpha[X, \alpha^{-1}(Y)] \\ &= \alpha(adX(\alpha^{-1}(Y))) \\ &= (\alpha \circ adX \circ \alpha^{-1})(Y), \end{aligned}$$

for any $Y \in \mathfrak{g}$. On the other hand, taking into account this equality, we have successively

$$\begin{aligned} K(Ad(g)X, Ad(g)Y) &= -trace(ad(Ad(g)X) \circ ad(Ad(g)Y)) \\ &= -trace\left(Ad(g) \circ adX \circ adY \circ (Ad(g))^{-1}\right) \\ &= -trace(adX \circ adY) \\ &= K(X,Y), \end{aligned}$$

for any $X, Y \in \mathfrak{g}$ and $g \in G$, as required.

If K is nondegenerate the Lie group G is called **semi-simple**. Moreover, if G is compact, K is positive definite. Indeed, let $(\cdot,\cdot)$ be any inner product on $\mathfrak{g}$. By integration we obtain an inner product $\langle \cdot,\cdot \rangle$ on $\mathfrak{g}$ invariant under Ad:

$$\langle X, Y \rangle = \int_G (Ad(g)X, Ad(g)Y)\, dg,$$

for each $X, Y \in \mathfrak{g}$. Consequently, $Ad(g)$ is an orthogonal transformation of $(\mathfrak{g}, \langle \cdot,\cdot \rangle)$ for each $g \in G$, i.e. $Ad(G) \subseteq O(\mathfrak{g})$ (= the orthogonal group of $\mathfrak{g}$). Since

$$Ad \circ \exp_G = \exp_{O(\mathfrak{g})} \circ ad$$

it follows that $ad(X)$ is skew-symmetric for all $X \in \mathfrak{g}$. Thus, if $[a_{ij}]$ is the matrix of $ad(X)$ relative to an orthonormal basis of $\mathfrak{g}$, one deduces that

$$\begin{aligned} K(X,Y) &= -trace(adX \circ adY) \\ &= -\sum_{i,j} a_{ij} a_{ji} \\ &= \sum_{i,j} a_{ij}^2 \\ &\geq 0. \end{aligned}$$

Since G is semi-simple, K is non-degenerate, and so K is positive definite. Hence if G is compact and semi-simple, K gives rise to a Riemannian metric on G by left translation.

Let M and N be smooth manifolds. Then the cartesian product $M \times N$ has a canonical manifold structure. For $(x,y) \in M \times N$, the tangent space $T_{(x,y)}(M \times N)$ is naturally isomorphic to the cartesian product of the vector spaces $T_x M$ and $T_y N$.

Definition 2.2 Let (M,g) and (N,h) be two Riemannian manifolds. Then the product metric tensor field on $M \times N$ is defined by:

$$[(g \times h)(X,Y)](x,y) = g_x(X_1(x), Y_1(x)) + h_y(X_2(y), Y_2(y)),$$

where $X(x,y) = (X_1(x), X_2(y))$, $Y(x,y) = (Y_1(x), Y_2(y))$.

Note that $(M \times N, g \times h)$ is a Riemannian manifold.

Example 2.7 Let us consider the Riemannian manifold (S^1, g_{can}). Then $T^2 = S^1 \times S^1$ has the product metric $g_{can} \times g_{can}$, which in polar coordinates θ, ϕ can be written as

$$g_{can} \times g_{can} = d\theta^2 + d\phi^2 .$$

Definition 2.3 Let (M,g) and (N,h) be two Riemannian manifolds. A map $\Phi : M \to N$ is an **isometry** if Φ is a diffeomorphism and if $g = \Phi^*(h)$, i.e.

$$h_{\Phi(x)}(d\Phi_x(u), d\Phi_x(v)) = g_x(u,v),$$

for each $x \in M$ and each $u, v \in T_x M$, where $d\Phi_x : T_x M \to T_{\Phi(x)} N$ is **the differential of Φ at $x \in M$** given by

$$d\Phi_x(X_x) = (\Phi_*(X))_{\Phi(x)}$$

for each $X \in \mathfrak{X}(M)$, that is

$$[(d\Phi_x)(u)](\tilde{f}) = u(\tilde{f} \circ \Phi), \qquad u \in T_x M,$$

for each smooth real valued function $\tilde{f}$ defined on a neighborhood of $\Phi(x) \in N$.

If Φ is as in Definition 2.3 one says that Φ is an isometry between the Riemannian manifolds (M,g) and (N,h) and one writes $\Phi : (M,g) \to (N,h)$.

Note that if $\Phi : (M,g) \to (N,h)$ is an isometry, then

$$h(\Phi_*(X),\Phi_*(Y))\circ\Phi = g(X,Y)$$

for all $X,Y\in\mathfrak{X}(M)$.

Example 2.8 Let M[resp. N] be a smooth n[resp. $n+m$] dimensional manifold and $\Phi: M\to N$ an immersion, i.e. Φ is smooth $(=C^\infty)$ and $d\Phi_x: T_xM\to T_{\Phi(x)}N$ is injective for each $x\in M$. If N has a Riemannian structure h, Φ induces a Riemannian structure g on M by defining:

$$g_x(u,v)=h_{\Phi(x)}(d\Phi_x(u),d\Phi_x(v)),$$

for each $u,v\in T_xM$. Since $d\Phi_x$ is injective (Φ is immersion) g_x is positive definite. The other conditions of Definition 2.1 are easily verified, i.e. in this way we obtain a mapping

$$g:\mathfrak{X}(M)\times\mathfrak{X}(M)\to C^\infty(M),$$

which defines canonically a Riemannian structure on M such that $d\Phi_x:(T_xM,g_x)\to(T_{\Phi(x)}N,h_{\Phi(x)})$ is an orthogonal mapping for each $x\in M$, hence $\Phi:(M,g)\to(N,h)$ is a **Riemannian immersion**. This metric tensor field is just the metric tensor field induced by Φ, i.e. $g=\Phi^*(h)$. In particular, this remark is also true if $\Phi: M\to N$ is a smooth local diffeomorphism, i.e. $d\Phi_x: T_xM\to T_{\Phi(x)}N$ is bijective for each $x\in M$. For instance, if $\Phi: M\to N$ is a smooth covering projection, that is Φ is a smooth map onto N provided each point $y\in N$ has a connected open neighborhood V such that Φ maps each component of $\Phi^{-1}(V)$ diffeomorphically onto V, then Φ is a local diffeomorphism, so that each Riemannian structure on N gives rise to a canonical Riemannian structure on M. If (N,h) is a Riemannian manifold, $\Phi: M\to N$ a covering projection and g the canonical Riemannian structure induced by Φ on M, one says that $\Phi:(M,g)\to(N,h)$ is a **Riemannian covering**. Conversely, if g is a Riemannian structure on M, then there does not exist in general a canonical technique in order to obtain a Riemannian structure on N. Sufficient conditions are given later in Theorem 2.3.

More generally, given a Riemannian manifold (M,g), we shall make obvious sufficient conditions in order that an equivalence relation R on M to be compatible with the differential structure of M as well as with the Riemannian structure g, that is the quotient manifold M/R should carry a natural Riemannian structure (see Theorem 2.2 and Corollary 2.1).

Remark 2.2 Let (M,h) be a Riemannian manifold and $\Phi:\tilde{M}\to M$ a smooth covering projection. Recall that a covering transformation of Φ is a smooth diffeomorphism $\Psi:\tilde{M}\to\tilde{M}$ such that $\Phi\circ\Psi=\Phi$. The set of all covering transformations of a smooth covering projection forms a group, with composition of mappings as the group operation. If g is the Riemannian structure on $\tilde{M}$ induced by Φ (see Example 2.8), then any covering transformation $\Psi:\tilde{M}\to\tilde{M}$ is an isometry of $(\tilde{M},g)$ onto itself. Indeed, for any $x\in\tilde{M}$ and $u,v\in T_x\tilde{M}$ we have:

$$\begin{aligned}g_x(u,v) &\overset{def}{=} h_{\Phi(x)}\left(d\Phi_x(u), d\Phi_x(v)\right)\\ &= h_{\Phi(x)}\left(d(\Phi\circ\Psi)_x(u), d(\Phi\circ\Psi)_x(v)\right)\\ &= h_{\Phi(\Psi(x))}\left(d\Phi_{\Psi(x)}\circ d\Psi_x(u), d\Phi_{\Psi(x)}\circ d\Psi_x(v)\right)\\ &\overset{def}{=} g_{\Psi(x)}\left(d\Psi_x(u), d\Psi_x(v)\right).\end{aligned}$$

An isometry preserves all the structures of a Riemannian manifold. Two manifolds are equivalent from the Riemannian geometry point of view if there exists an isometry between them. Such manifolds are said to be **isometric**. Note that two Riemannian manifolds as smooth manifolds can be the same, yet as Riemannian manifolds be distinct. [For instance this is the case of T^2 with the metrics $d\theta^2 + d\phi^2$, i.e. the product metric, and $d\theta^2 + (2+\cos\theta)^2 d\phi^2$].

Note that given (M,g), the set of all isometries $\Phi:(M,g)\to(M,g)$ forms a group $I(M,g)$ under composition called the **group of isometries** of (M,g).

A very general result on the group of isometries is the following classical theorem of Myers and Steenrod: the group $I(M,g)$ of isometries of the Riemannian manifold (M,g) is a Lie group with respect to the compact-open topology [that is, the topology for which the sets of the form $\{\Phi\in I(M,g) | \Phi(K)\subset U\}$ with K compact and U open in M form a subbase]. Moreover, if M is closed, then $I(M,g)$ is compact.

Definition 2.4 $I(M,g)$ is **transitive** if for each $x, y\in M$, there exists an $\Phi\in I(M,g)$ such that

$$\Phi(x) = y.$$

Example 2.9 Each translation, each rotation and each line reflection of $\mathbb{R}^2$ is an isometry of $(\mathbb{R}^2, g_{can})$. Moreover, the group generated by translations, rotations and line reflections of $\mathbb{R}^2$ is the group $I(\mathbb{R}^2, g_{can})$ (see Berger [1987]) and this group is transitive, since in fact the group of translations is transitive. Note that $I(\mathbb{R}^n, g_{can})$ is in fact the semi-direct product of $\mathbb{R}^n$ and $O(n)$, i.e. $\mathbb{R}^n\times O(n)$ with the group law given by

$$(x, A)(y, B) = (x + Ay, AB),$$

where $x, y\in\mathbb{R}^n$, $A, B\in O(n)$.

Example 2.10 The group of isometries $I(S^n, g_{can})$ is naturally isomorphic to the orthogonal group $O(n+1)$, the isomorphism being given by restriction to S^n (see for details Berger [1987]).

Example 2.11 Let B^2 be the open unit ball in $\mathbb{R}^2$ (identified with the complex plane) endowed with the hyperbolic metric tensor field h (see Example 2.3). From elementary complex variable theory

$$\mathcal{C} = \left\{ f : B^2 \to B^2 \;\middle|\; f(z) = e^{i\theta} \frac{z - x}{1 - \bar{x}z} \text{ for some } x \in B^2, 0 \le \theta < 2\pi \right\}$$

is the conformal group of transformations of B^2 onto itself, i.e. every $f \in \mathcal{C}$ is smooth, bijective and its differential $df_z : \mathbb{C} \to \mathbb{C}$ at every $z \in B^2$ is a similarity with respect to the canonical real Euclidean vector space structure on $\mathbb{C}$ (we shall identify $T_z B^2$ and $T_{f(z)} B^2$ both with $\mathbb{C}$, i.e. f preserves infinitesimal angles (see for details Berger [1987]). Note that any $f \in \mathcal{C}$ is an isometry of (B^2, h). Indeed, since df_z is a similarity, i.e. preserves ratios between lengths, $df_z = \lambda_z g_z$ for some real number $\lambda_z \neq 0$ and some isometry $g_z : \mathbb{C} \to \mathbb{C}$. But

$$\begin{aligned} df_z(1) &= f'(z) \\ &= e^{i\theta} \frac{1 - \bar{x}x}{(1 - \bar{x}z)^2} \end{aligned}$$

and

$$|\lambda_z| = |f'(z)|.$$

On the other hand,

$$\begin{aligned} |f'(z)|(1 - \bar{z}z) &= \frac{(1 - \bar{x}x)(1 - \bar{z}z)}{|1 - \bar{x}z|^2} \\ &= 1 - \frac{\bar{z} - \bar{x}}{1 - x\bar{z}} \cdot \frac{z - x}{1 - \bar{x}z} \\ &= 1 - \overline{f(z)} f(z), \end{aligned}$$

which implies that

$$\lambda_z^2 = \frac{\left[1 - \overline{f(z)} f(z)\right]^2}{(1 - \bar{z}z)^2}.$$

Consequently,

$$\begin{aligned} h_{f(z)}(df_z(u), df_z(v)) &= h_{f(z)}(\lambda_z g_z(u), \lambda_z g_z(v)) \\ &= 4\lambda_z^2 \frac{\langle g_z(u), g_z(v) \rangle}{\left[1 - \bar{f}(z) f(z)\right]^2} \\ &= \frac{4\langle u, v \rangle}{(1 - \bar{z}z)^2} \\ &= h_z(u, v), \end{aligned}$$

for all $u, v \in T_z B^2$, i.e. $df_z : (T_z B^2, h_z) \to (T_{f(z)} B^2, h_{f(z)})$ is an isometry for each $z \in B^2$, and therefore $f:(B^2,h)\to(B^2,h)$ is an isometry. Moreover $\mathcal{G}$ is transitive because for $x\in B^2$, $f_x:B^2\to B^2$, defined by $f_x(z)=\dfrac{z-x}{1-\overline{x}z}$, $z\in B^2$, maps x onto the origin, hence for each pair $x,y\in B^2$, $f_y^{-1}\circ f_x$ maps x onto y. One can show that $\mathcal{G}$ is precisely the group of all orientation – preserving isometries of (B^2,h).

Let us describe $I(B^n,h)$. Identify $\mathbb{R}^n$ with $\mathbb{R}^n\times\{0\}$ in $\mathbb{R}^{n+1}$. Let $e_{n+1}=(0,\ldots,0,1)\in\mathbb{R}^{n+1}$, $\pi:\mathbb{R}^n\to S^n\setminus\{e_{n+1}\}$ the stereographic projection with e_{n+1} as pole, ∞ a point not in $\mathbb{R}^{n+1}$ and define $\widehat{\mathbb{R}^n}=\mathbb{R}^n\cup\{\infty\}$. Now extend π to a bijection $\hat{\pi}:\widehat{\mathbb{R}^n}\to S^n$ by setting $\hat{\pi}(\infty)=e_{n+1}$ and define the metric $d:\widehat{\mathbb{R}^n}\times\widehat{\mathbb{R}^n}\to[0,+\infty)$ by the formula $d(x,y)=\|\hat{\pi}(x)-\hat{\pi}(y)\|$, where $\|\ \|$ denotes the usual norm on $\mathbb{R}^{n+1}$. The metric space $\left(\widehat{\mathbb{R}^n},d\right)$ is compact and is called the one-point compactification of $\mathbb{R}^n$. Let $a\in\mathbb{R}^n$ be a fixed unit vector, $t\in\mathbb{R}$ and consider the hyperplane $P(a,t)$ of $\mathbb{R}^n$ with unit normal vector a passing through the point $ta\,(t\in\mathbb{R})$: $P(a,t)=\{x\in\mathbb{R}^n\,|\,\langle a,x\rangle=t\}$. The subspace $\widehat{P(a,t)}=P(a,t)\cup\{\infty\}$ of $\widehat{\mathbb{R}^n}$ is homeomorphic to S^{n-1}. Let $\sigma:\mathbb{R}^n\to\mathbb{R}^n,\sigma(x)=x+2(t-\langle a,x\rangle)a, x\in\mathbb{R}^n$, be the reflection of $\mathbb{R}^n$ in $P(a,t)$ and let $\hat{\sigma}:\widehat{\mathbb{R}^n}\to\widehat{\mathbb{R}^n}$ be the extension of σ obtained by setting $\hat{\sigma}(\infty)=\infty$, and

$\rho:\mathbb{R}^n\setminus\{x_0\}\to\mathbb{R}^n$, $\rho(x)=x_0+\left(\dfrac{r}{\|x-x_0\|}\right)^2(x-x_0), x\in\mathbb{R}^n$, the reflection (or inversion with $x_0\in\mathbb{R}^n$ as pole) of $\mathbb{R}^n$ in the hypersphere $S(x_0,r)=\{x\in\mathbb{R}^n\,\|x-x_0\|=r\}$, $r\in(0,+\infty)$, and extend ρ to $\hat{\rho}:\widehat{\mathbb{R}^n}\to\widehat{\mathbb{R}^n}$ by setting $\hat{\rho}(x_0)=\infty$ and $\hat{\rho}(\infty)=x_0$. By definition, a sphere of $\widehat{\mathbb{R}^n}$ is either an usual sphere $S(x_0,r)$ of $\mathbb{R}^n$ or an extended hyperplane $\widehat{P(a,t)}$ (which is topologically an usual sphere). Recall that a **Möbius transformation** of $\widehat{\mathbb{R}^n}$ is a finite composition of reflections of $\widehat{\mathbb{R}^n}$ in spheres of $\widehat{\mathbb{R}^n}$. Since every isometry of $\mathbb{R}^n$ is a composition of at most $n+1$ reflections in hyperplanes, every isometry of $\mathbb{R}^n$ extends in a unique way to a Möbius transformation of $\widehat{\mathbb{R}^n}$. Thus one may regard the group $I(\mathbb{R}^n,g_{can})$ as a subgroup of the group $M\left(\widehat{\mathbb{R}^n}\right)$ of the Möbius transformation of $\widehat{\mathbb{R}^n}$. Let $k\in(0,+\infty)$ and $H_k:\mathbb{R}^n\to\mathbb{R}^n$, $H_k(x)=kx$, the vector homothety of ratio k. The natural extension $\widehat{H_k}:\widehat{\mathbb{R}^n}\to\widehat{\mathbb{R}^n}$ of H_k is a Möbius transformation, since H_k is the composite of the reflection in $S(0,1)$ followed by the reflection in $S(0,\sqrt{k})$. As every Euclidean similarity of $\mathbb{R}^n$ is the composite of an isometry followed by H_k for some k, every Euclidean similarity of $\mathbb{R}^n$ extends in a unique way to a Möbius transformation φ of $\widehat{\mathbb{R}^n}$ such that $\varphi(\infty)=\infty$. Conversely, if φ is a Möbius transformation of

$\widehat{\mathbb{R}^n}$ that fixes ∞, then ${}_{\mathbb{R}^n}|\varphi|_{\mathbb{R}^n}$ is a similarity of $\mathbb{R}^n$. Thus, one may regard the group of Euclidean similarities of $\mathbb{R}^n$ as a subgroup of $M\left(\widehat{\mathbb{R}^n}\right)$.

Under the identification of $\mathbb{R}^{n-1}$ with $\mathbb{R}^{n-1}\times\{0\}$ in $\mathbb{R}^n$, a point $x\in\mathbb{R}^{n-1}$ corresponds to the point $\tilde{x}=(x,0)$ of $\mathbb{R}^n$. Any Möbius transformation $\varphi:\widehat{\mathbb{R}^{n-1}}\to\widehat{\mathbb{R}^{n-1}}$ can be extended to a Möbius transformation $\tilde{\varphi}:\widehat{\mathbb{R}^n}\to\widehat{\mathbb{R}^n}$ as follows. If $\varphi:\widehat{\mathbb{R}^{n-1}}\to\widehat{\mathbb{R}^{n-1}}$ is the reflection in $\widehat{P(a,t)}$ [resp. $S(x_0,r)$], then $\tilde{\varphi}:\widehat{\mathbb{R}^n}\to\widehat{\mathbb{R}^n}$ is the reflection in $\widehat{P(\tilde{a},t)}$ [resp. $S(\tilde{x}_0,r)$]. In both cases $\tilde{\varphi}(x,0)=(\varphi(x),0)$ for all $x\in\mathbb{R}^{n-1}$, that is $\tilde{\varphi}$ extends φ. Moreover, $\tilde{\varphi}$ leaves $\widehat{\mathbb{R}^{n-1}}$ and $H^n=\left\{(x^1,\ldots,x^n)\in\mathbb{R}^n \,\middle|\, x^n>0\right\}$ invariant. If $\varphi:\widehat{\mathbb{R}^{n-1}}\to\widehat{\mathbb{R}^{n-1}}$ is an arbitrary Möbius transformation, then $\varphi=\sigma_1\circ\cdots\circ\sigma_m$, where $\sigma_j:\widehat{\mathbb{R}^{n-1}}\to\widehat{\mathbb{R}^{n-1}}$, $j=1,\ldots,m$, are reflections. Let $\tilde{\varphi}=\tilde{\sigma}_1\circ\cdots\circ\tilde{\sigma}_m$. By the previous remarks, $\tilde{\varphi}$ extends φ and leaves H^n invariant. One can show that $\tilde{\varphi}$ depends only on φ and not on the decomposition $\varphi=\sigma_1\circ\cdots\circ\sigma_m$. The map $\tilde{\varphi}$ is called **the Poincaré extension of** φ. One can prove that a Möbius transformation $\varphi:\widehat{\mathbb{R}^n}\to\widehat{\mathbb{R}^n}$ leaves H^n invariant if and only if φ is the Poincaré extension of a Möbius transformation of $\widehat{\mathbb{R}^{n-1}}$ (see Ratcliffe [1994], p. 121).

A **Möbius transformation of** S^n is a mapping $\varphi:S^n\to S^n$ such that $\hat{\pi}^{-1}\circ\varphi\circ\hat{\pi}:\widehat{\mathbb{R}^n}\to\widehat{\mathbb{R}^n}$ is a Möbius transformation, where $\hat{\pi}:\widehat{\mathbb{R}^n}\to S^n$ is the natural extension of the stereographic projection. The set $M(S^n)$ of all Möbius transformations of S^n forms a group under composition and the mapping $\psi\mapsto\hat{\pi}\circ\psi\circ\hat{\pi}^{-1}$ is an isomorphism from $M\left(\widehat{\mathbb{R}^n}\right)$ onto $M(S^n)$. Let $\varphi\in M(S^{n-1})$. **The Poincaré extension of** φ is the Möbius transformation $\tilde{\varphi}:\widehat{\mathbb{R}^n}\to\widehat{\mathbb{R}^n}$ defined by $\tilde{\varphi}=\eta\tilde{\psi}\eta^{-1}$, where $\tilde{\psi}:\widehat{\mathbb{R}^n}\to\widehat{\mathbb{R}^n}$ is the Poincaré extension of $\psi=\hat{\pi}^{-1}\circ\varphi\circ\hat{\pi}:\widehat{\mathbb{R}^{n-1}}\to\widehat{\mathbb{R}^{n-1}}$, $\hat{\pi}:\widehat{\mathbb{R}^{n-1}}\to S^{n-1}$ is the stereographic projection with $e_n=(0,\ldots,0,1)\in\mathbb{R}^n$ as pole, and $\eta:\widehat{\mathbb{R}^n}\to\widehat{\mathbb{R}^n}$ is defined by $\eta=\sigma\circ\rho$, where $\sigma:\widehat{\mathbb{R}^n}\to\widehat{\mathbb{R}^n}$ is the reflection in the sphere $S\left(e_n,\sqrt{2}\right)$, that is $\sigma(x)=e_n+\dfrac{2}{\|x-e_n\|^2}(x-e_n)$, $\sigma(e_n)=\infty$, $\sigma(\infty)=e_n$, and $\rho:\widehat{\mathbb{R}^n}\to\widehat{\mathbb{R}^n}$ is the reflection in $\widehat{\mathbb{R}^{n-1}}$. Since σ maps $\left\{(x^1,\ldots,x^n)\in\mathbb{R}^n \,\middle|\, x^n<0\right\}$ homeomorphically onto B^n, note that η maps H^n homeomorphically onto B^n. The Möbius transformation $\tilde{\varphi}:\widehat{\mathbb{R}^n}\to\widehat{\mathbb{R}^n}$ obviously extends φ and leaves the open ball $B^n=\left\{x\in\mathbb{R}^n \,\middle|\, \|x\|<1\right\}$ invariant; moreover, $\tilde{\varphi}$ is unique with this property: a Möbius transformation $\phi:\widehat{\mathbb{R}^n}\to\widehat{\mathbb{R}^n}$ leaves the open ball $B^n=\left\{x\in\mathbb{R}^n \,\middle|\, \|x\|<1\right\}$ invariant if and only if ϕ is the Poincaré extension of a Möbius transformation of S^{n-1}. By definition, **a Möbius transformation of** B^n is a Möbius transformation of $\widehat{\mathbb{R}^n}$ that leaves B^n invariant. Thus, the group $M(B^n)$ of the Möbius transformations of B^n, a subgroup of $M(\widehat{\mathbb{R}^n})$, is isomorphic to $M(S^{n-1})$. Every Möbius transformation of B^n restricts to an isometry of (B^n,h), and every isometry of (B^n,h) extends to a unique Möbius transformation of B^n (see Ratcliffe [1994] for details). In other words, $I(B^n,h)$ is isomorphic to $M(B^n)$ [and also to $M(S^{n-1})$].

Example 2.12 A **Möbius transformation of** $U^n = \{(x^1,\ldots,x^n)\in\mathbb{R}^n \mid x^n > 0\}$ is a Möbius transformation of $\widehat{\mathbb{R}^n}$ that leaves U^n invariant. As we have seen before, the group $M(U^n)$ of Möbius transformations of U^n is isomorphic to $M(\widehat{\mathbb{R}^{n-1}})$. Every Möbius transformation of U^n restricts to an isometry of (U^n,g), and every isometry of (U^n,g) extends to a unique Möbius transformation of U^n (see Ratcliffe [1994] for details). In other words, the group $I(U^n,g)$ is isomorphic to $M(U^n)$ [and also to $M(\widehat{\mathbb{R}^{n-1}})$]. In particular, every isometry of $(H^2=\{z=x+iy\in\mathbb{C}\mid y>0\},g)$ is of the form

$$z\mapsto\frac{az+b}{cz+d}\ \text{or}\ z\mapsto\frac{a(-\bar{z})+b}{c(-\bar{z})+b},$$

where $a,b,c,d\in\mathbb{R}$ and $ad-bc=1$. Consequently, the group of orientation preserving isometries of (H^2,g) is isomorphic to the projective special linear group $PSL(2,\mathbb{R})=\left\{\begin{bmatrix}a & b\\ c & d\end{bmatrix} \middle| a,b,c,d\in\mathbb{R}, ad-bc=1/\{\pm I_2\}\right\}$ (see Buser [1992] or Ratcliffe [1994] for details).

Example 2.13 The mapping

$$z\mapsto\frac{iz+1}{z+i}=\eta(z),$$

maps $H^2=\{z=x+iy\in\mathbb{C}\mid y>0\}$ biholomorphically onto the unit disc $B^2=\{u+iv\in\mathbb{C}\mid u^2+v^2<1\}$. Moreover, this is an isometry of (H^2,g) onto (B^2,h).

Let us mention that there exist also other techniques in order to obtain in a canonical way a new Riemannian structure from a given one. Firstly, recall that a smooth mapping $\Phi:M\to N$ is called a **submersion** if $d\Phi_x:T_xM\to T_{\Phi(x)}N$ is onto, for each $x\in M$.

Definition 2.5 If (M,g) and (N,h) are Riemannian manifolds, a **submersion** $\Phi:M\to N$ is called **Riemannian** if the restriction of $d\Phi_x$ to $[\ker d\Phi_x]^\perp$ (= the orthogonal subspace to $\ker d\Phi_x$ with respect to the inner product g_x) is an orthogonal isomorphism, i.e. preserves the inner products, of $[\ker d\Phi_x]^\perp$ onto $T_{\Phi(x)}N$ for each $x\in M$.

For example, if (M,g) and (N,h) are Riemannian manifolds, then with respect to the product metric tensor field $g\times h$ on $M\times N$ (see Definition 2.2), the canonical projections $p:M\times N\to M$ and $q:M\times N\to N$ are Riemannian submersions.

Theorem 2.2 Let $(\tilde{M}, g')$ be a Riemannian manifold, where $\tilde{M}$ is connected and without boundary. If G is a Lie group of isometries of $(\tilde{M}, g')$ which acts properly and freely on $\tilde{M}$, then there exists a unique canonical Riemannian structure g on the smooth orbit manifold $M = \tilde{M}/G$ such that the smooth submersion (= the canonical projection) $\pi : \tilde{M} \to M = \tilde{M}/G$ is Riemannian.

Proof. Recall that G acts freely on $\tilde{M}$ if it acts without fixed points (that is for $\tilde{x} \in \tilde{M}$ and $\varphi \in G \setminus \{1_{\tilde{M}}\}$, one has $\varphi(\tilde{x}) \neq \tilde{x}$), and that G acts properly on $\tilde{M}$ if, for any compact subset $K \subset \tilde{M}$, the set $\{\varphi \in G | \varphi(K) \cap K \neq \phi\}$ is relatively compact in G. Under the given assumptions concerning the action of G on $\tilde{M}$, it is standard that the orbit space $\tilde{M}/G = M$, which is connected, admits a natural smooth differential structure and $(\tilde{M}, G, M, \pi)$ is a smooth principal G-bundle, where $\pi : \tilde{M} \to M$ denotes the canonical projection. Thus π is a smooth submersion (see Bourbaki [1967] or Boothby [1975]). Let $x \in M$ and let us fix $\tilde{x} \in \pi^{-1}(x) \subset \tilde{M}$. Denote by $\tau_{\tilde{x}} : T_{\pi(\tilde{x})}M \to [\ker(d\pi_{\tilde{x}})]^{\perp}$ the inverse of the restriction of $d\pi_{\tilde{x}}$ to $[\ker(d\pi_{\tilde{x}})]^{\perp}$ and define the bilinear mapping $g_x : T_xM \times T_xM \to \mathbb{R}$ by:

$$g_x(u, v) = g'_{\tilde{x}}(\tau_{\tilde{x}}(u), \tau_{\tilde{x}}(v)),$$

for each $u, v \in T_xM$. This definition is independent on the choice of the point $\tilde{x}$ in the fiber $\pi^{-1}(x)$. Indeed, by definition, the action of G is transitive on each orbit, so that if $\tilde{x}_1$ is another point from $\pi^{-1}(x)$, then there exists $\Phi \in G$, i.e. an isometry $\Phi : (\tilde{M}, g') \to (\tilde{M}, g')$ (see Definition 2.3) such that $\Phi(\tilde{x}) = \tilde{x}_1$. But Φ is in the same time a covering transformation of $(\tilde{M}, \pi, M)$, i.e. $\pi \circ \Phi = \pi$, which implies that

$$d\pi_{\tilde{x}_1} \circ d\Phi_{\tilde{x}} = d\pi_{\tilde{x}}.$$

Since $\tilde{M}$ is connected, the above identity implies $(d\Phi_{\tilde{x}})[\ker d\pi_{\tilde{x}}] = \ker d\pi_{\tilde{x}_1}$, so that $d\Phi_{\tilde{x}}$ induces an orthogonal isomorphism of $[\ker d\pi_{\tilde{x}}]^{\perp}$ onto $[\ker d\pi_{\tilde{x}_1}]^{\perp}$. Consequently the same identity leads to $d\Phi_{\tilde{x}} \circ \tau_{\tilde{x}} = \tau_{\tilde{x}_1}$, where $\tau_{\tilde{x}_1}$ denotes the inverse of the restriction of $d\pi_{\tilde{x}_1}$ to $[\ker d\pi_{\tilde{x}_1}]^{\perp}$. Thus

$$\begin{aligned} g'_{\tilde{x}}(\tau_{\tilde{x}}(u), \tau_{\tilde{x}}(v)) &= g'_{\tilde{x}_1}(d\Phi_{\tilde{x}}(\tau_{\tilde{x}}(u)), d\Phi_{\tilde{x}}(\tau_{\tilde{x}}(v))) \\ &= g'_{\tilde{x}_1}(\tau_{\tilde{x}_1}(u), \tau_{\tilde{x}_1}(v)). \end{aligned}$$

Now let us check that g defines a Riemannian structure on M, that is if g is conceived as a section of the bundle of bilinear symmetric forms, then it is smooth (see also §1). Indeed, first note that the section $\tilde{x} \mapsto \tau_{\tilde{x}}$ of the bundle $\operatorname{End}(T\tilde{M})$ is smooth. On the other hand, the smooth principal G-bundle $(\tilde{M}, G, M, \pi)$ has local smooth sections, so that if U is an open subset of M over which such a smooth section s exists, by the above definition of g, $g|_U$ is given by

$$g_x = g'_{s(x)} \circ \left(\tau_{s(x)} \times \tau_{s(x)}\right),\ x \in U.$$

Q.E.D.

Since a compact group always acts properly, as a consequence we obtain the following result.

Corollary 2.1 If G is a compact Lie group of isometries of $(\tilde{M}, g')$ acting freely on $\tilde{M}$, then there exists a (unique) canonical Riemannian structure g on the orbit space $\tilde{M}/G = M$ such that $\pi : (\tilde{M}, g') \to (M, g)$ is a Riemannian submersion, where $\pi : \tilde{M} \to \tilde{M}/G = M$ denotes the canonical projection.

Example 2.14 Let $\mathbb{K}$ be the field $\mathbb{C}$ of complex numbers or the skew field $\mathbb{H}$ of quaternions and $d = \dim_{\mathbb{R}} \mathbb{K}$, i.e. $d = 2$ or 4. Regard the sphere $S^{d(n+1)-1}$ as the set of all $(z^1, \ldots, z^{n+1}) \in \mathbb{K}^{n+1}$ such that $|z^1|^2 + \cdots + |z^{n+1}|^2 = 1$ and identify $\mathbb{K}^{n+1}$ with $\mathbb{R}^{d(n+1)}$ via the $\mathbb{R}$-linear isomorphism

$$(z^1, \ldots, z^{n+1}) \mapsto (\text{real components of } z^1, \ldots, \text{real components of } z^{n+1}).$$

Regard the (d-1)-dimensional unit sphere $S^{d-1} \subset \mathbb{K}^* \subset \mathbb{R}^d$ as the group of numbers of $\mathbb{K}^*$ with absolute value 1. The action of S^{d-1} on $S^{d(n+1)-1}$ defined by multiplication, i.e.

$$z(z^1, \ldots, z^{n+1}) = (zz^1, \ldots, zz^{n+1}) \overset{def}{=} \phi_z(z^1, \ldots, z^{n+1}),\ z \in S^{d-1},$$

is isometric, i.e. $\phi_z : (S^{d(n+1)-1}, g_{can}) \to (S^{d(n+1)-1}, g_{can})$ is an isometry for each $z \in S^{d-1}$, and free, so that on the smooth orbit manifold $S^{d(n+1)-1}/S^{d-1}$ denoted by $P^n(\mathbb{K})$ and called the ***n*-dimensional projective space** over $\mathbb{K}$, there exists a canonical Riemannian structure g induced by g_{can}. $P^n(\mathbb{C})$ [resp. $P^n(\mathbb{H})$] is also called the ***n*-dimensional complex [resp. quaternionic] projective space**. In particular, the canonical projection $\pi : S^{d(n+1)-1} \to P^n(\mathbb{K})$ is a Riemannian submersion with respect to the Riemannian structures g_{can} and g and each fiber $\pi^{-1}(x)$, endowed with the Riemannian structure induced by g_{can}, is isometric to (S^{d-1}, g_{can}).

Next let us assume that Γ is a discrete group (that is, Γ is endowed with the discrete topology) acting properly on the n-dimensional smooth manifold $\tilde{M}$, that is $\{\gamma \in \Gamma | \gamma K \cap K \neq \varnothing\}$ is finite for any compact subset K of $\tilde{M}$. One can show that this condition holds if and only if for any $\tilde{x}, \tilde{y}$ in $\tilde{M}$ there exist open neighborhoods U and V of $\tilde{x}$ and $\tilde{y}$ respectively such that the set $\{\gamma \in \Gamma | \gamma U \cap V \neq \varnothing\}$ is finite. In particular, when $\tilde{x} = \tilde{y}$, by taking $W = U \cap V$ it follows that $\{\gamma \in \Gamma | \gamma W \cap W \neq \varnothing\}$ is finite. Conversely, if $(\tilde{M}, \tilde{g})$ is a Riemannian manifold and Γ is a discrete group of isometries of

$(\tilde{M},\tilde{g})$ such that for each $\tilde{x}\in\tilde{M}$ there exists an open set U in $\tilde{M}$ with $\tilde{x}\in U$ and $\{\gamma\in\Gamma|\gamma U\cap U\neq\varnothing\}$ finite, then Γ acts properly on $\tilde{M}$. Let us point out that if $(\tilde{M},\tilde{g})$ is a Riemannian manifold, where $\tilde{M}$ is connected and satisfies the second axiom of countability, then any discrete subgroup Γ of the isometry group $I(\tilde{M},\tilde{g})$ in the compact-open topology acts properly on $\tilde{M}$ (for details concerning the last two statements, see Satya Deo and Varadarajan [1997], p. 572 and p. 580).

If $\pi:(\tilde{M},g')\to(M,g)$ is a smooth Riemannian covering, the covering transformation group of $\pi:\tilde{M}\to M$ is a group of isometries of $(\tilde{M},g')$ (see Remark 2.2) acting properly and freely on $\tilde{M}$. Conversely, the following result is valid.

Theorem 2.3. Let Γ be a discrete group acting as isometries of the Riemannian manifold $(\tilde{M},g')$. If Γ acts properly and freely on $\tilde{M}$, then there exists a (unique) canonical Riemannian structure g on the orbit space $\tilde{M}/\Gamma=M$ such that the canonical Riemannian structure induced by g on $\tilde{M}$ via the canonical smooth covering projection $p:\tilde{M}\to\tilde{M}/\Gamma=M$ is just g', that is $p:(\tilde{M},g')\to(M,g)$ is a Riemannian covering. Moreover, if $\tilde{M}$ is connected, the covering transformation group of $p:\tilde{M}\to M$ is Γ. In particular, the same conclusion holds if Γ is a finite group of isometries of $(\tilde{M},g')$ acting freely on $\tilde{M}$.

Indeed, under these assumptions concerning the action of Γ on $\tilde{M}$, it is standard that there exists on the orbit space $\tilde{M}/\Gamma$ a unique smooth differential structure such that the canonical projection $p:\tilde{M}\to\tilde{M}/\Gamma=M$ is a smooth covering projection. In particular, $p:\tilde{M}\to M$ is a smooth local diffeomorphism. The same argument as in the proof of Theorem 2.2 yields the existence of g, which is the only possible.

Let (M,g) be a closed Riemannian manifold and Γ',Γ'' be two finite subgroups of $I(M,g)$ that act freely on M. If Γ' is conjugate to Γ'' in $I(M,g)$, then $(M/\Gamma',g')$ is isometric to $(M/\Gamma'',g'')$, where g' (resp. g'') is the Riemannian metric on M/Γ' (resp. M/Γ'') induced by g (see Theorem 2.3). Indeed, the conjugacy condition $\tilde{\Phi}\circ\Gamma'\circ\tilde{\Phi}^{-1}=\Gamma''$, where $\tilde{\Phi}\in I(M,g)$, implies that the mapping $\Phi:M/\Gamma'\to M/\Gamma''$ given by $\Phi(\pi'(x))=\pi''(\tilde{\Phi}(x))$, for any $x\in M$, is well-defined, where $\pi':M\to M/\Gamma'$ (resp. $\pi'':M\to M/\Gamma''$) denotes the canonical projection. Moreover, Φ is an isometry of $(M/\Gamma',g')$ onto $(M/\Gamma'',g'')$. This follows immediately from the definitions of g' and g'', and the fact that $\tilde{\Phi}$ is an isometry:

$$\begin{aligned} g''_{\Phi(\pi'(x))} \circ \left(d\Phi_{\pi'(x)} \times d\Phi_{\pi'(x)}\right) &= g''_{\pi''(\tilde{\Phi}(x))} \circ \left[\left(d\pi''_{\tilde{\Phi}(x)} \circ d\tilde{\Phi}_x \circ (d\pi'_x)^{-1}\right) \times \left(d\pi''_{\tilde{\Phi}(x)} \circ d\tilde{\Phi}_x \circ (d\pi'_x)^{-1}\right)\right] \\ &= g_{\tilde{\Phi}(x)} \circ \left[\left(d\tilde{\Phi}_x \circ (d\pi'_x)^{-1}\right) \times \left(d\tilde{\Phi}_x \circ (d\pi'_x)^{-1}\right)\right] \\ &= g_x \circ \left[(d\pi'_x)^{-1} \times (d\pi'_x)^{-1}\right] = g'_{\pi'(x)} \end{aligned}$$

for any $x \in M$.

Example 2.15 Let us consider the Riemannian manifold (S^n, g_{can}) and the group of isometries $G = \{1_{S^n}, \Phi\}$ where $\Phi : S^n \to S^n, \Phi(x) = -x$, is the reflection of S^n through the origin of $\mathbb{R}^{n+1}$. The orbit space S^n/G is the real projective space $P^n(\mathbb{R})$ and therefore via Theorem 2.3 there exists a canonical Riemannian structure on it. Note that even dimensional Riemannian manifolds of type $\left(S^n/\Gamma, g\right)$ with Γ a finite subgroup of $O(n+1)$ acting freely on S^n are only the spheres and the real projective spaces [g is the canonical Riemannian metric on S^n/Γ induced by g_{can} (see Theorem 2.3)]. Indeed, first notice that a finite subgroup Γ of $O(n+1)$ acts freely on S^{n+1} if and only if for any orthogonal operator $\varphi \in \Gamma \setminus \{1_{\mathbb{R}^{n+1}}\}$ of $\mathbb{R}^{n+1}$, 1 is not an eigenvalue of φ. Now assume Γ to be a finite subgroup of $O(2n+1), n \geq 1$, acting freely on S^{2n}. Then for any $\varphi \in \Gamma, \det(\varphi^2) = (\det(\varphi))^2 = 1$. Since $2n+1$ is odd, 1 is an eigenvalue of $\varphi^2 : \mathbb{R}^{2n+1} \to \mathbb{R}^{2n+1}$. Since Γ acts freely on $\mathbb{R}^{2n+1}, \varphi^2 = 1_{\mathbb{R}^{2n+1}}$, what implies that $\varphi = 1_{\mathbb{R}^{2n+1}}$ or $\varphi = -1_{\mathbb{R}^{2n+1}}$. Therefore, $\Gamma = \{1_{\mathbb{R}^{2n+1}}\}$ or $\Gamma = \{\pm 1_{\mathbb{R}^{2n+1}}\}$, that is $S^{2n}/\Gamma = S^{2n}$ or $P^{2n}(\mathbb{R})$. On the other hand, note that a finite subgroup Γ of $O(2n)$ acting freely on S^{2n-1} is in fact contained in the special orthogonal group $SO(2n)$. Indeed, let $\varphi \in \Gamma \setminus \{1_{\mathbb{R}^n}\}$. Then the characteristic roots of φ are $\lambda_1, \bar{\lambda}_1, \ldots, \lambda_k, \bar{\lambda}_k$ and -1 with multiplicity $2(n-k)$, where $\lambda_1, \ldots, \lambda_k$ are unimodular complex numbers $(\neq \pm 1)$, since Γ acts freely on S^{2n-1}. Thus one gets $\det(\varphi) = (-1)^{2(n-k)} \lambda_1 \cdot \bar{\lambda}_1 \cdots \lambda_k \cdot \bar{\lambda}_k = 1$.

If Γ', Γ'' are two finite subgroups of $O(n+1)$ that act freely on S^n such that $\left(S^n/\Gamma', g'\right)$ is isometric to $\left(S^n/\Gamma'', g''\right)$, then Γ' is conjugate to Γ'' in $O(n+1)$, where g' (resp. g'') is the Riemannian metric on S^n/Γ' (resp. S^n/Γ'') induced by g_{can} (see Theorem 2.3). Indeed, if $\Phi : \left(S^n/\Gamma', g'\right) \to \left(S^n/\Gamma'', g''\right)$ is an isometry, then there exists an isometry $\tilde{\Phi} : (S^n, g_{can}) \to (S^n, g_{can})$ such that $\Phi \circ \pi' = \pi'' \circ \tilde{\Phi}$, where $\pi' : S^n \to S^n/\Gamma'$, (resp. $\pi'' : S^n \to S^n/\Gamma''$) denotes the canonical projection. The isometry $\tilde{\Phi}$ may be considered as an element of $O(n+1)$ and yields an conjugation between Γ' and Γ'' : for any $\varphi' \in \Gamma'$ and $\varphi'' \in \Gamma''$,

$$\pi'' \circ \tilde{\Phi} \circ \varphi' = \Phi \circ \pi' \circ \varphi' = \Phi \circ \pi' = \pi'' \circ \tilde{\Phi} = \pi'' \circ \varphi'' \circ \tilde{\Phi},$$

that is $\tilde{\Phi}\circ\varphi' = \varphi''\circ\tilde{\Phi}$.

Example 2.16 Let $n \geq 1$ and regard the sphere S^{2n+1} as the set of all $\left(z^1,\ldots,z^{n+1}\right)\in\mathbb{C}^{n+1}$ such that $\left|z^1\right|^2+\cdots+\left|z^{n+1}\right|^2=1$. Let $p\geq 2$ and $q_1,\ldots,q_{n+1}$ be integers relatively prime to p. Let $G=\left\{z\in\mathbb{C}\,\middle|\, z^p=1\right\}$ be the cyclic group of order p of the complex p-th roots of unity. The action of G on S^{2n+1} defined by $z\cdot\left(z^1,\ldots,z^{n+1}\right)=\left(z^{q_1}z^1,\ldots,z^{q_{n+1}}z^{n+1}\right)$ is isometric with respect to g_{can}, and free, so that by Theorem 2.3 on the smooth orbit manifold S^{2n+1}/G denoted by $L_{2n+1}\left(p;q_1,\ldots,q_{n+1}\right)$ and called the **(2*n*+1)-dimensional lens space of type** $\left(p;q_1,\ldots,q_{n+1}\right)$, there exists a canonical Riemannian structure g induced by g_{can}. Note that $\left(L_{2n+1}\left(2;1,\ldots,1\right),g\right)$ is just $\left(P^{2n+1}\left(\mathbb{R}\right),g\right)$. On the other hand, let us note that if Γ is a finite cyclic subgroup of $O(2n+2), n\in\mathbb{N}$, acting freely on S^{2n+1}, then the quotient Riemannian manifold $\left(S^{2n+1}/\Gamma,g\right)$, where g is induced by g_{can} (see Theorem 2.3) is isometric to an $(2n+1)$-dimensional lens space. Indeed, assume that Γ is of order p and take a generator φ of Γ. Then φ is conjugate in $O(2n+2)$ to the element $\tilde{\varphi}\in O(2n+2)$ that can be identified with the orthogonal matrix

$$\begin{pmatrix} R\left(\dfrac{q_1}{p}\right) & & O \\ & \ddots & \\ O & & R\left(\dfrac{q_{n+1}}{p}\right)\end{pmatrix}$$

where $q_1,\ldots,q_{n+1}$ are integers relatively prime to p such that the characteristic roots of φ are $\exp\left(2\pi i\dfrac{q_j}{p}\right)$ and $\exp\left(2\pi i\dfrac{-q_j}{p}\right), 1\leq j\leq n+1$, and

$$R(\theta)=\begin{pmatrix}\cos 2\pi\theta & \sin 2\pi\theta \\ -\sin 2\pi\theta & \cos 2\pi\theta\end{pmatrix}$$

(see also the Appendix). Since Γ is conjugated in $O(2n+2)$ to the finite subgroup $\Gamma'=\left\{1_{R^{2n+1}},\tilde{\varphi},\tilde{\varphi}^2,\ldots,\tilde{\varphi}^{p-1}\right\}$, by an above remark the Riemannian manifold $\left(S^{n+1}/\Gamma', g_{can}/\Gamma\right)$ is isometric to the lens space $\left(L_{2n+1}\left(p;q_1,\ldots,q_{n+1}\right),g\right)$, where g_{can}/Γ is the Riemannian metric on S^{2n+1}/Γ induced by g_{can} (see Theorem 2.3).

The following statements are equivalent for the lens spaces $L=L\left(p;q_1,\ldots q_{n+1}\right)$ and $L'=L\left(p;q_1',\ldots,q_{n+1}'\right)$, where the integers $q_1,\ldots,q_{n+1}$ (resp. $q_1',\ldots,q_{n+1}'$) are relatively prime to p:

(i) L is homeomorphic to L';
(ii) L is diffeomorphic to L';

(iii) (L,g) is isometric to (L',g'), where g (resp. g') is the canonical Riemannian metric on L (resp. L');

(iv) There exist an integer k and $\varepsilon_1,\dots\varepsilon_{n+1}\in\{-1,1\}$ such that $(q_1,\dots,q_{n+1})$ is a permutation of $(\varepsilon_1 kq_1',\dots,\varepsilon_{n+1}kq_{n+1}')$.

Example 2.17. Let $(\mathbb{R}^n,g_{can}),n\geq 1$, be the Riemannian manifold considered in Example 2.1 and $(e_1,\dots,e_n)$ a basis of the real vector space $\mathbb{R}^n$. The additive group

$$\Gamma=\left\{\sum_{k=1}^{n}\lambda^k e_k \mid \lambda^k\in\mathbb{Z}, k=1,\dots,n\right\}=\mathbb{Z}e_1+\dots+\mathbb{Z}e_n,$$

where $\mathbb{Z}$ denotes the integers, which is a lattice in $\mathbb{R}^n$ of rank n, acts on $\mathbb{R}^n$ by translations, that is each $\gamma\in\Gamma$ gives rise to the translation $\tilde{\gamma}:\mathbb{R}^n\to\mathbb{R}^n,\tilde{\gamma}(x)=x+\gamma,x\in\mathbb{R}^n$. A lattice in $\mathbb{R}^n$ is a discrete subgroup of $(\mathbb{R}^n,+)$ and conversely, any discrete subgroup of $(\mathbb{R}^n,+)$, which is not contained in an $(n-1)$-dimensional linear subspace of $\mathbb{R}^n$, is a lattice in $\mathbb{R}^n$ (see Lekkerkerker [1969], p. 18). Note that any lattice Γ of $\mathbb{R}^n$ can be put in the form $\Gamma=A\mathbb{Z}^n$, where $A\in GL(n,\mathbb{R})$. This modality or writting is not unique: $A\mathbb{Z}^n=A'\mathbb{Z}^n$ if and only if there is $B\in GL(n,\mathbb{Z})$ such that $A'=AB$. Consequently, the set of all lattices of $\mathbb{R}^n$ can be identified to $GL(n,\mathbb{R})/GL(n,\mathbb{Z})$. Since Γ is a discrete group of isometries of $(\mathbb{R}^n,g_{can})$ and acts properly and freely on $\mathbb{R}^n$, by Theorem 2.3 there is a unique Riemannian metric on the orbit manifold $\mathbb{R}^n/\Gamma$, denoted by g_{can}/Γ, such that the natural projection $p:(\mathbb{R}^n,g_{can})\to(\mathbb{R}^n/\Gamma,g_{can}/\Gamma)$ is a Riemannian covering. Note that $\mathbb{R}^n/\Gamma$ is diffeomorphic to the n-dimensional torus $S^1\times\dots\times S^1$ (n times). The Riemannian manifold $(\mathbb{R}^n/\Gamma,g_{can}/\Gamma)$ is called **the n-dimensional flat torus defined by** Γ.

Note that $(\mathbb{R}^n/\Gamma,g_{can}/\Gamma)$ is isometric to $(\mathbb{R}^n/\Gamma',g_{can}/\Gamma')$ if and only if the lattices Γ and Γ' are congruent, that is there exists an orthogonal automorphism of $\mathbb{R}^n$ carrying Γ onto Γ'. In other words, the flat tori $(\mathbb{R}^n/\Gamma,g_{can}/\Gamma)$ and $(\mathbb{R}^n/\Gamma',g_{can}/\Gamma')$ are isometric if and only if there exists $C\in O(n,\mathbb{R})$ such that $\Gamma'=C\Gamma$. Consequently, the set of all isometry classes of flat tori can be identified with $O(n,\mathbb{R})\backslash\left(GL(n,\mathbb{R})/GL(n,\mathbb{Z})\right)$.

Example 2.18. Let $(\mathbb{R}^2,g_{can})$ be as in Example 2.1, $\Gamma_{a,b}=\mathbb{Z}(a,o)+\mathbb{Z}(o,b)$ the lattice of rank two in $\mathbb{R}^2$, where $a>0$, $b>0$ are real numbers and $\breve{\Gamma}_{a,b}$ the group generated by the translations of $\mathbb{R}^2$ induced by $\Gamma_{a,b}$ and the improper motion

$\varphi:\mathbb{R}^2\to\mathbb{R}^2, \varphi(x,y)=\left(x+\frac{a}{2},-y+b\right)$ [φ is just the product $\tau\circ\sigma=\sigma\circ\tau$, where τ is the translation by the vector $\left(\frac{a}{2},o\right)$ and σ the reflection through the line $y=\frac{b}{2}$]. Since $\breve{\Gamma}_{a,b}$ is discrete group of isometries of $(\mathbb{R}^2, g_{can})$ and acts properly and freely on $\mathbb{R}^2$, by Theorem 2.3 there is a unique Riemannian metric on the orbit manifold $\mathbb{R}^2/\breve{\Gamma}_{a,b} \overset{def}{=} K_{a,b}$, denoted by $g_{can}/\breve{\Gamma}_{a,b}$, such that the natural projection $p:(\mathbb{R}^2,g_{can})\to\left(K_{a,b}, g_{can}/\breve{\Gamma}_{a,b}\right)$ is a Riemannian covering. $K_{a,b}$ is called **the Klein bottle** corresponding to the group $\breve{\Gamma}_{a,b}$. Note that $\left(K_{a,b}, g_{can}/\breve{\Gamma}_{a,b}\right)$ is isometric to $\left(K_{a',b'}, g_{can}/\breve{\Gamma}_{a',b'}\right)$ if and only if $a=a', b=b'$ (see Berger, Gauduchon and Mazet [1971]).

Example 2.19. The real Heisenberg group H_n is the Lie subgroup of $GL(n+2,\mathbb{R})$ consisting of all matrices of the form

$$a(x,y,t)=\begin{bmatrix}1 & x & t\\ 0 & I_n & y^T\\ 0 & 0 & 1\end{bmatrix},$$

where x,y are row vectors in $\mathbb{R}^n$, y^T the transpose of y and I_n the $n\times n$ identity matrix. This Lie group is diffeomorphic to $\mathbb{R}^{2n+1}$. The product operation in H_n is given by

$$a(x,y,t)\cdot a(x',y',t')=a(x+x',y+y',t+t'+x\cdot y'),$$

where "." denotes the canonical inner (dot) product in $\mathbb{R}^n$. Its Lie algebra h_n is the Lie subalgebra of $gl(n+2,\mathbb{R})$ consisting of all matrices of the form

$$A(x,y,t)=\begin{bmatrix}0 & x & t\\ 0 & 0 & y^T\\ 0 & 0 & 0\end{bmatrix}.$$

The Lie bracket in h_n is given by

$$\left[A(x,y,t),A(x',y',t')\right]=A(0,0,x\cdot y'-x'\cdot y).$$

A **Riemannian Heisenberg manifold** is a pair $\left(H_n/\Gamma, g\right)$, where Γ is a cocompact discrete subgroup of H_n ["cocompact" means that the (left) quotient manifold H_n/Γ is compact] and g is a Riemannian metric on H_n/Γ such that the Riemannian metric induced by g on H_n, also denoted by g, is left H_n-invariant. The cocompact discrete subgroups of H_n can be classified. Indeed, for $r=(r_1,\ldots,r_n)\in(\mathbb{Z}^+)^n$ such that r_j divides $r_{j+1}, 1\le j\le n$, let $r\mathbb{Z}^n$ [respectively $\left(1/r\right)\mathbb{Z}^n$] denote the n-tuples $x=(x_1,\ldots,x_n)$ for which

$x_i \in r_i\mathbb{Z}$ [respectively $x_i \in \left(\frac{1}{r_i}\right)\mathbb{Z}$], $1 \le i \le n$. Then

$$\Gamma_r = \left\{a(x,y,t) \mid x \in r\mathbb{Z}^n, y \in \mathbb{Z}^n, t \in \mathbb{Z}\right\}$$

is a cocompact discrete subgroup of H_n. Moreover, these subgroups Γ_r classify the cocompact discrete subgroups of H_n up to automorphism. More precisely, if Γ is any cocompact discrete subgroup of H_n, then there exists a unique r for which some automorphism of H_n maps Γ to Γ_r. moreover, for r and s as before, Γ_r and Γ_s are isomorphic groups if and only if $r = s$ (for the proof, see Gordon and Wilson [1986]). On the other hand note that the map

$$\Gamma a \mapsto \varphi(\Gamma)\varphi(a)$$

is an isometry from $\left(H_n/\Gamma, g\right)$ to $\left(H_n/\varphi(\Gamma), (\varphi^*)^{-1}(g)\right)$ for any automorphism $\varphi : H_n \to H_n$. These facts allow to derive the following result: Given any Riemannian Heisenberg manifold $\left(H_n/\Gamma, g\right)$, there is a unique r as above and a left-invariant metric $\tilde{g}$ on H_n such that $\left(H_n/\Gamma, g\right)$ is isometric to $\left(H_n/\Gamma_r, \tilde{g}\right)$. Note that if $r \ne r'$, the smooth manifolds H_n/Γ_r and $H_n/\Gamma_{r'}$ have distinct fundamental groups.

Proposition 2.1. Let Γ be a cocompact discrete subgroup of H_n and let g and g' be left-invariant Riemannian metrics on H_n. $\left(H_n/\Gamma, g\right)$ is isometric to $\left(H_n/\Gamma, g'\right)$ if and only if there exists an automorphism $\varphi : H_n \to H_n$ such that $\varphi^*(g) = g'$ and $\varphi(\Gamma) = a\Gamma a^{-1}$ for some $a \in H_n$.

For the proof, see Gordon and Wilson [1988].

Let us recall the definition of properly discontinuous actions in differentiable setting.

Definition 2.6. A group Γ of diffeomorphisms of a smooth manifold M is **properly discontinuous** if each point $x \in M$ has a neighborhood U such that if $\varphi(U)$ meets U for $\varphi \in \Gamma$, then $\varphi = 1_M$.

By a previous remark, notice that if (M,g) is a Riemannian manifold and Γ acts properly discontinuous on M by means of isometries of (M,g), then Γ acts properly on M.

Note that in the case when $M = G/K$ is a symmetric space, a subgroup $\Gamma \subset G$ is discrete if and only if it acts properly discontinously on M.

The next facts follow immediately.

Proposition 2.2. Let M be a connected smooth manifold and Γ a group of diffeomorphisms acting properly discontinuously on M with compact quotient M/Γ. Then the following two statements are valid:

(i) There is a relatively compact open set C in M such that $\Gamma C = M$;

(ii) The group Γ is finitely generated. In particular, Γ is a countable set.

One can apply Proposition 2.3(i) to construct a Γ-invariant Riemannian metric on M.

Let (M,g) be an n-dimensional Riemannian manifold and $\omega:\mathfrak{X}(M)\to C^\infty(M)$ a linear map with respect to $C^\infty(M)$, i.e. $\omega\in A^1(M)$. Then there exists a unique vector field $\omega^\#\in\mathfrak{X}(M)$ such that

$$\omega(X)=g\left(X,\omega^\#\right),$$

for each $X\in\mathfrak{X}(M)$. Moreover the mapping $\#:A^1(M)\to\mathfrak{X}(M)$ is a $C^\infty(M)$-isomorphism. If $E_1,\dots,E_n$ is an orthonormal frame on M, then $E_1^b,\dots,E_n^b$ is its dual coframe, where $b:\mathfrak{X}(M)\to A^1(M)$ denotes the inverse of #.

In particular, if $f\in C^\infty(M)$, then the vector field $(df)^\#$ is called the **gradient** of f and it is denoted by grad(f) (see also §2 of Chapter 2). This vector field is characterized, therefore, by the relation

$$\begin{aligned} g(grad(f),Y) &= df(Y) \\ &= Y(f), \end{aligned}$$

for each $Y\in\mathfrak{X}(M)$ (see also §2 of Chapter 2).

If $L:\mathfrak{X}(M)\to\mathfrak{X}(M)$ is a smooth $(=C^\infty)$ tensor field of type (1,1), then the map $\hat{L}:\mathfrak{X}_2(M)\to C^\infty(M)$ given by

$$\hat{L}(X,Y)=g(L(X),Y),$$

for each $X,Y\in\mathfrak{X}(M)$ is a smooth tensor field of type (0,2) on M. Conversely, if $F:\mathfrak{X}_2(M)\to C^\infty(M)$ is a smooth tensor field of type (0,2) on M, then there exists a unique smooth tensor field $\breve{F}:\mathfrak{X}(M)\to\mathfrak{X}(M)$ of type (1,1) such that

$$F(X,Y)=g\left(\breve{F}(X),Y\right),$$

for each $X,Y\in\mathfrak{X}(M)$. Indeed, for each $X\in\mathfrak{X}(M)$ the map $\omega_X:\mathfrak{X}(M)\to C^\infty(M)$ given by

$$\omega_X(Y)=F(X,Y),$$

for each $Y\in\mathfrak{X}(M)$ is $C^\infty(M)$-linear and so that $\omega_X\in A^1(M)$. Then the mapping $\breve{F}:\mathfrak{X}(M)\to\mathfrak{X}(M)$ given by

$$\breve{F}(X)=\omega_X^\#,$$

for each $X\in\mathfrak{X}(M)$ is a smooth tensor field of type (1,1) on M and for each $X,Y\in\mathfrak{X}(M)$ we have:

$$F(X,Y)=\omega_X(Y)=g\left(Y,\omega_X^\#\right)=g\left(\breve{F}(X),Y\right).$$

Therefore the mapping $F\mapsto\breve{F}$ defines an isomorphism between the tensor fields of type (0,2) on M and the tensor fields of type (1,1) on M. In particular we shall define

$$\mathrm{trace}(F) = \mathrm{trace}(\breve{F}) \in C^\infty(M).$$

If $\{e_1,\ldots,e_n\}$ is an orthonormal basis of T_xM, then

$$\begin{aligned}(\mathrm{trace}(F))(x) &= (\mathrm{trace}(\breve{F}))(x)\\ &= \mathrm{trace}(\breve{F}_x)\\ &= \sum_i g_x(\breve{F}_x(e_i), e_i)\\ &= \sum_i F_x(e_i, e_i)\end{aligned}$$

Definition 2.7 Let (M,g) be a Riemannian manifold. The **length** of a smooth curve $c:[a,b]\to M$ is the positive real number

$$\lambda(c) = \int_a^b \left[g_{c(t)}(\dot{c}(t),\dot{c}(t))\right]^{\frac{1}{2}} dt .$$

Let $d: M\times M \to [0,\infty)$ be defined by:

$$d(x,y) = \inf\{\lambda(c) \mid c \text{ is a broken smooth curve from } x \text{ to } y\}.$$

Then d is a metric on M and the metric topology on M generated by d is the same as the manifold topology on M.

For example, in the case of $(\mathbb{R}^n, g_{can})$ one can prove that $d(x,y) = \|x-y\|$ for all $x,y\in\mathbb{R}^n$, i.e. d is just the standard Euclidean metric. For (S^n, g_{can}),

$$d(x,y) = \arccos\langle x,y\rangle \quad \text{for all } x,y\in S^n \subset \mathbb{R}^{n+1},$$

where $\langle\cdot,\cdot\rangle$ denotes the standard inner product on $\mathbb{R}^{n+1}$, i.e. d is the **intrinsic metric** on S^n. For (B^n,h) (see Example 2.3),

$$\cosh d(x,y) = 1 + \frac{2\|x-y\|^2}{(1-\|x\|^2)(1-\|y\|^2)}$$

for all $x,y\in B^n\subset\mathbb{R}^n$, where cosh is the hyperbolic cosine function and $\|\ \|:\mathbb{R}^n\to[0,+\infty)$ the usual norm. In the case of $(P^n(\mathbb{C}),g)$ (see Example 2.16), $d(x,y) = \bar{d}(\pi^{-1}(x),\pi^{-1}(y))$ for all $x,y\in P^n(\mathbb{C})$, where $\bar{d}$ denotes the intrinsic metric on S^{2n+1} and $\pi: S^{2n+1}\to P^n(\mathbb{C}) = S^{2n+1}/S^1$ the canonical projection (see Example 2.16), i.e. d is just the **Fubini-Study metric** on $P^n(\mathbb{C})$. Note that in the last case, the fibres $\pi^{-1}(x)$ and $\pi^{-1}(y)$ are isometric to (S^1, g_{can}). The distance d in the case (H^n, g) is given by the following formula:

$$\cosh(d(x,y)) = 1 + \frac{\|x-y\|^2}{2x^n y^n},$$

for all $x=(x^1,\dots,x^n), y=(y^1,\dots,y^n)\in H^n\subset\mathbb{R}^n$ (see Beardon [1983] and Ratcliffe [1994] for details).

One can prove that $\Phi:(M',g')\to(M,g)$ is an isometry if and only if $\Phi:(M',d')\to(M,d)$ is a distance preserving mapping with $\Phi(M')=M$.

The arc length along c is the function $s:[a,b]\to\mathbb{R}$ given by

$$s(u)=\int_a^u\left[g_{c(t)}(\dot c(t),\dot c(t))\right]^{\frac{1}{2}}dt.$$

Note that the real function s is of class C^1 but it is not necessarily smooth because $t\mapsto\left[g_{c(t)}(\dot c(t),\dot c(t))\right]^{\frac{1}{2}}$ is not necessarily differentiable when $\dot c(t)=0$. If we have $g_{c(t)}(\dot c(t),\dot c(t))\neq 0$ for all $t\in[a,b]$, then s is smooth and monotonically increasing.

The concepts of length and arc length extend to broken smooth curves since in this case the function $t\mapsto\left[g_{c(t)}(\dot c(t),\dot c(t))\right]^{\frac{1}{2}}$ is defined at all with the exception of a finite number of points.

Definition 2.8 A curve $c:[a,b]\to M$ is said to be **parametrized by arc length** if $g_{c(t)}(\dot c(t),\dot c(t))=1,$ for all $t\in[a,b]$.

In this case $s(u)=u-a$ for all $u\in[a,b]$.

Definition 2.9 The Riemannian manifold (*M*,*g*) is called **complete** if the metric space (*M*,*d*) is complete.

Example 2.20 $(\mathbb{R}^n,g_{can}),(S^n,g_{can})$ and (H^n,g) are complete Riemannian manifolds.

Definition 2.10 The **diameter** of the Riemannian manifold (M,g), denoted by *D*(*M*,*g*), is the diameter of the metric space (*M*,*d*).

Remark 2.3 It is easy to see that:

(i) If the length of a curve *c* in (*M*,*g*) is *L*, then its length in (*M*,*ag*) is $L\sqrt{a}$, where *a* is a strictly positive real number;

(ii) $D(M,ag)=\sqrt{a}D(M,g)$, where *a* is a strictly positive real number.

3. THE LEVI-CIVITA CONNECTION

Whereas the differentiation of functions on a manifold is naturally determined by the differentiable structure, the differentiation of vector fields, except for $\mathbb{R}^n$, is not naturally determined. In order to do this, one needs the choice of a connection, that is, a rule which associates to each vector fields $X,Y\in\mathfrak{X}(M)$, a vector field ∇_XY, satisfying the conditions of Definition 3.1 below.

Using an approach due to Koszul, we define intrinsically a canonical linear connection on a Riemannian manifold.

As we shall see, a Riemannian metric *g* on *M* does determine a unique connection on *M*, called the **Levi-Civita connection**.

Let M be a smooth n-dimensional manifold.

Definition 3.1 A linear connection on M is an $\mathbb{R}$-bilinear mapping

$$\nabla : \mathfrak{X}(M) \times \mathfrak{X}(M) \to \mathfrak{X}(M)$$

denoted by

$$(X,Y) \mapsto \nabla(X,Y) \overset{not}{=} \nabla_X Y$$

satisfying the following conditions:

(i) $\nabla_{fX} Y = f \nabla_X Y$;

(ii) $\nabla_X (fY) = X(f) \cdot Y + f \nabla_X Y$,

for any $X, Y \in \mathfrak{X}(M)$ and for any $f \in C^\infty(M)$.

Note that ∇ is $C^\infty(M)$-linear in the first argument, so that taking into account the considerations of §1 it follows that if $X, \tilde{X} \in \mathfrak{X}(M)$ are such that $X_x = \tilde{X}_x, x \in M$, then $(\nabla_X Y)(x) = (\nabla_{\tilde{X}} Y)(x)$, for each $Y \in \mathfrak{X}(M)$.

Thus, a linear connection ∇ on M induces for each $x \in M$ the mapping denoted also with ∇, namely

$$\nabla : T_xM \times \mathfrak{X}(M) \to T_xM,\ \nabla(u,Y) \overset{not}{=} \nabla_u Y \overset{def}{=} (\nabla_X Y)(x),$$

where $X \in \mathfrak{X}(M)$ is such that $X_x = u \in T_xM$, with the following properties

$$\nabla_{\alpha u + \beta v}(Y) = \alpha \nabla_u Y + \beta \nabla_v Y,$$
$$\nabla_u (Y_1 + Y_2) = \nabla_u Y_1 + \nabla_u Y_2,$$
$$\nabla_u (fY) = u(f) Y_x + f(x) \nabla_u Y,$$

for all $\alpha, \beta \in \mathbb{R}, u, v \in T_xM, Y, Y_1, Y_2 \in \mathfrak{X}(M)$ and $f \in C^\infty(M)$.

On the other hand, with respect to the second argument, ∇ behaves as a derivative, so that if $Y, \tilde{Y} \in \mathfrak{X}(M)$ are such that $Y|_U = \tilde{Y}|_U$, where U is an open neighborhood of a fixed point $x \in M$, then $(\nabla_X Y)(x) = (\nabla_X \tilde{Y})(x)$. For let $f \in C^\infty(M)$ be such that $\operatorname{supp}(f) \subset U$ and $f(x)=1$. Then $fY = f\tilde{Y}$ and consequently $\nabla_X (fY) = \nabla_X (f\tilde{Y})$, that is

$$f \nabla_X Y + (Xf) Y = f \nabla_X \tilde{Y} + (Xf) \tilde{Y},$$

what implies the desired equality. In particular, if D is an arbitrary non-empty open set of M, the linear connection ∇ induces a canonical mapping of $\mathfrak{X}(D) \times \mathfrak{X}(D)$ into $\mathfrak{X}(D)$, denoted also by ∇, such that $(\nabla_X Y)\big|_D = \nabla_{X|_D}(Y|_D)$ for any $X, Y \in \mathfrak{X}(M)$. Moreover this mapping is obviously $\mathbb{R}$-bilinear and satisfies the conditions of type (i) and (ii), i.e. ∇ is a linear connection on D. In particular, for $x \in M$ and $u \in T_xM$, note that $\nabla_u Y$ is uniquely determined by the restriction of Y to any open set D containing x.

Definition 3.2 Let M be a smooth n-dimensional manifold and ∇ a linear connection on M. Then ∇ can be extended in a natural way to the space $\mathcal{T}_r^0(M)$ of smooth tensor fields of type $(0,r)$ on M, namely

$$\nabla:(X,S)\in\mathcal{X}(M)\times\mathcal{T}_r^0(M)\mapsto\nabla(X,S)\in\mathcal{T}_r^0(M),$$

where

$$\nabla(X,S)(X_1,\dots,X_r)\overset{not}{=}(\nabla_X S)(X_1,\dots,X_r)$$
$$\overset{def}{=}X(S(X_1,\dots,X_r))-\sum_{i=1}^{r}S(X_1,\dots,X_{i-1},\nabla_X X_i,X_{i+1},\dots,X_r)$$

in the case when $r\geq 1$ and

$$\nabla(X,f)=\nabla_X f=X(f)$$

for each $f\in C^\infty(M)=\mathcal{T}_0^0(M)$.

Similarly, ∇ can be extended in a natural way to the space $\mathcal{T}_r^1(M)$, $r\geq 1$, namely

$$\nabla:\mathcal{X}(M)\times\mathcal{T}_r^1(M)\ni(X,L)\mapsto\nabla(X,L)\in\mathcal{T}_r^1(M),$$

where

$$\nabla(X,L)(Y_1,\dots,Y_r)\overset{not}{=}(\nabla_X L)(Y_1,\dots,Y_r)=\nabla_X(L(Y_1,\dots,Y_r))-\sum_{j=1}^{r}L(Y_1,\dots,\nabla_X Y_j,\dots,Y_r),$$

for all $X,Y_1,\dots,Y_r\in\mathcal{X}(M)$.

Remark 3.1 In particular, a direct computation shows that if $\omega\in A^p(M)$, then $\nabla(X,\omega)\overset{not}{=}\nabla_X\omega\in A^p(M)$ for each $X\in\mathcal{X}(M)$. Note that for each $\omega\in A^p(M)$, $\nabla(\cdot,\omega):\mathcal{X}(M)\to A^p(M)$, $X\mapsto\nabla(X,\omega)$, is $C^\infty(M)$-linear. Moreover, one can show that the linear mapping $\tilde{\nabla}:A^p(M)\to L(\mathcal{X}(M),A^p(M))$ [= the space of $C^\infty(M)$-linear mappings of $\mathcal{X}(M)$ into $A^p(M)$] $=A^1(M)\otimes_{C^\infty(M)}A^p(M)$, $\omega\mapsto\tilde{\nabla}(\omega)\overset{def}{=}\nabla(\cdot,\omega)$, is a first order linear differential operator [see for details R.S. Palais (editor) [1965]].

Note that we have used the same symbol ∇ for the initial linear connection and its natural extension.

Definition 3.3. Given $f\in C^\infty(M)$, its **covariant differential** $\overset{\circ}{\nabla} f$ is just its usual differential df, that is

$$\overset{\circ}{\nabla} f=df\in\mathcal{T}_1^0(M).$$

More generally, given $S\in\mathcal{T}_r^0(M)$, $r\geq 1$, one defines its **covariant differential** $\overset{\circ}{\nabla} S$ to be the tensor field of type $(0,r+1)$ given by

$$\left(\overset{\circ}{\nabla} S\right)(X_0, X_1, \ldots, X_r) \overset{def}{=} \left(\nabla_{X_0} S\right)(X_1, \ldots, X_r)$$

$$= X_0\left(S(X_1, \ldots, X_r)\right) - \sum_{j=1}^{r} S\left(X_1, \ldots, X_{j-1}, \nabla_{X_0} X_j, X_{j+1}, \ldots, X_r\right),$$

where $X_0, X_1, \ldots, X_r \in \mathfrak{X}(M)$ (see also Definition 3.2). In particular, one can define inductively the ***k*-th order covariant differential** $\overset{\circ}{\nabla}{}^k f \in \mathcal{T}_k^0(M)$ of $f \in C^\infty(M)$ by

$$\overset{\circ}{\nabla}{}^k f = \overset{\circ}{\nabla}\left(\overset{\circ}{\nabla}{}^{k-1} f\right), \qquad k \geq 2,$$

where $\overset{\circ}{\nabla}{}^1 = \overset{\circ}{\nabla}$.

Similarly, given $L \in \mathcal{T}_r^1(M)$, $r \geq 0$, one defines its covariant differential $\overset{\circ}{\nabla} L$ to be the smooth tensor field of type $(1, r+1)$ on M given by

$$\left(\overset{\circ}{\nabla} L\right)(X_0, X_1, \ldots, X_r) \overset{def}{=} \left(\nabla_{X_0} L\right)(X_1, \ldots, X_r)$$

$$= \nabla_{X_0}\left(L(X_1, \ldots, X_r)\right) - \sum_{j=1}^{r} L\left(X_1, \ldots, \nabla_{X_0} X_j, \ldots, X_r\right),$$

for any $X_0, X_1, \ldots, X_r \in \mathfrak{X}(M)$. In particular, one can define the covariant differential $\overset{\circ}{\nabla} X \in \mathcal{T}_1^1(M)$ of $X \in \mathfrak{X}(M)$ by

$$\left(\overset{\circ}{\nabla} X\right)(Y) = \nabla_Y(X),$$

for any $Y \in \mathfrak{X}(M)$.

Theorem 3.1 (**Levi-Civita**). On any Riemannian manifold (M,g) there exists a unique symmetric linear connection consistent with the metric tensor field g, i.e. such that the following equalities hold:

(iii) $\nabla_X Y - \nabla_Y X = [X, Y]$ (= the Lie bracket of X and Y) for each $X, Y \in \mathfrak{X}(M)$;

(iv) $X\left(g(Y,Z)\right) = g(\nabla_X Y, Z) + g(Y, \nabla_X Z)$, for each $X, Y, Z \in \mathfrak{X}(M)$.

Proof. First assume that there exists a linear connection ∇ on M satisfying (iii) and (iv). Then by circular permutations we have successively:

$$X\left(g(Y,Z)\right) = g(\nabla_X Y, Z) + g(Y, \nabla_X Z),$$
$$Y\left(g(Z,X)\right) = g(\nabla_Y Z, X) + g(Z, \nabla_Y X),$$
$$Z\left(g(X,Y)\right) = g(\nabla_Z X, Y) + g(X, \nabla_Z Y).$$

Adding the first two equalities, substracting the last one and using (iii) we obtain:

$$X(g(Y,Z))+Y(g(Z,X))-Z(g(X,Y))$$
$$=g(\nabla_X Y,Z)+g(Y,\nabla_X Z)+g(\nabla_Y Z,X)+g(Z,\nabla_Y X)-g(\nabla_Z X,Y)-g(X,\nabla_Z Y)$$
$$=g(\nabla_X Y,Z)+g(\nabla_Y X,Z)+g(\nabla_X Z,Y)-g(\nabla_Z X,Y)+g(\nabla_Y Z,X)-g(\nabla_Z Y,X)$$
$$=2g(\nabla_X Y,Z)-g([X,Y],Z)+g([X,Z],Y)+g([Y,Z],X).$$

Thus

$$\begin{aligned}2g(\nabla_X Y,Z)=&X(g(Y,Z))+Y(g(Z,X))-Z(g(X,Y))\\&+g([X,Y],Z)-g([X,Z],Y)-g([Y,Z],X)\end{aligned} \tag{3.1}$$

for any $X,Y,Z\in\mathfrak{X}(M)$.

On the other hand, for fixed $X,Y\in\mathfrak{X}(M)$, let $\omega_{X,Y}:\mathfrak{X}(M)\to C^\infty(M)$ be given by

$$\omega_{X,Y}(Z)$$
$$=\frac{1}{2}\left[X(g(Y,Z))+Y(g(Z,X))-Z(g(X,Y))+g([X,Y],Z)-g([X,Z],Y)-g([Y,Z],X)\right]$$

for any $Z\in\mathfrak{X}(M)$. It is easy to see that the mapping $\omega_{X,Y}$ is $C^\infty(M)$-linear, i.e. $\omega_{X,Y}\in A^1(M)$ and the $\mathbb{R}$-bilinear mapping

$$(X,Y)\in\mathfrak{X}(M)\times\mathfrak{X}(M)\mapsto(\omega_{X,Y})^\#\in\mathfrak{X}(M)$$

is a linear connection that satisfies (iii) and (iv). This proves the existence of a linear connection on M satisfying (iii) and (iv). Finally , the equality (3.1) implies that $\nabla_X Y=(\omega_{X,Y})^\#$ for any $X,Y\in\mathfrak{X}(M)$ which proves the uniqueness of ∇.

Q.E.D.

Definition 3.4 The linear connection constructed above is called the **Levi-Civita connection** or the **canonical symmetric linear connection** associated to the metric tensor field g.

Corollary 3.1. For every $X\in\mathfrak{X}(M)$, the operator ∇_X induced by the Levi-Civita connection commutes with the canonical isomorphism

$$\#:A^1(M)\to\mathfrak{X}(M),\ g(\omega^\#,Y)=\omega(Y),\ \omega\in A^1(M),\ Y\in\mathfrak{X}(M).$$

Proof. Let $b:\mathfrak{X}(M)\to A^1(M)$ be the inverse of #. Taking into account Theorem 3.1(iv) and Remark 3.1, it follows that

$$(\nabla_X Y)^b(Z)\overset{def}{=}g(\nabla_X Y,Z)=X(g(Y,Z))-g(Y,\nabla_X Z)=X[Y^b(Z)]-Y^b(\nabla_X Z)=(\nabla_X(Y^b))(Z)$$

for all $X,Y,Z\in\mathfrak{X}(M)$, that is

$$b\circ\nabla_X=\nabla_X\circ b$$

for all $X\in\mathfrak{X}(M)$, or equivalently

$$\#\circ\nabla_X=\nabla_X\circ\#$$

for all $X\in\mathfrak{X}(M)$.

Q.E.D.

Corollary 3.2 If $\Phi:(M,g)\to(\tilde{M},\tilde{g})$ is an isometry and ∇ (resp. $\tilde{\nabla}$) is the Levi-Civita connection associated to g (resp. $\tilde{g}$), then

$$\Phi_*\circ\nabla=\tilde{\nabla}\circ(\Phi_*\times\Phi_*).$$

Proof. The $\mathbb{R}$-bilinear mapping

$$\mathfrak{X}(\tilde{M})\times\mathfrak{X}(\tilde{M})\ni(\tilde{X},\tilde{Y})\mapsto\Phi_*\left(\nabla_{\Phi_*^{-1}(\tilde{X})}\Phi_*^{-1}(\tilde{Y})\right)\in\mathfrak{X}(\tilde{M})$$

satisfies the conditions (i), (ii) of Definition 3.1 and (iii), (iv) of Theorem 3.1, which uniquely characterize the Levi-Civita connection $\tilde{\nabla}$ associated to $\tilde{g}$. For instance, let us check only (ii), leaving (i), (iii) and (iv) to the reader. The property (ii) of ∇ implies

$$\begin{aligned}\Phi_*\left(\nabla_{\Phi_*^{-1}(\tilde{X})}\Phi_*^{-1}(\tilde{f}\tilde{Y})\right)&=\Phi_*\left(\nabla_{\Phi_*^{-1}(\tilde{X})}(\tilde{f}\circ\Phi)\Phi_*^{-1}(\tilde{Y})\right)\\&=\Phi_*\left[\Phi_*^{-1}(\tilde{X})(\tilde{f}\circ\Phi)\Phi_*^{-1}(\tilde{Y})+(\tilde{f}\circ\Phi)\nabla_{\Phi_*^{-1}(\tilde{X})}\Phi_*^{-1}(\tilde{Y})\right]\\&=\Phi_*\left[(\tilde{X}(\tilde{f})\circ\Phi)\Phi_*^{-1}(\tilde{Y})\right]+\Phi_*\left[(\tilde{f}\circ\Phi)\nabla_{\Phi_*^{-1}(\tilde{X})}\Phi_*^{-1}(\tilde{Y})\right]\\&=\tilde{X}(\tilde{f})\tilde{Y}+\tilde{f}\Phi_*\left(\nabla_{\Phi_*^{-1}(\tilde{X})}\Phi_*^{-1}(\tilde{Y})\right)\end{aligned}$$

for each $\tilde{f}\in C^\infty(\tilde{M})$ and $\tilde{X},\tilde{Y}\in\mathfrak{X}(\tilde{M})$.

Q.E.D.

Proposition 3.1 Let (M,g) be a Riemannian manifold, ∇ the Levi-Civita connection, $(U,x^1,\dots,x^n)$ a local coordinate system on M, $[g_{ij}(x)]$ the matrix representation of g with respect to this system and X,Y two vector fields on M whose local representations are given respectively by

$$X|_U=\sum_i X^i\frac{\partial}{\partial x^i};\ Y|_U=\sum_i Y^i\frac{\partial}{\partial x^i}.$$

Then in this coordinate system we have:

$$\nabla_X Y|_U=\sum_{i=1}^n\left(\sum_{j=1}^n X^j\frac{\partial Y^i}{\partial x^j}+\sum_{j,k=1}^n\Gamma^i_{jk}X^jY^k\right)\frac{\partial}{\partial x^i},$$

where the smooth functions $\Gamma^i_{jk}:U\to\mathbb{R}$ are defined by:

$$\nabla_{\frac{\partial}{\partial x^j}}\frac{\partial}{\partial x^k}=\sum_{i=1}^n\Gamma^i_{jk}\frac{\partial}{\partial x^i}.$$

The Γ^i_{jk} are the so-called **second Christoffel symbols** and these can be expressed only in terms of the first order derivatives of the metric tensor field:

$$\Gamma^i_{jk} = \frac{1}{2}\sum_{l=1}^{n} g^{il}\left(\frac{\partial g_{kl}}{\partial x^j} + \frac{\partial g_{lj}}{\partial x^k} - \frac{\partial g_{jk}}{\partial x^l}\right),$$

where as always $[g^{il}]$ is the inverse of the matrix $[g_{il}]$.

Proof. Since

$$\left[\frac{\partial}{\partial x^i}, \frac{\partial}{\partial x^j}\right] = 0, \tag{3.2}$$

for any $i,j = 1,2,...,n$ it follows that

$$\Gamma^i_{jk} = \Gamma^i_{kj},$$

for each $i,j,k = 1,2,...,n$.

The first part of the Proposition 3.1 is a direct consequence of the properties (i) and (ii) of the Levi-Civita connection ∇. To get the expression of the Γ^i_{jk} in terms of the first order derivatives of the g_{ij}, first note that

$$2g\left(\nabla_{\frac{\partial}{\partial x^j}}\frac{\partial}{\partial x^k}, \frac{\partial}{\partial x^l}\right) = \frac{\partial g_{kl}}{\partial x^j} + \frac{\partial g_{lj}}{\partial x^k} - \frac{\partial g_{jk}}{\partial x^l}. \tag{3.3}$$

Indeed, if one puts in the equality (3.1)

$$X = \frac{\partial}{\partial x^j},\ Y = \frac{\partial}{\partial x^k},\ Z = \frac{\partial}{\partial x^l}$$

and one takes into account (3.2) one derives (3.3).

On the other hand,

$$2g\left(\nabla_{\frac{\partial}{\partial x^j}}\frac{\partial}{\partial x^k}, \frac{\partial}{\partial x^l}\right) = 2g\left(\sum_{i=1}^{n}\Gamma^i_{jk}\frac{\partial}{\partial x^i}, \frac{\partial}{\partial x^l}\right) = 2\sum_{i=1}^{n}\Gamma^i_{jk}\, g\left(\frac{\partial}{\partial x^i}, \frac{\partial}{\partial x^l}\right)$$
$$= 2\sum_{i=1}^{n}\Gamma^i_{jk}\, g_{il}.$$

Therefore

$$\Gamma^i_{jk} = \frac{1}{2}\sum_{l=1}^{n} g^{il}\left(\frac{\partial g_{kl}}{\partial x^j} + \frac{\partial g_{lj}}{\partial x^k} - \frac{\partial g_{jk}}{\partial x^l}\right),$$

as required.

Q.E.D.

Example 3.1 Let $(\mathbb{R}^n, g_{can})$ be the Riemannian manifold considered in Example 2.1. In this case any smooth vector field X on $\mathbb{R}^n$ may be considered as a C^∞-mapping $X:\mathbb{R}^n \to \mathbb{R}^n$. Now let us define

$$(\nabla_X Y)(x) = (d_x Y)(X_x),$$

where $x \in \mathbb{R}^n$, $X, Y \in \mathfrak{X}(\mathbb{R}^n)$ and $d_x Y$ denotes the Fréchet differential of Y at x.

Then ∇ satisfies the conditions (i), (ii) of Definition 3.1 as well as (iii), (iv) of Theorem

3.1 and gives the Levi-Civita connection associated to $\left(\mathbb{R}^n, g_{can}\right)$. Let $\left(\partial/\partial x^1,\ldots,\partial/\partial x^n\right)$ be the natural basis with respect to the system of coordinates $\left(x^1,\ldots,x^n\right)$. Let

$$X=\sum_{i=1}^{n} X^i \partial/\partial x^i,\ Y=\sum_{j=1}^{n} Y^j \partial/\partial x^j,$$

$X^i, Y^j \in C^\infty\left(\mathbb{R}^n\right),\ i,j=1,\ldots,n$. Then

$$(\nabla_X Y)(x)=d_x Y\left(\sum_{j=1}^{n} X^j(x)\left(\frac{\partial}{\partial x^j}\right)_x\right)=\sum_{j,k=1}^{n} X^j(x)\frac{\partial Y^k}{\partial x^j}(x)\left(\frac{\partial}{\partial x^k}\right)_x$$
$$=\sum_{k=1}^{n} X_x\left(Y^k\right)\left(\frac{\partial}{\partial x^k}\right)_x.$$

With respect to this system of coordinates one gets

$$\sum_{k=1}^{n}\Gamma_{ij}^{k}\partial/\partial x^k=\nabla_{\partial/\partial x^i}\partial/\partial x^j=\sum_{l,k=1}^{n}\delta_i^l\,\partial/\partial x^l\left(\delta_j^k\right)\partial/\partial x^k=0,$$

that is $\Gamma_{ij}^k=0$ for any $i,j,k=1,\ldots,n$.

Example 3.2 If (θ,ϕ) are the spherical coordinates on S^2, i.e.

$$\begin{cases} x^1=\cos\theta\cos\phi, \\ x^2=\cos\theta\sin\phi, \\ x^3=\sin\theta, \end{cases}$$

then we have:

$$\nabla_{\frac{\partial}{\partial\theta}}\frac{\partial}{\partial\theta}=0,\ \nabla_{\frac{\partial}{\partial\phi}}\frac{\partial}{\partial\phi}=\frac{1}{2}\sin 2\theta\frac{\partial}{\partial\theta},\ \nabla_{\frac{\partial}{\partial\phi}}\frac{\partial}{\partial\theta}=\nabla_{\frac{\partial}{\partial\theta}}\frac{\partial}{\partial\phi}=-tg\theta\frac{\partial}{\partial\phi}.$$

Proposition 3.2 Let (M,g) be a Riemannian manifold and ∇ the Levi-Civita connection on M. Then $\nabla_X g=0$ for each $X\in\mathfrak{X}(M)$, that is $\overset{\circ}{\nabla} g=0$.

Proof. Indeed, for each $X, X_1, X_2\in\mathfrak{X}(M)$ we have:

$$(\nabla_X g)(X_1,X_2)=X\left(g(X_1,X_2)\right)-g(\nabla_X X_1,X_2)-g(X_1,\nabla_X X_2)=0$$

in virtue of (iv).

Q.E.D.

Remark 3.2 If (M,g) is a closed Riemannian manifold, $A^p(M)$ and $L\left(\mathfrak{X}(M),A^p(M)\right)$ have natural inner products (see Chapter 2, §1, C), so that the linear differential operator $\tilde{\nabla}$ corresponding to the Levi-Civita connection ∇ (see Remark 3.1) has a formal adjoint $\tilde{\nabla}^*: L\left(\mathfrak{X}(M),A^p(M)\right)\to A^p(M)$, i.e. $\left\langle\tilde{\nabla}\omega,S\right\rangle=\left\langle\omega,\tilde{\nabla}^*S\right\rangle$ for any $\omega\in A^p(M)$ and $S\in L\left(\mathfrak{X}(M),A^p(M)\right)$ (see for details R.S. Palais (editor) [1965] or J. Roe [1988]).

Proposition 3.3 Let (M,g) be a smooth n-dimensional Riemannian manifold, $f\in C^{\infty}(M)$ be such that $f>0$ and $\tilde{g}=fg$. If ∇ [resp. $\tilde{\nabla}$] is the Levi-Civita connection associated to g [resp. $\tilde{g}$], then for any $X,Y\in\mathfrak{X}(M)$ we have:

$$\tilde{\nabla}_X Y=\nabla_X Y+\frac{1}{2}\left[X(\ln f)Y+Y(\ln f)X-g(X,Y)\operatorname{grad}(\ln f)\right],$$

where ln denotes the natural real logarithmic function.

Proof. Taking into account (3.1) and the definition of the gradient of a smooth function on M, for each $X,Y,Z\in\mathfrak{X}(M)$ we can write successively:

$$\begin{aligned}
&2\tilde{g}(\tilde{\nabla}_X Y,Z)\\
&=X(\tilde{g}(Y,Z))+Y(\tilde{g}(Z,X))-Z(\tilde{g}(X,Y))+\tilde{g}([X,Y],Z)-\tilde{g}([X,Z],Y)-\tilde{g}([Y,Z],X)\\
&=X(f)g(Y,Z)+Y(f)g(Z,X)-Z(f)g(X,Y)+f\{X(g(Y,Z))+Y(g(Z,X))-Z(g(X,Y))+\\
&\quad+g([X,Y],Z)-g([X,Z],Y)-g([Y,Z],X)\}\\
&=X(f)g(Y,Z)+Y(f)g(Z,X)-g(\operatorname{grad}(f),Z)g(X,Y)+2fg(\nabla_X Y,Z)\\
&=2\tilde{g}\left(\nabla_X Y+\frac{1}{2}X(\ln f)Y+\frac{1}{2}Y(\ln f)X-\frac{1}{2}g(X,Y)\cdot\operatorname{grad}(\ln f),Z\right)
\end{aligned}$$

which implies the desired equality.

Q.E.D.

Definition 3.5 Let (M,g) be a smooth n-dimensional Riemannian manifold. If $f\in C^{\infty}(M)$, then its **Hessian form** $H_f:\mathfrak{X}(M)\times\mathfrak{X}(M)\to C^{\infty}(M)$ is defined by

$$H_f(X,Y)=[\nabla(X,df)](Y),$$

where $X,Y\in\mathfrak{X}(M)$, df is the differential of f and $\nabla:\mathfrak{X}(M)\times A^1(M)\to A^1(M)$ the natural extension of the Levi-Civita connection (see Remark 3.1), that is $H_f=\overset{\circ}{\nabla}(df)$ (see Definition 3.3).

H_f is a $C^{\infty}(M)$-bilinear form, that is a smooth tensor field of type (0,2) on M. Note that by the Definitions 3.5 and 3.2 we have also:

$$H_f(X,Y)=X(df(Y))-df(\nabla_X Y)=X(Y(f))-(\nabla_X Y)(f), \tag{3.4}$$

for each $X,Y\in\mathfrak{X}(M)$. Thus H_f is a symmetric form, since in virtue of the property (iii) of the Levi-Civita connection (see Theorem 3.1)

$$H_f(X,Y)-H_f(Y,X)=[X,Y]f-(\nabla_X Y-\nabla_Y X)f=0,$$

for any $X,Y\in\mathfrak{X}(M)$.

Proposition 3.4 Locally in a coordinate system $(U,x^1,\ldots,x^n)$ we have:

$$(H_f)_{ij}=\frac{\partial^2 f}{\partial x^i\partial x^j}-\sum_{k=1}^{n}\Gamma_{ij}^k\frac{\partial f}{\partial x^k},\quad i,j=1,2,\ldots,n.$$

Proof. Indeed, in virtue of (3.2), for any $i,j = 1,2,...,n$,

$$\begin{aligned}(H_f)_{ij} &= (H_f)\left(\frac{\partial}{\partial x^i},\frac{\partial}{\partial x^j}\right)\\ &= \frac{\partial}{\partial x^i}\left(\frac{\partial f}{\partial x^j}\right)-\left(\nabla_{\frac{\partial}{\partial x^i}}\frac{\partial}{\partial x^j}\right)(f)\\ &= \frac{\partial^2 f}{\partial x^i\partial x^j}-\sum_{k=1}^{n}\Gamma_{ij}^k\frac{\partial f}{\partial x^k},\end{aligned}$$

as required.

Q.E.D.

Definition 3.6 If (M,g) is a Riemannian manifold and $f\in C^\infty(M)$, the $C^\infty(M)$-linear mapping $h_f:\mathfrak{X}(M)\to\mathfrak{X}(M)$ given by

$$h_f(X)=\nabla_X(\text{grad } f),$$

$X\in\mathfrak{X}(M)$, that is $h_f=\overset{\circ}{\nabla}(\text{grad}(f))$ (see Definition 3.3), where ∇ is the Levi-Civita connection, defines a smooth tensor field of type $(1,1)$ on M called the **Hesse tensor field** associated to f.

Note that $H_f=\hat{h}_f$, since for all $X,Y\in\mathfrak{X}(M)$,

$$\begin{aligned}H_f(X,Y)&=X(df(Y))-df(\nabla_X Y)\\ &=X(g(\text{grad}(f),Y))-g(\text{grad}(f),\nabla_X Y)\\ &=g(\nabla_X(\text{grad}(f)),Y)\\ &=g(h_f(X),Y),\end{aligned}$$

the third equality being a consequence of the property (iv) of ∇ (see Theorem 3.1). In particular, since H_f is symmetric, h_f is selfadjoint with respect to g.

In particular,

$$trace(H_f)=trace(h_f)=\sum_{i=1}^{n}H_f(E_i,E_i)=\sum_{i=1}^{n}(E_i\circ E_i-\nabla_{E_i}E_i)(f) \tag{3.5}$$

for any local field of orthonormal frames $(E_1,\ldots,E_n)$ on (M,g).

4. THE CURVATURE OF A RIEMANNIAN MANIFOLD

There are many different ways of defining the curvature of a Riemannian manifold. In this section we give an intrinsic one but not so geometric.

Let (M,g) be a smooth n-dimensional Riemannian manifold, ∇ the Levi-Civita connection associated to g and the mapping $R:\mathfrak{X}(M)\times\mathfrak{X}(M)\times\mathfrak{X}(M)\to\mathfrak{X}(M)$ given by:

$$R(X,Y,Z)\overset{not}{=}R(X,Y)Z\overset{def}{=}\nabla_X(\nabla_Y Z)-\nabla_Y(\nabla_X Z)-\nabla_{[X,Y]}Z,$$

for each $X,Y,Z\in\mathfrak{X}(M)$.

Definition 4.1 R is a smooth tensor field of type (1,3) on M called the **curvature tensor field** of ∇.

It is easy to see that for each $X,Y,Z \in \mathfrak{X}(M)$ we have:

$$R(X,Y)Z = -R(Y,X)Z. \tag{4.1}$$

A straightforward computation leads us to:

Proposition 4.1 For each $X,Y,Z,U \in \mathfrak{X}(M)$ we have:

(i) $R(X,Y)Z + R(Y,Z)X + R(Z,X)Y = 0$ (first Bianchi's identity);

(ii) $g(R(X,Y)Z,U) = -g(R(X,Y)U,Z)$;

(iii) $g(R(X,Y)Z,U) = g(R(Z,U)X,Y)$;

(iv) $(\nabla_X R)(Y,Z,U) + (\nabla_Y R)(Z,X,U) + (\nabla_Z R)(X,Y,U) = 0$, that is $\left(\overset{\circ}{\nabla} R\right)(X,Y,Z,U) + \left(\overset{\circ}{\nabla} R\right)(Y,Z,X,U) + \left(\overset{\circ}{\nabla} R\right)(Z,X,Y,U) = 0$ (second Bianchi's identity).

Proof. For instance, let us verify only (i), leaving (ii), (iii) and (iv) to the reader. The symmetry of the Levi-Civita connection [see (iii) of Theorem 3.1] implies

$$\begin{aligned}
&R(X,Y)Z + R(Y,Z)X + R(Z,X)Y \\
&= \nabla_X(\nabla_Y Z) - \nabla_Y(\nabla_X Z) - \nabla_{[X,Y]}Z + \nabla_Y(\nabla_Z X) - \nabla_Z(\nabla_Y X) \\
&\quad - \nabla_{[Y,Z]}X + \nabla_Z(\nabla_X Y) - \nabla_X(\nabla_Z Y) - \nabla_{[Z,X]}Y \\
&= \nabla_X[Y,Z] + \nabla_Y[Z,X] + \nabla_Z[X,Y] - \nabla_{[X,Y]}Z - \nabla_{[Y,Z]}X - \nabla_{[Z,X]}Y \\
&= [X,[Y,Z]] + [Y,[Z,X]] + [Z,[X,Y]] = 0,
\end{aligned}$$

for any $X,Y,Z \in \mathfrak{X}(M)$, where the last equality is just the Jacobi identity for vector fields.

Q.E.D.

If $\Phi:(M,g) \to (\tilde{M},\tilde{g})$ is an isometry and R (resp. $\tilde{R}$) is the curvature tensor field of ∇ (resp. $\tilde{\nabla}$), then Corollary 3.2 implies that

$$\Phi_* \circ R = \tilde{R} \circ (\Phi_* \times \Phi_* \times \Phi_*).$$

For a smooth tensor field S of type $(0,r)$ on $M\,(r \geq 1)$, one has

$$\begin{aligned}
&\left(\overset{\circ}{\nabla}{}^2 S\right)(X,Y,Z_1,\ldots,Z_r) = \left[\nabla_X\left(\overset{\circ}{\nabla} S\right)\right](Y,Z_1,\ldots,Z_r) = \\
&= \nabla_X\left(\left(\overset{\circ}{\nabla} S\right)(Y,Z_1,\ldots,Z_r)\right) - \left(\overset{\circ}{\nabla} S\right)(\nabla_X Y,Z_1,\ldots,Z_r) - \sum_{j=1}^{r}\left(\overset{\circ}{\nabla} S\right)(Y,Z_1,\ldots,\nabla_X Z_j,\ldots,Z_r) \\
&= \nabla_X\left[(\nabla_Y S)(Z_1,\ldots,Z_r)\right] - (\nabla_{\nabla_X Y} S)(Z_1,\ldots,Z_r) - \sum_{j=1}^{r}(\nabla_Y S)(Z_1,\ldots,\nabla_X Z_j,\ldots,Z_r) \\
&= \left[(\nabla_X\nabla_Y - \nabla_{\nabla_X Y})(S)\right](Z_1,\ldots,Z_r)
\end{aligned}$$

for any $X,Y,Z_1,\ldots,Z_r \in \mathfrak{X}(M)$ (see Definition 3.3), which implies that

$$\left(\overset{\circ}{\nabla}{}^2 S\right)(X,Y,Z_1,\ldots,Z_r)-\left(\overset{\circ}{\nabla}{}^2 S\right)(Y,X,Z_1,\ldots,Z_r)$$
$$=\left[\left(\nabla_X\nabla_Y-\nabla_Y\nabla_X-\nabla_{[X,Y]}\right)(S)\right](Z_1,\ldots,Z_r).$$

On the other hand, the computation above shows that

$$\begin{aligned}\left(\overset{\circ}{\nabla}{}^2 S\right)&(X,Y,Z_1,\ldots,Z_r)=X\left(Y\left(S\left(Z_1,\ldots,Z_r\right)\right)\right)\\ &-\sum_{j=1}^{r}X\left(S\left(Z_1,\ldots,\nabla_Y Z_j,\ldots,Z_r\right)\right)-\left(\nabla_X Y\right)\left(S\left(Z_1,\ldots,Z_r\right)\right)\\ &+\sum_{j=1}^{r}S\left(Z_1,\ldots,\nabla_{\nabla_X Y}Z_j,\ldots,Z_r\right)-\sum_{j=1}^{r}Y\left(S\left(Z_1,\ldots,\nabla_X Z_j,\ldots,Z_r\right)\right)\\ &+\sum_{j=1}^{r}S\left(Z_1,\ldots,\nabla_Y\nabla_X Z_j,\ldots,Z_r\right)+\sum_{j\neq k}^{r}S\left(Z_1,\ldots,\nabla_X Z_j,\ldots,\nabla_Y Z_k,\ldots,Z_r\right),\end{aligned}$$

which yields to the following commutation law

$$\left(\overset{\circ}{\nabla}{}^2 S\right)(X,Y,Z_1,\ldots,Z_r)-\left(\overset{\circ}{\nabla}{}^2 S\right)(Y,X,Z_1,\ldots,Z_r)=-\sum_{j=1}^{r}S\left(Z_1,\ldots,R(X,Y)Z_j,\ldots,Z_r\right)$$

for any $X,Y,Z_1,\ldots,Z_r \in \mathfrak{X}(M)$ in the case when $r\geq 1$. Note that

$$\left(\overset{\circ}{\nabla}{}^2 f\right)(X,Y)-\left(\overset{\circ}{\nabla}{}^2 f\right)(Y,X)=[X,Y](f)$$

for each $f\in C^\infty(M)$ and $X,Y\in\mathfrak{X}(M)$.

In particular, one gets the following commutation formula

$$\begin{aligned}\left(\overset{\circ}{\nabla}{}^2 \omega\right)(X,Y,Z)-\left(\overset{\circ}{\nabla}{}^2 \omega\right)(Y,X,Z)&=\left[\left(\nabla_X\nabla_Y-\nabla_Y\nabla_X-\nabla_{[X,Y]}\right)(\omega)\right](Z)\\ &=-\omega\left(R(X,Y)Z\right)\end{aligned} \tag{4.2}$$

for any $X,Y,Z\in\mathfrak{X}(M)$ and $\omega\in A^1(M)$.

If the one-form ω is closed, note that

$$\left(\overset{\circ}{\nabla}{}^2 \omega\right)(X,Y,Z)=\left(\overset{\circ}{\nabla}{}^2 \omega\right)(X,Z,Y) \tag{4.3}$$

for any $X,Y,Z\in\mathfrak{X}(M)$. Indeed, by the property (iii) of the Levi-Civita connection (see Theorem 3.1), it follows that

$$\begin{aligned}(d\omega)(X,Y)&=X\left(\omega(Y)\right)-Y\left(\omega(X)\right)-\omega\left([X,Y]\right)\\ &=X\left(\omega(Y)\right)-Y\left(\omega(X)\right)-\omega\left(\nabla_X Y\right)+\omega\left(\nabla_Y X\right)\end{aligned}$$

for any $X,Y \in \mathfrak{X}(M)$ and $\omega \in A^1(M)$ (see also § 1).

Since

$$\left(\overset{\circ}{\nabla}\omega\right)(X,Y) = X(\omega(Y)) - \omega(\nabla_X Y)$$

for any $X,Y \in \mathfrak{X}(M)$ and $\omega \in A^1(M)$, one gets

$$\left(\overset{\circ}{\nabla}\omega\right)(X,Y) - \left(\overset{\circ}{\nabla}\omega\right)(Y,X) = (d\omega)(X,Y).$$

Hence $\overset{\circ}{\nabla}\omega \in \mathcal{T}_2^0(M)$ is symmetric if and only if $\omega \in A^1(M)$ is closed. Consequently, in order to prove (4.3) for a closed one form ω, it is sufficient to check

$$(\nabla_X S)(Y,Z) = (\nabla_X S)(Z,Y), \quad X,Y,Z \in \mathfrak{X}(M),$$

for any symmetric tensor field S of type (0,2) on M. But for such an S one has

$$\begin{aligned}(\nabla_X S)(Y,Z) &= X(S(Y,Z)) - S(\nabla_X Y, Z) - S(Y, \nabla_X Z)\\ &= X(S(Z,Y)) - S(\nabla_X Z, Y) - S(Z, \nabla_X Y) = (\nabla_X S)(Z,Y)\end{aligned}$$

for each $X,Y,Z \in \mathfrak{X}(M)$, as claimed.

Proposition 4.2 Let (M,g) be a Riemannian manifold, $f \in C^\infty(M)$ be such that $f >$ 0 and $\tilde{g} = fg$. If R [resp. $\tilde{R}$] is the curvature tensor field of the Levi-Civita connection ∇ [resp. $\tilde{\nabla}$] associated to g [resp. $\tilde{g}$], then

$$\begin{aligned}&\tilde{R}(X,Y)Z\\ &= R(X,Y)Z + \frac{1}{2}\left[H_{\ln f}(X,Z)Y - H_{\ln f}(Y,Z)X + g(X,Z)h_{\ln f}(Y) - g(Y,Z)h_{\ln f}(X)\right]\\ &\quad + \frac{1}{4}\left\{\begin{aligned}&\left[Y(\ln f)Z(\ln f) - g(Y,Z)g(\operatorname{grad}(\ln f),\operatorname{grad}(\ln f))\right]X\\ &-\left[X(\ln f)\cdot Z(\ln f) - g(X,Z)g(\operatorname{grad}(\ln f),\operatorname{grad}(\ln f))\right]Y\\ &+\left[X(\ln f)g(Y,Z) - Y(\ln f)g(X,Z)\right]\operatorname{grad}(\ln f)\end{aligned}\right\}\end{aligned}$$

for any $X,Y,Z \in \mathfrak{X}(M)$, where $H_{\ln f}$ (resp. $h_{\ln f}$) denotes the Hesse form (resp. Hesse tensor field) associated to $\ln f$.

Proof. Taking into account Proposition 3.3, the properties (iii) and (iv) of ∇ (see Theorem 3.1), the definitions of the Hesse tensor field and of the gradient vector field associated to $\ln f$ one obtains successively:

$$\begin{aligned}
\tilde{R}(X,Y)Z &= \tilde{\nabla}_X\left(\tilde{\nabla}_Y Z\right) - \tilde{\nabla}_Y\left(\tilde{\nabla}_X Z\right) - \tilde{\nabla}_{[X,Y]}Z \\
&= \tilde{\nabla}_X\left[\nabla_Y Z + \frac{1}{2}Y(\ln f)Z + \frac{1}{2}Z(\ln f)Y - \frac{1}{2}g(Y,Z)\operatorname{grad}(\ln f)\right] \\
&\quad - \tilde{\nabla}_Y\left[\nabla_X Z + \frac{1}{2}X(\ln f)Z + \frac{1}{2}Z(\ln f)X - \frac{1}{2}g(X,Z)\operatorname{grad}(\ln f)\right] \\
&\quad - \nabla_{[X,Y]}Z - \frac{1}{2}[X,Y](\ln f)Z - \frac{1}{2}Z(\ln f)[X,Y] + \frac{1}{2}g([X,Y],Z)\operatorname{grad}(\ln f) \\
&= \nabla_X(\nabla_Y Z) + \frac{1}{2}X(Y(\ln f))Z + \frac{1}{2}Y(\ln f)\nabla_X Z + \frac{1}{2}X(Z(\ln f))Y + \frac{1}{2}Z(\ln f)\nabla_X Y \\
&\quad - \frac{1}{2}X(g(Y,Z))\operatorname{grad}(\ln f) - \frac{1}{2}g(Y,Z)\nabla_X(\operatorname{grad}(\ln f)) + \frac{1}{2}X(\ln f)\nabla_Y Z \\
&\quad + \frac{1}{4}X(\ln f)Y(\ln f)Z + \frac{1}{4}X(\ln f)Z(\ln f)Y - \frac{1}{4}X(\ln f)g(Y,Z)\operatorname{grad}(\ln f) \\
&\quad + \frac{1}{2}(\nabla_Y Z)(\ln f)X + \frac{1}{4}Y(\ln f)Z(\ln f)X + \frac{1}{4}Z(\ln f)Y(\ln f)X \\
&\quad - \frac{1}{4}g(Y,Z)(\operatorname{grad}(\ln f))(\ln f)X - \frac{1}{2}g(X,\nabla_Y Z)\operatorname{grad}(\ln f) \\
&\quad - \frac{1}{4}Y(\ln f)g(X,Z)\operatorname{grad}(\ln f) - \frac{1}{4}Z(\ln f)g(X,Y)\operatorname{grad}(\ln f) \\
&\quad + \frac{1}{4}g(Y,Z)g(X,\operatorname{grad}(\ln f))\operatorname{grad}(\ln f) - \nabla_Y(\nabla_X Z) - \frac{1}{2}Y(X(\ln f))Z \\
&\quad - \frac{1}{2}X(\ln f)\nabla_Y Z - \frac{1}{2}Y(Z(\ln f))X - \frac{1}{2}Z(\ln f)\nabla_Y X + \frac{1}{2}Y(g(X,Z))\operatorname{grad}(\ln f) \\
&\quad + \frac{1}{2}g(X,Z)\nabla_Y(\operatorname{grad}(\ln f)) - \frac{1}{2}Y(\ln f)\nabla_X Z - \frac{1}{4}Y(\ln f)X(\ln f)Z \\
&\quad - \frac{1}{4}Y(\ln f)Z(\ln f)X + \frac{1}{4}Y(\ln f)g(X,Z)\operatorname{grad}(\ln f) - \frac{1}{2}(\nabla_X Z)(\ln f)Y \\
&\quad - \frac{1}{4}X(\ln f)Z(\ln f)Y - \frac{1}{4}Z(\ln f)X(\ln f)Y + \frac{1}{4}g(X,Z)(\operatorname{grad}(\ln f))(\ln f)Y \\
&\quad + \frac{1}{2}g(Y,\nabla_X Z)\operatorname{grad}(\ln f) + \frac{1}{4}X(\ln f)g(Y,Z)\operatorname{grad}(\ln f) \\
&\quad + \frac{1}{4}Z(\ln f)g(Y,X)\operatorname{grad}(\ln f) - \frac{1}{4}g(X,Z)g(Y,\operatorname{grad}(\ln f))\operatorname{grad}(\ln f) \\
&\quad - \nabla_{[X,Y]}Z - \frac{1}{2}[X,Y](\ln f)Z - \frac{1}{2}Z(\ln f)[X,Y] + \frac{1}{2}g([X,Y],Z)\operatorname{grad}(\ln f) \\
&= R(X,Y)Z + \frac{1}{2}\left[X(Z(\ln f)) - (\nabla_X Z)(\ln f)\right]Y - \frac{1}{2}\left[Y(Z(\ln f)) - (\nabla_Y Z)(\ln f)\right]X \\
&\quad + \frac{1}{2}\left[g(X,Z)h_{\ln f}(Y) - g(Y,Z)h_{\ln f}(X)\right] \\
&\quad + \frac{1}{4}\left[Y(\ln f)Z(\ln f) - g(Y,Z)g(\operatorname{grad}(\ln f),\operatorname{grad}(\ln f))\right]X \\
&\quad - \frac{1}{4}\left[X(\ln f)Z(\ln f) - g(X,Z)g(\operatorname{grad}(\ln f),\operatorname{grad}(\ln f))\right]Y
\end{aligned}$$

$$+\frac{1}{4}\left[X(\ln f)g(Y,Z)-Y(\ln f)g(X,Z)\right]\mathrm{grad}(\ln f),$$

which, via (3.4) implies the desired equality.

Q.E.D.

Let $n \geq 2$, $G_x^2 M$ be the $2(n-2)$-dimensional compact Grassmannian manifold of all 2-planes of T_xM and

$$G^2(M) = \underset{x\in M}{\cup} G_x^2 M .$$

$G^2(M)$ has a canonical smooth differential structure of dimension $3n-4$ such that $(G^2(M),\pi,M)$ is a smooth fiber bundle, where π is the canonical projection of $G^2(M)$ onto M.

Definition 4.2 The mapping $\sigma : G^2(M) \to \mathbb{R}$ given by

$$\sigma(x,P) = \frac{g_x(R_x(u,v)v,u)}{g_x(u,u)g_x(v,v)-\left[g_x(u,v)\right]^2}, \tag{4.4}$$

where (u,v) is a basis of the 2-plane P of T_xM, is called the **sectional curvature** of (M,g). The real number $\sigma(x,P)$ is called the sectional curvature of (M,g) at x in the "plane direction" $P\subset T_xM$.

It is easy to see that the above definition does not depend on the choice of the basis of P and in the case $n = 2$, when P is just T_xM, σ can be canonically identified with the Gauss curvature from the theory of surfaces, so that in this case $\sigma\in C^\infty(M)$. Indeed, under the linear automorphisms of P

$$u \mapsto au + bv \text{ and } v \mapsto cu + dv,$$

both the numerator and the denominator in the definition of $\sigma(x,P)$ change by the scaling factor $(ad-bc)^2$, thus $\sigma(x,P)$ depends only on the 2-plane P and on the point $x\in M$.

On the other hand, in a coordinate system (U,x^1,x^2) on a 2-dimensional manifold M, we can write

$$u = u^1\left(\frac{\partial}{\partial x^1}\right)_x + u^2\left(\frac{\partial}{\partial x^2}\right)_x,$$

$$v = v^1\left(\frac{\partial}{\partial x^1}\right)_x + v^2\left(\frac{\partial}{\partial x^2}\right)_x, \quad x\in U.$$

The above scaling properties imply:

$$\begin{aligned} g_x(R_x(u,v)v,u) &= \left(u^1v^2-u^2v^1\right)^2 g_x\left(R_x\left(\left(\frac{\partial}{\partial x^1}\right)_x,\left(\frac{\partial}{\partial x^2}\right)_x\right)\left(\frac{\partial}{\partial x^2}\right)_x,\left(\frac{\partial}{\partial x^1}\right)_x\right) \\ &= \left(u^1v^2-u^2v^1\right)^2 R_{1212}(x) \end{aligned}$$

and

$$g_x(u,u)g_x(v,v)-[g_x(u,v)]^2 = (u^1v^2-u^2v^1)^2\left[g_{11}(x)g_{22}(x)-g_{12}^2(x)\right]$$
$$= (u^1v^2-u^2v^1)^2 \det\left[g_{ij}(x)\right].$$

Then

$$\sigma(x,P) = \frac{R_{1212}(x)}{\det\left[g_{ij}(x)\right]}$$
$$= \frac{\left(\frac{\partial^2 g_{12}}{\partial x^1 \partial x^2} - \frac{1}{2}\frac{\partial^2 g_{22}}{\partial x^1 \partial x^1} - \frac{1}{2}\frac{\partial^2 g_{11}}{\partial x^2 \partial x^2}\right)(x)}{\det\left[g_{ij}(x)\right]},$$

which is just the Gaussian curvature function *K*.

Note that for $n \geq 3$, $\sigma \in C^\infty(G^2(M))$, see for details Bishop and Crittenden [1964].

For a fixed metric tensor field *g* on *M*, the curvature tensor field *R* determines uniquely σ via (4.4) and conversely *R* is completely determined by σ (for instance see for details Craioveanu and Puta [1988]). In particular, $R = 0$ if and only if $\sigma = 0$.

Definition 4.3 A Riemannian manifold (*M*,*g*) is called **a space of constant curvature** if there exists $k \in \mathbb{R}$ such that $\sigma(x,P) = k$, for any $(x,P) \in G^2(M)$. In this case *k* is called the **curvature of the space**. If $k > 0$ [resp. $k < 0$, resp. $k = 0$] the space is called **elliptic** [resp. **hyperbolic**, resp. **flat**.]

A Riemannian manifold (*M*,*g*) of dimension $n \geq 2$ is of constant curvature *k* if and only if

$$R(X,Y)Z = k\left[g(Y,Z)X - g(X,Z)Y\right] \tag{4.5}$$

for any $X,Y,Z \in \mathfrak{X}(M)$.

For, if (4.5) holds, then by (4.4), $\sigma(x,P) = k$ for any $(x,P) \in G^2(M)$. Conversely, assume that (*M*,*g*) if of constant curvature *k* and let us define

$$L : \mathfrak{X}(M) \times \mathfrak{X}(M) \times \mathfrak{X}(M) \to \mathfrak{X}(M)$$

by

$$L(X,Y)Z = R(X,Y)Z - k\left[g(Y,Z)X - g(X,Z)Y\right].$$

Then *L* is a smooth tensor field of type (1,3) on *M*, which satisfies the same algebraic properties (4.1) and (i), (ii), (iii) in Proposition 4.1 as the curvature tensor field *R*. Now note that

$$g_x(L_x(u,v)v,u) = 0$$

for any $x \in M$ and $u,v \in T_xM$ or equivalently

$$g(L(X,Y)Y,X) = 0 \tag{4.6}$$

for any $X,Y \in \mathfrak{X}(M)$. The equality is obvious if $\{u,v\}$ are linearly dependent. If $\{u,v\}$

are linearly independent $(n \geq 2)$, then $g(u,u)g(v,v)-g^2(u,v) \neq 0$ and $\sigma(x,P)=k$, where (u,v) is a basis of P, implies that

$$g_x\left(R_x(u,v)v - kg_x(v,v)u + kg_x(u,v)v, u\right) = 0,$$

that is (4.6) is valid. Now letting $Y+Z$ instead of Y in (4.6) and using (4.1), (ii), (iii) for L and (4.6) one gets

$$g(L(X,Y)X,Z)=0$$

for any $Z \in \mathfrak{X}(M)$, that is

$$L(X,Y)X = 0$$

for any $X,Y \in \mathfrak{X}(M)$. Setting $X+Z$ instead of X in the last equality, one obtains

$$L(X,Y)Z + L(Z,Y)X = 0$$

for each $X,Y,Z \in \mathfrak{X}(M)$. On the other hand, first Bianchi's identity implies

$$L(X,Y)Z = -L(Y,Z)X - L(Z,X)Y$$

for each $X,Y,Z \in \mathfrak{X}(M)$, so that the last two equalities imply

$$2L(Z,Y)X = L(Z,X)Y$$

for any $X,Y,Z \in \mathfrak{X}(M)$. Hence

$$2L(Z,Y)X = L(Z,X)Y = \frac{1}{2}L(Z,Y)X,$$

that is $L(Z,Y)X=0$ for any $X,Y,Z \in \mathfrak{X}(M)$, which implies (4.5).

Example 4.1 Since the curvature tensor field of the Levi-Civita connection associated to $(\mathbb{R}^n, g_{can})$ is zero, the Riemannian manifold $(\mathbb{R}^n, g_{can})$ is a flat space.

Example 4.2 Let us consider the case of the surface of revolution given parametrically by:

$$\begin{cases} x = a(\phi)\cos\theta, \\ y = a(\phi)\sin\theta, \\ z = b(\phi), \end{cases}$$

where $0 \leq \theta \leq 2\pi$ and ϕ is in some interval of $\mathbb{R}$. Then we have:

$$K = \frac{\dot{b}\left(\dot{a}\ddot{b} - \ddot{a}\dot{b}\right)}{a\left(\dot{a}^2 + \dot{b}^2\right)^2}.$$

In the particular case of the torus of revolution we have

$$\begin{cases} a = R + r\cos\phi \\ b = r\sin\phi, \end{cases}$$

where $R > r$, $0 \le \phi \le 2\pi$ and then

$$K = \frac{\cos\phi}{r(R + r\cos\phi)}.$$

For the 2-sphere of radius r we have

$$\begin{cases} a = r\sin\phi, \\ b = r\cos\phi, \end{cases}$$

where $0 \le \phi \le \pi$ and then

$$K = \frac{1}{r^2}.$$

Proposition 4.3 Let (M,g) be a smooth n-dimensional Riemannian manifold, $f \in C^\infty(M)$ be such that $f > 0$ and $\tilde{g} = fg$. If σ [resp. $\tilde{\sigma}$]$: G^2(M) \to \mathbb{R}$ is the sectional curvature of (M,g) [resp. $(M,\tilde{g})$], then

$$f(x)\tilde{\sigma}(x,P) = \sigma(x,P) - \frac{1}{2}\begin{bmatrix} (H_{\ln f})_x(u,u) + (H_{\ln f})_x(v,v) \\ +\frac{1}{2}g_x\left((\operatorname{grad}(\ln f))_x, (\operatorname{grad}(\ln f))_x\right) - \frac{1}{2}(u(\ln f))^2 \\ -\frac{1}{2}(v(\ln f))^2 \end{bmatrix},$$

for any $x \in M$, where (u,v) is an orthonormal basis of the 2-plane P of T_xM with respect to g_x.

Proof. In virtue of (4.4), the hypothesis $\tilde{g} = fg$, Proposition 4.2, the equality $H_{\ln f} = \hat{h}_{\ln f}$ and the definition of the vector field grad($\ln f$), it follows that

$$\tilde{\sigma}(x,P) = \frac{\tilde{g}_x\left(\tilde{R}_x(u,v)v,u\right)}{\tilde{g}_x(u,u)\tilde{g}_x(v,v) - \left[\tilde{g}_x(u,v)\right]^2} = \frac{1}{f(x)}g_x\left(\tilde{R}_x(u,v)v,u\right)$$

$$= \frac{1}{f(x)}\begin{bmatrix} \sigma(x,P) + \frac{1}{2}g_x(v,u)(H_{\ln f})_x(u,v) - \frac{1}{2}g_x(u,u)(H_{\ln f})_x(v,v) \\ +\frac{1}{2}g_x(u,v)(H_{\ln f})_x(v,u) - \frac{1}{2}g_x(v,v)(H_{\ln f})_x(u,u) + \frac{1}{4}g_x(u,u)(v(\ln f))^2 \\ -\frac{1}{4}g_x(v,v)g_x(u,u)g_x\left((\operatorname{grad}(\ln f))_x, (\operatorname{grad}(\ln f))_x\right) - \frac{1}{4}g_x(v,u)u(\ln f)v(\ln f) \\ +\frac{1}{4}g_x(u,v)g_x(v,u)g_x\left((\operatorname{grad}(\ln f))_x, (\operatorname{grad}(\ln f))_x\right) \\ +\frac{1}{4}g_x(v,v)u(\ln f)\cdot g_x\left((\operatorname{grad}(\ln f))_x, u\right) - \frac{1}{4}g_x(u,v)v(\ln f)g_x\left((\operatorname{grad}(\ln f))_x, u\right) \end{bmatrix}$$

$$= \frac{1}{f(x)}\left\{\sigma(x,P) - \frac{1}{2}\left[\begin{array}{l}(H_{\ln f})_x(u,u) + (H_{\ln f})_x(v,v) + \frac{1}{2}g_x\left((\operatorname{grad}(\ln f))_x, (\operatorname{grad}(\ln f))_x\right) \\ -\frac{1}{2}(u(\ln f))^2 - \frac{1}{2}(v(\ln f))^2\end{array}\right]\right\}$$

as desired.

Q.E.D.

Corollary 4.1 If (B^n,h) is the Riemannian manifold considered in Example 2.3, then (B^n,h) is of constant curvature -1.

Proof. If g_{can} denotes the canonical metric tensor field on the unit ball $B^n \subset \mathbb{R}^n$, then $h = f \cdot g_{can}$ with $f : B^n \to \mathbb{R}$, $f(x) = 4\Big/\left(1-\|x\|^2\right)^2$, $x \in B^n$, where $\| \ \|$ denotes the usual Euclidean norm on $\mathbb{R}^n$. Let $\tilde{\sigma}$ [resp. σ] be the sectional curvature of (B^n,h) [resp. (B^n,g_{can})]. Then $\sigma = 0$ (see also Example 4.1). Let P be a 2-plane of T_xB^n, $x \in B^n$, and

$$\left(u = \sum_{i=1}^{n} u^i \left(\frac{\partial}{\partial x^i}\right)_x, v = \sum_{i=1}^{n} v^i \left(\frac{\partial}{\partial x^i}\right)_x\right)$$

be an orthonormal basis of P with respect to $(g_{can})_x$. Since for any $x = (x^1,\ldots,x^n) \in B^n$ and $i, j = 1,\ldots,n,\ i \neq j$,

$$\left(\frac{\partial}{\partial x^i}\right)_x (\ln f) = \frac{\partial \ln f}{\partial x^i}(x) = \frac{4x^i}{1-\|x\|^2},$$

$$\frac{\partial^2 \ln f}{\partial x^{i^2}}(x) = 4\frac{1-\|x\|^2 + 2(x^i)^2}{\left(1-\|x\|^2\right)^2},$$

$$\frac{\partial^2 \ln f}{\partial x^i \partial x^j}(x) = \frac{8x^i x^j}{\left(1-\|x\|^2\right)^2},$$

it follows that

$$\begin{aligned}(H_{\ln f})_x(u,u) - \frac{1}{2}(u(\ln f))^2 &= \sum_{i,j=1}^{n} \frac{\partial^2 \ln f}{\partial x^i \partial x^j}(x) u^i u^j - \frac{1}{2}\left(\sum_{i=1}^{n} u^i \frac{\partial \ln f}{\partial x^i}(x)\right)^2 \\ &= \frac{4}{1-\|x\|^2}\sum_{i=1}^{n}\left(u^i\right)^2 = \frac{4}{1-\|x\|^2}\end{aligned}$$

and similarly,

$$(H_{\ln f})_x(v,v) - \frac{1}{2}(v(\ln f))^2 = \frac{4}{1-\|x\|^2}.$$

But

$$g_x\left(\left(\operatorname{grad}(\ln f)\right)_x, \left(\operatorname{grad}(\ln f)\right)_x\right) = \sum_{i=1}^{n}\left(\frac{\partial \ln f}{\partial x^i}(x)\right)^2$$
$$= \frac{16\|x\|^2}{\left(1-\|x\|^2\right)^2},$$

so that Proposition 4.3 implies

$$f(x)\tilde{\sigma}(x,P) = -\frac{1}{2}\left[\frac{8}{1-\|x\|^2} + \frac{8\|x\|^2}{\left(1-\|x\|^2\right)^2}\right]$$
$$= \frac{-4}{\left(1-\|x\|^2\right)^2}.$$

Hence $\tilde{\sigma}(x,P) = -1$, for each $(x,P) \in G^2(B^n)$, i.e. (B^n,h) is of constant curvature and the curvature of this space is -1.

Q.E.D.

Remark 4.1 Let $\Phi:(\tilde{M},\tilde{g}) \to (M,g)$ be a Riemannian local diffeomorphism, i.e. $\tilde{g}$ is the Riemannian structure induced by the smooth local diffeomorphism $\Phi:\tilde{M} \to M$ on $\tilde{M}$ (see Example 2.9). Assume that $\Phi:\tilde{M} \to M$ is onto. If $\tilde{\sigma}:G^2(\tilde{M}) \to \mathbb{R}$ [resp. $\sigma:G^2(M) \to \mathbb{R}$] is the sectional curvature of $(\tilde{M},\tilde{g})$ [resp. (M,g)], then for each $\tilde{x} \in \tilde{M}$,

$$\tilde{\sigma}(\tilde{x},\tilde{P}) = \sigma\left(\Phi(\tilde{x}), d\Phi_{\tilde{x}}(\tilde{P})\right), \tag{4.7}$$

for any $(\tilde{x},\tilde{P}) \in G^2(\tilde{M})$. In particular, $(\tilde{M},\tilde{g})$ is of constant curvature k if and only if (M,g) is of constant curvature k.

For, if $\tilde{R}$ (resp. R) is the curvature tensor field of the Levi-Civita connection $\tilde{\nabla}$ (resp. ∇) associated to $\tilde{g}$ (resp. g), then for each $\tilde{x} \in \tilde{M}$,

$$d\Phi_{\tilde{x}}\left(\tilde{R}_{\tilde{x}}(\tilde{u},\tilde{v})\tilde{w}\right) = R_{\Phi(\tilde{x})}\left(d\Phi_{\tilde{x}}(\tilde{u}), d\Phi_{\tilde{x}}(\tilde{v})\right)d\Phi_{\tilde{x}}(\tilde{w}),$$

for any $\tilde{u},\tilde{v},\tilde{w} \in T_{\tilde{x}}\tilde{M}$ (see also §3), so that (4.7) is an immediate consequence of the last equality and of the fact that $d\Phi_{\tilde{x}}:T_{\tilde{x}}\tilde{M} \to T_{\Phi(\tilde{x})}M$ is an orthogonal mapping.

Remark 4.2 Let (M,g) be an n-dimensional Riemannian manifold $(n \geq 2)$ and suppose that the group $I(M,g)$ of isometries of (M,g) is transitive on $G^2(M)$, i.e. for each $(x,P),(x',P') \in G^2(M)$ there exists $\phi \in I(M,g)$ such that $\phi(x) = x'$ and $d\phi_x(P) = P'$. Then (M,g) has constant curvature. For, let $(x_0,P_0) \in G^2(M)$. Then for any $(x,P) \in G^2(M)$, there exists $\phi \in I(M,g)$ such that $\phi(x_0) = x$ and $d\phi_{x_0}(P_0) = P$. Thus, in virtue of Remark 4.1,

$$\sigma(x_0,P_0) = \sigma\left(\phi(x_0), d\phi_{x_0}(P_0)\right) = \sigma(x,P).$$

For example, (S^n, g_{can}), $n \geq 2$, has constant curvature, since the group $O(n+1)$ of isometries of (S^n, g_{can}) (see Example 2.11) is transitive on $G^2(S^n)$.
More generally, let (M,g) be an n-dimensional Riemannian manifold $(n \geq 3)$ and suppose that the group $I(M,g)$ of isometries of (M,g) is transitive on each fiber $G_x^2 M$ of $G^2(M)$, i.e. for each $x \in M$ and each 2-planes P, P' of $G_x^2 M$ there exists $\Phi \in I(M,g)$ such that $\Phi(x) = x$ and $d\Phi_x(P) = P'$. Then, the same argument as before shows that σ depends only on the points of M, that is for each $x \in M, \sigma(x,P) = K(x)$, for any $P \in G_x^2 M$, where $K \in C^\infty(M)$, i.e. if M is connected, (M,g) is of constant curvature (see Remark 4.3).

Example 4.3 The space of constant curvature (S^n, g_{can}) (see Remark 4.2) has the curvature +1. In particular, $(P^n(\mathbb{R}), g)$ [resp. $(L_{2n+1}(p; q_1, \ldots, q_{n+1}), g)$] has constant curvature and its value is the same as that of (S^n, g_{can}) [resp. (S^{2n+1}, g_{can})], namely +1. More generally, let Γ be a finite subgroup of $O(n+1)$ acting freely on $S^n (n \geq 2)$. Then the quotient Riemannian manifold $\left(S^n/\Gamma, g\right)$ (see Theorem 2.3) has constant sectional curvature 1. Conversely, it is well-known that any closed, connected, n-dimensional Riemannian manifold of constant sectional curvature 1 is isometric to a quotient Riemannian manifold $\left(S^n/\Gamma, g\right), n \geq 2$, where Γ is a finite subgroup of $O(n+1)$ acting freely on S^n.

For let $\Phi : \mathbb{R}^n \to S^n \subset \mathbb{R}^{n+1}$ given by

$$\begin{aligned} \Phi(x) &= \left(\Phi^1(x), \ldots, \Phi^{n+1}(x)\right) \\ &= \frac{1}{1+\|x\|^2}\left(2x, \|x\|^2 - 1\right), \end{aligned}$$

$x \in \mathbb{R}^n$, be the inverse mapping of the stereographic projection of $S^n \setminus \{N\}$ to $\mathbb{R}^n$ from the north pole $N \in S^n$. Then for any $i, j = 1, 2, \ldots, n,\ i \neq j$, and each $x \in \mathbb{R}^n$,

$$\frac{\partial \Phi^i}{\partial x^i}(x) = 2\frac{\|x\|^2 - 2(x^i)^2 + 1}{\left(1+\|x\|^2\right)^2},$$

$$\frac{\partial \Phi^j}{\partial x^i}(x) = -4\frac{x^i x^j}{\left(1+\|x\|^2\right)^2},$$

$$\frac{\partial \Phi^{n+1}}{\partial x^i}(x) = 4\frac{x^i}{\left(1+\|x\|^2\right)^2},$$

so that for each $i = 1, 2, \ldots, n$,

$$\left(g_{can}\right)_{\Phi(x)}\left(d\Phi_x\left(\frac{\partial}{\partial x^i}\right)_x, d\Phi_x\left(\frac{\partial}{\partial x^i}\right)_x\right)=\sum_{a=1}^{n+1}\left(\frac{\partial \Phi^a}{\partial x^i}(x)\right)^2$$
$$=\frac{4}{\left(1+\|x\|^2\right)^4}\left[\left(\|x\|^2-2\left(x^i\right)^2+1\right)^2+4x^i\sum_{j\neq i}\left(x^j\right)^2+4\left(x^i\right)^2\right]$$
$$=\frac{4}{\left(1+\|x\|^2\right)^2}.$$

On the other hand, for any $i,j=1,2,\ldots,n,\ i\neq j$,

$$\left(g_{can}\right)_{\Phi(x)}\left(d\Phi_x\left(\frac{\partial}{\partial x^i}\right)_x, d\Phi_x\left(\frac{\partial}{\partial x^j}\right)_x\right)=\sum_{a=1}^{n+1}\frac{\partial \Phi^a}{\partial x^i}(x)\frac{\partial \Phi^a}{\partial x^j}(x)$$
$$=\frac{8}{\left(1+\|x\|^2\right)^4}\left[2\sum_{k\neq i,j}\left(x^k\right)^2x^ix^j-\|x\|^2x^ix^j+2\left(x^i\right)^3x^j-x^ix^j-\|x\|^2x^ix^j+2x^i\left(x^j\right)^3-x^ix^j+2x^ix^j\right]$$
$$=0.$$

Consequently,

$$\left(g_{can}\right)_{\Phi(x)}\left(d\Phi_x\left(\frac{\partial}{\partial x^i}\right)_x, d\Phi_x\left(\frac{\partial}{\partial x^j}\right)_x\right)=\frac{4}{\left(1+\|x\|^2\right)^2}\delta_{ij}$$
$$=\frac{4}{\left(1+\|x\|^2\right)^2}\left(g_{can}\right)_x\left(\left(\frac{\partial}{\partial x^i}\right)_x,\left(\frac{\partial}{\partial x^j}\right)_x\right),$$

for each $i,j = 1,2,\ldots,n$, i.e. Φ is an isometry of $\left(\mathbb{R}^n,\tilde{g}\right)$ onto $\left(S^n\setminus\{N\},g_{can}\right)$, where $\tilde{g}=fg_{can}$ with $f:\mathbb{R}^n\to\mathbb{R}$ given by:

$$f(x)=\frac{4}{\left(1+\|x\|^2\right)^2},\quad x\in\mathbb{R}^n.$$

But a similar computation with that of the proof of Corollary 4.1 shows that $\left(\mathbb{R}^n,\tilde{g}\right)$ is of constant curvature and its curvature is $+1$. Hence by Remarks 4.1 and 4.2 $\left(S^n,g_{can}\right)$ is of constant curvature and its curvature is the same as that of $\left(\mathbb{R}^n,\tilde{g}\right)$, that is $+1$. Finally, the last statement is an immediate consequence of Remark 4.2.

Example 4.4 The Riemannian manifold (H^n,g) considered in Example 2.4 is of constant curvature -1. Conversely, any complete Riemannian manifold of dimension n of constant sectional curvature -1 is isometric to the quotient Riemannian manifold $\left(H^n/\Gamma, g_{can}\right)$, where Γ is a torsion-free discrete subgroup of $I\left(H^n,g\right)$. Note that a discrete group Γ of isometries of (H^n, g) [resp. $(\mathbb{R}^n, g_{can})$] acts freely on H^n (resp. $\mathbb{R}^n$) if and only if Γ is torsion-free.

On the other hand, one can show that any simply-connected complete n-dimensional Riemannian manifold of constant sectional curvature $+1$ (resp. 0, resp. -1) is isometric to (S^n, g_{can}) [resp. $(\mathbb{R}^n, g_{can})$, resp. (H^n, g)], $n \geq 2$.

For reasons that will be clear later, let us point out, some geometrical objects that can be derived from the curvature tensor field.

Definition 4.4 The **Ricci tensor field** on M is the contraction c_2R of R, i.e. the tensor field of type (0,2) on M given by:

$$(c_2R)_x(u,v) = \operatorname{trace}[w \mapsto R_x(u,w)v],$$

where $u, v, w \in T_xM$ and $x \in M$. If $(e_1,\dots,e_n)$ is an orthonormal basis of T_xM, then

$$(c_2R)_x(u,v) = \sum_{i=1}^{n} g_x(R_x(u,e_i)v, e_i), \tag{4.8}$$

which implies the symmetry of c_2R.

Definition 4.5 If $v \in T_xM \setminus \{0\}$, then the real number

$$\rho(x,v) = \frac{(c_2R)_x(v,v)}{g_x(v,v)}$$

is called **Ricci curvature** of (M,g) at x in the direction v.

Note that the Ricci curvature determines the Ricci tensor field: for $u, v \in T_xM \setminus \{0\}$,

$$(c_2R)_x(u,v) = \begin{cases} \frac{1}{2}[g_x(u+v,u+v)\rho(x,u+v) - g_x(u,u)\rho(x,u) - g_x(v,v)\rho(x,v)] & \text{if } u \neq -v, \\ -g_x(u,u)\rho(x,u) & \text{if } u = -v. \end{cases}$$

For a unit tangent vector $u \in U_xM = \{v \in T_xM \mid g_x(v,v) = 1\} \subset T_xM$ one has $\rho(x,u) = (c_2R)_x(u,u)$, so that ρ can be thought as a function on the unit tangent bundle $UM = \bigcup_{x \in M} U_xM$. Moreover, $\rho : UM \to \mathbb{R}$ is a smooth function.

If the Ricci curvature $\rho : UM \to \mathbb{R}$ of (M,g) is constant, then (M,g) is called an **Einstein manifold**. From the above formula for $(c_2R)_x$ it follows that (M,g) is an Einstein manifold if and only if there exists a constant $k \in \mathbb{R}$ such that

$$c_2R(X,Y) = kg(X,Y)$$

for any $X, Y \in \mathfrak{X}(M)$.

Remark 4.3 Let (M,g) be a Riemannian manifold of dimension $n \geq 3$ such that M is connected. If there exists $f \in C^\infty(M)$ such that $\rho(x,u) = f(x)$ for any $(x,u) \in UM$, then (M,g) is an Einstein manifold (for an elegant proof of this statement, see Besse [1987]). In particular, if there exists $f \in C^\infty(M)$ such that

$\sigma(x,P) = f(x)$ for any $(x,P) \in G^2(M)$, then

$$\rho(x,u) = -\sum_{j=2}^{n} \sigma\left(x, P_{u,e_j}\right) = -(n-1) f(x),$$

where $(e_1 = u, e_2, \ldots, e_n)$ is an orthonormal basis for $T_x M$ and P_{u,e_j} is the 2-plane generated by u and e_j, $j \geq 2$ (see the formula below for ρ), and consequently (M,g) is of constant curvature.

Definition 4.6 The smooth real valued function $\operatorname{trace}(c_2 R) \in C^\infty(M)$ is called the **scalar curvature** of (M,g) and it is denoted by τ. Hence,

$$\tau : x \in M \mapsto \tau(x) = \operatorname{trace}(c_2 R)(x) \in \mathbb{R}.$$

If $(e_1,\ldots,e_n)$ is an orthonormal basis for $T_x M$, in virtue of (4.8) we have:

$$\begin{aligned}\rho(x,e_i) &= (c_2 R)_x (e_i, e_i) \\ &= \sum_{j=1}^{n} g_x\left(R_x(e_i,e_j)e_i, e_j\right) \\ &= -\sum_{j \neq i} \sigma\left(x, P_{e_i,e_j}\right),\end{aligned}$$

where P_{e_i,e_j} is the 2-plane of $T_x M$ generated by e_i and e_j, $i \neq j$, so that

$$\begin{aligned}\tau(x) &= \sum_{i=1}^{n} (c_2 R)_x (e_i, e_i) \\ &= \sum_{i=1}^{n} \rho(x, e_i) \\ &= -2 \sum_{1 \leq i < j \leq n} \sigma\left(x, P_{e_i,e_j}\right).\end{aligned} \tag{4.9}$$

Thus, if (M,g) is Einsteinian, that is the Ricci curvature ρ is equal to the constant $k \in \mathbb{R}$, then $\tau = nk$ and consequently $c_2 R = \frac{\tau}{n} g$. In particular, if (M,g) is a space of constant curvature k, then $\tau = -n(n-1)k$.

Example 4.5. Let G be an n-dimensional compact semisimple Lie group whose Lie algebra is $\mathfrak{g}$. The tangent bundle TG of G can be trivialized by using left translation, namely $TG = G \times \mathfrak{g}$, and the Killing form on $\mathfrak{g}$ gives rise, by left translation, to a natural Riemannian metric g on G (see Example 2.6), which is left invariant. Using the above trivialization of TG, the left invariant vector fields will be identified with the elements of $\mathfrak{g}$. Thus, if one restricts the Levi-Civita connection ∇ on G to left invariant vector fields, ∇ can be conceived as a mapping of $\mathfrak{g} \times \mathfrak{g}$ into $\mathfrak{g}$.

For left invariant vector fields X, Y on G, we have:

$$\nabla_X Y = \frac{1}{2}[X,Y]. \tag{4.10}$$

Indeed, the left invariance of g means that $L_Z g = 0$ for any left invariant vector field Z on G (L_Z denotes the Lie derivative with respect to Z), or equivalently:

$$g([Z,X],Y)+g(X,[Z,Y])=Z(g(X,Y)) \tag{4.11}$$

for all left invariant vector fields X,Y on G. Now $g(X,Y)$ is constant on G, thus (4.11) implies

$$g([X,Z],Y)=-g([Y,Z],X). \tag{4.12}$$

On the other hand, if X,Y,Z are arbitrary left invariant vector fields on G, then (3.1) leads to

$$2g(\nabla_X Y,Z)=g([X,Y],Z)-g([Y,Z],X)+g([Z,X],Y)$$

and therefore via (4.12) to the equality:

$$g(2\nabla_X Y-[X,Y],Z)=0,$$

which implies (4.10).

Now, note that (4.10) and Jacobi's identity imply:

$$R(X,Y)Z=-\frac{1}{4}[[X,Y],Z]$$

for left invariant vector fields X,Y,Z on G.

In particular, (4.12) yields

$$g(R(X,Y)Y,X)=-\frac{1}{4}g([[X,Y],Y],X)=\frac{1}{4}g([X,Y],[X,Y])$$

for any left invariant vector fields X,Y on G, so that (4.4) implies that (G,g) has non-negative sectional curvature.

If $\{E_1,\ldots,E_n\}$ is an orthonormal basis for the left invariant vector fields on G, the formula (4.8) leads to

$$\begin{aligned}(c_2R)(X,Y)&=\sum_{i=1}^{n}g(R(X,E_i)Y,E_i)\\&=-\frac{1}{4}\sum_{i=1}^{n}g([[X,E_i],Y],E_i)\\&=\frac{1}{4}\sum_{i=1}^{n}g([E_i,Y],[X,E_i]),\end{aligned}$$

where the last equality follows from (4.12). Hence

$$(c_2R)(X,Y)=-\frac{1}{4}\sum_{i=1}^{n}g([X,E_i],[Y,E_i]),$$

for left invariant vector fields X,Y on G.

Finally, using the first equality of (4.9) we get the expression of the scalar curvature τ of (G,g):

$$\tau=\sum_{j=1}^{n}(c_2R)(E_j,E_j)=-\frac{1}{4}\sum_{i,j=1}^{n}g([E_i,E_j],[E_i,E_j]). \tag{4.13}$$

Proposition 4.4 Let $(U,x^1,\ldots,x^n)$ be a local chart on M and ∇ the Levi-Civita connection on M associated to g. Then the following relations hold:

(i) $R^l_{ijk}=\dfrac{\partial \Gamma^l_{ik}}{\partial x^j}-\dfrac{\partial \Gamma^l_{ij}}{\partial x^k}+\sum_{s=1}^{n}(\Gamma^s_{ik}\Gamma^l_{js}-\Gamma^s_{ij}\Gamma^l_{ks})$, where

$$R^l_{ijk}\overset{def}{=}\left[R\left(\frac{\partial}{\partial x^j},\frac{\partial}{\partial x^k}\right)\frac{\partial}{\partial x^i}\right](x^l)\in C^\infty(U),\ i,j,k,l=1,\ldots,n;$$

(ii) $R_{ij}=\sum_{s=1}^{n}R^s_{ijs}$, where $R_{ij}\overset{def}{=}(c_2R)\left(\dfrac{\partial}{\partial x^i},\dfrac{\partial}{\partial x^j}\right)\in C^\infty(U)$;

(iii) $\tau=\sum_{i,j}R_{ij}g^{ij}$.

Proof. The proof reduces to a straightforward computation. We shall check only (i). Since

$$R\left(\frac{\partial}{\partial x^j},\frac{\partial}{\partial x^k}\right)\frac{\partial}{\partial x^i}=\sum_l R^l_{ijk}\frac{\partial}{\partial x^l},$$

and

$$\begin{aligned}R\left(\frac{\partial}{\partial x^j},\frac{\partial}{\partial x^k}\right)\frac{\partial}{\partial x^i}&=\nabla_{\frac{\partial}{\partial x^j}}\nabla_{\frac{\partial}{\partial x^k}}\frac{\partial}{\partial x^i}-\nabla_{\frac{\partial}{\partial x^k}}\nabla_{\frac{\partial}{\partial x^j}}\frac{\partial}{\partial x^i}\\&=\nabla_{\frac{\partial}{\partial x^j}}\left(\sum_s\Gamma^s_{ki}\frac{\partial}{\partial x^s}\right)-\nabla_{\frac{\partial}{\partial x^k}}\left(\sum_s\Gamma^s_{ji}\frac{\partial}{\partial x^s}\right)\\&=\sum_l\left(\frac{\partial}{\partial x^j}\Gamma^l_{ik}+\sum_s\Gamma^s_{ik}\Gamma^l_{js}-\frac{\partial}{\partial x^k}\Gamma^l_{ij}-\sum_s\Gamma^s_{ij}\Gamma^l_{ks}\right)\frac{\partial}{\partial x^l}.\end{aligned}$$

it follows that

$$R^l_{ijk}=\frac{\partial}{\partial x^j}\Gamma^l_{ik}-\frac{\partial}{\partial x^k}\Gamma^l_{ij}+\sum_s(\Gamma^s_{ik}\Gamma^l_{js}-\Gamma^s_{ij}\Gamma^l_{ks}),$$

and this is the required equality.

Q.E.D.

Finally, let us define two tensor fields of type (0,4) associated to the curvature tensor field $R:\mathfrak{X}_3(M)\to\mathfrak{X}(M)$. The first one, denoted also by R and called the **Riemannian tensor field** is defined by

$$R:\mathfrak{X}_4(M)\to C^\infty(M),\ R(X,Y,Z,V)=g(R(Z,V)Y,X), \tag{4.14}$$

for any $X,Y,Z,V\in\mathfrak{X}(M)$. This tensor field satisfies the following properties, which are reformulations of (4.1) and Proposition 4.1: For each $X,Y,Z,V,W\in\mathfrak{X}(M)$ one has:

(i) $R(X,Y,Z,V)=-R(X,Y,V,Z)$;

(ii) $R(X,Y,Z,V)=-R(Y,X,Z,V)$;

(iii) $R(X,Y,Z,V)+R(X,Z,V,Y)+R(X,V,Y,Z)=0$ (the **first Bianchi's identity**);

(iv) $R(X,Y,Z,V)=R(Z,V,X,Y)$;

(v) $(\nabla_X R)(W,V,Y,Z)+(\nabla_Y R)(W,V,Z,X)+(\nabla_Z R)(W,V,X,Y)=0$ (the **second Bianchi's identity**).

Note that by (4.5), a Riemannian manifold (*M*,*g*) of dimension $n \geq 2$ is of constant curvature *k* if and only if

$$R(X_1,X_2,X_3,X_4)=k\left[g(X_1,X_3)g(X_2,X_4)-g(X_1,X_4)g(X_2,X_3)\right]$$

for any $X_1,X_2,X_3,X_4 \in \mathfrak{X}(M)$.

The second one, denoted by *C* and called the **Weyl conformal curvature tensor field** is defined in the case $n \geq 3$ by:

$$\begin{aligned}&C(X,Y,Z,V)=\\&=R(X,Y,Z,V)-\\&-\frac{1}{n-2}\left[g(Y,Z)c_2R(X,V)-g(Y,V)c_2R(X,Z)+g(X,V)c_2R(Y,Z)-g(X,Z)c_2R(Y,V)\right]\\&+\frac{\tau}{(n-1)(n-2)}\left[g(Y,Z)\cdot g(X,V)-g(Y,V)\cdot g(X,Z)\right].\end{aligned}$$

Note that

$$|C|^2=|R|^2-\frac{4}{n-2}|c_2R|^2+\frac{2}{(n-1)(n-2)}\tau^2, \tag{4.15}$$

where $|\ |$ denotes the pointwise norm on the corresponding space of tensor fields induced by *g* (for more details concerning these norms, see Chapter 2, §1, C). Indeed, let $(U,x^1,\dots,x^n)$ be a local coordinate neighbourhood and (C_{ijkl}) [resp. (R_{ijkl}), resp. (R_{ij})] be the corresponding local coordinates of *C* (resp. *R*, resp. c_2R) (see also Proposition 4.4). Then

$$|C|^2=\sum_{i,j,k,l}C_{ijkl}C^{ijkl},$$

where

$$C_{ijkl}=R_{ijkl}-\frac{1}{n-2}\left(g_{il}R_{jk}-g_{ik}R_{jl}+g_{jk}R_{il}-g_{jl}R_{ik}\right)+\frac{1}{(n-1)(n-2)}\left(g_{jk}g_{il}-g_{jl}g_{ik}\right)\tau$$

and

$$C^{ijkl}=R^{ijkl}-\frac{1}{n-2}\left(g^{il}R^{jk}-g^{ik}R^{jl}+g^{jk}R^{il}-g^{jl}R^{ik}\right)+\frac{1}{(n-1)(n-2)}\left(g^{jk}g^{il}-g^{jl}g^{ik}\right)\tau.$$

Consequently,

$$|C|^2 = \sum_{i,j,k,l} R_{ijkl}R^{ijkl} - \frac{1}{n-2}\sum_{i,j,k,l} R_{ijkl}\left(g^{il}R^{jk} - g^{ik}R^{jl} + g^{jk}R^{il} - g^{jl}R^{ik}\right)$$

$$-\frac{1}{n-2}\sum_{i,j,k,l} R^{ijkl}\left(g_{il}R_{jk} - g_{ik}R_{jl} + g_{jk}R_{il} - g_{jl}R_{ik}\right) + \frac{\tau}{(n-1)(n-2)}\sum_{i,j,k,l} R_{ijkl}\left(g^{jk}g^{il} - g^{jl}g^{ik}\right)$$

$$+\frac{\tau}{(n-1)(n-2)}\sum_{i,j,k,l} R^{ijkl}\left(g_{jk}g_{il} - g_{jl}g_{ik}\right)$$

$$+\frac{1}{(n-2)^2}\sum_{i,j,k,l}\left(g_{il}R_{jk} - g_{ik}R_{jl} + g_{jk}R_{il} - g_{jl}R_{ik}\right)\left(g^{il}R^{jk} - g^{ik}R^{jl} + g^{jk}R^{il} - g^{jl}R^{ik}\right)$$

$$-\frac{\tau}{(n-1)(n-2)^2}\sum_{i,j,k,l}\left(g^{il}R^{jk} - g^{ik}R^{jl} + g^{jk}R^{il} - g^{jl}R^{ik}\right)\left(g_{jk}g_{il} - g_{jl}g_{ik}\right)$$

$$-\frac{\tau}{(n-1)(n-2)^2}\sum_{i,j,k,l}\left(g_{il}R_{jk} - g_{ik}R_{jl} + g_{jk}R_{il} - g_{jl}R_{ik}\right)\left(g^{jk}g^{il} - g^{jl}g^{ik}\right)$$

$$+\frac{\tau^2}{(n-1)^2(n-2)^2}\sum_{i,j,k,l}\left(g_{jk}g_{il} - g_{jl}g_{ik}\right)\left(g^{jk}g^{il} - g^{jl}g^{ik}\right)$$

$$= |R|^2 - \frac{8}{n-2}\sum_{j,k} R_{jk}R^{jk} + \frac{4\tau^2}{(n-1)(n-2)} + \frac{4n-8}{(n-2)^2}\sum_{j,k} R_{jk}R^{jk}$$

$$+\frac{4}{(n-2)^2}\tau^2 - \frac{\tau(8n\tau - 8\tau)}{(n-1)(n-2)^2} + \frac{\tau^2(2n^2-2n)}{(n-1)^2(n-2)^2}$$

$$= |R|^2 - \frac{4}{n-2}|c_2R|^2 + \frac{2}{(n-1)(n-2)}\tau^2.$$

Definition 4.7 A Riemannian manifold (M,g) is said to be **conformally flat**, if for any point $x \in M$ there exist an open neighborhood U and $f \in C^\infty(U)$, $f > 0$, such that $\left(U, fg|_U\right)$ is a flat space.

The Weyl conformal curvature tensor field of a conformally flat Riemannian manifold (M,g) vanishes everywhere and the converse statement also holds if the dimension of M is ≥ 4.

If $n = 1$, note that $R = 0$.

Example 4.6 Let (M,g) be a two dimensional Riemannian manifold, $x \in M$ and (e_1,e_2) an orthonormal basis of T_xM. Then we have:

$$\begin{aligned} K(x) &= g_x\left(R_x(e_1,e_2)e_2,e_1\right) \\ &= R_x(e_1,e_2,e_1,e_2) \\ &= R_{1212}(x), \end{aligned}$$

where $K : M \to \mathbb{R}$ denotes the **Gauss curvature** of (M,g). This implies that

$$\begin{aligned} |R|_x^2 &= \sum_{i,j,k,l} R_{ijkl}^2(x) \\ &= 4\left(K(x)\right)^2. \end{aligned}$$

On the other hand, in virtue of (4.8),

$$\begin{aligned}(c_2R)_x(e_j,e_j) &= \sum_{i=1}^{2} g_x(R_x(e_j,e_i)e_j,e_i)\\ &= R_{1212}(x)\\ &= K(x)\end{aligned}$$

for each $j = 1,2$ and $(c_2R)_x(e_1,e_2)=0$, so that $|c_2R|_x^2 = 2(K(x))^2$. Finally, $(\tau(x))^2 = 4(K(x))^2$ since $\tau(x) = -2K(x)$ [see (4.9)]. In particular, $|c_2R|_x^2 = \frac{1}{2}(\tau(x))^2$ and $|R|_x^2 = 2|c_2R|_x^2$, for each $x \in M$.

Example 4.7 Let (M,g) be a three dimensional Riemannian manifold, $x \in M$ and let us take an orthonormal basis (e_1,e_2,e_3) of T_xM such that the corresponding matrix of the symmetric bilinear form $(c_2R)_x$ has diagonal form. Then $(c_2R)_x(e_i,e_j)=0,\ i \neq j$, and

$$(c_2R)_x(e_i,e_i) = g_x(R_x(e_i,e_j)e_i,e_j) + g_x(R_x(e_i,e_k)e_i,e_k),$$

where (i,j,k) is a cyclic permutation of (1,2,3). Thus one gets

$$g_x(R_x(e_i,e_j)e_j,e_k)=0,$$

where i,j,k are different, and

$$2g_x(R_x(e_j,e_k)e_j,e_k) = (c_2R)_x(e_j,e_j)+(c_2R)_x(e_k,e_k)-(c_2R)_x(e_i,e_i), \tag{4.16}$$

where (i,j,k) is a cyclic permutation of (1,2,3). Note that the equality (4.16) holds for any orthonormal basis (e_1,e_2,e_3) of T_xM. Hence the only non zero components of R_x are $R_x(e_1,e_2,e_1,e_2) \overset{not}{=} \alpha_{12}$, $R_x(e_1,e_3,e_1,e_3) \overset{not}{=} \alpha_{13}$, $R_x(e_2,e_3,e_2,e_3) \overset{not}{=} \alpha_{23}$ and those deduced from these by symmetry. Note that

$$(c_2R)_x(e_1,e_1) = -R_x(e_1,e_2,e_1,e_2) - R_x(e_1,e_3,e_1,e_3) = -\alpha_{12}-\alpha_{13}$$

and similarly $(c_2R)_x(e_2,e_2) = -\alpha_{12}-\alpha_{23}$, $(c_2R)_x(e_3,e_3) = -\alpha_{13}-\alpha_{23}$, so that

$$\tau(x) = \sum_{j=1}^{3}(c_2R)_x(e_j,e_j) = -2(\alpha_{12}+\alpha_{13}+\alpha_{23}).$$

Consequently,

$$|R_x|^2 = 4(\alpha_{12}^2+\alpha_{13}^2+\alpha_{23}^2),\ |(c_2R)_x|^2 = (\alpha_{12}+\alpha_{13})^2+(\alpha_{12}+\alpha_{23})^2+(\alpha_{13}+\alpha_{23})^2,$$

$$(\tau(x))^2 = 4(\alpha_{12}+\alpha_{13}+\alpha_{23})^2,$$

which implies the relation $|R_x|^2 - 4|(c_2R)_x|^2 + (\tau(x))^2 = 0$. Moreover,

$$C_x(e_1,e_2,e_1,e_2) = R_x(e_1,e_2,e_1,e_2) + (c_2R)_x(e_1,e_1) + (c_2R)_x(e_2,e_2) - \frac{1}{2}\tau$$
$$= \alpha_{12} - (\alpha_{12} + \alpha_{13} + \alpha_{12} + \alpha_{23}) + \alpha_{12} + \alpha_{13} + \alpha_{23} = 0,$$

and similarly $C_x(e_1,e_3,e_1,e_3) = C_x(e_2,e_3,e_2,e_3) = 0$, that is $C = 0$. In particular, note that 3-dimensional Einstein manifolds are of constant curvature. For, assume that $c_2R = kg$ for some constant $k \in \mathbb{R}$. Let $(x,P) \in G^2(M)$, (u,v) an orthonormal basis for P and choose an orthonormal basis $(e_1 = u, e_2 = v, e_3)$ for T_xM. Then (4.16) implies that

$$\sigma(x,P) = g_x(R_x(u,v)u,v) = g_x(R_x(e_1,e_2)e_1,e_2) = k$$

for each $(x,P) \in G^2(M)$.

Example 4.8 For the Riemannian manifold (S^n, g_{can}) note that
(i) $|R|^2 = 2n(n-1)$;
(ii) $|c_2R|^2 = n(n-1)^2$;
(iii) $\tau^2 = n^2(n-1)^2$;
(iv) $C = 0$.

In order to motivate the definition of the Weyl conformal curvature tensor field associated to (M,g), let us describe the irreducible decomposition of the algebraic curvature tensor space $C(n)$ of an n-dimensional real Euclidean vector space (E,g) with respect to the action of the orthogonal group $G = O(E,g)$, $n \geq 2$.

First, note that the inner product $g : E \times E \to \mathbb{R}$ induces a canonical identification between E and E^*.

Let $\mathcal{T}_4^0(E) = \otimes^4 E^*$ be the space of tensors of type (0,4) of E, that is the real vector space of multilinear mappings of $E \times E \times E \times E$ into $\mathbb{R}$, and $C(n)$ the vector subspace of $\mathcal{T}_4^0(E)$ consisting of tensors R of type (0,4) of E, which satisfy the same algebraic identities as the curvature tensor field of a Riemannian manifold at one point:

(i) $R(u_1,u_2,u_3,u_4) = -R(u_1,u_2,u_4,u_3)$,
(ii) $R(u_1,u_2,u_3,u_4) = -R(u_2,u_1,u_3,u_4)$,
(iii) $R(u_1,u_2,u_3,u_4) + R(u_1,u_3,u_4,u_2) + R(u_1,u_4,u_2,u_3) = 0$,
(iv) $R(u_1,u_2,u_3,u_4) = R(u_3,u_4,u_1,u_2)$,

for any $u_1,u_2,u_3,u_4 \in E$. For example, the tensor

$$R_0 : E \times E \times E \times E \to \mathbb{R}$$

defined by

$$R_0(u_1,u_2,u_3,u_4) = g(u_1,u_4)g(u_2,u_3) - g(u_1,u_3)g(u_2,u_4)$$

for any $u_1,u_2,u_3,u_4 \in E$, is an element of $C(n)$.

Note that if $R_1, R_2 \in C(n)$ and $R_1(u_1,u_2,u_1,u_2) = R_2(u_1,u_2,u_1,u_2)$ for any

$u_1, u_2 \in E$, then $R_1 = R_2$ (one can use the same arguments as in the considerations before Example 4.1).

The properties (i) and (ii) of the elements of $C(n)$ show that $C(n)$ lies in the subspace $\wedge^2 E^* \otimes \wedge^2 E^*$ of $\otimes^4 E^*$, where $\wedge^2 E^*$ denotes the second exterior power of E^*, while the properties (i), (ii) and (iv) of these elements tell us that $C(n)$ lies in the subspace $S^2\left(\wedge^2 E^*\right)$ of $\otimes^4 E^*$, where $S^2\left(\wedge^2 E^*\right)$ denotes the second symmetric power of $\wedge^2 E^*$. The property (iii) suggests to define the mapping

$$\beta : \mathcal{T}_4^0(E) \to \mathcal{T}_4^0(E)$$

by

$$\left(\beta(T)\right)(u_1,u_2,u_3,u_4) = \frac{1}{3}\left[T(u_1,u_2,u_3,u_4) + T(u_1,u_3,u_4,u_2) + T(u_1,u_4,u_2,u_3)\right],$$

where $T \in \mathcal{T}_4^0(E)$ and $u_1,u_2,u_3,u_4 \in E$. Obviously, β maps $S^2\left(\wedge^2 E^*\right)$ into itself and $\beta^2 = \beta$, so that we have the direct decomposition

$$S^2\left(\wedge^2 E^*\right) = \ker(\beta) \oplus \operatorname{Im}(\beta).$$

Moreover, the property (iii) tell us that

$$C(n) = \ker(\beta)\left[\text{in } S^2\left(\wedge^2 E^*\right)\right],$$

and one can show that $\operatorname{Im}(\beta)$ is isomorphic to the 4-th exterior power of E^*, which proves that $\beta = 0$ if $n = 2$ or 3.

G acts on $\mathcal{T}_4^0(E)$ by

$$(tL)(u_1,u_2,u_3,u_4) = L\left(t^{-1}(u_1), t^{-1}(u_2), t^{-1}(u_3), t^{-1}(u_4)\right),$$

$t \in G$, $L \in \mathcal{T}_4^0(E)$, $u_1,u_2,u_3,u_4 \in E$, and $C(n)$ is a G-invariant subspace of $\mathcal{T}_4^0(E)$. Note that $tR_0 = R_0$ for any $t \in G$ and $C(2)$ is generated by R_0.

Let us define the Ricci contraction

$$c_2 : C(n) \to S^2\left(E^*\right)$$

by

$$\left(c_2(R)\right)(u_1,u_2) = \sum_{j=1}^{n} R\left(u_1, e_j, e_j, u_2\right),$$

for any $R \in C(n)$ and $u_1, u_2 \in E$, where $(e_1,\ldots,e_n)$ is an orthonormal basis for E and $S^2\left(E^*\right)$ the real vector space of symmetric bilinear maps of $E \times E$ into $\mathbb{R}$. Note that $c_2(R_0) = (n-1)g$. Conversely, there is a natural way to build an element of $C(n)$ from two elements of $S^2\left(E^*\right)$. The **Kulkarni-Nomizu product** of the symmetric bilinear forms $h,k : E \times E \to \mathbb{R}$ is the tensor $h \owedge k$ of type (0,4) of E given by

$$(h \circledwedge k)(u_1,u_2,u_3,u_4)$$
$$= h(u_1,u_4)k(u_2,u_3)+h(u_2,u_3)k(u_1,u_4)-h(u_1,u_3)k(u_2,u_4)-h(u_2,u_4)k(u_1,u_3)$$

for any $u_1,u_2,u_3,u_4 \in E$. Note that for any $h,k \in S^2(E^*)$,

$$h \circledwedge k = k \circledwedge h,$$
$$h \circledwedge k \in C(n),$$
$$g \circledwedge g = 2R_0$$

and

$$c_2(k \circledwedge g) = (n-2)k + \text{trace}(k)g \tag{4.17}$$

Indeed, the first three properties are obvious. For the last one, an easy computation gives

$$\left[c_2(k \circledwedge g)\right](u_1,u_2) = \sum_{j=1}^{n}(k \circledwedge g)(u_1,e_j,e_j,u_2)$$
$$= \sum_{j=1}^{n}\left[k(u_1,u_2)+k(e_j,e_j)g(u_1,u_2)-g(e_j,u_2)k(u_1,e_j)-g(u_1,e_j)k(e_j,u_2)\right]$$
$$= \left[(n-2)k+\text{trace}(k)g\right](u_1,u_2)$$

for any $u_1,u_2 \in E$, that is (4.17) holds.

For the elements of $C(n)$, let us also introduce the notion equivalent to the scalar curvature associated to the curvature tensor field. Namely, let us define

$$\tau : C(n) \to \mathbb{R}$$

by

$$\tau(R) = \text{trace}(c_2(R)) = -\sum_{i,j=1}^{n} R(e_i,e_j,e_i,e_j)$$

for any $R \in C(n)$, where, as before, $(e_1,\dots,e_n)$ is an orthonormal basis of E. Note that τ is a G-invariant linear form on $C(n)$, that is $\tau(tR) = \tau(R)$ for any $t \in G$ and $R \in C(n)$, so that ker(τ) is a G-invariant subspace of $C(n)$. Obviously c_2 is a G-equivariant map, that is $c_2(tR) = tc_2(R)$ for any $t \in G$ and $R \in C(n)$, and $\ker(c_2) \subset \ker(\tau)$.

Then one has the following fundamental result.

Proposition 4.5 (**Singer-Thorpe**) If $n \geq 4$, $C(n)$ admits the following orthogonal direct sum decomposition into (unique) irreducible G-invariant subspaces

$$C(n) = W(n) \oplus \left(S_0^2(E^*) \circledwedge g\right) \oplus \mathbb{R}(g \circledwedge g),$$

where $W(n) = \ker(\beta) \cap \ker(c_2)$ and $S_0^2(E^*)$ denotes the space of traceless symmetric bilinear forms on E.

For a proof of this result, see Besse [1987], pp. 47-48.

The space $W(n)$ is called **the space of Weyl tensors** of (E,g). Note that $W(3) = \{0\}$, so that

$$C(3) = \left(S_0^2\left(E^*\right) \bigcirc\!\!\!\!\wedge\, g\right) \oplus \mathbb{R}\left(g \bigcirc\!\!\!\!\wedge\, g\right).$$

Remark 4.4 For any $R \in C(n)$, $n \geq 4$, let $W(R)$ be its component in the Weyl space $W(n)$. Using (4.17), one can compute $W(R)$ explicitly. Indeed, let

$$R = W(R) + k \bigcirc\!\!\!\!\wedge\, g + \alpha g \bigcirc\!\!\!\!\wedge\, g, \qquad k \in S_0^2\left(E^*\right),\ \alpha \in \mathbb{R},$$

be the decomposition of R given by Proposition 4.5. By (4.17) one gets

$$c_2(R) = (n-2)k + 2(n-1)\alpha g. \tag{4.18}$$

Now taking the trace in (4.18) one obtains $\tau(R) = 2n(n-1)\alpha$, that is $\alpha = \dfrac{\tau(R)}{2n(n-1)}$, so that (4.18) leads to the formula

$$k = \frac{1}{n-2}\left[c_2(R) - \frac{\tau(R)}{n} g\right].$$

Consequently, one gets

$$W(R) = R - \frac{1}{n-2} c_2(R) \bigcirc\!\!\!\!\wedge\, g + \frac{\tau(R)}{2(n-1)(n-2)} g \bigcirc\!\!\!\!\wedge\, g.$$

Therefore, in some sense, $W(R)$ can be viewed as a "remainder" after two successive "divisions" of R by g.

Let (M, g) be now a Riemannian manifold and R its Riemannian curvature tensor field defined by (4.14). For any $x \in M$, the curvature tensor R_x belongs to the space of algebraic curvature tensors associated to $(T_x M, g_x)$. The Weyl conformal curvature tensor C_x of (M, g) (with $n \geq 4$) at $x \in M$ is just the Weyl component of R_x.

Note that (M, g) (with $n \geq 3$) is an Einstein manifold if and only if the component

$$\frac{1}{n-2}\left[(c_2 R)_x - \frac{\tau(x)}{n} g_x\right] \bigcirc\!\!\!\!\wedge\, g_x \text{ of } R_x \text{ in } S_0^2\left(T_x^* M\right) \bigcirc\!\!\!\!\wedge\, g_x$$

is zero for each $x \in M$.

We have noticed before that (M, g) (with $n \geq 2$) is of constant sectional curvature k if and only if $R = -\dfrac{k}{2} g \bigcirc\!\!\!\!\wedge\, g$. Equivalently, (M, g) is of constant sectional curvature if and only if the scalar curvature τ is constant and $R = \dfrac{\tau}{2n(n-1)} g \bigcirc\!\!\!\!\wedge\, g$.

For any 2-dimensional Riemannian manifold one has

$$R = \frac{\tau}{4} g \bigcirc\!\!\!\!\wedge\, g \tag{4.19}$$

and for any 3-dimensional Riemannian manifold, by a previous remark one has

$$R=\left(c_2R-\frac{\tau}{3}g\right)\owedge g+\frac{\tau}{12}g\owedge g.$$

On the other hand, for any n-dimensional Riemannian manifold (M,g) with $n\geq 4$, one has

$$R=C+\frac{1}{n-2}\left(c_2R-\frac{\tau}{n}g\right)\owedge g+\frac{\tau}{2n(n-1)}g\owedge g, \tag{4.20}$$

which is an immediate consequence of the direct sum decomposition given in Proposition 4.5. These remarks lead us to the following

Corollary 4.2. A 2 or 3-dimensional Riemannian manifold has constant sectional curvature if and only if it is Einsteinian. An n-dimensional Riemannian manifold, $n\geq 4$, has constant sectional curvature if and only if it is Einsteinian (i.e. $c_2R=\frac{\tau}{n}g$) and conformally flat (i.e. $C=0$).

There exists in general a close relation between the smooth real functions $|R|,|c_2R|$ and τ. For example, for a 2-dimensional Riemannian manifold the following equalities are true:

$$|R|^2=2|c_2R|^2=\tau^2$$

[see Example 4.6 and (4.19)]. More generally, the following statements are true.

Proposition 4.6. For any n-dimensional Riemannian manifold $(M,g),n\geq 3$, one has

$$|c_2R|^2\geq\frac{\tau^2}{n}$$

with equality sign if and only if (M,g) is an Einstein manifold. In particular, the equality sign occurs if (M,g) is of constant sectional curvature (see Remark 4.3).

Proof. The orthogonal direct sum decomposition (4.20) leads to the equality

$$|R|^2=|C|^2+\frac{1}{(n-2)^2}\left|\left(c_2R-\frac{\tau}{n}g\right)\owedge g\right|^2+\frac{\tau^2}{4n^2(n-1)^2}|g\owedge g|^2,$$

or equivalently

$$|R|^2=|C|^2+\frac{1}{(n-2)^2}\left|\left(c_2R-\frac{\tau}{n}g\right)\owedge g\right|^2+\frac{2\tau^2}{n(n-1)}, \tag{4.21}$$

because $|g\owedge g|^2=8n(n-1)$.

On the other hand, using (4.15) one gets

$$\begin{aligned}|R|^2 &= |C|^2 + \frac{4}{n-2}\left(|c_2R|^2 - \frac{\tau^2}{n}\right) + \left[\frac{4}{n(n-2)} - \frac{2}{(n-1)(n-2)}\right]\tau^2 \\ &= |C|^2 + \frac{4}{n-2}\left(|c_2R|^2 - \frac{\tau^2}{n}\right) + \frac{2}{n(n-1)}\tau^2,\end{aligned} \tag{4.22}$$

so that by comparing (4.21) and (4.22) one obtains

$$\left|\left(c_2R - \frac{\tau}{n}g\right) \owedge g\right|^2 = 4(n-2)\left(|c_2R|^2 - \frac{\tau^2}{n}\right),$$

which implies immediately the statements.

Q.E.D.

Proposition 4.7. For any n-dimensional Riemannian manifold $(M,g), n \geq 3$, one has

$$|R|^2 \geq \frac{2}{n-1}|c_2R|^2 \text{ or } |R|^2 \geq \frac{2}{n(n-1)}\tau^2$$

with equality sign if and only if (M,g) is of constant sectional curvature.

Proof. Using (4.15) and Proposition 4.6 one gets successively

$$\begin{aligned}|R|^2 &\geq \frac{4}{n-2}|c_2R|^2 - \frac{2}{(n-1)(n-2)}\tau^2 \geq \left[\frac{4}{n-2} - \frac{2n}{(n-1)(n-2)}\right]|c_2R|^2 \\ &= \frac{2}{n-1}|c_2R|^2,\end{aligned}$$

that is the desired inequality. If $|R|^2 = \frac{2}{n-1}|c_2R|^2$, then (4.22) leads to

$$|C|^2 + \frac{2n}{(n-1)(n-2)}\left(|c_2R|^2 - \frac{\tau^2}{n}\right) = 0,$$

and so Proposition 4.6 implies that $C = 0$ and $|c_2R|^2 = \frac{\tau^2}{n}$, that is (M,g) is conformally flat and Einsteinian, i.e. (M,g) is of constant sectional curvature (see Corollary 4.2). Conversely, if (M,g) is of constant sectional curvature, then (M,g) is conformally flat and Einsteinian (see Corollary 4.2), and so (4.21) implies that

$$|R|^2 = \frac{2\tau^2}{n(n-1)} = \frac{2}{n-1}|c_2R|^2.$$

Q.E.D.

For other proofs of Propositions 4.6 and 4.7, see Berger, Gauduchon, Mazet [1971] or Chen, Vanhecke [1981].

5. GEODESICS AND THE EXPONENTIAL MAP

Let M be a smooth n-dimensional manifold and $c:(a,b)\to M$ be a C^k-path in $M(k\geq 1)$.

Definition 5.1. A **C^k-vector field X along c** is a C^k-mapping $X:(a,b)\to TM$ such that $X(t)\in T_{c(t)}M$ for all $t\in(a,b)$.

Let (U,ϕ) be a local chart on M defining the coordinates $(x^1,\ldots,x^n)$ on U. If the path c is such that $\operatorname{Im}(c)\subseteq U$ and X is a C^k-vector field along c, then there exist C^k-functions $\lambda^i:(a,b)\to\mathbb{R}, \lambda^i(t)=(X(t))(x^i), i=1,\ldots,n$, such that

$$X(t)=\sum_{i=1}^{n}\lambda^i(t)\left(\frac{\partial}{\partial x^i}\right)_{c(t)} \quad \text{for all } t\in(a,b).$$

We now assume that on M is given a Riemannian metric g (see Definition 2.1). Let (U,ϕ) be a local chart on M and $c:(a,b)\to M$ be a C^k-path in $M(k\geq 1)$ such that $U\supseteq\operatorname{Im}(c)$. If $X=\sum_{i=1}^{n}\lambda^i\left(\frac{\partial}{\partial x^i}\circ c\right):(a,b)\to TM$ is a C^k-vector field along c, then the function

$$\nabla_D X:(a,b)\to TM$$

defined by

$$(\nabla_D X)(s)=\sum_{i=1}^{n}\left[\frac{d\lambda^i}{dt}(s)+\sum_{j,k=1}^{n}\Gamma^i_{jk}(c(s))\lambda^j(s)\frac{dc^k}{dt}(s)\right]\left(\frac{\partial}{\partial x^i}\right)_{c(s)} \tag{5.1}$$

for all $s\in(a,b)$, where $c^k=x^k\circ c$, $k=1,\ldots,n$, and $\Gamma^i_{jk}:U\to\mathbb{R}$, $i,j,k=1,\ldots,n$, are the second Christoffel symbols associated to g (see Proposition 3.1), is independent of the choice of the chart (U,ϕ) on M as before. Consequently $\nabla_D X$ is a well-defined C^{k-1}-vector field along c that it is possible to be defined even if the image of c is not contained in the domain of one chart on M. Namely, for fixed $s\in(a,b)$, let $\phi:U\to\mathbb{R}^n$ be a local chart on M about $c(s)$ and define $(\nabla_D X)(s)$ by

$$(\nabla_D \tilde{X})(s)=\sum_{i=1}^{n}\left[\frac{d\tilde{\lambda}^i}{dt}(s)+\sum_{j,k=1}^{n}\Gamma^i_{jk}(\tilde{c}(s))\tilde{\lambda}^j(s)\frac{d\tilde{c}^k}{dt}(s)\right]\left(\frac{\partial}{\partial x^i}\right)_{\tilde{c}(s)},$$

where $\tilde{c}$ (resp. $\tilde{X}$) denotes the restriction of c (resp. X) to $c^{-1}(\operatorname{Im}(c)\cap U)$. $\nabla_D X$ is called **the derivative of X along c** and satisfies

$$\nabla_D(X_1+X_2)=\nabla_D X_1+\nabla_D X_2, \tag{5.2}$$

$$\nabla_D(fX)=\frac{df}{dt}X+f\nabla_D X, \tag{5.3}$$

where X, X_1, X_2 are C^k-vector fields along c and $f:(a,b)\to\mathbb{R}$ is of class $C^k(k\geq 1)$. In particular, for fixed c, ∇_D is a $\mathbb{R}$-linear operator on the real vector space of all smooth vector fields along the smooth path c.

Definition 5.2. Let $c:(a,b)\to M$ be a C^k-path in $M\,(k\geq 1)$. One says that a C^k-vector field X along c is **parallel along c** if

$$(\nabla_D X)(t)=0 \text{ for all } t\in(a,b).$$

By (5.2), (5.3), the set of parallel smooth vector fields along the smooth path c is a $\mathbb{R}$-vector space. One can use the theory of linear ordinary differential equations in order to deduce for each $t_0\in(a,b)$ and $u\in T_{c(t_0)}M$ the existence of a unique parallel smooth vector field X along the smooth path c satisfying $X(t_0)=u$. In particular, the $\mathbb{R}$-vector space of parallel smooth vector fields along the smooth path c is finite dimensional and has the dimension equal to n.

Note that if X,Y are C^k-vector fields along $c:(a,b)\to M$, $k\geq 1$, then

$$\frac{d}{dt}\left(g_{c(t)}\left(X(t),Y(t)\right)\right)=g_{c(t)}\left((\nabla_D X)(t),Y(t)\right)+g_{c(t)}\left(X(t),(\nabla_D Y)(t)\right)$$

for all $t\in(a,b)$. Thus, if X,Y are parallel vector fields along c, then the real function

$$g_{c(\ )}\left(X(\ ),Y(\ )\right):(a,b)\to\mathbb{R}$$

is constant. In particular, the real function $|X|:(a,b)\to\mathbb{R}$,

$t\mapsto |X|(t)\overset{def}{=}\left[g_{c(t)}\left(X(t),X(t)\right)\right]^{\frac{1}{2}}$, is constant.

Definition 5.3. Let (M,g) be a smooth n-dimensional Riemannian manifold and $\gamma:(a,b)\to M$ be a C^k-path $(k\geq 2)$ in M. γ is called a **geodesic** if

$$(\nabla_D\dot{\gamma})(t)=0 \tag{5.4}$$

for all $t\in(a,b)$, where $\dot{\gamma}$ denotes the tangent field of γ. (Note that $\dot{\gamma}$ is a C^{k-1}-vector field along γ).

To write the equation for a geodesic $\gamma:(a,b)\to M$ in a chart, assume (U,ϕ) be the chart on M defining the coordinates $(x^1,\ldots,x^n)$ such that $\gamma((a,b))\subset U$ and set $\gamma^i=x^i\circ\gamma$, $i=1,\ldots,n$. Taking into account (5.1), the equation (5.4) reads as

$$\frac{d^2\gamma^i}{dt^2}+\sum_{j,k=1}^{n}\left(\Gamma^i_{jk}\circ\gamma\right)\frac{d\gamma^j}{dt}\frac{d\gamma^k}{dt}=0,\ i=1,\ldots,n, \tag{5.5}$$

where $\Gamma^i_{jk}:U\to\mathbb{R}, i,j,k=1,\ldots,n$, are the second Christoffel symbols (see Proposition 3.1). Note that if γ is C^2 and satisfies (5.5), then γ is C^∞.

Using as parameter the arc length s instead of t, an easy computation shows that (5.5) are equivalent to the Euler equations

$$\frac{d}{ds}\left(\frac{\partial L}{\partial \dot{\gamma}^i}\right)-\frac{\partial L}{\partial \gamma^i}=0, \quad i=1,\ldots,n, \tag{5.6}$$

where

$$\begin{aligned} L &= \sum_{i,j=1}^{n} g_{ij}\left(\gamma^1,\ldots,\gamma^n\right)\dot{\gamma}^i\dot{\gamma}^j, \\ \dot{\gamma}^i &= \frac{d\gamma^i}{ds}, \quad i=1,\ldots,n. \end{aligned} \tag{5.7}$$

Remark 5.1. If $F\left(\gamma^1,\ldots,\gamma^n,\ \dot{\gamma}^1,\ldots,\dot{\gamma}^n\right)$ is homogeneous of degree $m \neq 1$ in variables $\dot{\gamma}^1,\ldots,\dot{\gamma}^n$ and satisfies (5.6), then

$$F = \text{constant}$$

is a first integral of (5.6). In particular, if L is given by (5.7), then L = constant is a first integral of (5.6).

Indeed the equations

$$\frac{\partial F}{\partial \gamma^i}=\frac{d}{ds}\left(\frac{\partial F}{\partial \dot{\gamma}^i}\right), \quad i=1,\ldots,n, \tag{5.8}$$

and Euler's theorem for homogeneous functions imply that

$$\begin{aligned} \frac{dF}{ds} &= \sum_{i=1}^{n}\left(\frac{\partial F}{\partial \gamma^i}\dot{\gamma}^i+\frac{\partial F}{\partial \dot{\gamma}^i}\frac{d\dot{\gamma}^i}{ds}\right) \\ &= \sum_{i=1}^{n}\left[\frac{d}{ds}\left(\frac{\partial F}{\partial \dot{\gamma}^i}\right)\dot{\gamma}^i+\frac{\partial F}{\partial \dot{\gamma}^i}\frac{d\dot{\gamma}^i}{ds}\right] \\ &= \frac{d}{ds}\left(\sum_{i=1}^{n}\dot{\gamma}^i\frac{\partial F}{\partial \dot{\gamma}^i}\right)=m\frac{dF}{ds}. \end{aligned}$$

Hence, if $m \neq 1$, $\dfrac{dF}{ds}=0$ and F = constant is a first integral of (5.8).

Example 5.1. Let us describe explicitely the geodesics in $\left(S^2, g_{can}\right)$ (see Example 2.5) in the chart given by the spherical coordinates (θ,ϕ) on S^2 (see Example 3.2). In this chart, $g_{11}=1, g_{22}=\cos^2\theta, g_{12}=g_{21}=0$. We use the arc length as parameter. Since the function

$$L\left(\theta,\phi,\dot{\theta},\dot{\phi}\right)=\dot{\theta}^2+\cos^2\theta\,\dot{\phi}^2$$

is homogeneous of degree two with respect to the variables $\dot{\theta},\dot{\phi}$, it follows that

$$L = 1 \tag{5.9}$$

is a first integral of the Euler equations

$$\frac{d}{ds}\left(2\dot{\theta}\right)+\sin 2\theta\dot{\phi}^2=0, \quad \frac{d}{ds}\left(2\dot{\phi}\cos^2\theta\right)=0 \tag{5.10}$$

(see Remark 5.1), that is of the equations of a geodesic. Hence, for the integration of this system of equations, we can use only the second equation of (5.10), what implies

$$\cos^2\theta \frac{d\phi}{ds} = k \ (= \text{constant}). \tag{5.11}$$

Now (5.9) and (5.11) yield to

$$\left(\frac{d\phi}{d\theta}\right)^2 = \frac{k^2}{\cos^2\theta\left(\cos^2\theta - k^2\right)},$$

what imposes the restriction $k \in (-1,1)$. If one takes $t = tg\theta$, one obtains

$$\phi = \pm k \int \frac{dt}{\sqrt{1-k^2-k^2t^2}} = \pm \arcsin \frac{kt}{\sqrt{1-k^2}} + C,$$

where C is a real constant. Consequently,

$$\sin(\phi - C) = \pm \frac{kt}{\sqrt{1-k^2}},$$

that is

$$\sin\phi\cos C - \cos\phi\sin C = \pm \frac{k}{\sqrt{1-k^2}} tg\theta,$$

or equivalently

$$-x^1 \sin C + x^2 \cos C \mp \frac{k}{\sqrt{1-k^2}} x^3 = 0. \tag{5.12}$$

The geodesics are therefore the intersections of the sphere S^2, i.e.

$$\left(x^1\right)^2 + \left(x^2\right)^2 + \left(x^3\right)^2 = 1$$

with the planes through the origin (5.12), i.e. great circles (see also Example 5.3).

Example 5.2. Now let us describe explicitely the geodesics in (H^2,g) (see Example 2.4) in the canonical chart defining the coordinates $\left(x^1,x^2\right), x^2>0$.

In this chart, $g_{11} = g_{22} = \dfrac{1}{\left(x^2\right)^2}$, $g_{12} = g_{21} = 0$. We use as before the arc length as parameter. Since the function

$$L\left(x^1,x^2,\dot{x}^1,\dot{x}^2\right) = \frac{1}{\left(x^2\right)^2}\left[\left(\dot{x}^1\right)^2 + \left(\dot{x}^2\right)^2\right]$$

is homogeneous of degree two with respect to the variables $\dot{x}^1, \dot{x}^2$, it follows that

$$L = 1 \tag{5.13}$$

is a first integral of the Euler equations

$$\frac{d}{ds}\left[\frac{1}{\left(x^2\right)^2} 2\dot{x}^1\right] = 0, \ \frac{d}{ds}\left[\frac{1}{\left(x^2\right)^2} 2\dot{x}^2\right] + \frac{2}{\left(x^2\right)^3}\left[\left(\dot{x}^1\right)^2 + \left(\dot{x}^2\right)^2\right] = 0 \tag{5.14}$$

(see Remark 5.1), that is of the equations of a geodesic. For the integration of this system of equations, we use only the first equation of (5.14), what implies

$$\frac{dx^1}{ds} = C\left(x^2\right)^2, \tag{5.15}$$

where C is a real constant. In the case when $C = 0$, (5.15) leads to

$$x^1 = D, \tag{5.16}$$

where D is a real constant. If $C \neq 0$, (5.13) and (5.15) yield to

$$\frac{dx^2}{dx^1} = \frac{1}{\frac{dx^1}{ds}} \cdot \frac{dx^2}{ds} = \pm \frac{1}{Cx^2}\sqrt{1 - C^2\left(x^2\right)^2},$$

what implies

$$x^1 - x_0^1 = \mp \frac{1}{C}\sqrt{1 - C^2\left(x^2\right)^2},$$

or equivalently

$$\left(x^1 - x_0^1\right)^2 + \left(x^2\right)^2 = \frac{1}{C^2}, \tag{5.17}$$

where x_0^l is a real constant. The geodesics are therefore the intersections of H^2 (i.e. $x^2 > 0$) either with the straight lines (5.16) parallel with the x^2-axis or with the circles having the centres on the x^1-axis, i.e. either straight open half-lines parallel to the x^2-axis or open semicircles having the centers on the x^1-axis (see Figure 5.1).

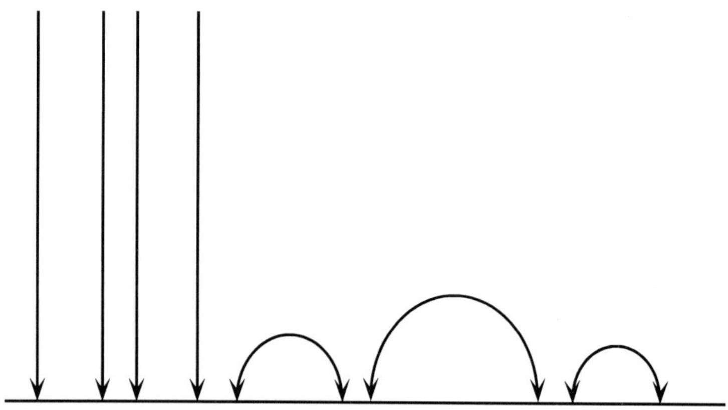

Figure 5.1

H^2 with these geodesics provides a model for the non-Euclidean geometry of Bolyai and Lobachevski for which all axioms of 2-dimensional Euclidean geometry hold except the parallel axiom: through any point P of H^2 not on a given geodesic (straight open half-line) γ there pass infinite many geodesics that do not intersect γ (see Figure 5.2).

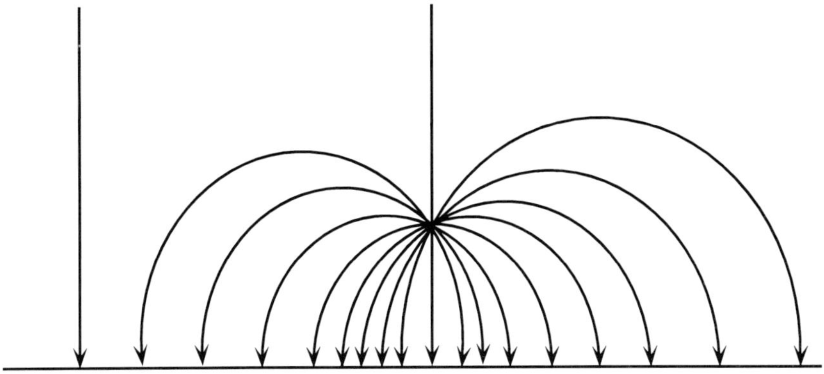

Figure 5.2

In the Euclidean plane E^2 this is not the case (see Figure 5.3): through any point of $E^2\backslash\gamma$, where γ is a geodesic (straight line) of E^2, there pass a unique geodesic that does not intersect γ (hence parallel with γ).

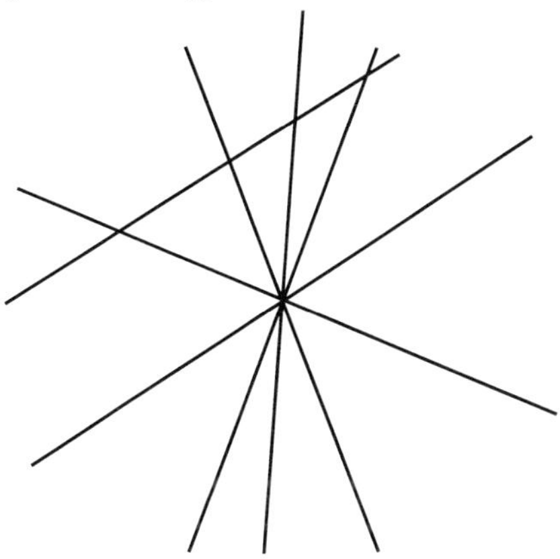

Figure 5.3

Let M be a smooth n-dimensional manifold. Recall that to a chart (U,ϕ) on M defining the coordinates $(x^1,...,x^n)$ on U one can associate the mapping

$$\bar{\phi} : \bigcup_{x\in U} T_xM \to \mathbb{R}^n \times \mathbb{R}^n = \mathbb{R}^{2n},\ \bar{\phi}(u) = \left(\phi(x), u(x^1), \ldots, u(x^n)\right),$$

$u \in T_xM$, that gives a natural chart on TM. This associated chart defines the coordinates $\left(\bar{x}^1, \ldots, \bar{x}^{2n}\right)$ on $\bigcup_{x\in U} T_xM = \pi^{-1}(U)$, where $\pi : TM \to M$ denotes the canonical projection, namely $\bar{x}^i(u) = x^i(x) = x^i(\pi(u))$, i.e. $\bar{x}^i = x^i \circ \pi$ and

$\bar{x}^{n+i}(u) = u(x^i),\ i = 1,\ldots,n,\ u \in T_xM,\ x \in U$. Note that if $A = \sum_{k=1}^{2n} A(\bar{x}^k)\frac{\partial}{\partial \bar{x}^k}$ is a smooth vector field on $\pi^{-1}(U)$, then

$$d\pi \circ A = \sum_{i=1}^{n} A\left(x^i \circ \pi\right)\left(\frac{\partial}{\partial x^i} \circ \pi\right) = \sum_{i=1}^{n} A\left(\overline{x}^i\right)\left(\frac{\partial}{\partial x^i} \circ \pi\right),$$

so that in particular for any $i = 1,\dots,n$,

$$d\pi \circ \frac{\partial}{\partial \overline{x}^i} = \frac{\partial}{\partial x^i} \circ \pi, \ d\pi \circ \frac{\partial}{\partial \overline{x}^{n+i}} = 0.$$

Now, let (M,g) be a smooth n-dimensional Riemannian manifold. The system of differential equations (5.5) of a geodesic in U can be written as a system of first-order differential equations in $\pi^{-1}(U)$:

$$\begin{aligned} &\frac{d\overline{x}^i}{dt} = \overline{x}^{n+i}, \\ &\frac{d\overline{x}^{n+i}}{dt} = -\sum_{j,k=1}^{n}\left(\Gamma_{jk}^i \circ \pi\right)\overline{x}^{n+j}\overline{x}^{n+k}. \end{aligned} \qquad i = 1,\dots,n. \tag{5.18}$$

The solutions of the system of differential equations (5.18) are therefore integral curves of the vector field S on $\pi^{-1}(U)$ given by

$$S = \sum_{i=1}^{n}\left[\overline{x}^{n+i}\frac{\partial}{\partial \overline{x}^i} - \sum_{j,k=1}^{n}\overline{x}^{n+j}\overline{x}^{n+k}\left(\Gamma_{jk}^i \circ \pi\right)\frac{\partial}{\partial \overline{x}^{n+i}}\right].$$

Since the geodesic equations are independent of the choice of local charts on M, it follows that S is a globally defined vector field on TM. This vector field $S : TM \to TTM$ is called **the geodesic spray** of the Levi-Civita connection.

Proposition 5.1. If the path $\gamma : (a,b) \to M$ is a geodesic in (M,g), then $\dot{\gamma} : (a,b) \to TM$ is an integral curve of the geodesic spray S of the Levi-Civita connection associated to g. Conversely, if $c : (a,b) \to TM$ is an integral curve of the geodesic spray S of the Levi-Civita connection associated to g, then $\gamma = \pi \circ c$ is a geodesic in (M,g) and $\dot{\gamma} = c$.

Proof. Let $\gamma : (a,b) \to M$ be a geodesic on (M,g). We show that $\dot{\gamma}$ is an integral curve of S, i.e. $\ddot{\gamma} = S \circ \dot{\gamma}$. For a chart (U,ϕ) on M, let us consider the associated natural chart on TM. With respect to these charts, one has

$$(\dot{\gamma})^{n+i} = \overline{x}^{n+i} \circ \dot{\gamma}, \quad i = 1,\dots,n,$$

and

$$\dot{\gamma}(t) = \sum_{i=1}^{n}(\dot{\gamma}(t))(x^i)\left(\frac{\partial}{\partial x^i}\right)_{\gamma(t)} = \sum_{i=1}^{n}(\dot{\gamma})^{n+i}\left(\frac{\partial}{\partial x^i}\right)_{\gamma(t)}.$$

On the other hand, by the definition of $\dot{\gamma}$,

$$\dot{\gamma} = \sum_{i=1}^{n}\frac{d}{dt}\left(x^i \circ \gamma\right)\left(\frac{\partial}{\partial x^i}\right) \circ \gamma = \sum_{i=1}^{n}\frac{d\gamma^i}{dt}\left(\frac{\partial}{\partial x^i}\right) \circ \gamma,$$

that is

$$\frac{d\gamma^i}{dt} = (\dot{\gamma})^{n+i}. \tag{5.19}$$

Since $\pi \circ \dot{\gamma} = \gamma$, it follows that

$$(\dot{\gamma})^i \overset{def}{=} \overline{x}^i \circ \dot{\gamma} = \gamma^i, \; i = 1, \ldots, n. \tag{5.20}$$

Consequently, (5.19) and (5.20) imply that

$$\frac{d(\dot{\gamma})^i}{dt} = (\dot{\gamma})^{n+i}, \; i = 1, \ldots, n. \tag{5.21}$$

The equality $\nabla_D \dot{\gamma} = 0$ leads to

$$\sum_{i=1}^{n} \left[\frac{d(\dot{\gamma})^{n+i}}{dt} + \sum_{j,k=1}^{n} \left(\Gamma^i_{jk} \circ \gamma\right)(\dot{\gamma})^{n+j} \frac{d\gamma^k}{dt} \right] \left(\frac{\partial}{\partial x^i} \right) \circ \gamma = 0,$$

so that in virtue of (5.21) one gets

$$\frac{d(\dot{\gamma})^{n+i}}{dt} = -\sum_{j,k=1}^{n} \left(\Gamma^i_{jk} \circ \gamma\right)(\dot{\gamma})^{n+j} (\dot{\gamma})^{n+k}, \; i = 1, \ldots, n. \tag{5.22}$$

(5.21) and (5.22) show that $\dot{\gamma} : (a,b) \to TM$ is an integral curve of *S*.

Conversely, let $c : (a,b) \to TM$ be an integral curve of *S*. We prove that $\gamma = \pi \circ c$ is a geodesic in (*M*,*g*), where $\pi : TM \to M$ denotes the canonical projection. To a chart on *M*, we associate the natural chart on *TM*. Setting

$$c^i = \overline{x}^i \circ c, \quad c^{n+i} = \overline{x}^{n+i} \circ c, \quad i = 1, \ldots, n,$$

the fact that *c* is an integral curve of *S*, that is $\dot{c} = S \circ c$, reads as

$$\frac{dc^i}{dt} = c^{n+i}, \quad \frac{dc^{n+i}}{dt} = -\sum_{j,k=1}^{n} \left(\Gamma^i_{jk} \circ \gamma\right) c^{n+j} c^{n+k}, \quad i = 1, \ldots, n. \tag{5.23}$$

The equality $\gamma = \pi \circ c$ implies that

$$\gamma^i \overset{def}{=} x^i \circ \gamma = x^i \circ \pi \circ c = \overline{x}^i \circ c = c^i, \quad i = 1, \ldots, n.$$

Consequently,

$$\frac{d\gamma^i}{dt} = \frac{dc^i}{dt} = c^{n+i}, \quad i = 1, \ldots, n, \tag{5.24}$$

so that in virtue of (5.23) one obtains

$$\frac{d^2\gamma^i}{dt^2} = -\sum_{j,k=1}^{n} \left(\Gamma^i_{jk} \circ \gamma\right) \frac{d\gamma^j}{dt} \frac{d\gamma^k}{dt}, \quad i = 1, \ldots, n,$$

that is γ is a geodesic. Note that

$$c=\sum_{i=1}^{n}\left(\overline{x}^{n+i}\circ c\right)\left[\left(\frac{\partial}{\partial x^i}\right)\circ\gamma\right]=\sum_{i=1}^{n}c^{n+i}\left(\frac{\partial}{\partial x^i}\right)\circ\gamma$$
$$=\sum_{i=1}^{n}\frac{d\gamma^i}{dt}\left(\frac{\partial}{\partial x^i}\right)\circ\gamma\overset{def}{=}\dot{\gamma},$$

that is $\dot{\gamma}=c$.

Q.E.D.

Let $\alpha:D(S)\to TM$ be the global flow of the geodesic spray S of the Levi-Civita connection, where

$$D(S)=\{(t,\xi)\in\mathbb{R}\times TM\,|\,t\in J(\xi)\},$$

$J(\xi)=(t^-(\xi),t^+(\xi))$ being the maximal interval in $\mathbb{R}$ on which the integral curve of the vector field S with the initial condition ξ is defined. In virtue of Proposition 5.1, the path

$$\gamma_\xi:J(\xi)\to M,\ \gamma_\xi(t)=(\pi\circ\alpha)(t,\xi),t\in J(\xi),$$

is the unique maximal geodesic in (M,g) satisfying

$$\gamma_\xi(0)=\pi(\xi),\ \dot{\gamma}_\xi(0)=\xi.$$

Note that $\gamma_\xi(t)$ depends C^∞ on t and ξ and $\dot{\gamma}_\xi=\alpha(\ ,\xi)$. Moreover, for any $a\in\mathbb{R}^*$,

$$J(a\xi)=\begin{cases}(at^+(\xi),at^-(\xi)) & \text{when } a<0,\\ (at^-(\xi),at^+(\xi)) & \text{when } a>0,\end{cases}\quad\text{and}\quad \gamma_{a\xi}(t)=\gamma_\xi(at). \tag{5.25}$$

Example 5.3 Fix a point $x\in S^n\subset\mathbb{R}^{n+1}$ and a unit vector $u\in T_xS^n=\{x\}^\perp$. Let γ_u be the unique maximal geodesic of (S^n,g_{can}) such that $\gamma_u(0)=x,\dot{\gamma}_u(0)=u$. Consider the reflection of $\mathbb{R}^{n+1}$ onto $\mathbb{R}^{n+1}$ through the vector 2-plane spanned by x and u. This induces a mapping $\Phi:(S^n,g_{can})\to(S^n,g_{can})$, which is an isometry (see Example 2.10) and so preserves the Levi-Civita connection associated to (S^n,g_{can}) (see Corollary 3.2). Thus, if γ is a geodesic so is $\Phi\circ\gamma$. Since Φ fixes x and u, the uniqueness part for the system of differential equations that defines a geodesic of (S^n,g_{can}) implies $\Phi\circ\gamma_u=\gamma_u$. Therefore, γ_u lies in the 2-plane spanned by x and u and so lies on a great circle of S^n. Since the real function $|\dot{\gamma}_u|$ is constant, one gets $|\dot{\gamma}_u(t)|=|\dot{\gamma}_u(0)|=|u|=1$ for all t, which implies $\gamma_u(t)=(\cos t)x+(\sin t)u$. A similar argument shows that the unique parallel vector field X along γ_u satisfying $X(0)=v\in\{u\}^\perp$ is given by the constant vector field $X=v$.

Let $\mathcal{E}(M)$ be the subset of $D(TM)$ defined by

$$\mathcal{E}(M)=\{\xi\in TM\,|\,1\in J(\xi)\}.$$

Since the domain $D(S)$ of the global flow α of S is open, it follows that $\mathcal{E}(M)$ is open. Moreover, (5.25) implies that for any $\xi\in\mathcal{E}(M)$ one has $a\xi\in\mathcal{E}(M)$ for all $a\in[0,1]$.

Definition 5.4. The map $\exp : \mathcal{E}(M) \to M$ defined by $\exp(\xi) = \gamma_\xi(1)$ is called the **exponential map associated to (*M*,*g*)**. For any given $x \in M$, one defines $\exp_x : T_xM \cap \mathcal{E}(M) \to M$ by $\exp_x = \exp\big|_{T_xM \cap \mathcal{E}(M)}$.

Proposition 5.2. The mappings exp and $\exp_x$ are differentiable of class C^∞, exp has maximal rank on the image of the canonical embedding of M into TM via the zero vector field, and $\exp_x$ has maximal rank at $0 \in T_xM$.

For the proof the reader can consult Chavel [1993].

Proposition 5.2 allows to construct charts of M in terms of geodesics of (M,g). Thus under the identification of $T_{0_x}(T_xM)$ with T_xM, where 0_x denotes the origin of the vector space T_xM, one can show that the differential of $\exp_x$ at 0_x is the identity of T_xM. Consequently there exists an $\varepsilon > 0$ such that $\exp_x$ is a diffeomorphism of the open ball $B_\varepsilon(0_x)$ of radius ε of T_xM centered at 0_x onto an open set U in M containing x. Taking an orthonormal basis $(e_1,\ldots,e_n)$ of T_xM one can define the diffeomorphism

$$\varphi : U \to \left\{ x \in \mathbb{R}^n \,\middle|\, \|x\| < \varepsilon \right\}, \varphi(y) = \left(x^1(y),\ldots,x^n(y)\right),$$

where $y \in U$ and $\exp_x\left(\sum_{i=1}^n x^i(y)e_i\right) = y$.

Therefore one obtains a chart (U,φ) around x and a corresponding coordinate system $\left(x^1,\ldots,x^n\right)$, which is called **a normal coordinate system at *x***.

Let (M,g) be a smooth Riemannian manifold, $\gamma:(a,b) \to M$ a smooth geodesic such that $\dot\gamma(t) \in T_{\gamma(t)}M$ is of unit length for all $t \in (a,b)$.

Definition 5.5 A smooth vector field $X:(a,b) \to TM$ along γ is called a **Jacobi field along** γ if

$$\left[\nabla_D(\nabla_D X)\right](t) + R_{\gamma(t)}\left(\dot\gamma(t), X(t)\right)\dot\gamma(t) = 0$$

for all $t \in (a,b)$.

Using again results from the theory of linear ordinary differential equations one can prove that for each $t_0 \in (a,b)$ and $u_0, u_1 \in T_{\gamma(t_0)}M$ there is a unique Jacobi field X along the smooth geodesic $\gamma:(a,b) \to M$ satisfying $X(t_0) = u_0$, $(\nabla_D X)(t_0) = u_1$.

It is easy to show that Jacobi fields along the geodesic $\gamma:(a,b) \to M$ are infinitesimal variations of γ through a family of geodesics (for example, see Chavel [1993]), that is if $H:(a,b)\times(-\varepsilon,\varepsilon) \to M$ is a variation of geodesics of γ [i.e. $(a,b) \ni t \mapsto H(t,s) \in M$ is a geodesic for each $s \in (-\varepsilon,\varepsilon)$ and $H(\ ,0) = \gamma$], then $Y = \frac{\partial H}{\partial s}\Big|_{s=0}$ is a Jacobi field along γ. Moreover, all Jacobi fields along γ arise this way. Using this remark, one can find Jacobi fields in normal coordinate systems. More precisely, let $x \in M$ and $u \in T_xM$ be fixed and $J(u) \subseteq \mathbb{R}$ be the maximal open interval on which there is the unique geodesic $\gamma_u : J(u) \to M$ such that $\gamma_u(0) = x$ and $\dot\gamma_u(0) = u$.

Then for each $v \in T_x M$ the Jacobi field $X : J(u) \to TM$ along γ_u with $X(0) = 0, (\nabla_D X)(0) = v$, is given by

$$X(t) = (d\exp_x)_{tu}(tv).$$

REFERENCES

BEARDON, A.F., *The Geometry of Discrete Groups*, Springer-Verlag, Berlin-Heidelberg-New York, 1983.

BERGER, M., GAUDUCHON, P. and MAZET, E., *Le Spectre d'une Variété Riemannienne*, Lecture Notes in Math., Vol. 194, Springer-Verlag, Berlin, 1971.

BERGER, M., *Geometry II*, Universitext, Springer-Verlag, Berlin-Heidelberg, 1987.

BESSE, A.L., *Einstein Manifolds*, Springer-Verlag, Berlin, 1987.

BISHOP, R. and CRITTENDEN, R., *Geometry of Manifolds*, Academic Press, New York, 1964.

BOOTHBY, W.M., *An Introduction to Differentiable Manifolds and Riemannian Geometry*, Second Edition, Academic Press, Inc., Boston, 1986.

BOURBAKI, N., *Variétés Différentielles et Analytiques*, Fascicule de Résultats/Paragraphes 1 à 7, Hermann, Paris, 1967.

BUSER, P., *Geometry and Spectra of Compact Riemann Surfaces*, Progress in Mathematics, Vol. 106, Birkhäuser, Boston, 1992.

CHAVEL, I., *Riemannian Geometry - A Modern Introduction*, Cambridge Tracts in Mathematics, Vol. 108, Cambridge University Press, 1993.

CHEN, B. –Y. and VANHECKE, L., *Differential geometry of geodesic spheres*, J. Reine Angew. Math., vol. 325(1981), 28-67.

CRAIOVEANU, M. and PUTA, M., *Introducere în Geometria Spectrală*, Editura Academiei Române, Bucuresti, 1988.

DEO, S. and VARADARAJAN, K., *Discrete groups and discontinuous actions*, Rocky Mountain Journal of Mathematics, 27(1997), 559-583.

GALLOT, S., HULIN, D. and LAFONTAINE, J., *Riemannian Geometry*, Second Edition, Universitext, Springer-Verlag, Berlin, 1993.

GHEORGHIEV, GH. and OPROIU, V., *Varietăti Finit si Infinit Dimensionale*, Vol. II, Editura Academiei Române, Bucuresti, 1979.

GORDON, C.S. and WILSON, E.N., *The spectrum of the Laplacian on Riemannian Heisenberg manifolds*, Michigan Math. J. 33(1986), No. 2, 253-271.

GORDON, C.S. and WILSON, E.N., *Isometry groups of Riemannian solvmanifolds*, Trans. Amer. Math. Soc., 307(1988), 245-269.

LEKKERKERKER, C.G., *Geometry of numbers*, Bibliotheca Mathematica, Vol. VIII, Wolters-Noordhoff Publ. – Groningen, North-Holland Publ. Co. – Amsterdam, 1969.

O' NEILL, B., *Semi-Riemannian Geometry with Applications to Relativity*, Academic Press, Inc., 1983.

PALAIS, R.S. (ed.), *Seminar on the Atiyah-Singer Index Theorem*, Princeton, 1965.

RATCLIFFE, J.G., *Foundations of Hyperbolic Manifolds*, Graduate Texts in Mathematics 149, Springer-Verlag, Berlin, 1994.

ROE, J., *Elliptic Operators, Topology and Asymptotic Methods*, Pitman Research Notes in Mathematics Series 179, Longman Group UK Limited, 1988.

SAKAI, T., *Riemannian Geometry*, Translations of Mathematical Monographs, Vol. 149, American Mathematical Society, Providence, Rhode Island, 1996.

Chapter 2
CANONICAL DIFFERENTIAL OPERATORS ASSOCIATED TO A RIEMANNIAN MANIFOLD

1. HILBERT SPACES ASSOCIATED TO A COMPACT RIEMANNIAN MANIFOLD

Let (M,g) be a n-dimensional Riemannian manifold. Then g allows us to define a **σ-regular Borel measure** μ_g on M determined by a positive linear functional $\lambda_g : C_0(M) \to \mathbb{R}$, where $C_0(M)$ is the space of continuous real valued functions defined on M with compact support, that is for each $f \in C_0(M)$ with $f \geq 0$ we have $\lambda_g(f) \geq 0$. Indeed, let (U,ϕ) be a local chart on M with coordinates $x^1,\ldots,x^n$, i.e. $x^i = \xi^i \circ \phi, 1 \leq i \leq n$, where $\xi^i : \mathbb{R}^n \to \mathbb{R}$ denotes the canonical projection. In particular, for each $x \in \mathbb{R}^n$ we have

$$x = \left(\xi^1(x), \ldots, \xi^n(x)\right).$$

On the domain U of our local chart (U,ϕ), let us consider the vector fields $\dfrac{\partial}{\partial x^1}, \ldots, \dfrac{\partial}{\partial x^n}$ and let us denote

$$g_{ij} = g\left(\frac{\partial}{\partial x^i}, \frac{\partial}{\partial x^j}\right)\Bigg|_U,$$

$1 \leq i, j \leq n$. Let $\xi_{ij} = g_{ij} \circ \phi^{-1}$ and let

$$g_{\phi(U)} = \left[\xi_{ij}\right]$$

be the corresponding matrix. Its determinant $\left|g_{\phi(U)}\right|$ is then a strictly positive real function. For each $f \in C_0(M)$ with $\operatorname{supp}(f) \subset U$, let us define

$$\lambda_{U,\phi}(f) = \int_{\phi(U)} \left(f|_U \circ \phi^{-1}\right) \left|g_{\phi(U)}\right|^{\frac{1}{2}} d\xi,$$

where $d\xi$ denotes the Lebesgue measure on $\mathbb{R}^n$. Let us note that the above integral defines a positive linear functional $\lambda_{U,\phi} : C_0(U) \to \mathbb{R}$, where $C_0(U) = \left\{f \in C_0(M) \middle| \operatorname{supp}(f) \subset U\right\}$. If (U',ϕ') is another chart with $U \cap U' \neq \varnothing$ we shall prove that for each $f \in C_0(M)$ with $\operatorname{supp}(f) \subset U \cap U'$, the value $\lambda_{U\cap U',\phi}(f)$ is independent of the choice of charts. Indeed, let us denote

$$\psi = \phi \circ \phi'^{-1} : \phi'(U \cap U') \to \phi(U \cap U')$$

and $J = J\left(\xi'^1, \ldots, \xi'^n\right)$ the Jacobian of this smooth diffeomorphism. Then using the chain rule we can write successively:

$$\lambda_{U\cap U',\phi}(f)=\int_{\phi(U\cap U')}\left(f\big|_{U\cap U'}\circ\phi^{-1}\right)\left|g_{\phi(U\cap U')}\right|^{\frac{1}{2}}d\xi$$

$$=\int_{\phi'(U\cap U')}\left(f\big|_{U\cap U'}\circ\phi'^{-1}\right)\left|g_{\phi(U\cap U')}\circ\psi\right|^{\frac{1}{2}}|J|\,d\xi.$$

On the other hand, using the change of coordinates for a tensor field of type (0,2) we obtain:

$$\left|g_{\phi'(U\cap U')}\right|=\left|g_{\phi(U\cap U')}\circ\psi\right|\cdot J^2$$

and then

$$\lambda_{U\cap U',\phi}(f)=\int_{\phi'(U\cap U')}\left(f\big|_{U\cap U'}\circ\phi'^{-1}\right)\left|g_{\phi'(U\cap U')}\right|^{\frac{1}{2}}d\xi'$$

$$=\lambda_{U\cap U',\phi'}(f),$$

as desired.

The existence and uniqueness of the canonical linear functional $\lambda_g:C_0(M)\to\mathbb{R}$ associated to the Riemannian manifold (M,g) such that

$$\lambda_g\big|_{C_0(U)}=\lambda_{U,\phi}$$

for each local chart (U,ϕ) is a consequence of the following general theorem:

Theorem 1.1 (the localization theorem). Let $\{U_\alpha\}$ be an open covering of the local compact, Hausdorff topological space M. For each index α, let $\lambda_\alpha:C_0(U_\alpha)\to\mathbb{R}$ be a linear functional such that for each α,β with $U_\alpha\cap U_\beta\neq\phi$ we have:

$$\lambda_\alpha\big|_{C_0(U_\alpha\cap U_\beta)}=\lambda_\beta\big|_{C_0(U_\alpha\cap U_\beta)}.$$

Then there exists a unique linear functional $\lambda:C_0(M)\to\mathbb{R}$ such that $\lambda\big|_{C_0(U_\alpha)}=\lambda_\alpha$. If each functional λ_α is positive, then λ has the same property.

Proof. Let $f\in C_0(M)$ and $K=\operatorname{supp}(f)$. Since each Hausdorff compact space is paracompact, it follows that K is paracompact. Let $\{\beta_i\}$ be a partition of unity subordinated to a cover of K by a finite number of open sets of type U_α. Then each $\beta_i f$ has the support contained in an $U_{\alpha(i)}$ and let us define

$$\lambda(f)=\sum_i\lambda_{\alpha(i)}(\beta_i f).$$

This sum is independent of the choice of $\alpha(i)$ and of the choice of the partition of unity. If we prove these two things it follows that λ satisfies all requirements of the theorem.

At the beginning let us note that if $U_{\alpha'(i)}$ is another open set such that $\operatorname{supp}(\beta_i f)\subset U_{\alpha'(i)}$, then $\operatorname{supp}(\beta_i f)\subset U_{\alpha(i)}\cap U_{\alpha'(i)}$ and the hypothesis concerning the functionals λ_α shows that the corresponding term of the above theorem does not depend of the choice of the index $\alpha(i)$.

On the other hand, if $\{\beta_k'\}$ is another partition of unity subordinated to a cover of K with a finite number of open sets of type U_α, then for each index i

$$\beta_i f = \sum_k \beta_k' \beta_i f$$

and this implies:

$$\sum_i \lambda_{\alpha(i)}(\beta_i f) = \sum_i \sum_k \lambda_{\alpha(i)}(\beta_k' \beta_i f).$$

If $\operatorname{supp}(\beta_k' \beta_i f)$ is contained in an U_α then the value $\lambda_\alpha(\beta_k' \beta_i f)$ is independent of the choice of the index α. The expression from the right hand side of the last equality is then symmetric with respect to our partitions of unity $\{\beta_i\}$ and $\{\beta_k'\}$, i.e.

$$\sum_i \sum_k \lambda_{\alpha(i)}(\beta_k' \beta_i f) = \sum_k \lambda_{\alpha(k)}(\beta_k' f).$$

Q.E.D.

Definition 1.1 The associated measure μ_g of the positive functional λ_g constructed above is called the **canonical measure of the Riemannian manifold** (M,g). Moreover the functional λ_g is given by the integral, that is

$$\lambda_g(f) = \int_M f d\mu_g ,$$

for all $f \in C_0(M)$.

Note that in the case of $(\mathbb{R}^n, g_{can})$, $\mu_{g_{can}}$ coincides with the Lebesgue measure.

In particular, if M is a closed manifold, then

$$\int_M 1 \cdot d\mu_g$$

is called the **volume** of (M,g) and will be denoted by Vol(M,g).

Remark 1.1 Let (M,g), (N,h) be Riemannian manifolds, $\Phi:(M,g) \to (N,h)$ an isometry and (U,ϕ) a local chart on M such that $\left(\Phi(U), \phi \circ \Phi^{-1}\big|_{\Phi(U)}\right)$ is a local chart on N. Then an easy computation shows that

$$\lambda_{\phi(U),\phi\circ\Phi^{-1}}\Big|_{\phi(U)} = \lambda_{U,\phi} \circ \alpha_{\Phi|U} ,$$

where $\alpha_{\Phi|U} : C_0(\Phi(U)) \to C_0(U)$ is the canonical $\mathbb{R}$-linear mapping induced by Φ through composition. This implies

$$\lambda_h = \lambda_g \circ \alpha_\Phi ,$$

where $\alpha_\Phi : C_0(N) \to C_0(M)$ is the canonical $\mathbb{R}$-linear mapping induced by Φ through composition. In particular, if M is closed, then:

$$\begin{aligned}\operatorname{Vol}(N,h) &= \int_N d\mu_h \\ &= \lambda_h(1) \\ &= \lambda_g \circ \alpha_\Phi(1) \\ &= \int_M d\mu_g \\ &= \operatorname{Vol}(M,g),\end{aligned}$$

that is $\operatorname{Vol}(M,g) = \operatorname{Vol}(N,h)$.

More generally, if $\Phi : M \to N$ is a smooth diffeomorphism and h a Riemannian metric on N, then one can define a positive linear functional $\Phi^*\lambda_h : C_0(M) \to \mathbb{R}$, called the pull-back of λ_h, by

$$(\Phi^*\lambda_h)(f) = \lambda_h(f \circ \Phi^{-1}), \qquad f \in C_0(M).$$

Note that in fact $\Phi^*\lambda_h = \lambda_{\Phi^*(h)}$, so that the following change of variable formula

$$\int_M f d\mu_{\Phi^*(h)} = \int_N f \circ \Phi^{-1} d\mu_h$$

is valid.

Remark 1.2 Let (M,g) be a Riemannian manifold, $\pi : \tilde{M} \to M$ a smooth covering projection, $\tilde{g}$ the canonical Riemannian structure on $\tilde{M}$ induced by g and π (see Chapter 1, Example 2.8). If $f \in C_0(\tilde{M})$, then the function $f : M \to \mathbb{R}$ given by:

$$f(x) = \sum_{\tilde{x} \in \pi^{-1}(x)} \tilde{f}(\tilde{x})$$

is with compact support and furthermore the following equality holds:

$$\int_{\tilde{M}} \tilde{f} d\mu_{\tilde{g}} = \int_M f d\mu_g,$$

(see for details Berger, Gauduchon and Mazet [1971]). In particular, if $\tilde{M}$ is a closed manifold and $\pi : \tilde{M} \to M$ an q-sheeted covering projection, then

$$\operatorname{Vol}(\tilde{M},\tilde{g}) = q \cdot \operatorname{Vol}(M,g).$$

Indeed, in this case the real valued function f which is canonically associated to the constant function 1 is the constant function $f(x) = q$, for each $x \in M$ and the above assertion follows easily.

Remark 1.3 Let M be a smooth manifold and let $\{(U_i,\phi_i)\}$ be an atlas. Recall that this atlas is said to be **oriented** if for all transition mappings $\phi_j \circ \phi_i^{-1}$ (defined whenever $U_i \cap U_j$ is not empty) the Jacobian determinant $\Delta_{\phi_j \circ \phi_i^{-1}}(x)$ is > 0 for all $x \in \phi_i(U_i \cap U_j)$. Two oriented atlases are said to define the **same orientation** if their union is oriented. This defines an equivalence relation and an orientation equivalence class of oriented atlases is said to define an **oriented manifold**.

Assume now that M is an n-dimensional smooth, oriented manifold. Let ω be a continuous differential n-form on M. Then there exists a linear functional $\lambda_\omega : C_0(M) \to \mathbb{R}$ defined as follows. If (U,ϕ) is a chart belonging to an oriented atlas with coordinates $(x^1,\ldots,x^n)$, i.e. $x^i = \xi^i \circ \phi, 1 \le i \le n$, and $F d\xi^1 \wedge \ldots \wedge d\xi^n$ with $F : \phi(U) \to \mathbb{R}$ continuous is the local representation of ω in this chart, then for any $f \in C_0(U)$, let us define:

$$\lambda_{\omega,U,\phi}(f) = \int_{\phi(U)} \left(f \circ \phi^{-1}\right) F d\xi ,$$

where $d\xi$ represents the Lebesgue measure on $\mathbb{R}^n$. This integral defines a linear functional on $C_0(U)$. As in the beginning of this section, the change of variables formula shows that if (U,ϕ) and (U',ϕ') belong to the oriented atlas, and if f has compact support in $U \cap U'$, then the value of the functional is independent of the choice of oriented charts. Thus we get a functional λ_ω by the localization theorem, using partition of unity. Note that this is the unique functional on $C_0(M)$ whose restriction to each $C_0(U)$ is equal to $\lambda_{\omega,U,\phi}$. We shall call λ_ω the functional associated with ω and for any $f \in C_0(M)$, $\lambda_\omega(f)$ will be denoted by $\int_M f\omega$.

Remark 1.4 Let us suppose now that (M,g) is an n-dimensional Riemannian manifold and M is oriented. In this case g gives us the possibility to construct a canonical nowhere zero smooth differential n-form v_g on M called the **volume form** on (M,g), namely if $(U,x^1,\ldots,x^n)$ is a locally oriented chart of M, then

$$v_g\big|_U = \sqrt{\det\left(g_{ij}\right)}dx^1 \wedge \ldots \wedge dx^n$$

is a nowhere zero local n-form on M. It is easy to see that the n-form $v_g\big|_U$ is independent of the oriented chart $(U,x^1,\ldots,x^n)$ and thus defines a global non-zero smooth differential n-form v_g on M. Moreover, the functional associated with v_g (see Remark 1.3) is equal to the functional λ_g defined in the beginning of this section. In particular, if M is compact, we have:

$$\operatorname{Vol}(M,g) = \int_M v_g .$$

Example 1.1 A straightforward computation shows us that the volume form $v_{g_{can}}$ for $\left(S^{n-1}, g_{can}\right)$ is given in Euclidean coordinates by

$$v_{g_{can}} = \sum_{i=1}^{n} (-1)^{n-i} x^i dx^1 \wedge \ldots \wedge \widehat{dx^i} \wedge \ldots \wedge dx^n ,$$

where $\widehat{\ }$ over a symbol means that symbol is defeated.

Let us determine $\operatorname{Vol}\left(S^{n-1}, g_{can}\right)$ for $n \ge 1$. In order to do this, let us compute the integral

$$I=\int_{-\infty}^{+\infty}\cdots\int_{-\infty}^{+\infty} e^{-\left[\left(x^1\right)^2+\cdots+\left(x^n\right)^2\right]} dx^1\wedge\ldots\wedge dx^n ,$$

in two different ways, where $(x^1,\ldots,x^n)$ are the usual coordinates on $\mathbb{R}^n$. On the one hand, note that

$$I=\left(2\int_0^{\infty} e^{-t^2}\,dt\right)^n$$
$$=\pi^{\frac{n}{2}}.$$

On the other hand, $(0,\infty)\times S^{n-1}$ is smooth diffeomorphic to $\mathbb{R}^n\setminus\{0\}$ via the smooth diffeomorphism $x^i=\rho v^i, i=1,2,\ldots,n,\ \rho\in(0,\infty)$ and $\left(v^1,\ldots,v^n\right)\in S^{n-1}$. Hence $dx^i=v^i d\rho+\rho dv^i$, so that an easy computation leads us to

$$dx^1\wedge\ldots\wedge dx^n=\rho^{n-1}\left[\sum_{i=1}^{n}(-1)^{n-i} v^i dv^1\wedge\ldots\wedge\widehat{dv^i}\wedge\ldots\wedge dv^n\right]\wedge d\rho$$
$$=\rho^{n-1}v_{g_{can}}\wedge d\rho.$$

Thus,

$$I=\int_0^{\infty}\int_{S^{n-1}} e^{-\rho^2}\rho^{n-1}v_{g_{can}}\wedge d\rho$$
$$=\int_0^{\infty} e^{-\rho^2}\rho^{n-1}d\rho\cdot\mathrm{Vol}\left(S^{n-1},g_{can}\right).$$

Making in the last integral the change of variable $\rho^2=t$ one obtains

$$\int_0^{\infty} e^{-\rho^2}\rho^{n-1}d\rho=\frac{1}{2}\int_0^{\infty} e^{-t}t^{\frac{n}{2}-1}\,dt$$
$$=\frac{1}{2}\Gamma\left(\frac{n}{2}\right),$$

where Γ denotes Euler's gamma function. Consequently,

$$\mathrm{Vol}\left(S^{n-1},g_{can}\right)=\frac{2\pi^{\frac{n}{2}}}{\Gamma\left(\frac{n}{2}\right)},$$

i.e.

$$\mathrm{Vol}\left(S^{n},g_{can}\right)=\frac{2\pi^{\frac{n+1}{2}}}{\Gamma\left(\frac{n+1}{2}\right)}.$$

In particular, taking into account the Remark 1.2 it follows that:

$$\operatorname{Vol}\left(P^n(\mathbb{R}), g\right) = \frac{\pi^{\frac{n+1}{2}}}{\Gamma\left(\frac{n+1}{2}\right)}$$

and

$$\operatorname{Vol}\left(L_{2n+1}(p, q_1, \ldots, q_{n+1}), g\right) = \frac{2\pi^{n+1}}{pn!}$$

(see Chapter 1, Examples 2.15 and 2.16).

Example 1.2. Let Γ be the lattice of rank n in $\mathbb{R}^n$ defining the n-dimensional flat torus $\left(\mathbb{R}^n/\Gamma, g_{can}/\Gamma\right)$ (see Chapter 1, Example 2.17). To each

$$u_0 = \sum_{k=1}^{n} \lambda_0^k e_k$$

in Γ one can associate the parallelepiped $P(u_0)$ defined by

$$P(u_0) = \left\{ x = \sum_{k=1}^{n} \xi^k e_k \;\middle|\; \xi^k \in \left(\lambda_0^k, \lambda_0^k + 1\right), k = 1, \ldots, n \right\}.$$

Then no two distinct points of $P(u_0)$ represent the same point of $\mathbb{R}^n/\Gamma$ and the image under the natural projection $p : \mathbb{R}^n \to \mathbb{R}^n/\Gamma$ of the closure $\overline{P(u_0)}$ is all of $\mathbb{R}^n/\Gamma$, that is $P(u_0)$ is a fundamental domain of $\mathbb{R}^n/\Gamma$, so that

$$\operatorname{Vol}\left(\mathbb{R}^n/\Gamma, g_{can}/\Gamma\right) = \operatorname{Vol}\left(P(u_0)\right) = \left|e_1 \wedge \ldots \wedge e_n\right|$$

for every $u_0 \in \Gamma$.

Example 1.3. Let $\Gamma_{a,b}$ and $\breve{\Gamma}_{a,b}$ be as in Chapter 1, Example 2.18. Since the Klein bottle $\left(K_{a,b}, g_{can}/\breve{\Gamma}_{a,b}\right)$ admits $\left(\mathbb{R}^2/\Gamma_{a,b}, g_{can}/\Gamma_{a,b}\right)$ as a double Riemannian covering, one gets

$$\operatorname{Vol}\left(K_{a,b}, g_{can}/\breve{\Gamma}_{a,b}\right) = \frac{1}{2}ab.$$

Example 1.4. Let H_n be the real $(2n+1)$-dimensional Heisenberg group. With the notations as in Chapter 1, Example 2.19, note that $\xi_n = \left\{A(0,0,t) \mid t \in \mathbb{R}\right\}$ is both the center and the derived subalgebra of h_n. If one identifies the subspace $\left\{A(x,y,0) \mid x, y \in \mathbb{R}^n\right\}$ of h_n with $\mathbb{R}^{2n}$, then h_n is the vector direct sum $h_n = \mathbb{R}^{2n} \oplus \xi_n$. The system

$$\left(A_1, \ldots, A_n, B_1, \ldots, B_n, A(0,0,1)\right),$$

where the first $2n$ elements form the standard basis of $\mathbb{R}^{2n}$, is called the **standard basis** of h_n. The non-zero brackets among the elements of this basis are given by

$$[A_k, B_k] = A(0,0,1) \text{ for } 1 \leq k \leq n.$$

A left invariant metric g on H_n is uniquely determined by the induced inner product $<,>$ on h_n, where h_n is viewed as the tangent space to H_n at the identity. Conversely, every inner product on h_n determines a left-invariant metric on H_n. Let us identify g with the matrix of $<,>$ relative to the standard basis of h_n. For any g, one can choose an inner automorphism φ such that $\mathbb{R}^{2n}$ is orthogonal to ξ_n relative to $\varphi^*(g)$. By Proposition 2.2 of Chapter 1, $\left(H_n/\Gamma, g\right)$ is isometric to $\left(H_n/\Gamma, \varphi^*(g)\right)$ for every cocompact discrete subgroup Γ of H_n. Replacing g by $\varphi^*(g)$ one may then assume that g has the form

$$g = \begin{bmatrix} h & 0 \\ 0 & g_{2n+1} \end{bmatrix}$$

with h a positive definite $2n \times 2n$ matrix and $g_{2n+1} > 0$. For a cocompact discrete subgroup Γ of H_n, let us define $|\Gamma| = r_1 r_2 \dots r_n$ for $r = (r_1, \dots, r_n)$ the unique n-tuple as in Example 2.19 of Chapter 1 for which Γ is isomorphic to Γ_r. Standard computation implies that

$$\operatorname{Vol}\left(H_n/\Gamma_r, g\right) = |\Gamma_r| (\det(g))^{\frac{1}{2}},$$

where g is identified with a positive definite matrix.

Remark 1.5 Let $\pi : (M, g) \to (N, h)$ be a Riemannian submersion with π onto and $\dim M > \dim N$. For each $x \in N$, $\pi^{-1}(x)$ is a smooth submanifold of M so that g induces a Riemannian structure g_x on $\pi^{-1}(x)$ (see Chapter 1, Example 2.5). Let μ_{g_x} be the canonical measure of the Riemannian manifold $(\pi^{-1}(x), g_x)$ (see Definition 1.1) and $\tilde{f} \in C_0(M)$. For any $x \in M$, let us define $\tilde{f}_x : \pi^{-1}(x) \to \mathbb{R}$ with $\tilde{f}_x = \tilde{f}\big|_{\pi^{-1}(x)}$ and $f : M \to \mathbb{R}$,

$$f(x) = \int_{\pi^{-1}(x)} \tilde{f}_x d\mu_{g_x}.$$

Then $\tilde{f}_x \in C_0(\pi^{-1}(x))$ for each $x \in M$, $f \in C_0(M)$ and moreover one can prove Fubini's property:

$$\int_{\tilde{M}} \tilde{f} d\mu_g = \int_M f d\mu_h$$

(see for details Berger, Gauduchon and Mazet [1971]). In particular, if $\tilde{M}$ is closed, then

$$\operatorname{Vol}(\tilde{\mathrm{M}},\mathrm{g}) = \int_M f d\mu_h ,$$

where $f : M \to \mathbb{R}$ is given by

$$f(x) = \operatorname{Vol}(\pi^{-1}(x), g_x).$$

Note that Fubini's property is also satisfied by local isometries.

For instance, if $\pi : (S^{2n+1}, g_{can}) \to (P^n(\mathbb{C}), g)$ is the Riemannian submersion defined in Example 2.14 of Chapter 1, then $(\pi^{-1}(x), (g_{can})_x)$ is isometric to (S^1, g_{can}) for any $x \in P^n(\mathbb{C})$, so that

$$\operatorname{Vol}(\pi^{-1}(x), (g_{can})_x) = 2\pi$$

for each $x \in P^n(\mathbb{C})$. Consequently,

$$\operatorname{Vol}(S^{2n+1}, g_{can}) = 2\pi \operatorname{Vol}(P^n(\mathbb{C}), g),$$

that is

$$\operatorname{Vol}(P^n(\mathbb{C}), g) = \frac{\pi^n}{\Gamma(n+1)} = \frac{\pi^n}{n!}.$$

Similarly, let us consider the Riemannian submersion $\pi : (S^{4n+3}, g_{can}) \to (P^n(\mathbb{H}), g)$ (see Chapter 1, Example 2.14). Since

$$\operatorname{Vol}(S^3, g_{can}) = 2\pi^2$$

it follows that

$$\operatorname{Vol}(P^n(\mathbb{H}), g) = \frac{\pi^{2n}}{\Gamma(2n+2)} = \frac{\pi^{2n}}{(2n+1)!}.$$

Now let us briefly recall some facts on integrable functions on M and integrable subsets of M. A function F on M with real values but that may take the value $+\infty$ is called a **lower semicontinuous function** if $p_n \to p$ implies $\liminf F(p_n) \geq F(p)$. For a lower semicontinuous function $F \geq 0$ on M, let us define

$$\bar{\lambda}_g(F) = \sup\{\lambda_g(f) \mid f \in C_0(M) \text{ satisfies } f \leq F\}$$

and for any function $f \geq 0$ on M, let us define

$$\bar{\lambda}_g(f) = \inf\{\bar{\lambda}_g(F) \mid F \geq f \text{ is lower semicontinuous}\}.$$

A function f on M is said to be **integrable** if there exists a sequence $\{f_n\} \subset C_0(M)$ such that $\bar{\lambda}_g(|f - f_n|) \to 0$. Then $\{\lambda_g(f_n)\}$ is a convergent sequence, and its limit does not depend on the choice of $\{f_n\}$. We denote the limit by $\int_M f d\mu_g$ and we call it the integral of f. In particular, any $f \in C_0(M)$ is integrable and its integral is just the above defined $\lambda_g(f)$. A subset $K \subset M$ is said to be **integrable** if its characteristic function $\chi_K : M \to \mathbb{R}$,

$$\chi_K(x) = \begin{cases} 1 & \text{for} \quad x \in K, \\ 0 & \text{for} \quad x \in M \setminus K, \end{cases}$$

is integrable. In this case, $\int_M \chi_K d\mu_g$ [also denoted by $\int_K d\mu_g$ or $\mathrm{Vol}(K,g)$] is called the **n-dimensional volume** of K. Any compact (or more generally, precompact) subset K of M is integrable with $\mathrm{Vol}(K,g) < +\infty$.

Suppose now that (M,g) is a closed Riemannian manifold. We shall point out some canonical Hilbert spaces that can be associated to (M,g) via the positive measure μ_g described above.

A. The Hilbert space $L^2(M)$

Let us define the inner product $\langle \cdot , \cdot \rangle$ on $C^\infty(M)$ by:

$$\langle f_1, f_2 \rangle = \int_M f_1 f_2 d\mu_g ,$$

for each $f_1, f_2 \in C^\infty(M)$. It is easy to see that $\langle \cdot , \cdot \rangle$ is a bilinear, symmetric map and moreover it is positive definite since

$$\langle f, f \rangle = \int_M f^2 d\mu_g \geq 0$$

and

$$\langle f, f \rangle = 0$$

if and only if $f = 0$. The real number

$$\|f\| = \sqrt{\langle f, f \rangle}$$

is called the L^2-norm of f. Thus $\left(C^\infty(M), \langle \cdot , \cdot \rangle\right)$ is a pre-Hilbert space and its completion via Cauchy sequences with respect to this L^2-norm is denoted by $L^2(M)$ and is called **the space of square integrable real valued functions** on M.

Thus, every $u \in L^2(M)$ is the equivalence class $[f_k]$ of a Cauchy sequence $\{f_k\}_{k \in N^*} \subset C^\infty(M)$ with respect to the L^2-norm, where by definition two sequences $\{f_k'\}$ and $\{f_k''\}$ are equivalent if and only if $\|f_k' - f_k''\| \to 0$ as $k \to \infty$. For $u = [f_k^1], v = [f_k^2] \in L^2(M)$, their inner product is

$$\langle u,v\rangle = \lim_{k\to\infty}\left\langle f_k^1, f_k^2\right\rangle.$$

Note that if $\Phi:(M,g)\to(N,h)$ is an isometry between closed Riemannian manifolds, then the mapping $\Phi^*:C^\infty(N)\to C^\infty(M)$ (see Chapter 1, §1) acts as an orthogonal transformation i.e. for any $f_1, f_2 \in C^\infty(N)$,

$$\left\langle \Phi^*(f_1),\Phi^*(f_2)\right\rangle = \int_M (f_1\circ\Phi)(f_2\circ\Phi)\,d\mu_g = \int_M (f_1\cdot f_2)\circ\Phi\,d\mu_g = \int_N f_1\cdot f_2\,d\mu_h = \langle f_1, f_2\rangle$$

(see also Remark 1.1).

For any $p>0$, one can use the Riemannian metric g to define the space $L^p(M)$ by saying a measurable function f to be an element of $L^p(M)$ if the integral

$$\left(\int_M |f|^p\,d\mu_g\right)^{1/p}$$

is finite. The L^p-norm $\|f\|_p$ of f is given by this expression, when it is finite. Hölder's inequality states that for $p,q\in(1,+\infty)$ satisfying $\frac{1}{p}+\frac{1}{q}=1$, one has

$$\int_M |f_1 f_2|\,d\mu_g \le \|f_1\|_p\,\|f_2\|_q$$

for any $f_1\in L^p(M)$, $f_2\in L^q(M)$.

B. The Hilbert space $L^2\mathfrak{X}(M)$

For each $X,Y\in\mathfrak{X}(M)$, $g(X,Y)$ is a smooth real valued function on M (see Chapter 1, Definition 2.1). Then we can define an inner product on $\mathfrak{X}(M)$ by

$$\langle X,Y\rangle = \int_M g(X,Y)\,d\mu_g\,,$$

for each $X,Y\in\mathfrak{X}(M)$. Therefore $\mathfrak{X}(M)$ has a natural structure of a pre-Hilbert space and its completion with respect to the L^2-norm given by

$$\|X\| = \sqrt{\langle X,X\rangle}$$

is denoted by $L^2\mathfrak{X}(M)$ and is called the **space of square integrable vector fields** on M.

Note that if $\Phi:(M,g)\to(N,h)$ is an isometry between closed Riemannian manifolds, the mapping $\Phi_*:\mathfrak{X}(M)\to\mathfrak{X}(N)$ acts as an orthogonal transformation, that is

$$\left\langle \Phi_*(X),\Phi_*(Y)\right\rangle = \langle X,Y\rangle \quad \text{for all } X,Y\in\mathfrak{X}(M).$$

Indeed,

$$\langle \Phi_*(X), \Phi_*(Y) \rangle = \int_N h(\Phi_*(X), \Phi_*(Y))\, d\mu_h = \int_M h(\Phi_*(X), \Phi_*(Y)) \circ \Phi\, d\mu_g$$
$$= \int_M g(X,Y)\, d\mu_g = \langle X, Y \rangle$$

for all $X, Y \in \mathfrak{X}(M)$.

C. The Hilbert spaces $L^2 A^p(M)$, $1 \le p \le n = \dim M$

As before, first we define a pointwise inner product for *p*-forms, $1 \le p \le n$.

The Riemannian structure $g : \mathfrak{X}(M) \times \mathfrak{X}(M) \to C^\infty(M)$ (see Chapter 1, Definition 2.1), that expresses the Riemannian metric on *M*, can be translated to a pointwise inner product $(\,|\,)$ on $A(M) = \sum_{p=1}^{n} A^p(M)$, the exterior algebra of smooth differential forms of degree ≥ 1, by requiring the following three conditions:

$$(\alpha | \beta) = g(\alpha^\#, \beta^\#) \tag{1.1}$$

for all $\alpha, \beta \in A^1(M)$ [for the definition of $\# : A^1(M) \to \mathfrak{X}(M)$, see §2 of Chapter 1];

$$A^p(M) \text{ is orthogonal to } A^q(M) \text{ for } p \ne q; \tag{1.2}$$

$$(\alpha^1 \wedge \cdots \wedge \alpha^p | \beta^1 \wedge \cdots \wedge \beta^p) = \det\left((\alpha^i | \beta^j)\right)_{1 \le i,j \le p} \tag{1.3}$$

for all $\alpha^1, \dots, \alpha^p, \beta^1, \dots, \beta^p \in A^1(M)$.

Thus, in a local chart (U, Φ) on *M* with coordinates $(x^1, \dots, x^n)$, one gets

$$g^{ij}(x) = (dx^i | dx^j)(x), \; 1 \le i, j \le n,$$

where $x \in U$ and $\left[g^{ij}(x)\right]$ is the inverse of the matrix $\left[g_{ij}(x)\right]$. If

$$\alpha = \sum_{1 \le i_1 < \ldots < i_p \le n} \alpha_{i_1 \ldots i_p}\, dx^{i_1} \wedge \ldots \wedge dx^{i_p},$$
$$\beta = \sum_{1 \le j_1 < \ldots < j_p \le n} \beta_{j_1 \ldots j_p}\, dx^{j_1} \wedge \ldots \wedge dx^{j_p}$$

are the expressions for $\alpha, \beta \in A^p(M)$ in this local chart, by linearity, one obtains the following expression for the smooth real function $(\alpha | \beta)$:

$$(\alpha | \beta)(x) = \sum_{\substack{1 \le i_1 < \cdots < i_p \le n \\ 1 \le j_1 < \cdots < j_p \le n}} \alpha_{i_1 \ldots i_p}(x) \beta_{j_1 \ldots j_p}(x) \left(dx^{i_1} \wedge \ldots \wedge dx^{i_p} \,\middle|\, dx^{j_1} \wedge \ldots \wedge dx^{j_p}\right)(x),$$

where

$$\left(dx^{i_1} \wedge \ldots \wedge dx^{i_p} \,\middle|\, dx^{j_1} \wedge \ldots \wedge dx^{j_p}\right)(x) = \sum_{k_1, \ldots, k_p} \varepsilon_{k_1 \ldots k_p} g^{i_1 j_{k_1}}(x) \cdots g^{i_p j_{k_p}}(x),$$

$x \in U$, the indices $k_1, \dots, k_p$ run over $(1, \dots, p)$, and

$$\varepsilon_{k_1\ldots k_p} = \begin{cases} 1 \text{ if the permutation } \begin{pmatrix} 1 & \ldots p \\ k_1 & \ldots k_p \end{pmatrix} \text{ is even,} \\ -1 \text{ if the permutation } \begin{pmatrix} 1 & \ldots p \\ k_1 & \ldots k_p \end{pmatrix} \text{ is odd.} \end{cases}$$

In particular,

$$(df_1 | df_2)(x) = \sum_{i,j=1}^{n} g^{ij}(x) \frac{\partial f_1}{\partial x^i}(x) \frac{\partial f_2}{\partial x^j}(x) \tag{1.4}$$

for all $x \in U$ and $f_1, f_2 \in C^\infty(M)$.

More generally, for all smooth tensor fields $S, T \in \mathcal{G}_r^0(M)$ one can define the smooth real valued function $(S|T) \in C^\infty(M)$.

More precisely, let $x \in M$. Then $T_x, S_x : \underbrace{T_xM \times \cdots \times T_xM}_{r \text{ times}} \to \mathbb{R}$ are tensors of type $(0,r)$ of T_xM, that is $\mathbb{R}$-linear with respect to each variable. Let $\mathcal{G}_r^0(T_xM)$ be the tensor space of type $(0,r)$ of T_xM and T_x^*M the dual space of T_xM. If $v_1^*, \ldots, v_r^* \in T_x^*M$ are given then the mapping

$$v_1^* \otimes \cdots \otimes v_r^* : \underbrace{T_xM \times \cdots \times T_xM}_{r \text{ times}} \to \mathbb{R}$$

defined by

$$(v_1^* \otimes \cdots \otimes v_r^*)(u_1, \ldots, u_r) = \prod_{j=1}^{r} v_j^*(u_j),$$

gives a tensor of type $(0,r)$ of T_xM. One may define the inner product $(\ ,\ )_x$ on $\mathcal{G}_r^0(T_xM)$ from the inner product g_x on T_xM so that $\{e^{j_1} \otimes \cdots \otimes e^{j_r} \mid j_1, \ldots, j_r = 1, \ldots, r\}$ is an orthonormal basis of $\mathcal{G}_r^0(T_xM)$, where $(e_1, \ldots, e_n)$ is an orthonormal basis of T_xM and $(e^1, \ldots, e^n)$ its dual basis. Therefore, one can define

$$(T|S)(x) = (T_x, S_x)_x, \ x \in M.$$

The inner product $\langle \cdot, \cdot \rangle$ on $A^p(M)$ is given by:

$$\langle \alpha, \beta \rangle = \int_M (\alpha \mid \beta) \, d\mu_g .$$

The completion of $A^p(M)$ with respect to the corresponding norm [also called the L^2-norm on $A^p(M)$]

$$\|\alpha\| = \sqrt{\langle \alpha, \alpha \rangle}$$

is denoted by $L^2 A^p(M)$ and is called **the space of square integrable *p*-forms** on M. Similarly, one can define the space $L^2\mathcal{G}_r^0(M)$ of square integrable tensor fields of type $(0,r)$ on M.

Remark 1.6. For $\alpha \in A^1(M)$, the adjoint of $i(\alpha^\#): A^p(M) \to A^{p-1}(M)$ relative to the inner product $\langle\, ,\rangle$ is the exterior multiplication operator $\varepsilon(\alpha): A^{p-1}(M) \to A^p(M)$ defined by $\varepsilon(\alpha)\beta = \alpha \wedge \beta$.

Indeed, first notice that it suffices to verify this property for elementary forms on *M*. On the other hand, using the definition of the inner product $\langle\, ,\rangle$, one has successively

$$\left\langle i(\alpha^\#)\theta^1 \wedge \ldots \wedge \theta^p, \omega^1 \wedge \ldots \wedge \omega^{p-1}\right\rangle = \int_M \left(i(\alpha^\#)\theta^1 \wedge \ldots \wedge \theta^p \middle| \omega^1 \wedge \ldots \wedge \omega^{p-1}\right) d\mu_g$$

$$= \int_M \sum_{k=1}^p (-1)^{k+1} \theta^k(\alpha^\#) \left(\theta^1 \wedge \ldots \wedge \hat{\theta}^k \wedge \ldots \wedge \theta^p \middle| \omega^1 \wedge \ldots \wedge \omega^{p-1} \right) d\mu_g$$

$$= \int_M \sum_{k=1}^p (-1)^{k+1} g\left((\theta^k)^\#, \alpha^\#\right) \begin{vmatrix} (\theta^1|\omega^1) & \cdots & (\theta^1|\omega^{p-1}) \\ \vdots & & \vdots \\ (\theta^{k-1}|\omega^1) & \cdots & (\theta^{k-1}|\omega^{p-1}) \\ (\theta^{k+1}|\omega^1) & \cdots & (\theta^{k+1}|\omega^{p-1}) \\ \vdots & & \vdots \\ (\theta^p|\omega^1) & \cdots & (\theta^p|\omega^{p-1}) \end{vmatrix} d\mu_g$$

$$= \int_M \sum_{k=1}^p (-1)^{k+1} (\theta^k|\alpha) \begin{vmatrix} (\theta^1|\omega^1) & \cdots & (\theta^1|\omega^{p-1}) \\ \vdots & & \vdots \\ (\theta^{k-1}|\omega^1) & \cdots & (\theta^{k-1}|\omega^{p-1}) \\ (\theta^{k+1}|\omega^1) & \cdots & (\theta^{k+1}|\omega^{p-1}) \\ \vdots & & \vdots \\ (\theta^p|\omega^1) & \cdots & (\theta^p|\omega^{p-1}) \end{vmatrix} d\mu_g$$

$$= \int_M \begin{vmatrix} (\theta^1|\alpha) & (\theta^1|\omega^1) & \cdots & (\theta^1|\omega^{p-1}) \\ (\theta^2|\alpha) & (\theta^2|\omega^1) & \cdots & (\theta^2|\omega^{p-1}) \\ \vdots & \vdots & & \vdots \\ (\theta^p|\alpha) & (\theta^p|\omega^1) & \cdots & (\theta^p|\omega^{p-1}) \end{vmatrix} d\mu_g$$

$$= \int_M \left(\theta^1 \wedge \ldots \wedge \theta^p \middle| \alpha \wedge \omega^1 \wedge \ldots \wedge \omega^{p-1}\right) d\mu_g$$

$$= \left\langle \theta^1 \wedge \ldots \wedge \theta^p, \varepsilon(\alpha)(\omega^1 \wedge \ldots \wedge \omega^{p-1}) \right\rangle$$

for all $\theta^1, \ldots, \theta^p, \omega^1, \ldots, \omega^{p-1} \in A^1(M)$.

In all that follows we shall use the following notations:

(i) $|f|^2 = f^2,\ f \in C^\infty(M)$;

(ii) $|X|^2 = g(X,X),\ X \in \mathfrak{X}(M)$;

(iii) $|\alpha|^2 = (\alpha|\alpha),\ \alpha \in A^p(M)$;

(iv) $|S|^2 = (S|S),\ S \in \mathcal{T}_r^0(M)$.

2. SOME CANONICAL DIFFERENTIAL OPERATORS ON A RIEMANNIAN MANIFOLD

Let (M,g) be an n-dimensional smooth Riemannian manifold. Then there exist some canonical linear differential operators defined on M as follows:

A. The gradient operator

This operator denoted

$$grad : C^\infty(M) \ni f \mapsto grad(f) \in \mathfrak{X}(M)$$

is defined by

$$X(f) = g(X, grad(f)),$$

for all $X \in \mathfrak{X}(M)$.

In particular note that

$$(grad(f_1))(f_2) = g(grad(f_1), grad(f_2))$$

for all $f_1, f_2 \in C^\infty(M)$.

Let $g' = ag$, where $a \in C^\infty(M), a > 0$. If $grad_g$ (resp. $grad_{g'}$) denotes the gradient operator associated to the Riemannian metric g (resp. g'), the previous definition implies immediately that

$$grad_{ag} = \frac{1}{a} grad_g. \tag{2.1}$$

Proposition 2.1. Let $(U, x^1, \ldots, x^n)$ be a local chart on M. Then we have:

$$grad(f)|_U = \sum_{i,j} g^{ij} \frac{\partial f}{\partial x^j} \frac{\partial}{\partial x^i}.$$

In particular, the equality

$$(grad(f_1)) f_2 = (df_1 | df_2)$$

is true for all $f_1, f_2 \in C^\infty(M)$.

Proof. Indeed, let

$$X|_U = \sum_i X^i \frac{\partial}{\partial x^i},\ grad(f)|_U = \sum_j (grad(f)|_U)^j \frac{\partial}{\partial x^j}.$$

Then

$$g(X, grad(f))|_U = \sum_{i,j} g_{ij} X^i (grad(f)|_U)^j ,$$

so that the equality defining *grad(f)* implies

$$\sum_i X^i \frac{\partial f}{\partial x^i} = \sum_{i,j} X^i g_{ij} (grad(f)|_U)^j$$

or equivalently

$$(grad(f)|_U)^j = \sum_i g^{ij} \frac{\partial f}{\partial x^i},$$

which leads to the desired equality.

In particular, one gets

$$[(grad(f_1))(f_2)](x) = \sum_{i,j=1}^{n} g^{ij}(x) \frac{\partial f_1}{\partial x^i}(x) \frac{\partial f_2}{\partial x^j}(x)$$

for any $x \in U$ and all $f_1, f_2 \in C^\infty(M)$, so that the last statement is an immediate consequence of (1.4).

Q.E.D.

Let us point out some of its properties. Thus, for any smooth functions $f_1, f_2 \in C^\infty(M)$ one has

$$\begin{aligned} grad(f_1 + f_2) &= grad(f_1) + grad(f_2), \\ grad(f_1 f_2) &= f_2(grad(f_1)) + f_1(grad(f_2)). \end{aligned} \tag{2.2}$$

Both these equalities are straightforward consequences of the definition of the gradient operator.

If $\Phi : (M, g) \to (N, h)$ is an isometry, then

$$\Phi_*\left(grad_g\left(\Phi^*(\tilde{f})\right)\right) = grad_h(\tilde{f}) \tag{2.3}$$

for any $\tilde{f} \in C^\infty(N)$.

Indeed, for any $X \in \mathfrak{X}(M)$,

$$\begin{aligned} h\left(\Phi_*\left(grad_g\left(\Phi^*(\tilde{f})\right)\right), \Phi_*(X)\right) &= g\left(grad_g\left(\Phi^*(\tilde{f})\right), X\right) \circ \Phi^{-1} \\ &= X\left(\Phi^*(\tilde{f})\right) \circ \Phi^{-1} = [\Phi_*(X)](\tilde{f}) = h\left(grad_h(\tilde{f}), \Phi_*(X)\right) \end{aligned}$$

(see also Chapter 1, §1 and §2), which implies (2.3).

Note that (2.3) implies

$$g\left(grad_g\left(\Phi^*(\tilde{f}_1)\right), grad_g\left(\Phi^*(\tilde{f}_2)\right)\right) = h\left(grad_h(\tilde{f}_1), grad_h(\tilde{f}_2)\right) \circ \Phi$$

for any $\tilde{f}_1, \tilde{f}_2 \in C^\infty(N)$.

Indeed, taking into account the definitions of the gradient of $\Phi^*(\tilde{f}_2)$ (resp. $\tilde{f}_2$) and of Φ^*, one gets

$$g\left(grad_g\left(\Phi^*\left(\tilde{f}_1\right)\right), grad_g\left(\Phi^*\left(\tilde{f}_2\right)\right)\right) = \left[grad_g\left(\Phi^*\left(\tilde{f}_1\right)\right)\right]\left(\Phi^*\left(\tilde{f}_2\right)\right)$$
$$= \left[\Phi_*\left(grad_g\left(\Phi^*\left(\tilde{f}_1\right)\right)\right)\right]\left(\tilde{f}_2\right)\circ\Phi = \left[grad_h\left(\tilde{f}_1\right)\right]\left(\tilde{f}_2\right)\circ\Phi$$
$$= h\left(grad_h\left(\tilde{f}_1\right), grad_h\left(\tilde{f}_2\right)\right)\circ\Phi.$$

as claimed.

B. The divergence operator

The divergence operator, denoted

$$div : \mathfrak{X}(M) \ni X \mapsto div(X) \in C^\infty(M)$$

is defined by

$$div(X) = trace\left(\overset{\circ}{\nabla} X\right),$$

where $\overset{\circ}{\nabla} X : \mathfrak{X}(M) \to \mathfrak{X}(M)$ is the covariant differential of $X \in \mathfrak{X}(M)$ (see Chapter 1, Definition 3.3).

Hence, for any local field of orthonormal frames $(E_1, \ldots, E_n)$ on (M,g), one obtains

$$div(X) = \sum_{j=1}^{n} g\left(\nabla_{E_j} X, E_j\right).$$

For any $f \in C^\infty(M)$ and any $X, Y \in \mathfrak{X}(M)$, one has

$$div(X+Y) = div(X) + div(Y), \tag{2.4}$$

$$div(fX) = f div(X) + g(\mathrm{grad}(f), X). \tag{2.5}$$

Indeed, (2.4) and (2.5) are an immediate consequence of the above local formula for the divergence operator. For instance,

$$div(fX) = \sum_{j=1}^{n} g\left(\nabla_{E_j}(fX), E_j\right) = f\sum_{j=1}^{n} g\left(\nabla_{E_j} X, E_j\right) + \sum_{j=1}^{n} g\left(E_j(f)X, E_j\right)$$
$$= f div(X) + \sum_{j=1}^{n} g\left(\mathrm{grad}(f), E_j\right) g\left(X, E_j\right)$$
$$= f div(X) + g\left(\sum_{j=1}^{n} g\left(\mathrm{grad}(f), E_j\right) E_j, X\right)$$
$$= f div(X) + g(\mathrm{grad}(f), X).$$

Also, let us note that

$$div(X)\big|_U = \sum_{k=1}^{n}\left(\frac{\partial X^k}{\partial x^k} + \sum_{l=1}^{n} X^l \Gamma_{lk}^k\right),$$

where $(U,x^1,\dots,x^n)$ is a local chart on M, $X|_U=\sum_{i=1}^n X^i\frac{\partial}{\partial x^i}$ and (Γ^l_{jk}) are the second Christoffel symbols (see Chapter 1, §3). Indeed, using (2.4) and (2.5) one gets

$$div(X)|_U=div\left(\sum_{k=1}^n X^k\frac{\partial}{\partial x^k}\right)=\sum_{k=1}^n div\left(X^k\frac{\partial}{\partial x^k}\right)$$
$$=\sum_{k=1}^n X^k div\left(\frac{\partial}{\partial x^k}\right)+\sum_{k=1}^n g\left(\mathrm{grad}\left(X^k\right),\frac{\partial}{\partial x^k}\right).$$

On the other hand, by Proposition 2.1 one has

$$\mathrm{grad}\left(X^k\right)=\sum_{i,j=1}^n g^{ij}\frac{\partial X^k}{\partial x^j}\frac{\partial}{\partial x^i},$$

so that

$$\sum_{k=1}^n g\left(\mathrm{grad}\left(X^k\right),\frac{\partial}{\partial x^k}\right)=\sum_{i,j,k=1}^n g^{ij}\frac{\partial X^k}{\partial x^j}g_{ik}=\sum_{k=1}^n\frac{\partial X^k}{\partial x^k}.$$

Finally the equality

$$\left[\overset{\circ}{\nabla}\left(\frac{\partial}{\partial x^k}\right)\right]\left(\frac{\partial}{\partial x^j}\right)=\sum_{i=1}^n\Gamma^i_{jk}\frac{\partial}{\partial x^i}$$

shows that

$$div\left(\frac{\partial}{\partial x^k}\right)=trace\left(\overset{\circ}{\nabla}\left(\frac{\partial}{\partial x^k}\right)\right)=\sum_{j=1}^n\Gamma^j_{jk},$$

which leads to the above claimed formula.

Proposition 2.2. If $(U,x^1,\dots,x^n)$ is a local chart on M, then

$$div(X)|_U=\frac{1}{\sqrt{\det(g_{ij})}}\sum_{k=1}^n\frac{\partial}{\partial x^k}\left(\sqrt{\det(g_{ij})}\cdot X^k\right),$$

where $X|_U=\sum_{j=1}^n X^j\frac{\partial}{\partial x^j}$ and $g_{ij}=g\left(\frac{\partial}{\partial x^i},\frac{\partial}{\partial x^j}\right)$, $i,j=1,\dots,n$.

Proof. Indeed, using the above formula for $div(X)|_U$ one can write successively

$$div(X)|_U=\sum_{k=1}^n\left(\frac{\partial X^k}{\partial x^k}+\sum_{l=1}^n X^l\Gamma^k_{lk}\right)=\sum_{k=1}^n\left(\frac{\partial X^k}{\partial x^k}+\frac{1}{2}X^k\sum_{i,j=1}^n g^{ij}\frac{\partial g_{ij}}{\partial x^k}\right)$$
$$=\sum_{k=1}^n\left\{\frac{\partial X^k}{\partial x^k}+\frac{1}{2}X^k\frac{1}{\det(g_{ij})}\sum_{l,m=1}^n G^{lm}\frac{\partial g_{lm}}{\partial x^k}\right\}=\sum_{k=1}^n\left[\frac{\partial X^k}{\partial x^k}+X^k\frac{\partial}{\partial x^k}\left(\ln\sqrt{\det(g_{ij})}\right)\right]$$
$$=\frac{1}{\sqrt{\det(g_{ij})}}\sum_{k=1}^n\frac{\partial}{\partial x^k}\left(\sqrt{\det(g_{ij})}X^k\right),$$

as claimed, where $\left(G^{ij}\right)$ denotes the adjoint matrix of the matrix $\left(g_{ij}\right)$.

Q.E.D.

Using Proposition 2.2 one can immediately deduce an equivalent definition of $div(X)$, namely $div(X)$ can also be defined by

$$d\left(i(X)v_g\right)=div(X)v_g,$$

where v_g denotes the local volume n-form corresponding to a local orientation on M (see also Remark 1.4) and $i(X)$ the interior multiplication operator defined by $X\in\mathfrak{X}(M)$ (see §1 of Chapter 1). Note that this definition of $div(X)$ does not depend on the choice of the local orientation on M since a changing of local charts that changes the orientation has as a consequence the sign changing of v_g in both sides of the above equality.

Let $g'=ag$, where $a\in C^{\infty}(M), a>0$. If div_g (resp. $div_{g'}$) denotes the divergence operator corresponding to the Riemannian metric g (resp. g'), then one has

$$div_{ag}(X)=div_g(X)+\frac{n}{2a}X(a) \tag{2.6}$$

for any $X\in\mathfrak{X}(M)$.

Indeed, putting $g'=ag$, Proposition 2.2 leads to

$$\begin{aligned}
div_{ag}(X)&=\frac{1}{\sqrt{\det\left(g'_{ij}\right)}}\sum_{k=1}^{n}\frac{\partial}{\partial x^k}\left(\sqrt{\det\left(g'_{ij}\right)}X^k\right)\\
&=\frac{1}{a^{\frac{n}{2}}\sqrt{\det\left(g_{ij}\right)}}\sum_{k=1}^{n}\frac{\partial}{\partial x^k}\left(a^{\frac{n}{2}}\sqrt{\det\left(g_{ij}\right)}X^k\right)\\
&=\frac{1}{\sqrt{\det\left(g_{ij}\right)}}\sum_{k=1}^{n}\frac{\partial}{\partial x^k}\left(\sqrt{\det\left(g_{ij}\right)}X^k\right)+\frac{n}{2a}\sum_{k=1}^{n}X^k\frac{\partial a}{\partial x^k}\\
&=div_g(X)+\frac{n}{2a}X(a).
\end{aligned}$$

If $\Phi:(M,g)\to(N,h)$ is an isometry, then

$$\Phi^*\left(div_h\left(\Phi_*(X)\right)\right)=div_g(X) \tag{2.7}$$

for any $X\in\mathfrak{X}(M)$.

Indeed, let $\left(E_1,\ldots,E_n\right)$ be a local field of orthonormal frames on (M,g) defined around a given point $x\in M$. Then $\left(\Phi_*\left(E_1\right),\ldots,\Phi_*\left(E_n\right)\right)$ is a local field around $\Phi(x)$ of orthonormal frames on (N,h), so that by Corollary 3.2 of Chapter 1 and the definition of an isometry one gets

$$\left[\Phi^*\left(div_h\left(\Phi_*(X)\right)\right)\right](x)=\left[div_h\left(\Phi_*(X)\right)\circ\Phi\right](x)$$

$$=\left\{\left[\sum_{j=1}^n h\left(\nabla_{\Phi_*(E_j)}\Phi_*(X),\Phi_*(E_j)\right)\right]\circ\Phi\right\}(x)$$

$$=\sum_{j=1}^n\left[g\left(\nabla_{E_j}X,E_j\right)\right](x)=\left[div_g(X)\right](x)$$

for any $X\in\mathfrak{X}(M)$ and $x\in M$, i.e. (2.7) is valid.

Remark 2.1. Given $X\in\mathfrak{X}(M)$, let $\{\Phi_t\}$ be the induced flow on *M*. For any fixed compact set *K* in *M*, let us denote

$$V(t)=\int_{\Phi_t(K)} d\mu_g\,.$$

Then

$$V'(0)=\int_K div(X)\,d\mu_g\,,$$

i.e. *div*(*X*) measures the infinitesimal change of volume by the flow generated by *X*.

Theorem 2.1 (The divergence theorem). If $X\in\mathfrak{X}(M)$ is with compact support, then

$$\int_M div(X)\,d\mu_g=0\,. \tag{2.8}$$

Proof. Let $\{\Phi_t\}$ be the flow induced by *X* on *M* and *K* the support of *X*. Since *K* is compact, *X* is complete. Consequently, Φ_t is defined for all $t\in\mathbb{R}$ and leaves $M\setminus K$ invariant. In particular,

$$\Phi_t(K)\subseteq K=\Phi_0(K)$$

for all *t*, hence the real function

$$V(t)=\int_{\Phi_t(K)} d\mu_g$$

has a maximal value at $t=0$, so that applying Remark 2.1 to $K=\operatorname{supp}X$, it follows that this function is differentiable at $t=0$ and

$$0=V'(0)=\int_K div(X)\,d\mu_g=\int_M div(X)\,d\mu_g\,,$$

that is (2.8).

Q.E.D.

Corollary 2.1. If *M* is closed, then

$$\langle \operatorname{grad}(f),X\rangle=-\langle f,div(X)\rangle \tag{2.9}$$

for any $f\in C^\infty(M)$ and $X\in\mathfrak{X}(M)$.

Proof. Indeed, the equality (2.9) is an immediate consequence of (2.5), Theorem

2.1 and the definitions of the inner products on $\mathfrak{X}(M)$ and $C^\infty(M)$ respectively (see this Chapter, §1,A,B).

Q.E.D.

C. The Laplace-Beltrami operator

One of the most fundamental objects associated to a Riemannian manifold (*M*,*g*) is its **Laplace-Beltrami operator**, a second-order elliptic selfadjoint partial differential operator which acts on $C^\infty(M)$.

This operator, denoted

$$\Delta : C^\infty(M) \ni f \mapsto \Delta f \in C^\infty(M)$$

is defined by

$$\Delta f = -div(\mathrm{grad}(f)).$$

Consequently,

$$\Delta f = -trace\left(\overset{\circ}{\nabla}(\mathrm{grad}(f))\right) = -trace(h_f) = -trace(H_f),$$

where h_f (resp. H_f) is the Hesse tensor field (resp. Hesse form) associated to *f* (see Chapter 1, Definitions 3.5 and 3.6), so that

$$\Delta f = -\sum_{j=1}^{n}\left(E_j \circ E_j - \nabla_{E_j} E_j\right)(f)$$

for any local field of orthonormal frames $(E_1, \ldots, E_n)$ on (*M*,*g*) [see also Chapter 1, formula (3.5)].

For $f_1, f_2 \in C^\infty(M)$, one has

$$div\left(f_2(\mathrm{grad}(f_1))\right) = -f_2(\Delta f_1) + g(\mathrm{grad}(f_2), \mathrm{grad}(f_1)), \tag{2.10}$$

$$\Delta(f_1 f_2) = f_2(\Delta f_1) - 2g(\mathrm{grad}(f_1), \mathrm{grad}(f_2)) + f_1(\Delta f_2). \tag{2.11}$$

To derive (2.10), one simply substitutes $f = f_2$ and $X = \mathrm{grad}(f_1)$ in (2.5) and one takes into account the definition of Δ. On the other hand, the definition of Δ, (2.2), (2.4) and (2.10) imply that

$$\begin{aligned}\Delta(f_1 f_2) &= -div(\mathrm{grad}(f_1 f_2)) \\ &= -div\left(f_2(\mathrm{grad}\, f_1)\right) - div\left(f_1(\mathrm{grad}(f_2))\right) \\ &= f_2(\Delta f_1) - g(\mathrm{grad}(f_2), \mathrm{grad}(f_1)) + f_1(\Delta f_2) - g(\mathrm{grad}(f_1), \mathrm{grad}(f_2)) \\ &= f_2(\Delta f_1) - 2g(\mathrm{grad}(f_1), \mathrm{grad}(f_2)) + f_1(\Delta f_2),\end{aligned}$$

i.e. (2.11) is valid.

Proposition 2.3 (Green's formula)

If $f_1, f_2 \in C^\infty(M)$ are such that $f_2(\mathrm{grad}(f_1))$ has compact support, then

$$\int_M \left[f_2 \Delta f_1 - g\left(\operatorname{grad}(f_2), \operatorname{grad}(f_1)\right)\right] d\mu_g = 0. \tag{2.12}$$

Proof. One takes $X = f_2\left(\operatorname{grad}(f_1)\right)$ and one applies the divergence theorem (Theorem 2.1) using (2.10).

Q.E.D.

Corollary 2.2. If both $f_1, f_2 \in C^\infty(M)$ have compact support, then

$$\int_M (f_2 \Delta f_1 - f_1 \Delta f_2) d\mu_g = 0. \tag{2.13}$$

In particular, if M is closed, then

$$\int_M \Delta f d\mu_g = 0 \tag{2.14}$$

for any $f \in C^\infty(M)$.

Proof. Indeed, using (2.10) twice, one gets

$$div\left(f_2\left(\operatorname{grad}(f_1)\right)\right) = -f_2(\Delta f_1) + g\left(\operatorname{grad}(f_2), \operatorname{grad}(f_1)\right),$$
$$div\left(f_1\left(\operatorname{grad}(f_2)\right)\right) = -f_1(\Delta f_2) + g\left(\operatorname{grad}(f_1), \operatorname{grad}(f_2)\right),$$

what implies

$$div\left(f_2\left(\operatorname{grad}(f_1)\right)\right) - div\left(f_1\left(\operatorname{grad}(f_2)\right)\right) = f_1(\Delta f_2) - f_2(\Delta f_1).$$

On the other hand, under the above hypotheses, both $f_2\left(\operatorname{grad}(f_1)\right)$ and $f_1\left(\operatorname{grad}(f_2)\right)$ have compact support, so that (2.13) is a consequence of the divergence theorem. Concerning (2.14), one simply substitutes $f_2 = 1$ (the constant function 1) in (2.13).

Q.E.D.

The next formula, namely the Bochner-Lichnerowicz formula, will be used later.

Proposition 2.4 (the Bochner-Lichnerowicz formula). Let (M,g) be a smooth Riemannian manifold. Then

$$-\frac{1}{2}\Delta\left(|df|^2\right) = |H_f|^2 - \left(d(\Delta f)|df\right) - (c_2 R)\left(\operatorname{grad}(f), \operatorname{grad}(f)\right)$$

for any $f \in C^\infty(M)$, where $d : C^\infty(M) \to A^1(M)$ denotes the exterior differential operator (see also the notations in §1, C).

Proof. By Theorem 3.1 (iv) of Chapter 1, one gets

$$X\left(|df|^2\right) = X\left(df|df\right) = Xg\left(\operatorname{grad}(f), \operatorname{grad}(f)\right)$$
$$= 2g\left(\nabla_X\left(\operatorname{grad}(f)\right), \operatorname{grad}(f)\right),$$

so that by Theorem 3.1 (iv), Chapter 1, again and Corollary 3.1, Chapter 1, one obtains

$$\left(X \circ Y - \nabla_X Y\right)\left(|df|^2\right) = 2Xg\left(\nabla_Y\left(\operatorname{grad}(f)\right), \operatorname{grad}(f)\right) - 2g\left(\nabla_{\nabla_X Y}\left(\operatorname{grad}(f)\right), \operatorname{grad}(f)\right)$$

$$= 2g\left(\nabla_X\left(\operatorname{grad}(f)\right),\nabla_Y\left(\operatorname{grad}(f)\right)\right)+2g\left(\left(\nabla_X\nabla_Y-\nabla_{\nabla_X Y}\right)\left(\operatorname{grad}(f)\right),\operatorname{grad}(f)\right)$$
$$= 2\left[\left(\nabla_X\left(\operatorname{grad}(f)\right)\right)^b\middle|\left(\nabla_Y\left(\operatorname{grad}(f)\right)\right)^b\right]+2\left(\left(\left(\nabla_X\nabla_Y-\nabla_{\nabla_X Y}\right)\left(\operatorname{grad}(f)\right)\right)^b\middle|\left(\operatorname{grad}(f)\right)^b\right)$$
$$= 2\left(\nabla_X(df)\middle|\nabla_Y(df)\right)+2\left(\left(\nabla_X\nabla_Y-\nabla_{\nabla_X Y}\right)(df)\middle|df\right),$$

that is

$$\frac{1}{2}\left(X\circ Y-\nabla_X Y\right)\left(|df|^2\right)=\left(\nabla_X(df)\middle|\nabla_Y(df)\right)+\left(\left(\nabla_X\nabla_Y-\nabla_{\nabla_X Y}\right)(df)\middle|df\right)$$

for any $X,Y\in\mathfrak{X}(M)$ and $f\in C^\infty(M)$.

Let $(E_1,\ldots,E_n)$ be a local field of orthonormal frames on (M,g). The last formula implies

$$\begin{aligned}-\frac{1}{2}\Delta\left(|df|^2\right)&=\frac{1}{2}\sum_{i=1}^n\left(E_i\circ E_i-\nabla_{E_i}E_i\right)\left(|df|^2\right)\\&=\sum_{i=1}^n\left(\nabla_{E_i}(df)\middle|\nabla_{E_i}(df)\right)+\sum_{i=1}^n\left(\left(\nabla_{E_i}\nabla_{E_i}-\nabla_{\nabla_{E_i}E_i}\right)(df)\middle|df\right).\end{aligned}\tag{2.15}$$

Concerning (2.15), first notice that

$$\sum_{i=1}^n\left(\nabla_{E_i}(df)\middle|\nabla_{E_i}(df)\right)=\sum_{i=1}^n\sum_{j=1}^n\left[\left(\nabla_{E_i}(df)\right)(E_j)\right]^2=\sum_{i,j=1}^n\left(H_f(E_i,E_j)\right)^2=|H_f|^2\tag{2.16}$$

(see also Chapter 1, Definition 3.5). On the other hand,

$$\sum_{i=1}^n\left(\left(\nabla_{E_i}\circ\nabla_{E_i}-\nabla_{\nabla_{E_i}E_i}\right)(df)\middle|df\right)=\sum_{i=1}^n\sum_{j=1}^n\left[\left(\nabla_{E_i}\circ\nabla_{E_i}-\nabla_{\nabla_{E_i}E_i}\right)(df)\right](E_j)\,df(E_j)$$
$$=\sum_{i=1}^n\sum_{j=1}^n\left[\overset{\circ}{\nabla}{}^2(df)\right](E_i,E_i,E_j)\,g\left(\operatorname{grad}(f),E_j\right)$$
$$=\sum_{i=1}^n\left[\overset{\circ}{\nabla}{}^2(df)\right]\left(E_i,E_i,\sum_{j=1}^n g\left(\operatorname{grad}(f),E_j\right)E_j\right)$$
$$=\sum_{i=1}^n\left[\overset{\circ}{\nabla}{}^2(df)\right]\left(E_i,E_i,\operatorname{grad}(f)\right)=\sum_{i=1}^n\left[\overset{\circ}{\nabla}{}^2(df)\right]\left(E_i,\operatorname{grad}(f),E_i\right),$$

where the last equality is a direct consequence of formula (4.3) of Chapter 1. Using formula (4.2) of Chapter 1, the last formula gives

$$\begin{aligned}&\sum_{i=1}^n\left(\left(\nabla_{E_i}\circ\nabla_{E_i}-\nabla_{\nabla_{E_i}E_i}\right)(df)\middle|df\right)\\&=\sum_{i=1}^n\left[\overset{\circ}{\nabla}{}^2(df)\right]\left(\operatorname{grad}(f),E_i,E_i\right)-\sum_{i=1}^n df\left(R\left(E_i,\operatorname{grad}(f)\right)E_i\right).\end{aligned}\tag{2.17}$$

Next let us compute the first term of the right-hand side of (2.17). Since

$$\left[\overset{\circ}{\nabla}{}^2(df)\right](X,Y,Z)=X\left[Y(Z(f))-(\nabla_Y Z)(f)\right]-(\nabla_X Y)(Z(f))+(\nabla_{\nabla_X Y}Z)(f)$$

for any $X,Y,Z\in\mathfrak{X}(M)$, by the definition of the second covariant differential $\overset{\circ}{\nabla}{}^2$, one gets

$$\begin{aligned}&\sum_{i=1}^{n}\left[\overset{\circ}{\nabla}{}^2(df)\right](Y,E_i,E_i)\\&=Y\left(\sum_{i=1}^{n}\left((E_i\circ E_i-\nabla_{E_i}E_i)(f)\right)\right)-\left[\sum_{i=1}^{n}\left((\nabla_Y E_i)\circ E_i-\nabla_{\nabla_Y E_i}E_i\right)\right](f)\end{aligned}\tag{2.18}$$

for any $Y\in\mathfrak{X}(M)$ and $f\in C^\infty(M)$.

Now let us compute the second term of the right-hand side of (2.18). Assume that

$$\nabla_{E_j}E_i=\sum_{k=1}^{n}\alpha_{ji}^k E_k,\qquad i,j=1,\ldots,n,$$

where α_{ji}^k are real smooth functions which are locally defined. Since

$$g(E_i,E_k)=\delta_{ik}$$

for any $i,k=1,\ldots,n$, it follows that

$$g\left(\nabla_{E_j}E_i,E_k\right)+g\left(E_i,\nabla_{E_j}E_k\right)=0$$

[see Chapter 1, Theorem 3.1 (iv)], which implies

$$\alpha_{ji}^k+\alpha_{jk}^i=0$$

for any $i,j,k=1,\ldots,n$. Let $Y=\sum_{j=1}^{n}Y^jE_j$. Then by (3.4) of Chapter 1 and the symmetry of the Hessian form H_f one gets

$$\begin{aligned}\sum_{i=1}^{n}\left((\nabla_Y E_i)\circ E_i-\nabla_{\nabla_Y E_i}E_i\right)(f)&=\sum_{i,j=1}^{n}Y^j\left(\left(\nabla_{E_j}E_i\right)\circ E_i-\nabla_{\nabla_{E_j}E_i}E_i\right)(f)\\&=\sum_{i,j,k=1}^{n}Y^j\alpha_{ji}^k\left(E_k\circ E_i-\nabla_{E_k}E_i\right)(f)=\sum_{j=1}^{n}Y^j\sum_{i,k=1}^{n}\alpha_{ji}^kH_f(E_k,E_i)=0,\end{aligned}$$

for any $Y\in\mathfrak{X}(M)$, and consequently (2.17) yields to the following formula

$$\sum_{i=1}^{n}\left[\overset{\circ}{\nabla}{}^2(df)\right](Y,E_i,E_i)=Y\left(\sum_{i=1}^{n}\left(E_i\circ E_i-\nabla_{E_i}E_i\right)(f)\right)=-Y(\Delta f)=-g\left(\operatorname{grad}(\Delta f),Y\right).$$

In particular,

$$\sum_{i=1}^{n}\left[\overset{\circ}{\nabla}{}^2(df)\right]\left(\operatorname{grad}(f),E_i,E_i\right)=-g\left(\operatorname{grad}(\Delta f),\operatorname{grad}(f)\right)=-\left(d(\Delta f)\middle|df\right)\tag{2.19}$$

for any $f \in C^{\infty}(M)$.

Concerning the second term of the right-hand side of (2.17), notice that the definition of the gradient operator and of the Riemann tensor field [see Chapter 1, formula (4.14)] lead to

$$\begin{aligned} &-\sum_{i=1}^{n} df\left(R\left(E_i, \operatorname{grad}(f)\right)E_i\right) = \sum_{i} df\left(R\left(\operatorname{grad}(f), E_i\right)E_i\right) \\ &= \sum_{i=1}^{n} g\left(R\left(\operatorname{grad}(f), E_i\right)E_i, \operatorname{grad}(f)\right) = \sum_{i=1}^{n} R\left(\operatorname{grad}(f), E_i, \operatorname{grad}(f), E_i\right) \qquad (2.20) \\ &= -(c_2 R)\left(\operatorname{grad}(f), \operatorname{grad}(f)\right) \end{aligned}$$

for any $f \in C^{\infty}(M)$. Finally, note that (2.15), (2.16), (2.17), (2.19) and (2.20) yield to the desired equality.

Q.E.D.

Let $g' = ag$, where $a \in C^{\infty}(M), a > 0$. If Δ_g (resp. $\Delta_{g'}$) denotes the Laplace-Beltrami operator associated to the Riemannian metric g(resp. g'), note that

$$\Delta_{ag} = \frac{1}{a}\Delta_g + \left(1 - \frac{n}{2}\right)\frac{1}{a^2} grad_g(a). \qquad (2.21)$$

Indeed, taking into account the definition of the Laplace-Beltrami operator and the properties (2.1), (2.5) and (2.6) one gets successively:

$$\begin{aligned} \Delta_{ag}(f) &= -div_{ag}\left(grad_{ag}(f)\right) = -div_{ag}\left(\frac{1}{a} grad_g(f)\right) \\ &= -\frac{1}{a} div_{ag}\left(grad_g(f)\right) - ag\left(grad_{ag}\left(\frac{1}{a}\right), grad_g(f)\right) \\ &= -\frac{1}{a} div_g\left(grad_g(f)\right) - \frac{n}{2a^2}\left[grad_g(f)\right](a) - ag\left(\frac{1}{a} grad_g\left(\frac{1}{a}\right), grad_g(f)\right) \\ &= \frac{1}{a}\Delta_g(f) - \frac{n}{2a^2}\left[grad_g(a)\right](f) + \frac{1}{a^2} g\left(grad_g(a), grad_g(f)\right) \\ &= \frac{1}{a}\Delta_g(f) + \left(1 - \frac{n}{2}\right)\frac{1}{a^2}\left[grad_g(a)\right](f) \end{aligned}$$

for any $f \in C^{\infty}(M)$.

If $\Phi : (M, g) \to (N, h)$ is an isometry, then

$$\Delta_g \circ \Phi^* = \Phi^* \circ \Delta_h, \qquad (2.22)$$

where Δ_g (resp. Δ_h) denotes the Laplace-Beltrami operator associated to (M,g) [resp. (N,h)].

Indeed, using the definition of Δ, (2.7) and (2.3), one gets

$$\begin{aligned}\Delta_g\left(\Phi^*\left(\tilde{f}\right)\right) &= -div_g\left(grad_g\Phi^*\left(\tilde{f}\right)\right) = -\Phi^*\left(div_h\left(\Phi_*\left(grad_g\Phi^*\left(\tilde{f}\right)\right)\right)\right)\\ &= -\Phi^*\left(div_h\left(grad_h\left(\tilde{f}\right)\right)\right) = \Phi^*\left(\Delta_h\tilde{f}\right).\end{aligned}$$

for any $\tilde{f} \in C^\infty(N)$.

In fact, one can show for a Riemannian manifold (M,g) that the only diffeomorphisms $\Phi : M \to M$ which commute with the Laplace-Beltrami operator Δ are the isometries (see Helgason [1962], p. 387).

It is natural to ask if there exist other smooth maps between Riemannian manifolds which commute with the corresponding Laplace-Beltrami operators. Concerning this question, let us mention the following result:

Proposition 2.5. Let $(M,g),(N,h)$ be closed, connected, smooth Riemannian manifolds, Δ_g and Δ_h the corresponding Laplace-Beltrami operators and $\Phi:(M,g)\to(N,h)$ a surjective Riemannian submersion. The following conditions are equivalent:

(i) Each fiber $\Phi^{-1}(y)$ of $\Phi, y \in N$, is a minimal submanifold of (M,g), that is the mean curvature vector field of the submanifold of M defined by the fibers of Φ vanishes.

(ii) $\Delta_g \circ \Phi^* = \Phi^* \circ \Delta_h$,

where $\Phi^* : C^\infty(N) \to C^\infty(M), \Phi^*\left(\tilde{f}\right) = \tilde{f}\circ\Phi, \tilde{f}\in C^\infty(N)$, is the canonical mapping induced by Φ.

In particular, if Φ is a Riemannian covering, then the property (ii) holds.

For a proof of this result, see Watson [1973] or Gilkey and Park [1996].

Conversely, if $(M,g),(N,h)$ are closed, connected, smooth oriented Riemannian manifolds and the surjective smooth mapping $\Phi : M \to N$ satisfies (ii), then Φ is a Riemannian submersion satisfying (i). Moreover, if M and N have the same dimension, then Φ is a Riemannian covering.

Note that any Riemannian immersion $\Phi:(M,g)\to(N,h)$ such that $\dim M < \dim N$ does not commute with the corresponding Laplace-Beltrami operators.

Proposition 2.6. If $\left(U,x^1,\ldots,x^n\right)$ is a local chart on M, then

$$\Delta f\big|_U = -\frac{1}{\sqrt{\det\left(g_{ij}\right)}}\sum_{p,k=1}^{n}\frac{\partial}{\partial x^k}\left(g^{kp}\frac{\partial f}{\partial x^p}\sqrt{\det\left(g_{ij}\right)}\right).$$

Proof. Indeed, we have

$$\Delta f = -div\left(grad\left(f\right)\right),$$

$$grad\left(f\right)\big|_U = \sum_{k,p=1}^{n} g^{kp}\frac{\partial f}{\partial x^p}\frac{\partial}{\partial x^k},$$

$$div\left(X\right)\big|_U = \frac{1}{\sqrt{\det\left(g_{ij}\right)}}\sum_{k=1}^{n}\frac{\partial}{\partial x^k}\left(X^k\sqrt{\det\left(g_{ij}\right)}\right)$$

and consequently

$$\Delta f|_U = -div\left(\sum_{k,p=1}^{n} g^{kp}\frac{\partial f}{\partial x^p}\frac{\partial}{\partial x^k}\right) = -\frac{1}{\sqrt{\det(g_{ij})}}\sum_{k,p=1}^{n}\frac{\partial}{\partial x^k}\left(g^{kp}\frac{\partial f}{\partial x^p}\sqrt{\det(g_{ij})}\right),$$

as required.

Q.E.D.

Example 2.1. Let $(\mathbb{R}^n, g_{can})$ be the n-dimensional Euclidean space with its canonical Riemannian structure. Then for each $f \in C^\infty(\mathbb{R}^n)$, we have

$$\Delta f = -\sum_{i=1}^{n}\frac{\partial^2 f}{\partial x^{i2}}.$$

For an arbitrary Riemannian manifold (M,g), if one takes a normal coordinate system $(x^1,\dots,x^n)$ around $x \in M$ (see Chapter 1, §5), from Proposition 2.5 one gets

$$(\Delta f)(x) = -\sum_{i=1}^{n}\frac{\partial^2 f}{\partial x^{i2}}(x).$$

Indeed, if g_{ij} are the components of the Riemannian metric g with respect to the normal coordinate system $(x^1,\dots,x^n)$ around the fixed point $x \in M$, then

$$g_{ij}(x) = \delta_{ij},\ \frac{\partial g_{ij}}{\partial x^k}(x) = 0,\ \Gamma_{ij}^k(x) = 0$$

for every $i,j,k \in \{1,\dots,n\}$ (for details, see for example Sakai [1996]). If $(e_1,\dots,e_n)$ is an orthonormal basis of T_xM, let γ_i be the geodesic satisfying the conditions $\gamma_i(0) = x, \dot{\gamma}_i(0) = e_i, i = 1,\dots,n,$ and $(x^1,\dots,x^n)$ the normal coordinate system at x determined by this basis. Since $x^j(\gamma_i(t)) = \delta_i^j t, i,j = 1,\dots,n$, it follows that

$$\frac{d^2}{dt^2}\Big|_{t=0}(f\circ\gamma_i)(t) = \frac{\partial^2 f}{\partial x^{i2}}(x),$$

which implies the following more geometric (local) definition of Δ:

$$\Delta f(x) = -\sum_{i=1}^{n}\frac{d^2}{dt^2}\Big|_{t=0}(f\circ\gamma_i)(t).$$

Example 2.2. Let H^2 be the upper half plane, i.e.

$$H^2 = \{(x,y)\in\mathbb{R}^2 \,|\, y>0\}$$

with the Poincaré metric, i.e.

$$[g_{ij}] = \begin{bmatrix}\frac{1}{y^2} & 0\\ 0 & \frac{1}{y^2}\end{bmatrix}$$

(see Chapter 1, Example 2.4). Then, for each $f \in C^\infty(H^2)$ we have

$$\Delta f = -y^2\left(\frac{\partial^2 f}{\partial x^2} + \frac{\partial^2 f}{\partial y^2}\right).$$

Remark 2.2. If M is closed, the Laplace-Beltrami operator Δ is formally symmetric, i.e.

$$\langle \Delta f_1, f_2 \rangle = \langle f_1, \Delta f_2 \rangle \text{ for all } f_1, f_2 \in C^\infty(M),$$

and formally positive, i.e.

$$\langle \Delta f, f \rangle \geq 0 \quad \text{for all } f \in C^\infty(M).$$

Indeed, the first property is an immediate consequence of (2.13), while the second one is a direct consequence of the equality

$$\langle \Delta f, f \rangle = \|grad(f)\|^2 = \|df\|^2 \text{ for all } f \in C^\infty(M), \tag{2.23}$$

which follows immediately from (2.12) and Proposition 2.1. The equality (2.23) implies that any eigenvalue of Δ is necessarily positive and $\Delta f = 0$ (i.e. f is harmonic) if and only if f is a real constant function on M (connected). Hence 0 is an eigenvalue of Δ and the corresponding eigenspace is of dimension one.

If M is closed, note that the condition $\Delta f \geq 0$, $f \in C^\infty(M)$, implies in fact, via (2.14), $\Delta f = 0$, that is f is a constant function.

As a consequence of the previous considerations, let us give a simple description of the minimal submanifolds in the Euclidean spaces. Let M be an n-dimensional submanifold in $\mathbb{R}^m$ equipped with the Riemannian metric g_{can}. For $x_0 \in \mathbb{R}^m$, let us define the function $\tilde{f}_{x_0} : \mathbb{R}^m \to \mathbb{R}$, $\tilde{f}_{x_0}(x) = \langle x_0, x \rangle$, and $f_{x_0} = \tilde{f}_{x_0}|_M$. If Δ is the Laplace-Beltrami operator on (M,g), where g is the induced metric by g_{can} (see Chapter 1, Example 2.5) and μ_x is the mean curvature vector of M at $x \in M$ defined by

$$\mu_x = -\frac{1}{n}\sum_{j=1}^{n} S_x(e_j, e_j) \in (T_x M)^\perp,$$

where $(e_1, \ldots, e_n)$ is an orthonormal basis of $T_x M$, $S : \mathfrak{X}(M) \times \mathfrak{X}(M) \to C^\infty(\nu_M)$ the second fundamental form of M and ν_M the normal bundle of the submanifold M, then

$$\left[\Delta(f_{x_0})\right](x) = n\langle \mu_x, x_0 \rangle,\ x \in M.$$

In particular, M is minimal, that is $\mu = 0$, if and only if the coordinate functions restricted to M, are harmonic.

Indeed, let ∇ (resp. $\tilde{\nabla}$) be the Levi-Civita connection associated to (M,g) [resp. $(\mathbb{R}^m, g_{can})$], $X, Y \in \mathfrak{X}(M)$ and $x \in M$. Then there exist an open neighborhood U of x in $\mathbb{R}^m$ and smooth vector fields $\tilde{X}, \tilde{Y} \in \mathfrak{X}(U)$ such that

$\tilde{X}\big|_{M\cap U}=X\big|_{M\cap U}, \tilde{Y}\big|_{M\cap U}=Y\big|_{M\cap U}$. Let $\tilde{\nabla}_X Y\in\mathfrak{X}(M)$ given by $\left(\tilde{\nabla}_X Y\right)(x)=\left(\tilde{\nabla}_{\tilde{X}}\tilde{Y}\right)(x)$. This tangent vector at x does not depend on the extension $\tilde{X}$ (resp. $\tilde{Y}$) of X (resp. Y). Since the vector field $grad\left(\tilde{f}_{x_0}\right):\mathbb{R}^m\to\mathbb{R}^m$ takes the constant value x_0, one gets

$$\begin{aligned}\tilde{X}\left(\tilde{Y}\left(\tilde{f}_{x_0}\right)\right)&=\tilde{X}\left(\left\langle\tilde{Y},grad\left(\tilde{f}_{x_0}\right)\right\rangle\right)=\left\langle\tilde{\nabla}_{\tilde{X}}\tilde{Y},grad\left(\tilde{f}_{x_0}\right)\right\rangle+\left\langle\tilde{Y},\tilde{\nabla}_{\tilde{X}}\left(grad\left(\tilde{f}_{x_0}\right)\right)\right\rangle\\&=\left\langle\tilde{\nabla}_{\tilde{X}}\tilde{Y},grad\left(\tilde{f}_{x_0}\right)\right\rangle\end{aligned}$$

(for the last equality, see Chapter 1, Example 3.1). On the other hand, by definition $\tilde{\nabla}_X Y=\nabla_X Y+S(X,Y)$, so that

$$\begin{aligned}\left[X\left(Y\left(f_{x_0}\right)\right)\right](x)&=\left[\tilde{X}\left(\tilde{Y}\left(\tilde{f}_{x_0}\right)\right)\right](x)=\left\langle\left(\tilde{\nabla}_{\tilde{X}}\tilde{Y}\right)(x),x_0\right\rangle=\left\langle\left(\tilde{\nabla}_X Y\right)(x),x_0\right\rangle\\&=\left\langle\left(\nabla_X Y\right)(x),x_0\right\rangle+\left\langle S_x\left(X(x),Y(x)\right),x_0\right\rangle\\&=\left\langle\left(\nabla_X Y\right)(x),\left(grad\left(f_{x_0}\right)\right)_x\right\rangle+\left\langle S_x\left(X(x),Y(x)\right),x_0\right\rangle\\&=\left[\left(\nabla_X Y\right)\left(f_{x_0}\right)\right](x)+\left\langle S_x\left(X(x),Y(x)\right),x_0\right\rangle,\end{aligned}$$

that is

$$\left[\left(X\circ Y-\nabla_X Y\right)\left(f_{x_0}\right)\right](x)=\left\langle S_x\left(X(x),Y(x)\right),x_0\right\rangle,$$

and consequently

$$\begin{aligned}\left[\Delta\left(f_{x_0}\right)\right](x)&=-\sum_{j=1}^{n}\left[\left(E_j\circ E_j-\nabla_{E_j}E_j\right)\left(f_{x_0}\right)\right](x)\\&=-\sum_{j=1}^{n}\left\langle S_x\left(e_j,e_j\right),x_0\right\rangle=-\left\langle\sum_{j=1}^{n}S_x\left(e_j,e_j\right),x_0\right\rangle\\&=n\left\langle\mu_x,x_0\right\rangle,\end{aligned}$$

where $\left(E_1,\ldots,E_n\right)$ is a local field of orthonormal frames on (M,g) such that $E_j(x)-e_j, j=1,\ldots,n$.

Now, if M is minimal, its mean curvature vector μ is zero and, for any $x_0\in\mathbb{R}^m$, $f_{x_0}:M\to\mathbb{R}$ is harmonic, so that in particular the coordinate functions restricted to M are harmonic. Conversely, if the coordinate functions restricted to M are harmonic, then all the $f_{x_0}:M\to\mathbb{R}$ are also harmonic, so that μ is zero, and consequently M is minimal. In particular, using the fact that the coordinate functions restricted to a closed, connected, minimal submanifold $M\subset\mathbb{R}^m$ should be harmonic, hence constant by Remark 2.2, one can show that there are no non-trivial (i.e. not reduced to a point) closed, connected, minimal submanifolds in the Euclidean spaces.

Remark 2.3. One can prove that Δ is an elliptic operator of order two. In fact Δ is the only natural linear differential operator whose symbol is $-\|\ \|^2$ and with no zero-order term.

From now on let us assume that (M,g) is an n-dimensional closed Riemannian manifold and let us define the **Sobolev space** $H^1(M)$ as the subspace of $L^2(M)$ consisting of those functions in $L^2(M)$ possessing weak derivatives. More precisely, for $a,b \in C^\infty(M)$ we define

$$\langle a,b\rangle_1 = \langle a,b\rangle + B(a,b),$$

where

$$B(a,b) = \int_M g(grad(a), grad(b))\, d\mu_g .$$

B is a positive semi-definite symmetric bilinear form and $\langle\, ,\rangle_1$ an inner product on $C^\infty(M)$. With these notations, Green's formula (see Proposition 2.3) can be re-written as

$$\langle \Delta a,b\rangle = B(a,b) = \langle a,\Delta b\rangle. \tag{2.24}$$

$C^\infty(M)$ is a pre-Hilbert space with respect to $\langle\, ,\rangle_1$ and we shall denote by $H^1(M)$ its completion. The corresponding norm for $\langle\, ,\rangle_1$ will be denoted by $\|\ \|_1$. Recall that every $u \in H^1(M)$ is the equivalence class $[a_k]$ of a Cauchy sequence $\{a_k \mid k \in \mathbb{N}^*\} \subset C^\infty(M)$ with respect to the norm $\|\ \|_1$ and, by definition,two sequences $\{a_k\}$ and $\{b_k\}$ are equivalent if and only if

$$\lim_{k\to\infty} \|a_k - b_k\|_1 = 0 .$$

In particular, the equivalence class $[a_k]$ defines the zero element of $H^1(M)$ if and only if $\lim_{k\to\infty} \|a_k\|_1 = 0$.

For $u = [a_k], v = [b_k] \in H^1(M)$, their inner product is given by

$$\langle u,v\rangle_1 = \lim_{k\to\infty} \langle a_k, b_k\rangle_1 .$$

Note also that B can be extended to $H^1(M)$ by letting

$$B(u,v) = \lim_{k\to\infty} B(a_k, b_k)$$

for $u = [a_k], v = [b_k] \in H^1(M)$.

Since, by definition, $L^2(M)$ is the completion of the pre-Hilbert space $(C^\infty(M), \langle\, ,\rangle)$ (see §1, C) and $\|f\| \le \|f\|_1$ for all $f \in C^\infty(M)$, every Cauchy sequence of $C^\infty(M)$ with respect to $\|\ \|_1$ is a Cauchy sequence with respect to $\|\ \|$ and equivalence of Cauchy sequences with respect to $\|\ \|_1$ implies equivalence of Cauchy sequences with respect to $\|\ \|$. So we get a well-defined mapping

$$i : H^1(M) \to L^2(M), i([a_k]) = [a_k].$$

Moreover, the following statement is true.

Proposition 2.7. $i: H^1(M) \to L^2(M)$ is a linear injection.

Proof. For any $\alpha, \beta \in \mathbb{R}$ and $[a_k],[b_k] \in H^1(M)$, one has

$$i(\alpha[a_k]+\beta[b_k]) = i([\alpha a_k + \beta b_k]) = [\alpha a_k + \beta b_k] = \alpha[a_k]+\beta[b_k] = \alpha i([a_k]) + \beta i([b_k]),$$

i.e. i is linear.

Now assume that $\{a_k\} \subset C^\infty(M)$ is a Cauchy sequence with respect to $\| \ \|_1$ such that $\lim_{k\to\infty} \|a_k\| = 0$ and $\alpha = \lim_{k\to\infty} \|a_k\|_1$. Since $\{a_k\}$ is a Cauchy sequence with respect to $\| \ \|_1$, for fixed $\varepsilon > 0$, there exists $N(\varepsilon) \in \mathbb{N}$ such that

$$\|a_k - a_N\|_1 < \varepsilon \tag{2.25}$$

for all $k > N(\varepsilon)$. Since by (2.24) and by the Cauchy-Schwarz inequality

$$|B(a_k, a_N)| = |\langle a_k, \Delta a_N \rangle| \le \|a_k\| \|\Delta a_N\|$$

and by hypothesis

$$\lim_{k\to\infty} \|a_k\| = 0,$$

it follows that

$$\lim_{k\to\infty} B(a_k, a_N) = 0.$$

Similarly, using

$$|\langle a_k, a_N \rangle| \le \|a_k\| \|a_N\|$$

one gets

$$\lim_{k\to\infty} \langle a_k, a_N \rangle = 0.$$

On the other hand, (2.25) implies

$$\varepsilon^2 > \|a_k - a_N\|_1^2 = \|a_k\|_1^2 + \|a_N\|_1^2 - 2[\langle a_k, a_N\rangle + B(a_k, a_N)]$$

for all $k \ge N(\varepsilon)$, so that the right-hand side of the last inequality converges to $\alpha^2 + \|a_N\|_1^2$ as $k \to \infty$. Hence $\alpha < \varepsilon$ for all ε, that is $\alpha = 0$, which proves the injectivity of i.

Q.E.D.

We shall identify $H^1(M)$ with its image under i in $L^2(M)$. This gives the inclusions

$$C^\infty(M) \subset H^1(M) \subset L^2(M),$$

where the first inclusion is conceived as the canonical isometric embedding, and $H^1(M)$ may be considered as a dense subspace of the Hilbert space $L^2(M)$. One can show that the space $H^1(M)$ do not depend on the Riemannian metric g. Note that since $H^1(M) \subset L^2(M)$, $H^1(M)$ is contained in $L^p(M)$ for all $p \in [1,2]$.

One can define inductively the norm

$$\|f\|_k^2 = \|f\|_{k-1}^2 + \int_M \left(\overset{\circ}{\nabla}{}^k f \middle| \overset{\circ}{\nabla}{}^k f \right) d\mu_g, \quad k \geq 1,$$

where $\| \, \|_0 = \| \, \|$ (see §1, A) and then the **Sobolev space** $H^k(M)$ to be the completion of $C^\infty(M)$ with respect to the norm $\| \, \|_k$.

Remark 2.4. A specific case of the Sobolev embedding theorem says that functions in H^k for k sufficiently large are actually continuous, and their supremum norm can be estimated from their H^k-norm. More precisely, if $k > \frac{n}{2}$, where n is the dimension of M, then $H^k(M) \subset C^0(M)$ and

$$\sup_{x \in M} |f(x)| \leq K \|f\|_k,$$

where the constant K depends only on k, M and g but not on f (for a proof of this result, see for example Gilbarg and Trudinger [1983] or Gilkey [1974]).

Another fundamental analytic inequality is the **Sobolev inequality**: if (M, g) is an n-dimensional closed Riemannian manifold, $n \geq 2$, then there exists a constant $c > 0$ such that

$$c \inf_{\alpha \in \mathrm{R}} \|f - \alpha\|_{\frac{n}{n-1}}^n \leq \|grad(f)\|_1^n$$

for any f in the Sobolev space of functions which have L^1-derivatives. The supremum $s(M)$ of all $c > 0$ for which this inequality holds is called the **Sobolev constant of *M***. It is known that the Sobolev inequality is equivalent to the isoperimetric inequality, namely $I(M) \leq s(M) \leq 2I(M)$, where $I(M)$ is the isoperimetric constant of M (see Bombieri [1970]). If (M, g) is an n-dimensional compact Riemannian manifold with boundary the Sobolev inequality

$$s(M) \|f\|_{\frac{n}{n-1}}^n \leq \|grad(f)\|_1^n,$$

that is

$$s(M) \left(\int_M |f|^{\frac{n}{n-1}} d\mu_g \right)^{n-1} \leq \left(\int_M |grad(f)| d\mu_g \right)^n$$

is valid for all f in the Sobolev space of functions which have L^1-derivatives satisfying $f|_{\partial M} = 0$. In particular, if $f \in H^1(M)$ satisfies $f|_{\partial M} = 0$, then

$$c \|f\|_{\frac{2n}{n-2}}^2 \leq \|grad(f)\|_2^2$$

when $n \geq 3$ with $c = (s(M))^{\frac{2}{n}} \dfrac{(n-2)^2}{2(n-1)^2}$, and

$$\frac{s(M)}{4\sqrt{\text{Area}(M,g)}}\|f\|_4^2 \le \|grad(f)\|_2^2$$

when $n=2$ (see Li [1980]). Indeed, in the case when $n\ge 3$, consider the function $f_1 = |f|^{\frac{2(n-1)}{n-2}}$. By the Sobolev inequality applied to f_1 and Hölder's inequality one obtains

$$\begin{aligned}
s(M)\|f\|_{\frac{2n}{n-2}}^{\frac{2n(n-1)}{n-2}} &= s(M)\left(\int_M |f|^{\frac{2n}{n-2}} d\mu_g\right)^{n-1} = s(M)\left(\int_M |f_1|^{\frac{n}{n-1}} d\mu_g\right)^{n-1} \le \left(\int_M |grad(f_1)| d\mu_g\right)^n \\
&= \left[\frac{2(n-1)}{n-2}\right]^n \left(\int_M |f|^{\frac{n}{n-2}} |grad(f)| d\mu_g\right)^n \\
&\le \left[\frac{2(n-1)}{n-2}\right]^n \left(\int_M |f|^{\frac{2n}{n-2}} d\mu_g\right)^{\frac{n}{2}} \left(\int_M |grad(f)|^2 d\mu_g\right)^{\frac{n}{2}} \\
&= \left[\frac{2(n-1)}{n-2}\right]^n \|f\|_{\frac{2n}{n-2}}^{\frac{n^2}{n-2}} \|grad(f)\|_2^n .
\end{aligned}$$

Dividing both sides by $\|f\|_{\frac{2n}{n-2}}^{\frac{n^2}{n-2}}$ one gets the desired inequality. If $n=2$, apply Sobolev's inequality to f^2 and Hölder's inequality twice in order to get successively:

$$\begin{aligned}
s(M)\|f\|_4^4 &= s(M)\int_M f^4 d\mu_g \le \left(\int_M |grad(f^2)| d\mu_g\right)^2 \\
&= 4\left(\int_M |f||grad(f)| d\mu_g\right)^2 \le 4\int_M f^2 d\mu_g \int_M |grad(f)|^2 d\mu_g \\
&\le 4\left(\int_M 1 d\mu_g\right)^{\frac{1}{2}} \left(\int_M f^4 d\mu_g\right)^{\frac{1}{2}} \|grad(f)\|_2^2 = 4\sqrt{\text{Area}(M,g)}\,\|f\|_4^2 \|grad(f)\|_2^2 .
\end{aligned}$$

Dividing both sides by $\|f\|_4^2$ one obtains the desired inequality.

Note that for a general Riemannian manifold $(M,g), \dim M \ge 3$, the assertion for any $f\in C_0^\infty(M), f\ge 0$,

$$c\left(\int_M f^{\frac{2n}{n-2}} d\mu_g\right)^{\frac{n-2}{n}} \le \int_M |grad(f)|^2 d\mu_g ,$$

may not be true. It is well-known that this inequality holds for $(\mathbb{R}^n, g_{can})$. It is a consequence of the isoperimetric property of a ball in $\mathbb{R}^n$: any region of the same volume as the ball has larger boundary area unless it is the ball.

D. The Hodge-de Rham operators

Let (M,g) be a closed, connected, smooth oriented, n-dimensional Riemannian manifold.

Definition 2.1. The star Hodge operator

$$S^{(p)}: A^p(M) \to A^{n-p}(M)$$

is the zeroth order differential operator defined as follows:

(i) $S^{(p)}$ is a $C^\infty(M)$-linear mapping, i.e.

$$S^{(p)}(f_1\omega_1 + f_2\omega_2) = f_1 S^{(p)}(\omega_1) + f_2 S^{(p)}(\omega_2),$$

for any $f_1, f_2 \in C^\infty(M)$ and any $\omega_1, \omega_2 \in A^p(M)$;

(ii) If (U,ϕ) is a local chart on M with local coordinates $(x^1,\dots,x^n)$, then

$$S^{(p)}\left(dx^{i_1}\wedge\ldots\wedge dx^{i_p}\right) = \sqrt{\det(g_{ij})}\sum_{k_1,\dots,k_p}\varepsilon_{j_{k_1}\cdots j_{k_p} i_{p+1}\cdots i_n} g^{i_1 j_{k_1}}\cdots g^{i_p j_{k_p}} dx^{i_{p+1}}\wedge\ldots\wedge dx^{i_n},$$

where $1\le j_1<\cdots<j_p\le n$, $1\le i_{p+1}<\cdots<i_n\le n$, $(j_1,\dots,j_p,i_{p+1},\dots,i_n)$ is a permutation of $(1,\dots,n)$, $(k_1,\dots,k_p)$ a permutation of $(1,\dots,p)$ and

$$\varepsilon_{l_1\ldots l_n} = \begin{cases} 1 & \text{if the permutation } \begin{pmatrix}1\ldots n\\ l_1\ldots l_n\end{pmatrix} \text{ is even,} \\ -1 & \text{if the permutation } \begin{pmatrix}1\ldots n\\ l_1\ldots l_n\end{pmatrix} \text{ is odd.}\end{cases}$$

From Definition 2.1 it follows that:

$$S^{(0)}(1) = v_g,\ S^{(n)}(v_g) = 1, \tag{2.26}$$

where v_g denotes the canonical volume form of (M,g) (see Remark 1.4) and 1 is the real constant function on M having value 1, and

$$S^{(n-p)}\circ S^{(p)}(\omega) = (-1)^{p(n-p)}\omega, \text{ for any } \omega\in A^p(M). \tag{2.27}$$

Proposition 2.8. The following equality holds:

$$(\alpha|\beta)v_g = \alpha\wedge\left(S^{(p)}(\beta)\right),$$

or equivalently

$$(\alpha|\beta) = \left(S^{(n)}\right)^{-1}\left(\alpha\wedge\left(S^{(p)}(\beta)\right)\right)$$

for any $\alpha,\beta\in A^p(M)$, where $(\alpha|\beta)\in C^\infty(M)$ denotes the pointwise inner product of α and β (see this Chapter, §1, C).

Proof. It is sufficient to verify this equality for $\alpha = dx^{i_1}\wedge\ldots\wedge dx^{i_p}$ and $\beta = dx^{j_1}\wedge\ldots\wedge dx^{j_p}$, where $1\le i_1<\cdots<i_p\le n$, $1\le j_1<\cdots<j_p\le n$. By the definition of the star-Hodge operator it follows that:

$$\begin{aligned}\alpha\wedge\left(S^{(p)}(\beta)\right)&=\sqrt{\det\left(g_{ij}\right)}\varepsilon_{i_1\ldots i_n}\sum_{k_1,\ldots,k_p}\varepsilon_{k_1\ldots k_p}g^{j_1 i_{k_1}}\ldots g^{j_p i_{k_p}}dx^{i_1}\wedge\ldots\wedge dx^{i_n}\\&=\sqrt{\det\left(g_{ij}\right)}\sum_{k_1,\ldots,k_p}\varepsilon_{k_1\ldots k_p}g^{i_1 j_{k_1}}\cdots g^{i_p j_{k_p}}dx^1\wedge\ldots\wedge dx^n\\&=\left(\alpha|\beta\right)v_g,\end{aligned}$$

the last equality being a consequence of the definition of $\left(\alpha|\beta\right)$.

Q.E.D.

As a consequence of Proposition 2.8, note that the inner product on $A^p(M)$ (see this Chapter, §1, C) can be also defined by:

$$\langle\alpha,\beta\rangle=\int_M\alpha\wedge S^{(p)}(\beta).$$

Definition 2.2. The **codifferential operator** $\delta^{p+1}:A^{p+1}(M)\to A^p(M)$ is given by

$$\delta^{p+1}=(-1)^{np+1}S^{(n-p)}\circ d^{n-p-1}\circ S^{(p+1)},$$

where $d^{n-p-1}:A^{n-p-1}(M)\to A^{n-p}(M)$ denotes the exterior differential.

Remark 2.5. Both operators d^p and δ^p can be expressed in terms of the covariant differential $\overset{\circ}{\nabla}$ associated to (*M,g*) (see Chapter 1, Definition 3.3). More precisely, for any $\omega\in A^p(M)$ one has

$$\left(d^p\omega\right)\left(X_0,X_1,\ldots,X_p\right)=\sum_{j=0}^{p}(-1)^j\left(\overset{\circ}{\nabla}\omega\right)\left(X_j,X_0,\ldots,X_{j-1},X_{j+1},\ldots,X_p\right),$$

$X_0,\ldots,X_p\in\mathfrak{X}(M)$. On the other hand, if $E_1,\ldots,E_n$ is a local field of orthonormal frames on (*M,g*), then

$$\left(\delta^p\omega\right)\left(E_2,\ldots,E_p\right)=-\sum_{j=1}^{n}\left(\overset{\circ}{\nabla}\omega\right)\left(E_j,E_j,E_2,\ldots,E_p\right).$$

In particular, note that for any $\omega\in A^1(M)$,

$$\left(\delta^1\omega\right)=-trace\left(\overset{\circ}{\nabla}\omega\right),$$

where $\overset{\circ}{\nabla}\omega:\mathfrak{X}(M)\times\mathfrak{X}(M)\to C^\infty(M)$ is the smooth tensor field of the type (0,2) defined by

$$\left(\overset{\circ}{\nabla}\omega\right)(X,Y)=(\nabla_X\omega)(Y)=X\left(\omega(Y)\right)-\omega(\nabla_XY),$$

$X,Y\in\mathfrak{X}(M)$ (see Chapter 1, Definition 3.3). Indeed, by definition,

$$trace\left(\overset{\circ}{\nabla}\omega\right)=\sum_{j=1}^{n}\left(\overset{\circ}{\nabla}\omega\right)\left(E_j,E_j\right)=\sum_{j=1}^{n}\left(\nabla_{E_j}\omega\right)\left(E_j\right)=-\delta^1\omega$$

for any $\omega \in A^1(M)$.

Proposition 2.9. The operators d^p and δ^{p+1} are adjoint with respect to $\langle \cdot , \cdot \rangle$.

Proof. By (2.27) and Definition 2.2 one gets

$$d^{n-p-1} \circ S^{(p+1)} = (-1)^{(p-1)(p+1)} S^{(p)} \circ \delta^{p+1}.$$

Consequently

$$\begin{aligned} 0 = d^n \left(\alpha \wedge S^{(p+1)}(\beta) \right) &= d^p(\alpha) \wedge S^{(p+1)}(\beta) + (-1)^p \alpha \wedge \left(d^{n-p-1} \circ S^{(p+1)} \right)(\beta) \\ &= d^p(\alpha) \wedge S^{(p+1)}(\beta) - \alpha \wedge S^{(p)}\left(\delta^{(p+1)}(\beta) \right) \end{aligned}$$

for any $\alpha \in A^p(M)$, $\beta \in A^{p+1}(M)$, that is

$$\left(d^p(\alpha) \middle| \beta \right) = \left(\alpha \middle| \delta^{p+1}(\beta) \right),$$

what implies that

$$\left\langle d^p(\alpha), \beta \right\rangle = \left\langle \alpha, \delta^{p+1}(\beta) \right\rangle$$

for any $\alpha \in A^p(M)$ and $\beta \in A^{p+1}(M)$.

Q.E.D.

Corollary 2.3. The adjoint of the Lie derivative operator $L_X : A^p(M) \to A^p(M)$ with respect to the inner product $\langle , \rangle$ is given by

$$L_X^* = \delta \varepsilon \left(X^b \right) + \varepsilon \left(X^b \right) \delta .$$

Proof. Indeed, by formula (1.3) of Chapter 1 and Remark 1.6 one has

$$L_X^* = \left(d \circ i(X) \right)^* + \left(i(X) \circ d \right)^* = \left(i(X) \right)^* \circ \delta + \delta \circ \left(i(X) \right)^* = \varepsilon\left(X^b \right) \circ \delta + \delta \circ \varepsilon\left(X^b \right).$$

Q.E.D.

Remark 2.6. Let us denote the operator $i(X) : A^k(M) \to A^{k-1}(M)$ by i_X^k, $0 \le k \le \dim M$. The formal adjoints $\left(i_X^{p-1} \right)^* : A^{p-1}(M) \to A^p(M)$ and $\left(L_X^p \right)^* : A^p(M) \to A^p(M)$ of $i_X^p : A^p(M) \to A^{p-1}(M)$ (see also Remark 1.6) and $L_X^p = d^{p-1} \circ i_X^p + i_X^{p+1} \circ d^p : A^p(M) \to A^p(M)$ respectively satisfy:

$$\begin{aligned} &\left(i_X^{p-1} \right)^* = (-1)^{np-1} S^{(n-p)} \circ i_X^{n-p+1} \circ S^{(p-1)}, \\ &\left(L_X \right)^* = (-1)^{(n+1)p+1} S^{(n-p)} \circ L_X^{n-p} \circ S^{(p)}, \\ &\left(i_X^p(\alpha_1) \middle| \alpha_2 \right) = \left(\alpha_1 \middle| \left(i_X^{p-1} \right)^* (\alpha_2) \right), \\ &\left(L_X^p(\alpha_1) \middle| \alpha_3 \right) = \left(\alpha_1 \middle| \left(L_X^p \right)^* (\alpha_3) \right), \end{aligned}$$

for any $\alpha_1, \alpha_3 \in A^p(M)$ and $\alpha_2 \in A^{p-1}(M)$. Note that $L_X^p + \left(L_X^p \right)^* : A^p(M) \to A^p(M)$ is a zeroth order differential operator.

Corollary 2.4. The following two equalities are true:

(i) $f_1 \Delta f_2 = (df_1 | df_2) + \delta(f_1 df_2)$ for all $f_1, f_2 \in C^\infty(M)$;

(ii) $\langle \mathrm{grad}(f) f_1, f_2 \rangle = \langle f, \delta(f_2 df_1) \rangle$ for all $f, f_1, f_2 \in C^\infty(M)$.

Proof. (i) Using Propositions 2.3, 2.1 and the equalities (2.2) one gets successively

$$\begin{aligned}\langle f_1 \Delta f_2, f \rangle &= \int_M (f f_1) \Delta f_2 \, d\mu_g = \int_M g(\mathrm{grad}(f f_1), \mathrm{grad}(f_2)) \, d\mu_g \\ &= \int_M (d(f f_1) | df_2) \, d\mu_g = \int_M (df_1 | df_2) f \, d\mu_g + \int_M (f_1 df_2 | df) \, d\mu_g \\ &= \langle (df_1 | df_2), f \rangle + \langle f_1 df_2, df \rangle = \langle (df_1 | df_2) + \delta(f_1 df_2), f \rangle\end{aligned}$$

for all $f \in C^\infty(M)$.

(ii) Using Proposition 2.1 one gets successively

$$\begin{aligned}\langle \mathrm{grad}(f) f_1, f_2 \rangle &= \int_M (\mathrm{grad}(f) f_1) \cdot f_2 \, d\mu_g = \int_M (df | df_1) f_2 \, d\mu_g = \int_M (df | f_2 df_1) \, d\mu_g \\ &= \langle df, f_2 df_1 \rangle = \langle f, \delta(f_2 df_1) \rangle.\end{aligned}$$

Q.E.D.

Definition 2.3. The **Hodge-de Rham operator** (or the **Laplace operator on *p*-forms**)

$$\Delta^{(p)} : A^p(M) \to A^p(M)$$

is given by:

$$\Delta^{(p)} : \alpha \in A^p(M) \mapsto \Delta^{(p)} \alpha = d^{p-1} \circ \delta^p(\alpha) + \delta^{p+1} \circ d^p(\alpha) \in A^p(M).$$

Note that $\Delta^{(0)}$ is just the Laplace-Beltrami operator Δ defined in §2, C.

Proposition 2.10. The operator $\Delta^{(p)}$, for each $0 \le p \le n$, has the following properties:

(i) $\Delta^{(p)}$ is formally self-adjoint;

(ii) $\Delta^{(p)}$ is formally positive, i.e. $\langle \Delta^{(p)} \alpha, \alpha \rangle \ge 0$, for all $\alpha \in A^p(M)$;

(iii) $\Delta^{(p)} \alpha = 0$ if and only if $d^p \alpha = 0$ and $\delta^p \alpha = 0$;

(iv) $d^p \Delta^{(p)} = \Delta^{(p+1)} d^p$, $\delta^p \Delta^{(p)} = \Delta^{(p-1)} \delta^p$;

(v) $\Delta^{(p)} S^{(n-p)} = S^{(n-p)} \Delta^{(n-p)}$.

Proof.

(i)
$$\begin{aligned}\langle \Delta^{(p)} \alpha, \beta \rangle &= \langle d^{p-1} \circ \delta^p \alpha + \delta^{p+1} \circ d^p \alpha, \beta \rangle = \langle \delta^p \alpha, \delta^p \beta \rangle + \langle d^p \alpha, d^p \beta \rangle \\ &= \langle \alpha, d^{p-1} \circ \delta^p \beta \rangle + \langle \alpha, \delta^{p+1} \circ d^p \beta \rangle = \langle \alpha, \Delta^{(p)} \beta \rangle,\end{aligned}$$

for each $\alpha, \beta \in A^p(M)$.

(ii) $\langle \Delta^{(p)} \alpha, \alpha \rangle = \langle d^p \alpha, d^p \alpha \rangle + \langle \delta^p \alpha, \delta^p \alpha \rangle = \|d^p \alpha\|^2 + \|\delta^p \alpha\|^2 \ge 0$ for each $\alpha \in A^p(M)$.

(iii) If $\Delta^{(p)}\alpha = 0$, the equality: $\langle \Delta^{(p)}\alpha,\alpha\rangle = \langle d^p\alpha, d^p\alpha\rangle + \langle \delta^p\alpha,\delta^p\alpha\rangle$ implies that $\langle d^p\alpha, d^p\alpha\rangle + \langle \delta^p\alpha,\delta^p\alpha\rangle = 0$, or equivalently $d^p\alpha = 0$ and $\delta^p\alpha = 0$. Conversely, if $d^p\alpha = 0$ and $\delta^p\alpha = 0$, then by the definition of $\Delta^{(p)}$, $\Delta^{(p)}\alpha = 0$.

(iv) It is an easy computation and we shall omit any details.

(v) We shall consider two cases:

Case 1: *n* is even. Then we can write

$$\Delta^{(p)} = -\left(d^{p-1}\circ S^{(n-p+1)}\circ d^{n-p}\circ S^{(p)} + S^{(n-p)}\circ d^{n-p-1}\circ S^{(p+1)}\circ d^p\right)$$

and therefore using (2.27),

$$\begin{aligned} S^{(n-p)}\circ\Delta^{(n-p)} &= -\left(S^{(n-p)}\circ d^{n-p+1}\circ S^{(p+1)}\circ d^p\circ S^{(n-p)} + S^{(n-p)}\circ S^{(p)}\circ d^{p-1}\circ S^{(n-p+1)}\circ d^{n-p}\right) \\ &= -\left[-\delta^{p+1}\circ d^p\circ S^{(n-p)} + (-1)^{2p(n-p)}(-1)^{(p-1)n+1} d^{p-1}\circ\delta^p\circ S^{(n-p)}\right] \\ &= \Delta^{(p)}\circ S^{(n-p)}. \end{aligned}$$

Case 2: *n* is odd. Then

$$\Delta^{(p)}\circ S^{(n-p)} = d^{p-1}\circ\delta^p\circ S^{(n-p)} + \delta^{p+1}\circ d^p\circ S^{(n-p)}$$

and therefore using Definition 2.2 and (2.27),

$$\begin{aligned} &\Delta^{(p)}\circ S^{(n-p)} = (-1)^p d^{p-1}\circ S^{(n-p+1)}\circ d^{n-p} + (-1)^{p+1} S^{(n-p)}\circ d^{n-p+1}\circ S^{(p+1)}\circ d^p\circ S^{(n-p)} \\ &= (-1)^p(-1)^{n(n-p)+p(n-p)+1} S^{(n-p)}\circ\delta^{n-p+1}\circ d^{n-p} + (-1)^{p+1}(-1)^{n-p} S^{(n-p)}\circ d^{n-p+1}\circ\delta^{n-p} \\ &= S^{(n-p)}\circ\left(\delta^{n-p+1}\circ d^{n-p} + d^{n-p+1}\circ\delta^{n-p}\right) = S^{(n-p)}\circ\Delta^{(n-p)} \end{aligned}$$

as desired.

Q.E.D.

Remark 2.7. If $\omega\in A^p(M)$ is **parallel**, that is $\overset{\circ}{\nabla}\omega = 0$, then $\Delta^{(p)}(\omega) = 0$.

Indeed, if $\overset{\circ}{\nabla}\omega = 0$, by Remark 2.5 it follows that $d^p\omega = 0, \delta^p\omega = 0$, so that the statement is an immediate consequence of the equality

$$\langle \Delta^{(p)}(\omega),\omega\rangle = \|d^p\omega\|^2 + \|\delta^p\omega\|^2$$

[see the proof of Proposition 2.9 (ii)].

Remark 2.8. It is easy to see that the operators $\delta^p, \left(i_X^p\right)^*, \left(L_X^p\right)^*$ and $\Delta^{(p)}$ are independent of the orientation of *M*. Therefore they are defined (first locally and then being differential operators globally) for an arbitrary Riemannian manifold, not necessary orientable.

Proposition 2.5 can be generalized for the Hodge-de Rham operators as follows.

Proposition 2.11. Let $(M,g),(N,h)$ be as in Proposition 2.5, $\Delta_g^{(q)}\left(0\le q\le \dim M\right)$ and $\Delta_h^{(p)}\left(0\le p\le\dim N\right)$ the corresponding Hodge-de Rham operators, $\Phi:(M,g)\to(N,h)$ a surjective Riemannian submersion and $H: M\ni x\mapsto\left[\mathrm{Ker}\left(d\Phi_x\right)\right]^{\perp}\subseteq T_xM$ the horizontal distribution of Φ. The following conditions are equivalent:

(i) The fibers of Φ are minimal submanifolds of (M,g) and H is integrable.

(ii) For all p, $1\le p\le \dim N$, $\Delta_g^{(p)}\circ\Phi_p^*=\Phi_p^*\circ\Delta_h^{(p)}$, where $\Phi_p^*:A^p(N)\to A^p(M)$, $\left(\Phi_p^*(\omega)\right)_x=\omega_{\Phi(x)}\circ\underbrace{\left(d\Phi_x\times\cdots\times d\Phi_x\right)}_{p\text{-times}}$, $\omega\in A^p(N)$, is the canonical mapping induced by Φ.

In particular, if Φ is a Riemannian covering or an isometry, then the property $\Delta_g^{(p)}\circ\Phi_p^*=\Phi_p^*\circ\Delta_h^{(p)}$ holds for any p, $0\le p\le\dim M=\dim N$.

For a proof of this result, see Goldberg and Ishihara [1978] or Gilkey and Park [1996].

We shall mention also the following result which will be crucial in our considerations.

Proposition 2.12. For each p, $0\le p\le n$, the Hodge-de Rham operator $\Delta^{(p)}$ is an elliptic one.

Proof. It is easy to see that for each $\xi\in T_x^*M$ there is a well-defined linear transformation $\sigma_{\Delta^{(p)}}(\xi)$ on $\bigwedge^{(p)}T_x^*M$ defined by:

$$\sigma_{\Delta^{(p)}}(\xi)(v)=\Delta^{(p)}\left(\phi^2\alpha\right)(x),$$

where $v\in\bigwedge^p T_x^*M$, α is any smooth p-form such that $\alpha(x)=v$, and ϕ is any smooth real valued function on M such that $\phi(x)=0$ and $d\phi_x=\xi$. This linear transformation $\sigma_{\Delta^{(p)}}(\xi)$ is known as the symbol of $\Delta^{(p)}$. Now, the ellipticity of $\Delta^{(p)}$ is equivalent to the property that the symbol $\sigma_{\Delta^{(p)}}(\xi)$ is an isomorphism at each point $x\in M$, for each non-zero $\xi\in T_x^*M$, and this follows by a straightforward computation, see for details Gallot and Meyer [1975].

Q.E.D.

In Chapter 5, §1, we shall need a workable formula for the behaviour of the operator $\Delta^{(p)}:A^p(M)\to A^p(M)$ with respect to $C^\infty(M)$.

Proposition 2.13. For all $f\in C^\infty(M)$ and $\omega\in A^p(M)$,

$$\Delta^{(p)}(f\omega)=(\Delta f)\omega+f\Delta^{(p)}(\omega)-2\nabla_{\mathrm{grad}(f)}\omega.$$

Proof. First note that

$$\delta(f\omega)=f\delta(\omega)-i(\mathrm{grad}(f))\omega, \tag{2.28}$$

for any $f\in C^\infty(M)$ and any $\omega\in A^p(M)$.

Indeed, for every $\theta\in A^{p-1}(M)$,

$$\begin{aligned}\langle\theta,\delta(f\omega)\rangle&=\langle d\theta,f\omega\rangle=\langle fd\theta,\omega\rangle=\langle d(f\theta),\omega\rangle-\langle df\wedge\theta,\omega\rangle\\&=\langle f\theta,\delta\omega\rangle-\langle\varepsilon(df)\theta,\omega\rangle=\langle\theta,f\delta\omega\rangle-\left\langle\theta,i\left((df)^\#\right)\omega\right\rangle\\&=\left\langle\theta,f\delta\omega-i(\mathrm{grad}(f))\omega\right\rangle.\end{aligned}$$

On the other hand, one has

$$\Delta^{(p)}(f\omega)-f\Delta^{(p)}(\omega)=\left[\left(L_{\mathrm{grad}(f)}\right)^{*}-L_{\mathrm{grad}(f)}\right](\omega) \tag{2.29}$$

for every $f\in C^{\infty}(M)$ and $\omega\in A^{p}(M)$.

Indeed, using the definition of $\Delta^{(p)}$, (2.28), formula (1.3) of Chapter 1 and Corollary 2.3, one has successively

$$\begin{aligned}\Delta^{(p)}(f\omega)-f\Delta^{(p)}(\omega)&=d\delta(f\omega)+\delta d(f\omega)-f(d\delta\omega+\delta d\omega)\\&=d\left[f(\delta\omega)-i(\mathrm{grad}(f))\omega\right]+\delta(fd\omega+df\wedge\omega)-f(d\delta\omega+\delta d\omega)\\&=fd\delta\omega+df\wedge\delta\omega-d\left[i(\mathrm{grad}(f))\omega\right]+f\delta d\omega\\&\quad-i(\mathrm{grad}(f))d\omega+\delta(df\wedge\omega)-fd\delta\omega-f\delta d\omega\\&=-\left[d\left(i(\mathrm{grad}(f))\omega\right)+i(\mathrm{grad}(f))d\omega\right]+\varepsilon(df)(\delta\omega)+\delta\left(\varepsilon(df)\omega\right)\\&=\left[\left(L_{\mathrm{grad}(f)}\right)^{*}-L_{\mathrm{grad}(f)}\right](\omega).\end{aligned}$$

Since

$$(\alpha|\beta)=g(\alpha^{\#},\beta^{\#})=\beta(\alpha^{\#})=\alpha(\beta^{\#})$$

for $\alpha,\beta\in A^{1}(M)$, from formula (1.2) of Chapter 1 it follows that

$$\begin{aligned}&\left(L_{\mathrm{grad}(f)}\alpha|\beta\right)-\left(\alpha|L_{\mathrm{grad}(f)}\beta\right)=g\left(\left(L_{\mathrm{grad}(f)}(\alpha)\right)^{\#},\beta^{\#}\right)-g\left(\alpha^{\#},\left(L_{\mathrm{grad}(f)}(\beta)\right)^{\#}\right)\\&=\left(L_{\mathrm{grad}(f)}\alpha\right)(\beta^{\#})-\left(L_{\mathrm{grad}(f)}\beta\right)(\alpha^{\#})\\&=\mathrm{grad}(f)\left(\alpha(\beta^{\#})\right)-\alpha\left([\mathrm{grad}(f),\beta^{\#}]\right)-\mathrm{grad}(f)\left(\beta(\alpha^{\#})\right)+\beta\left([\mathrm{grad}(f),\alpha^{\#}]\right)\\&=g\left(\beta^{\#},[\mathrm{grad}(f),\alpha^{\#}]\right)-g\left(\alpha^{\#},[\mathrm{grad}(f),\beta^{\#}]\right),\end{aligned}$$

that is

$$\left(L_{\mathrm{grad}(f)}\alpha|\beta\right)-\left(\alpha|L_{\mathrm{grad}(f)}\beta\right)=g\left(\beta^{\#},[\mathrm{grad}(f),\alpha^{\#}]\right)-g\left(\alpha^{\#},[\mathrm{grad}(f),\beta^{\#}]\right). \tag{2.30}$$

Taking $X=\mathrm{grad}(f)$ in formula (3.1) of Chapter 1 one gets

$$\begin{aligned}&2g\left(\nabla_{\mathrm{grad}(f)}Y,Z\right)=\mathrm{grad}(f)\left(g(Y,Z)\right)+g\left([\mathrm{grad}(f),Y],Z\right)-g\left(Y,[\mathrm{grad}(f),Z]\right)\\&=g\left(\mathrm{grad}\left(g(Y,Z)\right),\mathrm{grad}(f)\right)+g\left([\mathrm{grad}(f),Y],Z\right)-g\left(Y,[\mathrm{grad}(f),Z]\right),\end{aligned} \tag{2.31}$$

for every $Y,Z\in\mathfrak{X}(M)$.

Corollary 3.1 of Chapter 1 and (2.31) implies that

$$\begin{aligned}&2\left(\nabla_{\mathrm{grad}(f)}\alpha|\beta\right)=2g\left(\left(\nabla_{\mathrm{grad}(f)}\alpha\right)^{\#},\beta^{\#}\right)=2g\left(\nabla_{\mathrm{grad}(f)}(\alpha^{\#}),\beta^{\#}\right)\\&=g\left(\mathrm{grad}\left(g(\alpha^{\#},\beta^{\#})\right),\mathrm{grad}(f)\right)+g\left([\mathrm{grad}(f),\alpha^{\#}],\beta^{\#}\right)-g\left(\alpha^{\#},[\mathrm{grad}(f),\beta^{\#}]\right)\end{aligned} \tag{2.32}$$

for every $\alpha, \beta \in A^1(M)$.

Using (2.32) and (2.30), it follows that

$$\begin{aligned}&\left(L_{\operatorname{grad}(f)}(\alpha)|\beta\right)-\left(\alpha\left|L_{\operatorname{grad}(f)}(\beta)\right.\right)-2\left(\nabla_{\operatorname{grad}(f)}\alpha|\beta\right)\\&=-g\left(\operatorname{grad}(\alpha|\beta),\operatorname{grad}(f)\right)=-\left(d(\alpha|\beta)|df\right),\end{aligned}\tag{2.33}$$

for every $\alpha, \beta \in A^1(M)$.

Now let $\omega = \alpha^1 \wedge \cdots \wedge \alpha^p, \theta = \beta^1 \wedge \cdots \wedge \beta^p$ be elementary p-forms on M. Since $L_{\operatorname{grad}(f)}, L^*_{\operatorname{grad}(f)}$ and $\nabla_{\operatorname{grad}(f)}$ are derivations of degree zero on $A(M)$, (1.3), (2.33) and the classical formula for differentiating a determinant lead to

$$\begin{aligned}
&\left(\left[L_{\operatorname{grad}(f)}-\left(L_{\operatorname{grad}(f)}\right)^*-2\nabla_{\operatorname{grad}(f)}\right](\omega)|\theta\right)\\
&=\sum_{j=1}^{p}\left(\alpha^1\wedge\cdots\wedge\left[L_{\operatorname{grad}(f)}-\left(L_{\operatorname{grad}(f)}\right)^*-2\nabla_{\operatorname{grad}(f)}\right](\alpha^j)\wedge\cdots\wedge\alpha^p\,|\theta\right)\\
&=\sum_{j=1}^{p}\begin{vmatrix}(\alpha^1|\beta^1) & \cdots & (\alpha^1|\beta^p)\\ \vdots & & \vdots\\ -\left(df\left|d\left(\alpha^j|\beta^1\right)\right.\right) & \cdots & -\left(df\left|d\left(\alpha^j|\beta^p\right)\right.\right)\\ \vdots & & \vdots\\ (\alpha^p|\alpha^1) & \cdots & (\alpha^p|\beta^p)\end{vmatrix}\\
&=-\left(df\left|\sum_{j=1}^{p}\begin{vmatrix}(\alpha^1|\beta^1) & \cdots & (\alpha^1|\beta^p)\\ \vdots & & \vdots\\ d\left(\alpha^j|\beta^1\right) & \cdots & d\left(\alpha^j|\beta^p\right)\\ \vdots & & \vdots\\ (\alpha^p|\beta^1) & \cdots & (\alpha^p|\beta^p)\end{vmatrix}\right.\right)\\
&=-\left(df\left|d\begin{vmatrix}(\alpha^1|\beta^1) & \cdots & (\alpha^1|\beta^p)\\ \vdots & & \vdots\\ (\alpha^p|\beta^1) & \cdots & (\alpha^p|\beta^p)\end{vmatrix}\right.\right)\\
&=-\left(df\left|d\left(\alpha^1\wedge\cdots\wedge\alpha^p\left|\beta^1\wedge\cdots\wedge\beta^p\right.\right)\right.\right)\\
&=-\left(df\left|d(\omega|\theta)\right.\right).
\end{aligned}$$

Integrating over M one obtains

$$\begin{aligned}\left\langle \left(L_{\operatorname{grad}(f)} - L^*_{\operatorname{grad}(f)} - 2\nabla_{\operatorname{grad}(f)}\right)(\omega), \theta \right\rangle &= -\left\langle df, d(\omega|\theta) \right\rangle \\ &= -\left\langle \delta df \middle| (\omega|\theta) \right\rangle = -\left\langle \Delta f \middle| (\omega|\theta) \right\rangle = -\int_M \Delta f (\omega|\theta) d\mu_g \\ &= -\int_M ((\Delta f)\omega|\theta) d\mu_g = -\langle (\Delta f)\omega, \theta \rangle\end{aligned}$$

for every elementary p-form θ, and therefore for every $\theta \in A^p(M)$, that is

$$\left(L_{\operatorname{grad}(f)} - L^*_{\operatorname{grad}(f)} - 2\nabla_{\operatorname{grad}(f)}\right)(\omega) = -(\Delta f)\omega \tag{2.34}$$

for each elementary p-form ω on M. But then (2.34) holds for every $\omega \in A^p(M)$ and the desired formula follows directly from (2.29) and (2.34).

Q.E.D.

REFERENCES

BERGER, M., GAUDUCHON, P. and MAZET, E., *Le Spectre d'une Variété Riemannienne*, Lecture Notes in Math., Vol. 194, Springer-Verlag, Berlin, 1971.

BOMBIERI, E., *Theory of minimal surfaces and a counter-example to the Bernstein conjecture in high dimension*, Lecture Notes, Courant Institute, 1970.

BUSER, P., *Geometry and Spectra of Compact Riemann Surfaces*, Progress in Mathematics, Vol. 106, Birkhäuser, Boston, 1992.

CHAVEL, I., *Eigenvalues in Riemannian Geometry*, Academic Press, New York, 1984.

CRAIOVEANU, M. and PUTA, M., *Introducere în Geometria Spectrală*, Editura Academiei Române, Bucuresti, 1988.

GALLOT, S. and MEYER, D., *Opérateur de courbure et Laplacien des formes différentielles d' une variété riemanniennes*, J. Math. Pures et Appl., 54(1975), 259-284.

GILBARG, D. and TRUDINGER, N., *Elliptic Partial Differential Equations of Second Order*, 2nd Ed., Die Grundlehren der Mathematischen Wissenschaften, B. 224(1983), Springer-Verlag, Berlin.

GILKEY, P.B., *The Index Theorem and the Heat Equation*, Publish and Perish Lecture Notes 4(1974).

GILKEY, P.B. and PARK, J.H., *Riemannian submersions which preserve the eigenforms of the Laplacian*, Illinois J. Math. 40(1996), 194-201.

GOLDBERG, S.I. and ISHIHARA, T., *Riemannian submersions commuting with the Laplacian*, J. Differential Geometry 13(1978), 139-144.

GORDON, C.S. and WILSON, E.N., *Isospectral deformations of compact solvmanifolds*, J. Diff. Geometry 19(1984), 241-256.

HELGASON, S., *Differential Geometry and Symmetric Spaces*, Academic Press, New York, 1962.

LANG, S., *Analysis II*, Addison-Wesley Publishing Co., Inc., Reading, Massachusetts,

1969.

LI, P., *On the Sobolev constant and the p-spectrum of a compact Riemannian manifold*, Ann. Scient. Éc. Norm. Sup., 4-*e* série, t. 13(1980), 451-469.

SAKAI, T., *Riemannian Geometry*, Translations of Mathematical Monographs, Vol. 149, American Mathematical Society, Providence, Rhode Island, 1996.

WATSON, B., *Manifold maps commuting with the Laplacian*, J. Differential Geometry, 8(1973), 85-94.

Chapter 3
SPECTRAL PROPERTIES OF THE LAPLACE-BELTRAMI OPERATOR AND APPLICATIONS

In this Chapter we shall be concerned with the eigenvalues of the Laplace-Beltrami operator associated to a closed Riemannian manifold (M,g). The problem of the extent to which these eigenvalues determine the geometry of (M,g) (see also Chapter 4, §1) has been translated into the colloquial question "Can one hear the shape of a manifold?" by several authors, most notably Kac [1966], because of the analogy with **the wave equation**, a partial differential equation which models the transverse vibrations of ideal stretched objects. Such objects vibrate due to the elastic restoring force they experience when they are deformed from their equilibrium position. If one thinks of M as being embedded in $M\times\mathbb{R}$, let $u(x,t), x\in M, t\in\mathbb{R}$, be the amount of deformation in the normal direction of the point $x\in M$ at time t. The function u is constrained to satisfy

$$\frac{\partial^2 u}{\partial t^2}+a^2\Delta u=0,$$

where Δ is the Laplace-Beltrami operator of (M,g) and a^2 a constant depending on the physical properties of M, along with some initial conditions. The classical method for solving this equation is by separation of variables: One seeks the solution as a sum of special solutions of the form

$$u(x,t)=v(x)T(t).$$

On substituting this particular form of u into the wave equation one gets

$$\frac{\Delta v(x)}{v(x)}=-\frac{T''(t)}{a^2T(t)}$$

so that there exists a constant $\lambda\in\mathbb{R}$ such that

$$\Delta v=\lambda v$$

and

$$T''+\lambda a^2T=0.$$

As we will see in §1, if M is closed there are only countably many values of λ for which the eigenvalue problem $\Delta v=\lambda v$ for Δ has nontrivial solutions, and this set of λ's forms a sequence

$$0=\lambda_0<\lambda_1\leq\lambda_2\leq\cdots,$$

which accumulates only at infinity. Here λ_k denotes the k-th eigenvalue of Δ (the λ's are repeated with their multiplicity). Moreover, if f_k is the L^2-normalized eigenfunction corresponding to λ_k, then $f\in L^2(M)$ can be expanded as $f=\sum_{k=0}^{\infty}c_kf_k$. This last fact enables us to solve the wave equation via Fourier series:

$$u(x,t)=\sum_{k=0}^{\infty}\left(\alpha_k\cos a\sqrt{\lambda_k}t+\beta_k\sin a\sqrt{\lambda_k}t\right)f_k(x),$$

where the α_k and β_k are determined by the initial conditions. Note that for a fixed $t = t_0$, $u(x,t_0)$ is a snapshot of the vibrating manifold at a specific time. An appropriate average in x gives the equation of the sound wave emitted by the manifold:

$$s(t) = \sum_{k=0}^{\infty} \left(\gamma_k \cos a\sqrt{\lambda_k}\, t + \delta_k \sin a\sqrt{\lambda_k}\, t \right).$$

Thus, in a certain sense, the λ_k's can be detected by "listening". The entire collection of λ's with their multiplicities is called **the spectrum** of (M,g) (or of Δ).

1. THE FUNDAMENTAL SOLUTION OF THE HEAT EQUATION ON RIEMANNIAN MANIFOLDS

In general it is difficult to study the spectrum of the Laplace-Beltrami operator directly. Instead one introduces certain functions of the eigenvalues which can be used to extract geometric information from the spectrum. A set of very useful functions are the heat coefficients which arise from the study of the heat equation on (M,g). These coefficients are locally computable from the metric. In fact, they are universal polynomials in the curvature and their covariant derivatives.

On a Riemannian manifold (M,g), the heat equation

$$\frac{\partial u}{\partial t} + \Delta u = 0,$$

where $u(x,t)$ represents the temperature at $x \in M$ at time $t \in [0,+\infty)$, models the diffusion of radiant energy through the manifold. As with the wave equation, if M is closed, the initial-value problem for the heat equation can be solved by separation of variables and Fourier series provided the spectrum and eigenfunctions of Δ on (M,g) are known. One is led to consider a particular family of initial conditions, whose solutions when taken together form what is called the "**fundamental solution**", or **heat kernel** of (M,g):

$$E(x,y,t) = \sum_{k=0}^{\infty} e^{-\lambda_k t} f_k(x) f_k(y),$$

where $x, y \in M$ and $t \in (0,+\infty)$. We remark that $E(x,y,t)$ represents the temperature at $x \in M$ at time t, if exactly one unit of energy is input at the point y at time $t = 0$ (a "spark").

For the beginning let us consider a more general situation, namely assume that (M,g) is a connected Riemannian manifold without boundary.

Definition 1.1. A fundamental solution of the heat equation (or abbreviated FSHE) on (M,g), or **a heat kernel** on (M,g) is a continuous function

$$E : M \times M \times (0,+\infty) \to \mathbb{R}$$

that satisfies the following conditions:

(i) E is C^2 with respect to the first variable and C^1 with respect to the third one;

(ii) $\left(\Delta + \frac{\partial}{\partial t} \right) E = 0$, where Δ is the Laplace-Beltrami operator associated to (M,g) acting on the first variable;

(iii) $\lim_{t \downarrow 0} E(\cdot, y, t) = \delta_y$

for every $y \in M$, where δ_y is the **Dirac delta function**, that is, for all bounded continuous functions f on M, one has

$$\lim_{t \downarrow 0} \int_M E(x,y,t) f(x) d\mu_g(x) = f(y)$$

for every $y \in M$.

For example,

$$E : \mathbb{R}^n \times \mathbb{R}^n \times (0,+\infty) \to \mathbb{R} \text{, defined by } E(x,y,t) = (4\pi t)^{-\frac{n}{2}} e^{-\|y-x\|^2/4t} \tag{1.1}$$

is an explicit formula for the unique FSHE on $(\mathbb{R}^n, g_{can})$ (for a proof, cf. Chavel [1984], pp. 142-148). A direct verification of the conditions (i)-(iii) of Definition 1.1 does not reveal how the expression like that in (1.1) arises, thus let us outline the formal computation based on the Fourier transform which motivates the candidate for a FSHE on $(\mathbb{R}^n, g_{can})$.

First, let us recall some basic facts on Fourier transform and convolutions.

Let $\mathscr{S}(\mathbb{R}^n)$ be the **Schwartz space** of complex valued functions on $\mathbb{R}^n$, that is $\mathscr{S}(\mathbb{R}^n)$ consists of those $f \in C^\infty(\mathbb{R}^n;\mathbb{C})$ which tend to 0 rapidly at infinity, as well as their partial derivatives of all orders. This means that for every polynomial P on $\mathbb{R}^n$ and multi-index $\alpha = (\alpha_1,\dots,a_n)$, the function $P \cdot D^\alpha f$ is bounded on $\mathbb{R}^n$, where

$$D^\alpha = (-i)^{\alpha_1+\dots+\alpha_n} \left(\partial/\partial x_1\right)^{\alpha_1} \cdots \left(\partial/\partial x_n\right)^{\alpha_n}.$$

$\mathscr{S}(\mathbb{R}^n)$ is a vector space over $\mathbb{C}$ and $C_0^\infty(\mathbb{R}^n,\mathbb{C}) \subset \mathscr{S}(\mathbb{R}^n) \subset L^p(\mathbb{R}^n)$ for all $p \geq 1$, where $C_0^\infty(\mathbb{R}^n,\mathbb{C})$ is the space of smooth complex valued functions on $\mathbb{R}^n$ with compact support. Thus $\mathscr{S}(\mathbb{R}^n)$ is dense in $L^p(\mathbb{R}^n)$.

We now define the **Fourier transform** of a function $f \in \mathscr{S}(\mathbb{R}^n)$ by

$$\hat{f}(y) = (2\pi)^{-\frac{n}{2}} \int_{\mathbb{R}^n} f(x) e^{-i\langle x,y\rangle} dx,$$

where $\langle , \rangle$ denotes the canonical inner product on $\mathbb{R}^n$. By induction, one gets

$$\widehat{D^\alpha f} = M^\alpha \hat{f}, D^\alpha(\hat{f}) = \widehat{M^\alpha f},$$

where

$$(M^\alpha f)(y) = y_1^{\alpha_1} \cdots y_n^{\alpha_n} f(y), \quad y = (y_1,\dots,y_n).$$

Thus the Fourier transform $f \mapsto \hat{f}$ is a linear map of $\mathscr{S}(\mathbb{R}^n)$ into itself.

Let $\lambda \in (0,+\infty)$ and for any function $f \in \mathscr{S}(\mathbb{R}^n)$, let

$$g(x) = f(\lambda x).$$

Then

$$\hat{g}(y) = \lambda^{-n}\hat{f}\left(\frac{y}{\lambda}\right).$$

One also checks that for the function $f \in \mathscr{S}(\mathbb{R}^n)$ defined by

$$f(x) = e^{-\|x\|^2/2},$$

one has

$$f = \hat{f}.$$

For $f_1, f_2 \in \mathscr{S}(\mathbb{R}^n)$, one defines the **convolution** of f_1 and f_2, $f_1 * f_2$, by

$$(f_1 * f_2)(x) = (2\pi)^{-\frac{n}{2}} \int_{\mathbb{R}^n} f_1(x-y) f_2(y)\, dy.$$

This integral is absolutely convergent, $f_1 * f_2 \in \mathscr{S}(\mathbb{R}^n)$, $f_1 * f_2 = f_2 * f_1$ and

$$D^\alpha(f_1 * f_2) = (D^\alpha f_1) * f_2 = f_1 * (D^\alpha f_2).$$

Furthermore,

$$\widehat{f_1 * f_2} = \hat{f}_1 \cdot \hat{f}_2.$$

Let us define the **Hermitian product**

$$\langle f_1, f_2 \rangle = \int_{\mathbb{R}^n} f_1(x)\overline{f_2(x)}\, dx, \quad f_1, f_2 \in \mathscr{S}(\mathbb{R}^n).$$

Thus, one obtains

$$\langle f_1, f_2 \rangle = \langle \hat{f}_1, \hat{f}_2 \rangle$$

and hence $\|f\| = \|\hat{f}\|$, $f \in \mathscr{S}(\mathbb{R}^n)$, where $\| \ \|$ denotes the corresponding L^2-norm. Consequently, the map $f \mapsto \hat{f}$ is an automorphism of $\mathscr{S}(\mathbb{R}^n)$ preserving the Hermitian product and thus the L^2-norm.

For a function $V : \mathbb{R}^n \times [0, +\infty) \to \mathbb{R}$, we denote

$$\left(\widehat{V}\right)_1(y,t) = (2\pi)^{-\frac{n}{2}} \int_{\mathbb{R}^n} V(x,t) e^{-i\langle x,y\rangle}\, dx,$$

that is $\left(\widehat{V}\right)_1$ is the Fourier transform with respect to the first variable $x \in \mathbb{R}^n$ of V.

Now, if E is a FSHE on $(\mathbb{R}^n, g_{can})$, then

$$H(y,t) = \begin{cases} \int_{\mathbb{R}^n} f(x) E(x,y,t)\, dx, & y \in \mathbb{R}^n, t \in (0,+\infty), \\ f(y) & , \ y \in \mathbb{R}^n, t = 0, \end{cases}$$

is a solution of the initial value problem for the heat equation on $(\mathbb{R}^n, g_{can})$ with the initial value f, where f is a bounded continuous function on $\mathbb{R}^n$, that is

$$\left(\Delta + \partial/\partial t\right) H = 0 \text{ and } H(\cdot, 0) = f$$

(for details, cf. Chavel [1984]).

On the other hand, given a solution $H : \mathbb{R}^n \times [0, +\infty) \to \mathbb{R}$ of the heat equation on $(\mathbb{R}^n, g_{can})$ with initial condition f, one has formally

$$\|y\|^2 \left(\widehat{H}\right)_1 (y,t) = \left(\widehat{\Delta H}\right)_1 (y,t) = -\left(\widehat{\frac{\partial H}{\partial t}}\right)_1 (y,t) = -\frac{\partial}{\partial t}\left(\widehat{H}\right)_1 (y,t),$$

where $\Delta = -\left(\frac{\partial^2}{\partial x_1^2} + \cdots + \frac{\partial^2}{\partial x_n^2}\right)$. The ordinary differential equation (with respect to the time variable)

$$\frac{\partial}{\partial t}\left(\widehat{H}\right)_1 (y,t) = -\|y\|^2 \left(\widehat{H}\right)_1 (y,t)$$

implies the existence of a function F such that

$$\left(\widehat{H}\right)_1 (y,t) = F(y)\, e^{-\|y\|^2 t}$$

from which one deduces

$$F(y) = \left(\widehat{H}\right)_1 (y,0) = \hat{f}(y).$$

Hence

$$\left(\widehat{H}\right)_1 (y,t) = \hat{f}(y) e^{-\|y\|^2 t} = \hat{f}(y) e^{-\left\|\sqrt{2t}\,y\right\|^2/2} = \hat{f}(y)\cdot \left\{\widehat{(2t)^{-\frac{n}{2}} e^{-\|\ \|^2/4t}}\right\}(y)$$

$$= \left\{\widehat{f * \left[(2t)^{-\frac{n}{2}} e^{-\|\ \|^2/4t}\right]}\right\}(y),$$

so that

$$H(y,t) = \left\{f * \left[(2t)^{-\frac{n}{2}} e^{-\|\ \|^2/4t}\right]\right\}(y) = (4\pi t)^{-\frac{n}{2}} \int_{\mathbb{R}^n} f(x) e^{-\|y-x\|^2/4t} dx,$$

which, together with a previous remark, suggests to take E as in (1.1).

Also, let us mention that there exists an explicit formula for the FSHE on the hyperbolic plane (H^2, g) (see Chapter 1, Example 2.4). More precisely, the function $E : H^2 \times H^2 \times (0, +\infty) \to \mathbb{R}$ given by

$$E(x,y,t) = \frac{\sqrt{2}}{(4\pi t)^{3/2}} e^{-t/4} \int_{d(x,y)}^{\infty} \frac{r e^{-r^2/4t}}{\sqrt{\cosh(r) - \cosh(d(x,y))}} dr$$

is a FSHE on (H^2, g), where cosh is the hyperbolic cosine function and $d(x,y)$ is the Riemannian distance from x to y. Let us recall

$$\cosh(d(x,y)) = 1 + \frac{|x-y|^2}{2\operatorname{Im} x \operatorname{Im} y},$$

where Im denotes the imaginary part of a complex number (see also Chapter 1, §2). A deductive proof of this fact based on the Abel transform is given in the book of Buser [1992], pp. 197-209. A similar deductive approach based on the Mehler transform can be found in the book of Chavel [1984], pp. 242-246 (see also the book of Davies [1989]).

The following result gives a sufficient condition for the existence of the FSHE on general Riemannian manifolds.

Theorem 1.1. Let $(\tilde{M},\tilde{g})$ be any connected Riemannian manifold admitting a properly discontinuous group of isometries Γ (see Chapter 1, Definition 2.6) with compact orbit space $\tilde{M}/\Gamma$. Then there is a unique FSHE $\tilde{E}:\tilde{M}\times\tilde{M}\times(0,+\infty)\to\mathbb{R}$ on $(\tilde{M},\tilde{g})$. In fact $\tilde{E}$ belongs to $C^\infty(\tilde{M}\times\tilde{M}\times(0,+\infty))$ and $\tilde{E}$ satisfies

$$\tilde{E}(\tilde{x},\tilde{y},t) = \tilde{E}(\tilde{y},\tilde{x},t)$$

for all $(\tilde{x},\tilde{y},t)\in\tilde{M}\times\tilde{M}\times(0,+\infty)$. Moreover, $\tilde{E}$ satisfies the semigroup property

$$\tilde{E}(\tilde{x},\tilde{y},t+s) = \int_{\tilde{M}} \tilde{E}(\tilde{x},\tilde{z},t)\,\tilde{E}(\tilde{z},\tilde{y},s)\,d\mu_{\tilde{g}}(\tilde{z}),$$

for all $\tilde{x},\tilde{y}\in\tilde{M}$ and $t,s\in(0,+\infty)$.

The main point is that $\tilde{M}=\Gamma C$ with C relatively compact open set in $\tilde{M}$ according to Proposition 2.3 in Chapter 1. This fact together with small modifications of the method given by Berger, Gauduchon, Mazet [1971] (see also Chavel [1984] or Craioveanu, Puta [1988]) can be utilized to obtain the necessary estimates. The uniqueness and symmetry can be concluded by Duhamel's principle (see Chavel [1984], p. 137). C^∞-smoothness of $\tilde{E}$ and semigroup property follow from uniqueness. For more details we refer to Donnelly [1979].

Let us remark that for arbitrary $(\tilde{M},\tilde{g})$, a FSHE, should it exist, need not be unique (see Chavel [1984]).

Now suppose that $(\tilde{M},\tilde{g})$ satisfies the assumptions of Theorem 1.1. The orbit space $M=\tilde{M}/\Gamma$ may be not a manifold (see Chapter 1, §2). The canonical projection $p:\tilde{M}\to M$ is a regular (topological) covering projection and Γ is its group of covering transformations. However, one can define $C^m(M)$ as follows: a function $f:M\to\mathbb{R}$ is said to be of **class C^m** if and only if $f\circ p\in C^m(\tilde{M})$. Since Γ acts isometrically on $\tilde{M}$, the Laplacian $\tilde{\Delta}$ of $(\tilde{M},\tilde{g})$ is Γ-invariant. Thus $\tilde{\Delta}$ induces an operator Δ on $C^2(M)$ called the Laplacian of M. Indeed, if $f\in C^m(M)$, then $f\circ p\in C^m(\tilde{M})$ and $\gamma^*(f\circ p)=f\circ p$ for any $\gamma\in\Gamma$, where

$$\gamma^*:C^m(\tilde{M})\to C^m(\tilde{M}),\gamma^*(\tilde{f})=\tilde{f}\circ\gamma,\tilde{f}\in C^m(\tilde{M}),$$

is the canonical mapping induced by the isometry $\gamma:\tilde{M}\to\tilde{M}$ (see Chapter 1, §1). Consequently, by the property (2.22) in Chapter 2, one gets

$$\gamma^*\left[\tilde{\Delta}(f\circ p)\right]=\tilde{\Delta}\gamma^*(f\circ p)=\tilde{\Delta}(f\circ p),$$

that is

$$\gamma^*\left[\tilde{\Delta}(f\circ p)\right]=\tilde{\Delta}(f\circ p).$$

This last equality implies the existence of a unique function $\Delta(f)\in C^m(M)$ such that

$$\tilde{\Delta}(f\circ p)=\Delta(f)\circ p$$

for any $f\in C^m(M)(m\geq 2)$.

A function $\tilde{a}\in C^\infty(\tilde{M})$ with compact support such that for all $\tilde{x}\in\tilde{M}$ one has $\sum_{\gamma\in\Gamma}\tilde{a}(\gamma(\tilde{x}))=1$ is called **a partition of unity relative to** Γ. To obtain such a function $\tilde{a}$, let $\tilde{b}\in C^\infty(\tilde{M})$ be a non-negative function whose compact support contains the relatively compact open set C in $\tilde{M}$ with $\Gamma C=\tilde{M}$ (see Chapter 1, Proposition 2.3) and choose

$$\tilde{a}(\tilde{x})=\frac{\tilde{b}(\tilde{x})}{\sum_{\gamma\in\Gamma}\tilde{b}(\gamma(\tilde{x}))},\quad \tilde{x}\in\tilde{M}.$$

One can define a measure on M by using a partition of unity relative to Γ as follows. Let $f:M\to\mathbb{R}$ be continuous and $\tilde{a}$ a partition of unity relative to Γ. Then set

$$\int_M f(x)\,dx=\int_{\tilde{M}}\tilde{a}(\tilde{x})(f\circ p)(\tilde{x})\,d\mu_{\tilde{g}}(\tilde{x}),$$

where $p:\tilde{M}\to M$ is the canonical projection.

A continuous function $E:M\times M\times(0,+\infty)\to\mathbb{R}$ is said to be **a fundamental solution for the heat equation** on M if E satisfies properties analogous to (i), (ii), (iii) of Definition 1.1. To obtain E one takes the fundamental solution $\tilde{E}$ on $(\tilde{M},\tilde{g})$ and sums over Γ to obtain a Γ-invariant expression. More precisely, the following result holds.

Theorem 1.2. Let Γ act on $\tilde{M}$ as in Theorem 1.1 and C in $\tilde{M}$ with $\Gamma C=\tilde{M}$, as in Proposition 2.3 of Chapter 1. For $x,y\in M$ set

$$E(x,y,t)=\sum_{\gamma\in\Gamma}\tilde{E}(\tilde{x},\gamma(\tilde{y}),t), \tag{1.2}$$

where $\tilde{x},\tilde{y}\in C$ are any preimages of x and y, respectively, that is $x=p(\tilde{x})$, $y=p(\tilde{y})$. If $\tilde{E}$ is the FSHE on $(\tilde{M},\tilde{g})$, the sum on the right side converges uniformly on $C\times C\times[t_1,t_2], 0<t_1\leq t_2$, to the FSHE on M. The FSHE E is unique, symmetric with respect to the first two arguments and satisfies the semigroup property

$$E(x,y,t+s)=\int_M E(x,z,t)E(z,y,s)\,dz.$$

For the proof, see Donnelly [1979].

In particular, the following two consequences can be derived.

Corollary 1.1. If (M,g) is any closed connected Riemannian manifold, then there is a unique FSHE $E: M\times M\times(0,+\infty)\to\mathbb{R}$ on (M,g). Moreover, E is of class C^∞, symmetric with respect to the first two arguments and satisfies the semigroup property

$$E(x,y,t+s)=\int_M E(x,z,t)E(z,y,s)\,d\mu_g(z),$$

for all $(x,y)\in M\times M$ and $t,s\in(0,+\infty)$.

Remark 1.1. Let us point out another property of E under the assumptions of Corollary 1.1. First, note that it is easy to see that for fixed $f\in C^\infty(M)$ the smooth function $H_0: M\times[0,+\infty)\to\mathbb{R}$ defined by

$$H_0(x,t)=\begin{cases}\int_M E(x,y,t)f(y)\,d\mu_g(y) & x\in M, t>0,\\ f(x) & x\in M, t=0,\end{cases}$$

is a solution of the heat equation on (M,g) with initial condition f, that is $\left(\Delta+\frac{\partial}{\partial t}\right)H_0=0$ on $M\times(0,+\infty)$ and $H_0(x,0)=f(x)$ for all $x\in M$. Indeed, since M is closed and E is smooth, the operators Δ (with respect to the variable x) and $\frac{\partial}{\partial t}$ pass under the integral sign. Moreover, any smooth solution $H: M\times[0,+\infty)\to\mathbb{R}$ of the heat equation on (M,g) with initial condition f satisfies

$$\begin{aligned}\frac{\partial}{\partial t}\int_M H^2(x,y,t)\,d\mu_g(y)&=2\int_M H\frac{\partial H}{\partial t}\,d\mu_g=-2\int_M H\Delta H\,d\mu_g\\ &=-2\int_M g(\operatorname{grad}(H),\operatorname{grad}(H))\,d\mu_g=-2\|\operatorname{grad}(H)\|^2\le 0\end{aligned}$$

for all $t\in(0,+\infty)$, that is for fixed $x\in M$, the smooth function $\int_M H^2(x,y,\cdot)\,d\mu_g(y):[0,+\infty)\to\mathbb{R}$ is monotone decreasing, which implies that $H=H_0$. Since any constant function $c: M\to\mathbb{R}$ is a solution of the heat equation on (M,g) with the initial condition c, it follows that

$$\int_M E(x,y,t)\,d\mu_g(y)=1$$

for all $x\in M$ and $t\in(0,+\infty)$.

Given a mapping $\Phi:(M',g')\to(M,g)$ an interesting question is to compare the heat kernels of the two Riemannian manifolds, under suitable assumptions for Φ. It turns out that a good assumption is that Φ satisfies Fubini's property for every continuous function on M' with compact support. This property is automatically satisfied for instance by Riemannian submersions and coverings, or by local isometries (see Chapter 2,

§1). For example, one has (see also Remark 2.1):

Corollary 1.2. If $p:(\tilde{M},\tilde{g})\to(M,g)$ is a Riemannian covering (see Chapter 1, Example 2.8), where (M,g) is a closed Riemannian manifold, then the fundamental solutions of the heat equations on $(\tilde{M},\tilde{g})$ and (M,g), respectively, are related by

$$E(x,y,t)=\sum_{\gamma\in\Gamma}\tilde{E}(\tilde{x},\gamma(\tilde{y}),t),$$

where $\tilde{x},\tilde{y}$ are any preimages of x,y, respectively, and Γ is the deck transformation group of the covering p.

For example, the heat kernel E of the n-dimensional flat torus $(M,g)=\left(\mathbb{R}^n/\Gamma, g_{can}/\Gamma\right)$ defined by the lattice Γ of $(\mathbb{R}^n,+)$ (see Chapter 1, Example 2.17) is given by

$$E(x,y,t)=(4\pi t)^{-\frac{n}{2}}\sum_{\gamma\in\Gamma}e^{-\|\tilde{y}+\gamma-\tilde{x}\|^2/4t},$$

where $\tilde{x},\tilde{y}$ are any preimages of x,y, respectively with respect to the Riemannian covering $p:(\mathbb{R}^n,g_{can})\to\left(\mathbb{R}^n/\Gamma, g_{can}/\Gamma\right)$.

More generally, the following result is true: if $p:(\tilde{M},\tilde{g})\to(M,g)$ is a local isometry between closed Riemannian manifolds, then the equality

$$\sum_{\tilde{y}\in p^{-1}(y)}\tilde{E}(\tilde{x},\tilde{y},t)=E(p(\tilde{x}),y,t)$$

holds, in the sense of distributions, for any $t\in(0,+\infty)$, for any $\tilde{x}\in\tilde{M}$ and for almost every $y\in M$. For the proof of an even more general result, see Bordoni [2000].

Now it is possible to deduce:

Theorem 1.3. (Spectral theorem). Suppose that Γ acts properly discontinuously and isometrically on the Riemannian manifold $(\tilde{M},\tilde{g})$ with compact orbit space $M=\tilde{M}/\Gamma$. The Laplace-Beltrami operator Δ on M is a symmetric operator on $C^\infty(M)$. The eigenvalue problem $\Delta f=\lambda f$ has a complete orthonormal system of eigenfunctions $f_1,f_2,\ldots$ in $L^2(M)$ of class C^∞ with corresponding eigenvalues $\lambda_1,\lambda_2,\ldots$ such that

(i) these eigenvalues can be arranged in increasing order: $\lambda_1\le\lambda_2\le\cdots$, where $\lambda_j\to\infty$ as $j\to\infty$;

(ii) $E(x,y,t)=\sum_{j=1}^{\infty}e^{-\lambda_j t}f_j(x)f_j(y)$, where the convergence of the series is uniform on $M\times M$ for each $t\in(0,+\infty)$.

Proof. Recall that a linear operator on a Hilbert space H is called **compact** if it is a norm limit of finite rank operators. Such operators enjoy a particularly simple spectral theory: to any self-adjoint compact operator on H there is associated an orthogonal decomposition of H consisting of the infinite family of finite-dimensional eigenspaces, with eigenvalues tending to zero.

(i) For each fixed $t \in (0,+\infty)$, the heat kernel E defines a self-adjoint compact operator $\mathscr{P}_t : L^2(M) \to L^2(M)$ given by

$$[\mathscr{P}_t(f)](x) = \int_M E(x,y,t) f(y)dy, \quad x \in M,$$

since E is continuous and symmetric with respect to the first two arguments (see Lang [1975], p. 13). Moreover, the semigroup property of E (see Theorem 1.2) implies the following semigroup property of the family $\{\mathscr{P}_t\}_{t\in(0,+\infty)}$:

$$\mathscr{P}_s \circ \mathscr{P}_t = \mathscr{P}_{s+t}$$

for any $s,t \in (0,+\infty)$. In particular, $\mathscr{P}_s \circ \mathscr{P}_t = \mathscr{P}_t \circ \mathscr{P}_s$ for any $s,t \in (0,+\infty)$, i.e. $\{\mathscr{P}_t\}_{t\in(0,+\infty)}$ is a commuting family of compact operators, and can therefore be simultaneously diagonalized (see Lang [1975], p. 12). Let $f_1, f_2, \ldots$ be a complete orthonormal system in $L^2(M)$ such that

$$\mathscr{P}_t(f_j) = \mu_j(t) f_j, \quad j \in \mathbb{N}^*.$$

Let $\mu_1, \mu_2, \ldots \geq 0$ be the eigenvalues of $\mathscr{P}_1$, where $\mu_j \to 0$ as $j \to \infty$. Then the eigenvalues of $\mathscr{P}_{1/k}$ with $k \in \mathbb{N}^*$ are $\mu_1^{1/k}, \mu_2^{1/k}, \ldots$ since $\mathscr{P}_1 = \left(\mathscr{P}_{1/k}\right)^k$. If $t = \frac{l}{k}, l \in \mathbb{N}^*$, is a positive rational, since $\mathscr{P}_t = \left(\mathscr{P}_{1/k}\right)^l$, it follows that $\mathscr{P}_t$ has the eigenvalues $\mu_1^t, \mu_2^t, \ldots$. The density of all positive rationals in $(0,+\infty)$ and the continuity of the heat kernel E gives $\mu_j(t) = \mu_j^t$, $j \in \mathbb{N}^*$, for all $t \in (0,+\infty)$. Now, since E satisfies a property analogous to (iii) of Definition 1.1,

$$f_j = \lim_{t \downarrow 0} \mathscr{P}_t(f_j) = \left(\lim_{t\downarrow 0} \mu_j^t\right) f_j$$

for all $j \in \mathbb{N}^*$, so that $\lim_{t\downarrow 0} \mu_j^t = 1$ for all $j \in \mathbb{N}^*$, what leads to the fact that $\mu_j > 0$ for all $j \in \mathbb{N}^*$. Thus, $f_j = \mu_j^{-t} \mathscr{P}_t(f_j)$ for all $j \in \mathbb{N}^*$ so that $f_j \in C^\infty(M)$ for all $j \in \mathbb{N}^*$, since E is smooth in the first and third argument. Now let $\lambda_j = -\ln \mu_j$, $j \in \mathbb{N}^*$. Since E is the FSHE on M, one has

$$0 = \Delta \mathscr{P}_t(f_j) + \frac{\partial}{\partial t}\mathscr{P}_t(f_j) = \Delta(\mu_j^t f_j) + \frac{\partial}{\partial t}(\mu_j^t f_j) = \mu_j^t(\Delta f_j + \ln \mu_j f_j) = e^{-t\lambda_j}(\Delta f_j - \lambda_j f_j),$$

i.e.

$$\Delta f_j = \lambda_j f_j \quad \text{for any } j \in \mathbb{N}^*.$$

Hence the f_j are eigenfunctions of Δ and the property (i) in the theorem corresponds to $\mu_j \to 0$ for $j \to \infty$.

(ii) Since all eigenvalues $\mu_1^t = e^{-\lambda_1 t}, \mu_2^t = e^{-\lambda_2 t}, \ldots$ of $\mathscr{P}_t$ are non-negative, it follows from Mercer's theorem that the kernel $E(\cdot,\cdot,t)$ of $\mathscr{P}_t$ has the desired expansion.

Q.E.D.

Remark 1.2. (i) The proof of Theorem 1.3 shows that the eigenspaces of $\mathscr{P}_t$ are also the eigenspaces of Δ. In particular, eigenspaces of Δ belonging to distinct eigenvalues are orthogonal in $L^2(M)$ and the direct sum of all eigenspaces of Δ is dense in $L^2(M)$. Indeed, in virtue of the spectral theorem for compact operators, the family of eigenspaces of $\mathscr{P}_t$ is an orthogonal decomposition of $L^2(M)$.

(ii) Since the eigenfunctions

$$f_1, f_2, \dots$$

of Δ determine a complete orthonormal system of $L^2(M)$, it follows that for $f \in L^2(M)$ one has

$$f = \sum_{j=1}^{\infty} \langle f, f_j \rangle f_j \text{ in } L^2(M)$$

(the Fourier series expansion) and

$$\|f\|^2 = \sum_{j=1}^{\infty} \langle f, f_j \rangle^2$$

(Parseval's identity).

Corollary 1.3. For each $t \in (0, +\infty)$ the series $\sum_{j=1}^{\infty} e^{-\lambda_j t}$ is convergent and its sum is equal to $\int_M E(x, x, t)\, dx$.

Proof. Theorem 1.3 implies that

$$E(x, x, t) = \sum_{j=1}^{\infty} e^{-\lambda_j t} [f_j(x)]^2$$

where the convergence of the series is uniform on M, so that our statement is a consequence of Fatou's theorem and of the completeness and orthonormality of the system $\{f_j \mid j \in \mathbb{N}^*\}$.

Q.E.D.

Remark 1.3. Since

$$\sum_j \langle \mathscr{P}_t(f_j), f_j \rangle = \sum_j \mu_j^t = \sum_j e^{-\lambda_j t},$$

Corollary 1.3 shows that the sum

$$\sum_j \langle \mathscr{P}_t(f_j), f_j \rangle$$

converges. Since $\mathscr{P}_t$ is a self-adjoint compact operator (see the proof of Theorem 1.3), this implies that $\mathscr{P}_t$ is a trace class operator (i.e. it is the product of two Hilbert-Schmidt operators, see Lang [1975], p. 159) for each $t \in (0, +\infty)$. Thus, this being the case, by definition the trace of $\mathscr{P}_t$ is given by

$$trace(\mathcal{P}_t) = \sum_j \langle \mathcal{P}_t(f_j), f_j \rangle = \sum_j e^{-\lambda_j t}.$$

On the other hand,

$$\int_M E(x,x,t)\,dx = \sum_j e^{-\lambda_j t}$$

(see Corollary 1.3). Hence we find

$$trace(\mathcal{P}_t) = \sum_j e^{-\lambda_j t} = \int_M E(x,x,t)\,dx.$$

Another immediate consequence of Theorem 1.3 (see also Chapter 2, Remark 2.2) is the following spectral theorem.

Corollary 1.4. Let (M,g) be any connected, closed Riemannian manifold. The eigenvalue problem $\Delta f = \lambda f$ has a complete orthonormal system of eigenfunctions $f_0, f_1, \ldots$ in $L^2(M)$ of class C^∞ with corresponding eigenvalues

$$\lambda_0, \lambda_1, \ldots$$

such that

(i) these eigenvalues are non-negative and can be arranged in increasing order:

$$0 = \lambda_0 < \lambda_1 \le \lambda_2 \le \cdots, \text{ where } \lambda_j \to \infty \text{ as } j \to \infty;$$

(ii) $E(x,y,t) = \sum_{j=0}^{\infty} e^{-\lambda_j t} f_j(x) f_j(y)$, where the convergence of the series is uniform on $M \times M$ for each $t \in (0,+\infty)$. In particular, for each $t \in (0,+\infty)$ the series $\sum_{j=0}^{\infty} e^{-\lambda_j t}$ is convergent and its sum is equal to

$$\int_M E(x,x,t)\,d\mu_g(x).$$

Example 1.1. Let us consider the Riemannian manifold $(\mathbb{R}^n, g_{can})$ (see Chapter 1, Example 2.1). In the local chart corresponding to the polar coordinate system $(x^1 = r\cos\theta, x^2 = r\sin\theta)$ for $r \in (0,+\infty)$ and $\theta \in S^1$ on $\mathbb{R}^2$, g_{can} is given in matrix form by

$$\left[(g_{can})'_{ij} \right] = \begin{bmatrix} 1 & 0 \\ 0 & r^2 \end{bmatrix}$$

(see Chapter 1, Example 2.2), so that

$$\left[(g_{can})'_{ij} \right]^{-1} = \begin{bmatrix} 1 & 0 \\ 0 & \frac{1}{r^2} \end{bmatrix}.$$

Consequently, in this local chart one has

$$\Delta = -\frac{1}{r}\left[\frac{\partial}{\partial r}\left(r\frac{\partial}{\partial r}\right)+\frac{\partial}{\partial \theta}\left(\frac{1}{r}\frac{\partial}{\partial \theta}\right)\right] = -\frac{\partial^2}{\partial r^2}-\frac{1}{r}\frac{\partial}{\partial r}-\frac{1}{r^2}\frac{\partial^2}{\partial \theta^2}$$

(see Chapter 2, Proposition 2.5). On the other hand, note that the simplest closed manifold is the unit circle S^1. Considered as a smooth submanifold of $\mathbb{R}^2$, it inherits a canonical Riemannian structure, denoted also by g_{can} (see Chapter 1, Example 2.5). The length of (S^1, g_{can}) is 2π and (S^1, g_{can}) is isometric to $\mathbb{R}/2\pi\mathbb{Z}$. Since $r\big|_{S^1} = 1$, it follows that the Laplace-Beltrami operator on (S^1, g_{can}) in this coordinate system induced on S^1 by the polar one, can be identified with $-\frac{d^2}{d\theta^2}$. Let us find the eigenvalues of this operator. First, let us note that $C^\infty(S^1)$ can be identified with the real vector subspace of the smooth periodic functions u in $C^\infty(\mathbb{R})$ of period 2π, that is $u(t+2\pi) = u(t)$ for all $t \in \mathbb{R}$. Let $\lambda > 0$ be an eigenvalue of $-\frac{d^2}{d\theta^2}$ and $f \in C^\infty(S^1)\setminus\{0\}$ such that

$$\frac{d^2 f}{d\theta^2} + \lambda f = 0 .$$

The general integral of the differential equation

$$u'' + \lambda u = 0$$

is $u(t) = a\cos\sqrt{\lambda}t + b\sin\sqrt{\lambda}t$, where a,b are arbitrary real constants. This solution u is a periodic function of period 2π if and only if $\sqrt{\lambda} = k \in \mathbb{N}$, so that $\lambda = k^2$. Therefore, the non-zero eigenvalues of the Laplace-Beltrami operator on the Riemannian manifold (S^1, g_{can}) are $k^2, k \in \mathbb{N}^*$. A linear independent system of eigenfunctions corresponding to $k^2, k \in \mathbb{N}^*$, consists of the functions $f_{2k-1}, f_{2k} \in C^\infty(S^1)$ given by

$$f_{2k-1}(\theta) = \sin k\theta, f_{2k}(\theta) = \cos k\theta ,$$

that is the dimension of the eigenspace corresponding to $k^2, k \in \mathbb{N}^*$, is two. Consequently,

$$\left\{\frac{1}{\sqrt{2\pi}}, \frac{1}{\sqrt{\pi}} f_{2k-1}, \frac{1}{\sqrt{\pi}} f_{2k}\right\}_{k\in\mathbb{N}^*}$$

is a complete orthonormal system in $L^2(S^1)$ and the expansion

$$f(\theta) = \frac{1}{\sqrt{2\pi}} a_0 + \sum_{k=1}^{\infty} \frac{1}{\sqrt{\pi}}\left(a_k \sin k\theta + b_k \cos k\theta\right)$$

is the usual expansion in terms of Fourier series. From this sequence of eigenfunctions, we can see that the sound of a "circular" musical instrument is made up of sine tones with frequencies in arithmetic progression – a fact well-known to musicians.

The FSHE $E : S^1 \times S^1 \times (0, +\infty) \to \mathbb{R}$ on (S^1, g_{can}) is given by

$$E(x,y,t)=\frac{1}{2\pi}+\frac{1}{\pi}\sum_{k=1}^{\infty}e^{-tk^2}\left(\sin kx\sin ky+\cos kx\cos ky\right)=\frac{1}{2\pi}+\frac{1}{\pi}\sum_{k=1}^{\infty}e^{-tk^2}\cos k\left(x-y\right)$$

for all $(x,y,t)\in S^1\times S^1\times(0,+\infty)$ [see Corollary 1.4 (ii)].

The properties of the heat kernel *E* discussed above under the assumptions of Corollary 1.1 (or Corollary 1.4) follow from the self-adjointness of *Δ* and from its spectral properties. However, there is another important property of *E* which cannot be derived only from the spectral properties. Thus, it is known that

$$E(x,y,t)>0$$

for all $x,y\in M$ and $t>0$. Note that this property of *E* is not at all obvious from the eigenfunction expansion (ii) in Corollary 1.4, because the eigenfunctions f_k are changing the sign (except for f_0). By the maximum principle, it is sufficient to show that for *y* fixed, $E(x,y,t)\geq 0$ holds for all $x\in M$ and small $t>0$ (for a proof of this fact, see for instance Schoen and Yau [1994], p. 99).

From now on assume that (*M*,*g*) is a closed, connected Riemannian manifold. Let us denote by

$$Z_{(M,g)}(t)=\sum_{j=0}^{\infty}e^{-\lambda_j t},\quad t\in(0,+\infty)$$

the trace of $\mathscr{P}_t$ (see also the notations of Corollary 1.4). Note that this series converges uniformly on $[t_0,+\infty)$ for any $t_0\in(0,+\infty)$, so that

$$Z_{(M,g)}:(0,+\infty)\to\mathbb{R}$$

is a continuous function. It is decreasing, $\lim_{t\to\infty}Z_{(M,g)}(t)=1$ and $\lim_{t\searrow 0}Z_{(M,g)}(t)=+\infty$. $Z_{(M,g)}$ is called the **partition function** of (*M*,*g*).

For an eigenvalue $\lambda\geq 0$ of *Δ*, let $m_\lambda\in\mathbb{N}^*$ be the dimension of the corresponding eigenspace. m_λ is called **the multiplicity of** λ. Then $Z_{(M,g)}$ can be written as

$$Z_{(M,g)}(t)=\sum_{\lambda}m_\lambda e^{-\lambda t}$$

for each $t\in(0,+\infty)$, where λ runs over all distinct eigenvalues of *Δ*.

Proposition 1.1. The partition function $Z_{(M,g)}:(0,+\infty)\to\mathbb{R}$ of (*M*,*g*) determines the eigenvalues of *Δ* and their multiplicities.

Proof. Since *M* is connected, $\lambda=0$ is always an eigenvalue of *Δ* with multiplicity 1. Indeed, for $f\in C^\infty(M)$, $\Delta f=0$ implies $df=0$, i.e. *f* is a constant function on *M* (see Chapter 2, Remark 2.2). Assume now that

$$0<\lambda_1<\lambda_2<\cdots\to+\infty$$

are the nonzero distinct eigenvalues of *Δ*. Thus, λ_1 is the smallest nonzero eigenvalue of *Δ*.

Let us consider the real number $\mu>0$. Then for $\mu\neq\lambda_1$,

$$e^{\mu t}\left[Z_{(M,g)}(t)-1\right]=\sum_{i\geq 1} m_i e^{(\mu-\lambda_i)t},$$

which implies

$$\lim_{t\to\infty} e^{\mu t}\left[Z_{(M,g)}(t)-1\right]=\begin{cases}0 & \text{if } \mu<\lambda_1,\\ +\infty & \text{if } \mu>\lambda_1.\end{cases}$$

If $\mu=\lambda_1$, then

$$e^{\lambda_1 t}\left[Z_{(M,g)}(t)-1\right]=m_1+\sum_{j\geq 2} m_j e^{(\lambda_1-\lambda_j)t},$$

so that

$$\lim_{t\to\infty} e^{\lambda_1 t}\left[Z_{(M,g)}(t)-1\right]=m_1 .$$

Thus, λ_1 is the unique strictly positive real number μ such that the limit of $e^{\mu t}\left[Z_{(M,g)}(t)-1\right]$ as $t\to+\infty$ is a natural number ≥ 1, namely m_1. By recurence, it follows that λ_k $(k\geq 2)$ is the unique strictly positive real number μ such that the limit of

$$e^{\mu t}\left[Z_{(M,g)}(t)-1-\sum_{j=1}^{k-1} m_j e^{-\lambda_j t}\right]$$

as $t\to+\infty$ is a natural number ≥ 1, namely m_k.

Q.E.D.

If $p:(\tilde{M},\tilde{g})\to(M,g)$ is an q-sheeted Riemannian covering between closed Riemannian manifolds, then we have

$$Z_{(\tilde{M},\tilde{g})}(t)+\int_{\tilde{M}}\sum_{\substack{\tilde{y}\in p^{-1}(p(\tilde{x}))\\ \tilde{y}\neq\tilde{x}}}\tilde{E}(\tilde{x},\tilde{y},t)\,d\mu_{\tilde{g}}(\tilde{x})=qZ_{(M,g)}(t)$$

for all $t\in(0,+\infty)$ (see Bordoni [2000]). Indeed, this equality is obtained by setting $p(\tilde{x})=y$ in the previous formula relating the corresponding heat kernels and integrating over $\tilde{M}$. In particular, the positivity of the heat kernel $\tilde{E}$ implies the classical **Kato type inequality**:

$$Z_{(\tilde{M},\tilde{g})}(t)\leq qZ_{(M,g)}(t)$$

for all $t\in(0,+\infty)$. Equality case occurs here if and only if the covering $p:\tilde{M}\to M$ is trivial, i.e. if and only if $\tilde{M}$ is the disjoint union of q copies of M. Indeed, if the Kato type inequality is an equality for at least one $t_0\in(0,+\infty)$, then the above formula implies $\tilde{E}(\tilde{x},\tilde{y},t_0)=0$ for any $\tilde{x}\neq\tilde{y}$ in $p^{-1}(p(\tilde{x}))$. Thus $\tilde{x}$ and $\tilde{y}$ lie in two different connected components of $\tilde{M}$. One should note that the Kato type inequality has been generalized for any Riemannian submersion $p:(\tilde{M},\tilde{g})\to(M,g)$ between closed Riemannian manifolds whose fibers are totally geodesic submanifolds of $(\tilde{M},\tilde{g})$. More precisely, such a submersion is associated to a principal bundle whose typical fiber F is, in this case, endowed with a Riemannian metric g_0 (see Hermann [1960]). Then

$$Z_{(\tilde{M},\tilde{g})}(t) \le Z_{(M,g)}(t)\cdot Z_{(F,g_0)}(t) = Z_{(M\times F,g\times g_0)}(t)$$

for all $t\in(0,+\infty)$, with equality if and only if p is the trivial fibration with the same typical fiber F (see Bérard and Gallot [1984] and Besson [1986]). The name "**Kato type inequality**" is justified on the one hand by the analogy with the inequality proved by Kato (see Kato [1973], Hess, Schrader and Uhlenbrock [1980]) and on the other hand by the fact that the proof given by Bérard and Gallot [1984] uses it.

By Corollary 1.4, every $f\in L^2(M)$ has a Fourier series expansion

$$f=\sum_{j=0}^{\infty}\alpha_j f_j \text{ with } \alpha_j=\langle f,f_j\rangle, j\in\mathbb{N}, \text{ and } \sum_{j=0}^{\infty}\alpha_j^2=\|f\|^2<\infty$$

[see also Remark 1.2 (ii)]. Now let us give another characterization of the **Sobolev space** $H^1(M)$ (see Chapter 2, §2).

Proposition 1.2. $H^1(M)=\left\{f=\sum_{j=1}^{\infty}\alpha_j f_j\in L^2(M)\,\middle|\,\sum_{j=0}^{\infty}\lambda_j\alpha_j^2<\infty\right\}.$

Proof. Let us denote the right-hand side of the previous equality by Λ. For $f\in\Lambda$, the partial sums

$$s_k=\sum_{j=0}^{k}\alpha_j f_j, \text{ where } \alpha_j=\langle f,f_j\rangle,$$

belong to $C^\infty(M)$ and $\|f-s_k\|\to 0$ as $k\to\infty$, which means that f is the equivalence class $[s_k]$ of the Cauchy sequence $\{s_k\}$ with respect to the norm $\|\ \|$. On the other hand, for $k>l$, Green's formula [see (2.12) in Chapter 2] and the orthonormality of the system $\{f_j\}$ give that

$$B(s_k-s_l,s_k-s_l)\overset{def}{=}\int_M g(\operatorname{grad}(s_k-s_l),\operatorname{grad}(s_k-s_l))\,d\mu_g=\langle s_k-s_l,\Delta(s_k-s_l)\rangle=\sum_{j=l+1}^{k}\lambda_j\alpha_j^2.$$

[see also (2.24) in Chapter 2]. Since $f\in\Lambda$, this implies that

$$\|s_k-s_l\|_1^2=\|s_k-s_l\|^2+B(s_k-s_l,s_k-s_l)=\sum_{j=l+1}^{k}(1+\lambda_j)\alpha_j^2\to 0$$

as $k,l\to\infty$, i.e. $\{s_k\}\subset C^\infty(M)$ is also a Cauchy sequence with respect to $\|\ \|_1$. Hence $f=[s_k]\in H^1(M)$.

Now, we show that Λ is closed in $H^1(M)$ endowed with its natural metric topology $\mathcal{T}$ induced by $\|\ \|_1$. Thus, let $u=\sum_j\alpha_j f_j\in H^1(M)$ be in the closure $\bar{\Lambda}$ of Λ with respect to $\mathcal{T}$. Since $C^\infty(M)$ is dense in $H^1(M)$ with respect to $\mathcal{T}$, there exists a sequence $a_1,a_2,\ldots$ in $C^\infty(M)$ that converges to u with respect to the norm $\|\ \|_1$. On the other hand, each a_p has a Fourier series expansion

$$a_p = \sum_{j=0}^{\infty} \alpha_{p,j} f_j$$

in $L^2(M)$. Then $\Delta(a_p) \in C^\infty(M) \subset L^2(M)$ and since

$$\langle \Delta(a_p), f_j \rangle = \langle a_p, \Delta(f_j) \rangle = \lambda_j \langle a_p, f_j \rangle = \lambda_j \alpha_{p,j},$$

it follows that

$$\Delta(a_p) = \sum_{j=0}^{\infty} \lambda_j \alpha_{p,j} f_j . \tag{1.3}$$

By Green's formula and (1.3) one gets

$$\|a_p\|_1^2 = \|a_p\|^2 + \langle a_p, \Delta(a_p) \rangle = \sum_{j=0}^{\infty} (1+\lambda_j) \alpha_{p,j}^2 ,$$

which leads to

$$\|a_p\|_1^2 \geq \sum_{j=0}^{k} (1+\lambda_j) \alpha_{p,j}^2 \tag{1.4}$$

for any k. By hypothesis, $\|a_p - u\|_1 \to 0$ as $p \to \infty$, so that

$\|a_p - u\| = \left\| \sum_{j=0}^{\infty} (\alpha_{p,j} - \alpha_j) f_j \right\| \to 0$ as $p \to \infty$ and therefore $\alpha_{p,j} \to \alpha_j$ as $p \to \infty$, for each $j \in \mathbb{N}$. Since $\|a_p - u\| \to 0$ as $p \to \infty$, it follows that $\|a_p\|_1 \to \|u\|_1$ as $p \to \infty$, so that (1.4) implies that

$$\sum_{j=0}^{k} (1+\lambda_j) \alpha_j^2 \leq \|u\|_1^2 \tag{1.5}$$

for any k. Since $\lambda_j \geq 0$ for any $j \in \mathbb{N}$, (1.5) shows that the series $\sum_{j=0}^{\infty} (1+\lambda_j) \alpha_j^2$ converges, and because of the fact the series $\sum_{j=0}^{\infty} \alpha_j^2$ converges too, it follows that the series $\sum_{j=0}^{\infty} \lambda_j \alpha_j^2$ converges, that is $u \in \Lambda$. This shows that $\overline{\Lambda} \subseteq \Lambda$, and thus Λ is closed.

Finally, let us show that the orthogonal complement of Λ in $H^1(M)$ (with respect to $\langle , \rangle_1$) is zero. If $u \in C^\infty(M)$ is orthogonal to Λ with respect to $\langle , \rangle_1$, then $\langle u, f_j \rangle_1 = 0$ for all $j \in \mathbb{N}$. But, by Green's formula,

$$\langle u, f_j \rangle_1 = \langle u, f_j \rangle + B(u, f_j) = \langle u, f_j \rangle + \langle u, \Delta f_j \rangle = (1+\lambda_j) \langle u, f_j \rangle,$$

so that $\langle u, f_j \rangle = 0$ for all $j \in \mathbb{N}$. By the completeness of the sequence $\{f_j\}$ in $L^2(M)$ this implies that $u = 0$. Since $C^\infty(M)$ is dense in $H^1(M)$ and Λ is closed in $H^1(M)$ (in the topology $\mathscr{T}$), it follows that if $u \in H^1(M)$ is orthogonal to Λ with respect to $\langle\, , \rangle_1$, then there exists a sequence $\{a_k\} \subset C^\infty(M) \cap \Lambda^\perp$ such that $\|a_k - u\| \to 0$ as $k \to \infty$. The previous argument shows that $a_k = 0$ for any k, that is $u = 0$, or equivalently $\Lambda^\perp = \{0\}$. Hence $H^1(M) = \Lambda$.

Q.E.D.

Remark 1.4. (i) The direct sum of all eigenspaces of Δ is dense in $H^1(M)$. (ii) Similarly one can show that

$$H^2(M) = \left\{ u = \sum_{j=0}^{\infty} \alpha_j f_j \in L^2(M) \,\middle|\, \sum_{j=0}^{\infty} \lambda_j^2 \alpha_j^2 < \infty \right\},$$

where $H^2(M)$ denotes the Sobolev space defined in Chapter 2, §2. Note that

$$C^\infty(M) \subset H^2(M) \subset H^1(M) \subset L^2(M)$$

and $H^2(M)$ is a dense subspace of the Hilbert space $L^2(M)$.

If $f = \sum_{j=0}^{\infty} \alpha_j f_j \in C^\infty(M)$, then $\Delta f \in C^\infty(M)$ and

$$\langle \Delta f, f_j \rangle = \langle f, \Delta f_j \rangle = \lambda_j \langle f, f_j \rangle = \lambda_j \alpha_j$$

for any $j \in \mathbb{N}$. Since $C^\infty(M) \subset L^2(M)$, that is $\Delta f \in L^2(M)$, we have consequently

$$\Delta f = \sum_{j=0}^{\infty} \lambda_j \alpha_j f_j \text{ with } \sum_{j=0}^{\infty} \lambda_j^2 \alpha_j^2 < \infty .$$

This remark suggests to define the following linear operator $T : Dom(T) \to L^2(M)$, where $Dom(T) = H^2(M)$, by letting

$$T(u) = \sum_{j=0}^{\infty} \lambda_j \alpha_j f_j \text{ for } u = \sum_{j=0}^{\infty} \alpha_j f_j \in H^2(M).$$

Note that the above considerations show that T is an extension of the Laplace-Beltrami operator Δ, that is $T\big|_{C^\infty(M)} = \Delta$. This is called **the Friedrich extension of the operator** Δ.

Before making obvious some properties of this extension, we shall recall some facts on unbounded operators. Let $(H, \langle\, , \rangle)$ be a real Hilbert space. An **unbounded operator** T on H is a linear map from a dense subspace $Dom(T) \subseteq H$, the domain of T, to H. The set

$$\{ v \in H \mid \text{there exists } w \in H \text{ such that } \langle u, w \rangle = \langle T(u), v \rangle \text{ for all } u \in Dom(T) \}$$

is a real vector space. For each v in this vector space there is exactly one w, if it exists, having the above property. Indeed, if there exist at least two such w, say w_1 and w_2 such that

$$\langle u, w_1 \rangle = \langle T(u), v \rangle = \langle u, w_2 \rangle$$

for all $u \in Dom(T)$, then

$$\langle u, w_1 - w_2 \rangle = 0$$

for all $u \in Dom(T)$. Since $Dom(T)$ is a dense subspace of H, this implies $w_1 = w_2$. Hence one can define an operator

$$T^* : Dom(T^*) \to H \text{ by } T^*(v) = w,$$

where $Dom(T^*)$ is the above defined space of such vectors v. The pair $(T^*, Dom(T^*))$ is called the **adjoint of** T. We say that T is **symmetric** if

$$\langle T(u), v \rangle = \langle u, T(v) \rangle$$

for all $u, v \in Dom(T)$.

We say that T is **self-adjoint** if in addition $Dom(T) = Dom(T^*)$, that is $T = T^*$.

If $T : Dom(T) \to H, S : Dom(S) \to H$ are two unbounded operators such that $Dom(T) \subset Dom(S)$ and $S\big|_{Dom(T)} = T$, one writes $T \subset S$. For example, if T is symmetric, then $T \subset T^*$. Note that if $T \subset S$, then $S^* \subset T^*$. On the other hand, if T, S are self-adjoint and $T \subset S$, then $T = S$. Indeed, these assumptions imply $S = S^* \subset T^* = T$, whence $T = S$.

Here are some important properties of the Friedrich extension

$$T : H^2(M) \to L^2(M) \text{ of } \Delta.$$

Proposition 1.3. (i) $\langle T(u), v \rangle = B(u, v)$ for all $u \in H^2(M)$ and all $v \in H^1(M)$. In particular, $\langle T(u), v \rangle = \langle u, T(v) \rangle$ for all $u, v \in H^2(M)$, i.e. T is symmetric;

(ii) T is self-adjoint. Moreover, T is a maximal (hence unique) self-adjoint extension of Δ in $L^2(M)$ [or, how one also says, Δ is **essentially self-adjoint** on $C^\infty(M)$];

(iii) $H^2(M) = \{u \in H^1(M) |$ there exists $w \in L^2(M)$ such that

$$\langle w, v \rangle = B(u, v) \text{ for all } v \in H^1(M)\}.$$

Proof. (i) If $u = \sum_{j=0}^{\infty} \alpha_j f_j, v = \sum_{j=0}^{\infty} \beta_j f_j \in H^1(M)$, then the partial sums

$$u_k = \sum_{j=0}^{k} \alpha_j f_j, v_k = \sum_{j=0}^{k} \beta_j f_j \in C^\infty(M)$$

form Cauchy sequences with respect to the norm $\| \ \|_1$ (see also the proof of Proposition 1.2) and $u = [u_k], v = [v_k]$. Thus,

$$B(u, v) = \lim_{k \to \infty} B(u_k, v_k) = \lim_{k \to \infty} \langle \Delta(u_k), v_k \rangle = \sum_{j=0}^{\infty} \lambda_j \alpha_j \beta_j .$$

In particular, if $u \in H^2(M)$, then by definition of T,

$$T(u)=\sum_{j=0}^{\infty}\lambda_j\alpha_j f_j,$$

so that

$$\langle T(u),v\rangle=\sum_{j=0}^{\infty}\lambda_j\alpha_j\beta_j=B(u,v)$$

for all $v\in H^1(M)$. If $v\in H^2(M)$, the same kind of computation as before shows that

$$\langle u,T(v)\rangle=B(u,v)$$

for all $u\in H^1(M)$, that is

$$\langle u,T(v)\rangle=\langle T(u),v\rangle$$

for all $u,v\in H^2(M)$.

(ii) First note that by (i), T is symmetric, what implies $T\subset T^*$. Now, assume that $v=\sum_{j=0}^{\infty}\beta_j f_j\in Dom(T^*)$. Then, by definition, there exists

$$w=\sum_{j=0}^{\infty}\gamma_j f_j\in L^2(M)$$

such that

$$\langle T(u),v\rangle=\langle u,w\rangle$$

for all $u\in H^2(M)$. In particular, taking $u=f_j, j\in\mathbb{N}$, one gets

$$\lambda_j\beta_j=\langle T(f_j),v\rangle=\langle f_j,w\rangle=\gamma_j,$$

that is

$$\lambda_j\beta_j=\gamma_j$$

for any $j\in\mathbb{N}$. But, the last equality implies

$$\sum_{j=0}^{\infty}\lambda_j^2\beta_j^2=\sum_{j=0}^{\infty}\gamma_j^2<\infty,$$

i.e. the convergence of the series $\sum_{j=0}^{\infty}\lambda_j^2\beta_j^2$, which means that $v\in H^2(M)$. Hence $Dom(T^*)=H^2(M)$ and T is self-adjoint.

Finally, let us consider $v\in Dom(\Delta^*)$, where $\Delta:C^\infty(M)\to L^2(M)$. Then there exists $w\in L^2(M)$ such that

$$\langle\Delta(u),v\rangle=\langle u,w\rangle$$

for all $u \in C^{\infty}(M)$. But a similar argument as before shows that $v \in H^2(M)$, i.e. $Dom(\Delta^*) \subset H^2(M)$ and $\Delta^* \subset T$.

(iii) If $u = \sum_{j=0}^{\infty} \alpha_j f_j$ belongs to the right-hand side of the equality in (iii), there exists $w = \sum_{j=0}^{\infty} \gamma_j f_j \in L^2(M)$ such that

$$\langle w, v \rangle = B(u, v)$$

for all $v \in H^1(M)$, so that in particular

$$\gamma_j = \langle w, f_j \rangle = B(u, f_j)$$

for all $j \in \mathbb{N}$. Since the partial sums

$$u_k = \sum_{j=0}^{k} \alpha_j f_j$$

are in $C^{\infty}(M)$, it follows by Green's formula that

$$B(u, f_j) = \lim_{k\to\infty} B(u_k, f_j) = \lim_{k\to\infty} \langle u_k, \Delta f_j \rangle = \lambda_j \lim_{k\to\infty} \langle u_k, f_j \rangle = \lambda_j \alpha_j$$

for all $j \in \mathbb{N}$, which implies

$$\gamma_j = \lambda_j \alpha_j$$

for all $j \in \mathbb{N}$. Therefore

$$\sum_{j=0}^{\infty} \lambda_j^2 \alpha_j^2 = \sum_{j=0}^{\infty} \gamma_j^2 < \infty,$$

that is $u \in H^2(M)$.

For the converse inclusion, note that if $u \in H^2(M)$, then letting $w = T(u) \in L^2(M)$, by the first equality in (i),

$$\langle w, v \rangle = B(u, v)$$

for all $v \in H^1(M)$, and thus u is in the right-hand side of the equality in (iii).

Q.E.D.

Remark 1.5. (i) The above results show that the positive-definite symmetric Laplace-Beltrami operator $\Delta : C^{\infty}(M) \to C^{\infty}(M)$ has a unique self-adjoint extension $T : H^2(M) \to L^2(M)$, where $H^2(M)$ is a dense subspace of $L^2(M)$, namely the largest domain we might consider for Δ, i.e. $H^2(M)$ is the set of functions $f = \sum_{j=0}^{\infty} \alpha_j f_j$ such that there exists a sequence $u_k = \sum_{j=0}^{k} \alpha_j f_j$ in $C^{\infty}(M)$ with the property that

$u_k \to f$ in $L^2(M)$ as $k \to \infty$ and $\Delta(u_k) = \sum_{j=0}^{k} \lambda_j \alpha_j f_j$ converges to an element in $L^2(M)$ which is in fact $T(f)$.

(ii) From the elementary theory of partial differential equations we have the following basic elliptic estimate, sometimes called **Gärding's inequality**: The operator $\Delta + I$, where I is the identity operator, has strictly positive eigenvalues, and trivial kernel, and

$$\|f\|_2 \le C\|\Delta f + f\|$$

for all $f \in H^2(M)$, where $\| \ \|_2$ denotes the Sobolev H^2-norm (see Chapter 2, §2) and the constant $C > 0$ depends only on M and g but not on f. This inequality states essentially that one can estimate the second derivatives of f in the L^2-sense just by knowing the L^2-norm $\| \ \|$ (see Chapter 2, §1, A) of the Laplace-Beltrami operator of f. Similarly, for any positive integer m, one gets

$$\|f\|_{2m} \le C^m \left\|(\Delta + I)^m (f)\right\|$$

for all $f \in H^{2m}(M)$ (for more details, see Gilkey [1974] or Roe [1988]).

Now one can prove a more precise statement concerning the convergence to infinity of the sequence of distinct eigenvalues of Δ, namely

Proposition 1.4. Let $0 = \lambda_0 < \lambda_1 < \lambda_2 < \cdots$ be the increasing sequence of all distinct eigenvalues of Δ and n the dimension of M. For any integer $m > \frac{n}{4}$, there exists a real constant $C > 0$ such that

$$\lambda_k \ge C k^{\frac{1}{2m}}$$

for k sufficiently large.

Proof. Let f_j be a normalized eigenfunction of Δ corresponding to λ_j and $\mu_j = (\lambda_j + 1)^m$, $j \in \mathbb{N}$ that is μ_j is the j-th distinct eigenvalue of $(\Delta + I)^m$. Then f_j is also a normalized eigenfunction of $(\Delta + I)^m$ corresponding to μ_j. For $\alpha \in (0, +\infty)$, let $S(\alpha)$ be the linear span of the set $\{f_j \mid \mu_j \le \alpha\}$. If $f = \sum_j \beta_j f_j \in S(\alpha)$, the definition of $S(\alpha)$ and Pythagoras theorem lead to

$$\left\|(\Delta + I)^m (f)\right\|^2 = \left\|\sum_j \mu_j (\beta_j f_j)\right\|^2 = \sum_j \mu_j^2 \|\beta_j f_j\|^2 \le \alpha^2 \|f\|^2,$$

that is

$$\left\|(\Delta + I)^m (f)\right\| \le \alpha \|f\|, \tag{1.6}$$

so that for $2m > \frac{n}{2}$, Remark 2.4 in Chapter 2, Remark 1.5 (ii) and (1.6) yield

$$\sup_{x\in M}|f(x)|\le C_1\|f\|_{2m}\le C_2\left\|(\Delta+I)^m(f)\right\|\le C_2\alpha\|f\|,$$

where the constants C_1 and C_2 depend only on m, M and g, but not on f. If the first l of the f_j's are in $S(\alpha)$, then this last estimate and Pythagoras theorem yield

$$\left|\sum_{j=0}^{l-1}\beta_j f_j(x)\right|\le C_2\alpha\left\|\sum_{j=0}^{l-1}\beta_j f_j\right\|=C_2\alpha\left(\sum_{j=0}^{l-1}\|\beta_j f_j\|^2\right)^{\frac{1}{2}}=C_2\alpha\left(\sum_{j=0}^{l-1}\beta_j^2\right)^{\frac{1}{2}}$$

for any real numbers β_j and any $x\in M$. If we fix $x\in M$ and one takes $\beta_j=f_j(x)$, then one obtains the estimate

$$\sum_{j=0}^{l-1}f_j^2(x)\le C_2\alpha\left[\sum_{j=0}^{l-1}f_j^2(x)\right]^{\frac{1}{2}},$$

or equivalently

$$\left[\sum_{j=0}^{l-1}f_j^2(x)\right]^{\frac{1}{2}}\le C_2\alpha$$

for all $x\in M$, that is

$$\sum_{j=0}^{l-1}f_j^2(x)\le C_2^2\alpha^2 \tag{1.7}$$

for all $x\in M$. We integrate (1.7) over M to conclude

$$l\le C_2^2\alpha^2 Vol(M,g).$$

This shows that $\dim S(\alpha)$ is finite. Let $k=\dim S(\alpha)$. If one takes $\alpha=\mu_k$ one gets

$$(\lambda_k+1)^m=\mu_k=\alpha\ge\frac{\sqrt{k}}{C_2\sqrt{Vol(M,g)}},$$

from which the desired estimate follows.

Q.E.D.

Definition 1.2. Assume that we have ordered all the distinct eigenvalues of the Laplace-Beltrami operator Δ in an increasing sequence

$$0=\lambda_0<\lambda_1<\lambda_2<\cdots.$$

The **spectrum** (or more precisely the ***o*-spectrum**, see also Chapter 5, Definition 1.3) of the closed, connected Riemannian manifold (M,g) denoted by $\mathbf{Spec}(M,g)$ [or more precisely by $\mathbf{Spec}^{(0)}(M,g)$], is the sequence of real numbers

$$0=\lambda_0<\underbrace{\lambda_1,\ldots,\lambda_1}_{m_{\lambda_1}\text{-times}}<\underbrace{\lambda_2,\ldots,\lambda_2}_{m_{\lambda_2}\text{-times}}<\cdots,$$

where each eigenvalue λ_j of Δ is repeated according to its multiplicity m_{λ_j}, $j\in\mathbb{N}$.

Proposition 1.1 shows that the partition function of (M,g) determines Spec(M,g).

Remark 1.6. Let M be a closed, smooth manifold of dimension n. Colin de Verdière [1986], [1987] proved a remarkable fact that, if $n \geq 3$, then every preassigned finite sequence $0 = \lambda_0 < \lambda_1 \leq \lambda_2 \leq \cdots \leq \lambda_k$ is the sequence of first $k + 1$ eigenvalues (with multiplicities) of the Laplace-Beltrami operator associated to some Riemannian metric g on M. In particular, there are no restrictions concerning the multiplicities of the eigenvalues. In the case $n = 2$ there is a topological obstruction to prescribing finitely many eigenvalues with multiplicities ≥ 2. Indeed, let M_2 be a closed surface with Euler characteristic $\chi(M_2)$ and m_j the multiplicity of the j-th eigenvalue $(j \geq 1)$ of the Laplace-Beltrami operator associated to an arbitrary Riemannian metric on M. Then one has

(i) if $M_2 = S^2$ (= the two-dimensional unit sphere), then $m_j \leq 2j+1$;

(ii) if $M_2 = P^2(\mathbb{R})$ (= the real projective plane), then $m_j \leq 2j+3$;

(iii) if $M_2 = T^2$ (= the two-dimensional torus), then $m_j \leq 2j+4$;

(iv) if $M_2 = K^2$ (= the Klein bottle), then $m_j \leq 2j+3$;

(v) if $\chi(M_2) < 0$, then $m_j \leq 2j - 2\chi(M) + 3$.

(for details, see Nadirashvili [1988]). Note that the above inequalities (i)-(iii) become equalities for the case of the standard Riemannian metrics on S^2, $P^2(\mathbb{R})$, and on T^2 obtained by factoring $\mathbb{R}^2$ modulo the lattice with the generators (0,2) and $(\sqrt{3},1)$, respectively (see also, §2). It turns out however that the preceding result of Colin de Verdière holds in this case for finite sequences of the form $0 = \lambda_0 < \lambda_1 < \lambda_2 < \cdots < \lambda_k$.

Remark 1.7. Now assume that (M,g) is a compact, connected Riemannian manifold and M has boundary ∂M, with induced Riemannian metric. Let $X^{(n)}$ be the outward unit normal vector field on ∂M. Similar results are valid for the eigenvalue problem

$$\Delta f = \lambda f \tag{1.8}$$

providing one imposes suitable boundary conditions on the eigenfunctions of Δ. By suitable boundary conditions we shall mean

$$f = 0 \text{ on } \partial M$$

(Dirichlet eigenvalue problem)

or

$$X^{(n)}(f) = 0 \text{ on } \partial M$$

(Neumann eigenvalue problem).

In fact for each one of this eigenvalue problem there are only countably many values of λ for which (1.8) has nontrivial solutions, each set of λ's forms a sequence which accumulates only at infinity, and each associate eigenspace is finite dimensional. Eigenspaces belonging to distinct eigenvalues are orthogonal in $L^2(M)$, and the direct sum of all the eigenspaces is dense in $L^2(M)$. The corresponding spectra, i.e. the sequences of real numbers

$$0 < \mu_1 < \underbrace{\mu_2, \ldots, \mu_2}_{m_{\mu_2}\text{-times}} < \cdots$$

and

$$0 = \nu_0 < \underbrace{\nu_1, \dots, \nu_1}_{m_{\nu_1}\text{-times}} < \underbrace{\nu_2, \dots, \nu_2}_{m_{\nu_2}\text{-times}} < \cdots,$$

respectively, where m_{μ_j} (resp. m_{ν_l}) is the multiplicity of the eigenvalue μ_j (resp. ν_l), i.e. the dimension of the eigenspace associated to μ_j (resp. ν_l) are denoted by $\mathbf{Spec}_D(\boldsymbol{M},\boldsymbol{g})$ and $\mathbf{Spec}_N(\boldsymbol{M},\boldsymbol{g})$ respectively (for more details, see Chavel [1984]).

More precisely, let (M,g) be an *n*-dimensional connected smooth Riemannian manifold and let *K* be a precompact (totally bounded) open set of *M* with a smooth boundary ∂K and conceive $\bar{K}$ as a smooth compact Riemannian manifold with boundary. Define the spaces

$$L^2(K) = \left\{ f \,\middle|\, \int_K f^2 d\mu_g < \infty \right\},$$

$$H^1(K) = \left\{ f \in L^2(K) \,\middle|\, |\mathrm{grad}(f)| \in L^2(K) \right\},$$

where $\mathrm{grad}(f)$ is understood in the sense of distributions, and define $H_0^1(K)$ to be the closure of the space $C_0^\infty(K)$ of smooth real valued functions on *K* with compact support in $H^1(K)$.

Let us consider the Dirichlet eigenvalue problem in *K*:

$$\begin{cases} \Delta u = \mu u, \\ u|_{\partial K} = 0. \end{cases} \tag{1.9}$$

A weak solution of (1.9) in *K* is a function $u \in H_0^1(K)$ that satisfies the equation $\Delta u = \mu u$ in the sense of distributions, that is

$$\int_K g(\mathrm{grad}(u), \mathrm{grad}(v))\, d\mu_g = \mu \int_K uv\, d\mu_g$$

for all $v \in H_0^1(K)$.

Similarly as before one can prove that there exists a complete orthonormal system $\{f_j | j \in \mathbb{N}^*\}$ in $L^2(K)$ such that each f_j is a weak eigenfunction of Δ in *K* with f_j having the eigenvalue $\mu_j = \mu_j(K)$ satisfying

$$0 \le \mu_1 \le \mu_2 \le \cdots$$

with $\mu_j \to \infty$ as $j \to \infty$.

If *K* is connected and $M \setminus \bar{K}$ is non-empty, then $\mu_1(K) > 0$ is a simple eigenvalue, that is $0 < \mu_1(K) < \mu_2(K)$ and f_1 is $\neq 0$ in *K*. In particular, each eigenvalue has finite multiplicity. Each $f_j \in C^\infty(K)$ and extends to a continuous function on $\bar{K}$.

The operator Δ can be considered as an unbounded operator in $L^2(K)$, with the domain $C_0^\infty(K)$. It is essentially self-adjoint. Its closure, denoted by Δ_K and called the

Dirichlet extension of Δ, has the domain $\left\{f \in H_0^1(K) \middle| \Delta f \in L^2(K)\right\}$ and the spectrum $\left\{\mu_j \middle| j \in \mathbb{N}^*\right\}$. With respect to the basis $\left\{f_j \middle| j \in \mathbb{N}^*\right\}$, the operator Δ_K is represented by the infinite diagonal matrix

$$\begin{pmatrix} \mu_1 & & & & \\ & \mu_2 & & 0 & \\ & & \ddots & & \\ & & & \mu_j & \\ & 0 & & & \ddots \end{pmatrix}.$$

By the spectral theory, one can define $f(\Delta_K)$, where f is a function on the spectrum of Δ_K. In particular, one can define the operator $\exp(-t\Delta_K) \overset{not}{=} e^{-t\Delta_K}$, where $t \in \mathbb{R}$. With respect to the basis $\left\{f_j \middle| j \in \mathbb{N}^*\right\}$, it has the matrix

$$\begin{pmatrix} \exp(-t\mu_1) & & & 0 & \\ & \exp(-t\mu_2) & & & \\ & & \ddots & & \\ & & & \exp(-t\mu_j) & \\ & 0 & & & \ddots \end{pmatrix}. \tag{1.10}$$

If $t \geq 0$, then $\exp(-t\Delta_K)$ is a bounded self-adjoint operator in $L^2(K)$.

Let us consider the following initial-boundary problem on $K \times (0, +\infty)$:

$$\begin{cases} \dfrac{\partial u}{\partial t} + \Delta u = 0, \\ u(x,0) = f(x), f \in L^2(K), \\ u\big|_{\partial K \times (0,+\infty)} = 0, \end{cases} \tag{1.11}$$

that is u is interpreted as a function from $[0, +\infty)$ to $L^2(K)$, i.e. a path in $L^2(K)$, such that

(i) u is Fréchet differentiable at $t \in (0, +\infty)$ and its Fréchet derivative $\dot{u}$ is equal to $-\Delta u$; in particular, this means that $\Delta u \in L^2(K)$;

(ii) u is L^2-continuous at $t = 0$ and $u(0, \cdot) = f$;

(iii) for each $t > 0, u(t, \cdot) \in H_0^1(K)$.

It is easy to check that the equation $\dot{u} + \Delta u = 0$ has

$$u = \left(e^{-t\Delta_K}\right)(f)$$

as solution.

The Fourier expansion of the function $f \in L^2(K)$ is given by

$$f = \sum_{j=1}^{\infty} \alpha_j f_j ,$$

where

$$\alpha_j = \int_K f_j f d\mu_g .$$

Then, by (1.10), one gets successively

$$\begin{aligned}
\left(\exp(-t\Delta_K) f\right)(x) &= \sum_{j=1}^{\infty} \alpha_j \left[\exp(-t\Delta_K) f_j\right](x) = \sum_{j=1}^{\infty} \alpha_j \exp\left(-t\mu_j(K)\right) f_j(x) \\
&= \sum_{j=1}^{\infty} \exp\left(-t\mu_j(K)\right) f_j(x) \int_K f(y) f_j(y) d\mu_g(y) \\
&= \int_K \left[\sum_{j=1}^{\infty} \exp\left(-t\mu_j(K)\right) f_j(x) f_j(y)\right] f(y) d\mu_g(y).
\end{aligned}$$

The real number in the brackets [ ] is denoted by $E_K(x,y,t)$ and the function $E_K : K \times K \times (0,+\infty) \to \mathbb{R}$ is called the heat kernel associated to *K*. Hence

$$E_K(x,y,t) = \sum_{j=1}^{\infty} \exp\left(-t\mu_j(K)\right) f_j(x) f_j(y), \tag{1.12}$$

$(x,y,t) \in K \times K \times (0,+\infty)$, and the weak solution *u* for (1.11) is given by

$$u(x,t) = \left(e^{-t\Delta_K} f\right)(x) = \int_K E_K(x,y,t) f(y) d\mu_g(y). \tag{1.13}$$

The eigenfunctions $f_j, j \in \mathbb{N}^*$, are C^∞-smooth, by the local elliptic regularity. One can show that the sequence

$$\mu_1 \le \mu_2 \le \cdots$$

is growing fast enough to ensure the convergence of (1.12) locally in any $C^l(K)$, so that $E_K \in C^\infty\left(K \times K \times (0,+\infty)\right)$.

The solution *u* defined by (1.13) is then C^∞-smooth and satisfies the heat equation in the classical sense. If *f* is continuous, then it is possible to show that the solution *u* is continuous in the classical sense on $K \times [0,+\infty)$ and $u(\ ,0) = f$.

Since $f_j \in H_0^1(K)$, one gets that E_K is also in $H_0^1(K)$ as a function of the first (or second) variable. Since the boundary ∂K is smooth, then this implies that E_K extends continuously to $\bar{K} \times \bar{K} \times (0,+\infty)$ and that E_K vanishes on ∂K. In particular, the solution *u* defined by (1.13) is also continuous on $\partial K \times (0,+\infty)$ and $u(x,t) = 0$ for all $(x,t) \in \partial K \times (0,+\infty)$.

Other properties of E_K that can be derived from the spectral properties are as follows:

(i) As a function of the first and third argument, E_K satisfies the heat equation

$$\Delta E_K + \frac{\partial}{\partial t} E_K = 0$$

and

$$\lim_{t \searrow 0} \int_K E_K(x,y,t) f(y) d\mu_g(y) = f(x).$$

(ii) The semigroup property, namely for all $x, y \in K$ and $t, s \in (0, +\infty)$,

$$E_K(x,y,t+s) = \int_K E_K(x,z,t) E_K(z,y,s) d\mu_g(z) \tag{1.14}$$

that is

$$\exp(-(t+s)\Delta_K) = \exp(-t\Delta_K)\exp(-s\Delta_K).$$

(iii) The symmetry, i.e.

$$E_K(x,y,t) = E_K(y,x,t)$$

for all $(x,y,t) \in K \times K \times (0,+\infty)$, which is obvious from (1.12).

Another property of E_K is the positivity, that is

$$E_K(x,y,t) > 0$$

for all $(x,y,t) \in K \times K \times (0,+\infty)$. The latter is not at all obvious from the eigenfunction expansion (1.12), because the eigenfunctions f_j (except f_1) are changing the sign. The positivity of E_K is a consequence of a property of the heat equation which is called the maximum (minimum) principle.

The maximum (minimum) principle. Let $T > 0, K_T = K \times (0,T)$ and $\partial' K_T = \partial K_T \setminus (K \times \{T\})$ the part of the boundary ∂K_T of the cylinder K_T without the top of the cylinder. If $u \in C^2(K_T) \cap C^{\circ}(\bar{K})$ is a solution of the heat equation on K_T, then

$$\sup_{K_T} u = \sup_{\partial' K_T} u \tag{1.15}$$

and

$$\inf_{K_T} u = \inf_{\partial' K_T} u. \tag{1.16}$$

If the initial function f in the initial-boundary problem (1.11) is non-negative then the minimum principle (1.16) implies that a classical solution u of (1.11) should be non-negative, too. Since ∂K is smooth, the function $u = e^{-t\Delta_K} f$ is a classical solution of (1.11) whence

$$e^{-t\Delta_K} f \geq 0 \text{ for all } f \in C_0^\infty(K), f \geq 0.$$

Therefore, the heat kernel E_K must be non-negative. In fact, one can show that E_K must be strictly positive on $K \times K \times (0,+\infty)$.

Another consequence of the maximum principle (1.15) is the inequality

$$e^{-t\Delta_K} 1 \le 1. \tag{1.17}$$

Indeed, $u = e^{-t\Delta_K} 1$ is a solution of (1.11) with $f = 1$ and, obviously, $\sup_{\partial' K_T} u \le 1$. Hence, by (1.15), one gets

$$\sup_{K_T} u \le 1$$

which means $u \le 1$ everywhere. Clearly, (1.17) and (1.13) imply

$$\int_K E_K(x, y, t)\, d\mu_g(y) \le 1 \tag{1.18}$$

for all $x \in K$.

Another important consequence of the minimum principle (1.16) is the monotonicity of E_K with respect to *K*: if $K \subset K'$, where K' is also a precompact open subset of *M*, then

$$E_K(x, y, t) \le E_{K'}(x, y, t). \tag{1.19}$$

Under the convention that $E_K(x, y, t)$ is equal to 0 for $x, y \notin K$, (1.19) holds for all $(x, y, t) \in M \times M \times (0, +\infty)$. Indeed, for any function $f \in C_0^\infty(K), f \ge 0$, let us compare the functions $u = e^{-t\Delta_K} f$ and $u' = e^{-t\Delta_{K'}} f$ on K_T. Both have the same initial datum, but on the boundary ∂K, one has $u(x, t) = 0 \le u'(x, t)$. Therefore,

$$\inf_{\partial' K_T} (u' - u) \ge 0,$$

and the minimum principle (1.16) implies $u' - u \ge 0$ on K_T, which means $u' \ge u$ everywhere. But $e^{-t\Delta_{K'}} f \ge e^{-t\Delta_K} f$ for any $f \in C_0^\infty(K), f \ge 0$, implies $E_{K'} \ge E_K$, that is (1.19).

The monotonicity of the heat kernel E_K with respect to *K* allows to construct the heat kernel *E* on the entire manifold *M* as follows. Let us consider a sequence $\{K_j \mid j \in \mathbb{N}^*\}$ of precompact open sets $K_j \subset M$ such that ∂K_j is smooth, $\bar{K}_j \subset K_{j+1}$ for all $j \in \mathbb{N}^*$ and $M = \bigcup_{j \in \mathbb{N}^*} K_j$. Such a sequence can be constructed on any manifold. Then let us define

$$E(x, y, t) = \lim_{j \to \infty} E_{K_j}(x, y, t). \tag{1.20}$$

Since $E_{K_{j+1}} \ge E_{K_j}$, the limit exists (finite or infinite) and does not depend on the choice of the sequence $\{K_j \mid j \in \mathbb{N}^*\}$. As follows from (1.18),

$$\int_M E(x, y, t)\, d\mu_g(y) \le 1,$$

so that *E* is finite almost everywhere. By the convergence properties of solutions to the parabolic equations, *E* is finite everywhere and C^∞-smooth.

Remark 1.8. Clearly, *E* inherits all previously discussed properties of E_K except for the eigenfunction expansion (1.12). Moreover, it is possible to define the Dirichlet extension Δ_M of the Laplace-Beltrami operator Δ associated to (M,g) and to show that E is the kernel of the semigroup $\exp(-t\Delta_M)$ acting in $L^2(M)$. However, the spectrum of Δ_M is not necessarily discrete as for a precompact open subset of M. This is why it is not possible in general to define the heat kernel by the eigenfunction expansion (1.12).

Example 1.2. Let M be some closed interval of $\mathbb{R}$, say $[0,\alpha]$, with the usual Riemannian metric g_{can}. The Dirichlet eigenvalue problem on $([0,\alpha],g_{can})$ leads us to the associated boundary problem

$$u'' + \mu u = 0, \tag{1.21}$$

$$u(0) = u(\alpha) = 0, \tag{1.22}$$

where $u \in C^\infty([0,\alpha])$, $\mu > 0$. The general integral of the differential equation (1.21) is

$$u(x) = a\cos\sqrt{\mu}x + b\sin\sqrt{\mu}x,\ x \in [0,\alpha],$$

where a,b are arbitrary real constants. (1.22) with $u \neq 0$ means that $a = 0$ and $\mu = \frac{k^2\pi^2}{\alpha^2}$, $k \in \mathbb{N}^*$. Therefore the eigenvalues of the Laplace-Beltrami operator on $([0,\alpha],g_{can})$ are $\mu_k = \frac{k^2\pi^2}{\alpha^2}$, $k \in \mathbb{N}^*$, and a linear independent system of eigenfunctions corresponding to μ_k consists of one function, say $f_k \in C^\infty([0,\alpha])$ given by $f_k(x) = \sin\left(\frac{\pi k}{\alpha}x\right)$, that is the dimension of the eigenspace corresponding to μ_k is one for all $k \in \mathbb{N}^*$, or equivalently

$$Spec_D([0,\alpha], g_{can}) = \left\{\frac{k^2\pi^2}{\alpha^2} \middle| k \in \mathbb{N}^*\right\}.$$

Note that $\left\{\sqrt{\frac{2}{\alpha}}f_k \middle| k \in \mathbb{N}^*\right\}$ is a complete othonormal system in the corresponding L^2-space of functions. For the Neumann eigenvalue problem on $([0,\alpha],g_{can})$ the boundary conditions are $u'(0) = u'(\alpha) = 0$, from which one gets the simple eigenvalues $\nu_k = \frac{\pi^2k^2}{\alpha^2}$, $k \in \mathbb{N}$, that is

$$Spec_N([0,\alpha], g_{can}) = \left\{\frac{k^2\pi^2}{\alpha^2} \middle| k \in \mathbb{N}\right\},$$

with eigenfunctions $f_k \in C^\infty([0,\alpha])$, $k \in \mathbb{N}$, given by $f_k(x) = \cos\frac{\pi k}{\alpha}x$.

2. EXAMPLES OF EXPLICIT SPECTRA

In this section we present some fundamental examples concerning the explicit computation of spectra for standard closed Riemannian manifolds.

We start with some basic facts.

First let us recall the **Stone-Weierstrass theorem**.

Theorem 2.1. Let (M,g) be a closed smooth Riemannian manifold and $\mathcal{A} \subset C^\infty(M)$ a subalgebra satisfying the following two conditions:

(i) $\mathcal{A}$ contains all constant functions;

(ii) $\mathcal{A}$ separates points of M, namely, for any different points $x,y \in M$ there exists an $f \in \mathcal{A}$ such that $f(x) \neq f(y)$.

Then $\mathcal{A}$ is dense in $C^\infty(M)$ with respect to the topology induced by the norm $\| \ \|$ (see Chapter 2, §1, A), and consequently in $L^2(M)$.

The next statement will be used later.

Proposition 2.1. Let (M,g) be a closed Riemannian manifold and assume given a nontrivial vector subspace W_j of $C^\infty(M)$ for each $j \in \mathbb{N}$ such that the following two conditions are satisfied:

(i) For any $j \in \mathbb{N}$, there exists an eigenvalue $\lambda_j \in (0,+\infty)$ of the Laplace-Beltrami operator Δ such that $W_j \subseteq E(\Delta,\lambda_j)$ (= the eigenspace of Δ corresponding to λ_j);

(ii) The sum $\sum_{j\in\mathbb{N}} W_j$ is dense in $C^\infty(M)$ with respect to the topology induced by the norm $\| \ \|$.

Then $\lambda_j, j \in \mathbb{N}$, are the eigenvalues of Δ and $W_j = E(\Delta,\lambda_j)$ for each $j \in \mathbb{N}$.

Proof. By hypothesis, any λ_j is an eigenvalue of Δ. Conversely, assume that λ is an eigenvalue of Δ such that $\lambda \neq \lambda_j$ for each $j \in \mathbb{N}$. Then the corresponding eigenspace $E(\Delta,\lambda)$ is $\neq 0$, and $E(\Delta,\lambda)$ is orthogonal to $E(\Delta,\lambda_j)$ with respect to the inner product $\langle \ , \ \rangle$ for any $j \in \mathbb{N}$ [see Remark 1.1 (i)], which is in contradiction with our assumption (ii). Therefore, $\{\lambda_j \,|\, j \in \mathbb{N}\}$ is the set of all eigenvalues of Δ. Now, let us prove the equality $W_j = E(\Delta,\lambda_j)$ for each $j \in \mathbb{N}$. Indeed, since $W_j \subseteq E(\Delta,\lambda_j)$ it follows that W_j is of finite dimension, hence closed in $E(\Delta,\lambda_j)$ with respect to the topology induced by $\| \ \|$, for any $j \in \mathbb{N}$. If we suppose that $W_j \neq E(\Delta,\lambda_j)$, there exists $f \in E(\Delta,\lambda_j)\setminus\{0\}$ such that f is orthogonal on W_j. On the other hand, by Remark 1.1 (i) f should be orthogonal on all W_k with $k \neq j$. Therefore, f is orthogonal on $\sum_{j\in\mathbb{N}} W_j$, so that $f = 0$ by assumption (ii), which contradicts $f \neq 0$. Consequently, $W_j = E(\Delta,\lambda_j)$ for any $j \in \mathbb{N}$.

Q.E.D.

The following statement is useful in some specific examples.

Proposition 2.2. Let $(\tilde{M},\tilde{g})$ be complete, $p:(\tilde{M},\tilde{g}) \to (M,g)$ a Riemannian covering or a surjective Riemannian submersion (see Chapter 1, Definition 2.5) with totally geodesic fibers, and $E(\tilde{\Delta},\tilde{\lambda})$ [resp. $E(\Delta,\lambda)$] the eigenspace of the Laplace-Beltrami operator $\tilde{\Delta}$ (resp. Δ) associated to $(\tilde{M},\tilde{g})$ [resp. (M,g)] corresponding to the eigenvalue $\tilde{\lambda}$ (resp. λ). If $f \in E(\Delta,\lambda)$, then $f \circ p \in E(\tilde{\Delta},\lambda)$. Conversely, if $\tilde{f} \in E(\tilde{\Delta},\tilde{\lambda})$ is constant on each fiber of p, then there exists a unique function $f \in C^\infty(M)$ such that $\tilde{f} = f \circ p$ and $f \in E(\Delta, \tilde{\lambda})$.

Indeed, under this assumption, one has

$$\tilde{\Delta}(f \circ p) = \Delta(f) \circ p \tag{2.1}$$

for any $f \in C^\infty(M)$ (see Berger, Gauduchon and Mazet [1971] or Craioveanu and Puta [1988] for more details) and our statement is an immediate consequence of this equality.

Remark 2.1. Under the assumption of Proposition 2.2, (M, g) is complete, $(\tilde{M}, M, p)$ is a fibre bundle with structure group and the smooth functions on M may be naturally considered as smooth functions on $\tilde{M}$ that are constant on each fiber of p. In particular, if p is a Riemannian covering, this means that $C^\infty(M)$ may be naturally considered as those functions of $C^\infty(\tilde{M})$ that are invariant under the covering transformation group of p. Therefore, for M closed, the eigenvalues of Δ are precisely those eigenvalues of $\tilde{\Delta}$ in whose eigenspaces there are nontrivial eigenfunctions of $\tilde{\Delta}$ that are constant on each fiber of p. The nontrivial vector subspace of the eigenspace of $\tilde{\Delta}$ consisting of all functions that are constant on each fiber of p is isomorphic to the eigenspace of Δ corresponding to the given eigenvalue.

Example 2.1. Let (M_1, g_1), (M_2, g_2) be closed Riemannian manifolds, $p_j : M_1 \times M_2 \to M_j$, $j = 1, 2$, the canonical projections and consider the Riemannian manifold $(M_1 \times M_2, g_1 \times g_2)$ (see Chapter 1, Definition 2.2). Then the vector subspace $\mathscr{A}$ of $C^\infty(M_1 \times M_2)$ generated by $\{(f_1 \circ p_1) \cdot (f_2 \circ p_2) | f_j \in C^\infty(M_j), j = 1, 2\}$ is dense in $L^2(M_1 \times M_2)$ by Theorem 2.1. If Δ (resp. Δ_j) is the Laplace-Beltrami operator associated to $(M_1 \times M_2, g_1 \times g_2)$ [resp. (M_j, g_j)], then in virtue of (2.1) and (2.11) in Chapter 2 one gets

$$\Delta((f_1 \circ p_1) \cdot (f_2 \circ p_2)) = (\Delta_1(f_1) \circ p_1) \cdot (f_2 \circ p_2) + (f_1 \circ p_1) \cdot (\Delta_2(f_2) \circ p_2)$$

for any $f_j \in C^\infty(M_j)$, $j = 1, 2$ (for more details, see Berger, Gauduchon and Mazet [1971] or Craioveanu and Puta [1988]). Therefore, if λ (resp. μ) is an eigenvalue of Δ_1 (resp. Δ_2) and $f_1 \in E(\Delta_1, \lambda)$, $f_2 \in E(\Delta_2, \mu)$, then $(f_1 \circ p_1) \cdot (f_2 \circ p_2) \in E(\Delta, \lambda + \mu)$. Let $\mathscr{A}_{\lambda,\mu}$ be the vector subspace of $E(\Delta, \lambda + \mu)$ generated by $\{(f_1 \circ p_1)(f_2 \circ p_2) | f_1 \in E(\Delta_1, \lambda), f_2 \in E(\Delta_2, \mu)\}$ and $W_\nu = \sum_{\lambda+\mu=\nu} \mathscr{A}_{\lambda,\mu}$. Since the direct sum $E_1 = \sum_\lambda E(\Delta_1, \lambda)$ [resp. $E_2 = \sum_\mu E(\Delta_2, \mu)$] is dense in $C^\infty(M_1)$ [resp. $C^\infty(M_2)$], it follows that the vector subspace $\sum W_{\lambda+\mu}$ of $\mathscr{A}$ generated by $\{(f_1 \circ p_1)(f_2 \circ p_2) | f_j \in E_j,\ j = 1, 2\}$ is dense in $\mathscr{A}$, and hence in $C^\infty(M_1 \times M_2)$.
By Proposition 2.1, $\{\lambda + \mu | \lambda \text{ (resp. } \mu) \text{ is an eigenvalue of } \Delta_1 \text{ (resp. } \Delta_2)\}$ is the set of all eigenvalues of Δ and

$$E(\Delta,\nu)=\sum_{\lambda+\mu=\nu}\mathscr{A}_{\lambda,\mu}.$$

Taking $f_2=1$ (resp. $f_1=1$) in the above formula for $\Delta((f_1\circ p_1)\cdot(f_2\circ p_2))$ one gets $\Delta\circ p_j^*=p_j^*\circ\Delta_j$ for $j=1,2$, where $p_j^*:C^\infty(M_j)\to C^\infty(M_1\times M_2)$, $p_j^*(f_j)=f_j\circ p_j$, $f_j\in C^\infty(M_j)$, that is the canonical projections commute with the corresponding Laplace-Beltrami operators. More generally, let $(\tilde{M},\tilde{g})$, (M,g) be closed, connected, smooth Riemannian manifolds, $p:(\tilde{M},\tilde{g})\to(M,g)$ a surjective Riemannian submersion and $p^*:C^\infty(M)\to C^\infty(\tilde{M})$, $p^*(f)=f\circ p$, $f\in C^\infty(M)$, the canonical $\mathbb{R}$-linear mapping induced by p. Under these assumptions, $(\tilde{M},M,p)$ is a locally trivial fibre bundle (in the sense of Ehresmann) and p^* is injective. Thus the corestriction $p^*:C^\infty(M)\to\operatorname{Im}p^*$ is an $\mathbb{R}$-linear isomorphism and note that $\operatorname{Im}p^*$ coincides with the real vector space of all functions $\tilde{f}\in C^\infty(\tilde{M})$ that are compatible with the equivalence relation R_p induced by p on $\tilde{M}$, that is xR_px' [i.e. $p(x)=p(x')$] implies $\tilde{f}(x)=\tilde{f}(x')$; in other words, any such $\tilde{f}$ is constant on each fiber of p. Thus, $C^\infty(M)$ can be identified with this vector space. In particular, if p is a Riemannian covering, then $C^\infty(M)$ can be also identified with the real vector space of all smooth functions on $\tilde{M}$ that are invariant with respect to the covering transformation group of p (see also Remark 2.1). Indeed, since p is a surjective locally trivial fibration, it induces the smooth diffeomorphism $\hat{p}:\tilde{M}/R_p\to M$ such that $\hat{p}\circ\pi=p$, where $\tilde{M}/R_p$ denotes the smooth quotient manifold and $\pi:\tilde{M}\to\tilde{M}/R_p$ the canonical smooth submersion. If $\tilde{f}\in C^\infty(\tilde{M})$ is compatible with R_p, there exists a unique function $\hat{\tilde{f}}\in C^\infty\left(\tilde{M}/R_p\right)$ such that $\hat{\tilde{f}}\circ\pi=\tilde{f}$. Thus, any such $\tilde{f}$ induces a unique function $f\in C^\infty(M)$, namely $f=\hat{\tilde{f}}\circ\hat{p}^{-1}$, such that $f\circ p=\tilde{f}$. Let $E(\tilde{\Delta},\tilde{\lambda})$ [resp. $E(\Delta,\lambda)$] the eigenspace of the Laplace-Beltrami operator $\tilde{\Delta}$ (resp. Δ) associated to $(\tilde{M},\tilde{g})$ [resp. (M,g)] corresponding to the eigenvalue $\tilde{\lambda}$ (resp. λ) $\in[0,+\infty)$. The following two conditions are equivalent:

(i) $\tilde{\Delta}\circ p^*=p^*\circ\Delta$ (see Chapter 2, Proposition 2.5 for the existence of such a mapping p);

(ii) Any eigenvalue $\lambda\in[0,+\infty)$ of Δ is also an eigenvalue of $\tilde{\Delta}$ and $p^*(E(\Delta,\lambda))\subseteq E(\tilde{\Delta},\lambda)$.

It is immediate that (i) implies (ii), that is if p satisfies condition (i), $p^*(f)=f\circ p$ is an eigenfunction of $\tilde{\Delta}$, with the same eigenvalue, for every eigenfunction f of Δ. In particular, if p is a Riemannian covering, then any eigenvalue λ of Δ is also an eigenvalue of $\tilde{\Delta}$ and the multiplicity of λ as eigenvalue of Δ is $\leq$ than the multiplicity of λ as eigenvalue of $\tilde{\Delta}$. If p is an isometry, $\tilde{\Delta}$ and Δ have the same eigenvalues [since

$p^{-1}:(M,g)\to(\tilde{M},\tilde{g})$ is also an isometry], and $E(\tilde{\Delta},\lambda)$ is isomorphic to $E(\Delta,\lambda)$ for every eigenvalue $\lambda\in[0,+\infty)$, that is $\mathbf{Spec}(\tilde{M},\tilde{g})=\mathbf{Spec}(M,g)$.

On the other hand, note that if $p:(\tilde{M},\tilde{g})\to(M,g)$ satisfies condition (i), $\lambda\in[0,+\infty)$ is an eigenvalue of $\tilde{\Delta}$ and the vector subspace $E_p(\tilde{\Delta},\lambda)$ of $E(\tilde{\Delta},\lambda)$ consisting of all eigenfunctions that are constant on each fiber of *p* is nontrivial, then λ is also an eigenvalue of Δ and $E(\Delta,\lambda)$ is isomorphic to $E_p(\tilde{\Delta},\lambda)$. Consequently, if *p* satisfies condition (i), the eigenvalues of Δ are precisely those eigenvalues of $\tilde{\Delta}$ whose eigenspaces contain nontrivial eigenfunctions of $\tilde{\Delta}$ that are constant on each fiber of *p* and the multiplicity of the eigenvalue λ of Δ is just the dimension of $E_p(\tilde{\Delta},\lambda)$.

For the proof of the converse statement, namely (ii) implies (i), see Gilkey and Park [1996].

A. The spectrum of a flat torus

Let $\left(T^n=\mathbb{R}^n/\Gamma, g_{can}/\Gamma\right)$ be the *n*-dimensional flat torus defined by the lattice $\Gamma=\mathbb{Z}e_1+\cdots+\mathbb{Z}e_n$ in $\mathbb{R}^n$, where $(e_1,\cdots,e_n)$ is a basis of $\mathbb{R}^n$ (see Chapter 1, Example 2.17). The elements of Γ act on $\mathbb{R}^n$ as parallel translations, which are isometries of $(\mathbb{R}^n,g_{can})$. Since the canonical projection $p:(\mathbb{R}^n,g_{can})\to\left(T^n,g_{can}/\Gamma\right)$ is a Riemannian covering, by Proposition 2.2 (see also Remark 2.1) the eigenfunctions of the Laplace-Beltrami operator Δ of T^n lift to eigenfunctions $f\in C^\infty(\mathbb{R}^n)$ of the Laplace-Beltrami operator $\Delta^{\mathbb{R}^n}$ of $(\mathbb{R}^n,g_{can})$ that are Γ-periodic functions on $\mathbb{R}^n$, that is $f(x)=f(x+\gamma)$ for all $x\in\mathbb{R}^n$ and $\gamma\in\Gamma$. Eigenfunctions of $\Delta^{\mathbb{R}^n}$ of the form

$$\exp\left[2\pi i\left(y^1x^1+\cdots+y^nx^n\right)\right]=\cos 2\pi\left(y^1x^1+\cdots+y^nx^n\right)+i\sin 2\pi\left(y^1x^1+\cdots+y^nx^n\right)$$

will have the correct periodicity provided $\gamma^1y^1+\cdots+\gamma^ny^n\in\mathbb{Z}$ for all $\gamma=(\gamma^1,\ldots,\gamma^n)\in\Gamma$ [$\Delta^{\mathbb{R}^n}$ will act on smooth complex-valued functions by acting on their real and imaginary parts separately, namely for real-valued functions $u,v\in C^\infty(\mathbb{R}^n)$, one has $\Delta(u+iv)=\Delta u+i\Delta v$].

The set

$$\Gamma^*\overset{def}{=}\left\{(y^1,\ldots,y^n)\in\mathbb{R}^n\,\middle|\,\sum_{k=1}^n\gamma^ky^k\in\mathbb{Z}\text{ for any }(\gamma^1,\ldots,\gamma^n)\in\Gamma\right\}$$

is again a lattice of $\mathbb{R}^n$, called the **dual lattice to** Γ. More precisely, if $\Gamma=A\mathbb{Z}^n$, where $A\in GL(n,\mathbb{R})$, then $\Gamma^*={}^tA^{-1}\mathbb{Z}^n$ (see also Chapter 1, Example 2.17). In particular, it follows that $(\Gamma^*)^*=\Gamma$ and note that the lattices Γ and Γ_1 in $\mathbb{R}^n$ are congruent (see Chapter 1, Example 2.17) if and only if Γ^* and Γ_1^* are congruent. Note also that Γ^* carries a basis $(e_1^*,\cdots,e_n^*)$ of $\mathbb{R}^n$ satisfying

$$e_j^* \cdot e_k = \delta_{jk} \left(1 \le j,k \le n\right),$$

where "$\cdot$" denotes the canonical inner product in $\mathbb{R}^n$. On the other hand, note that for $y \in \Gamma^* \subset \mathbb{R}^n$, the functions $f_y^{(1)}, f_y^{(2)} \in C^\infty\left(\mathbb{R}^n\right)$ given by

$$f_y^{(1)}(x) = \cos 2\pi (x \cdot y), f_y^{(2)}(x) = \sin 2\pi (x \cdot y), x \in \mathbb{R}^n,$$

are invariant under the action of Γ and they induce smooth functions $\widehat{f_y^{(k)}}$, $k = 1,2$, on T^n, namely

$$\widehat{f_y^{(k)}}(x + \Gamma) = f_y^{(k)}(x), \ k = 1,2, x \in \mathbb{R}^n,$$

so that $\widehat{f_y^{(k)}} \circ p = f_y^{(k)}$. Indeed, if $x - \tilde{x} \in \Gamma$, $x, \tilde{x} \in \mathbb{R}^n$, then $\exp 2\pi i\left[y \cdot (x - \tilde{x})\right] = 1$ since $y \in \Gamma^*$, so that

$$f_y^{(k)}(x) = f_y^{(k)}(\tilde{x})$$

for $k = 1,2$.

Proposition 2.3. The set of eigenvalues of the Laplace-Beltrami operator Δ of a flat torus $\left(T^n, g_{can}/\Gamma\right)$ is $\left\{4\pi^2 \|y\|^2 \,\middle|\, y \in \Gamma^*\right\}$, i.e. $(4\pi^2)$ times the set of lengths-squared of points of the dual lattice Γ^* to Γ. The eigenfunctions corresponding to the eigenvalue $\lambda = 0$ are real constant functions, and the eigenspaces corresponding to $\lambda > 0$ are spanned by $\widehat{f_y^{(1)}}, \widehat{f_y^{(2)}} \in C^\infty\left(T^n\right)\left(y \in \Gamma^* \setminus \{0\}\right)$ such that $\|y\|^2 = \dfrac{\lambda}{4\pi^2}$.

Proof. An easy computation shows that the functions $f_y^{(k)}\left(y \in \Gamma^*\right)$, $k = 1,2$, satisfy

$$\frac{\partial^2 f_y^{(k)}}{\partial x^{j2}} = -4\pi^2 \left(y^j\right)^2 f_y^{(k)}$$

for any $j = 1, \ldots, n$. Therefore

$$\Delta^{\mathbb{R}^n} f_y^{(k)} = 4\pi^2 \|y\|^2 f_y^{(k)}, \ k = 1,2.$$

Since

$$\Delta^{\mathbb{R}^n}\left(f_y^{(k)}\right) = \Delta\left(\widehat{f_y^{(k)}}\right) \circ p, \ k = 1,2$$

[see the equality (2.1)], it follows that

$$\Delta\left(\widehat{f_y^{(k)}}\right) = 4\pi^2 \|y\|^2 \widehat{f_y^{(k)}}, \tag{2.2}$$

$k = 1,2$. Let $\mathscr{A}$ be the real linear subspace of $C^\infty\left(T^n\right)$ spanned by $\left\{1, \widehat{f_y^{(1)}}, \widehat{f_y^{(2)}} \,\middle|\, y \in \Gamma^*\right\}$. By the addition formulae for trigonometric functions, it follows that

$$\widehat{f_y^{(1)}}\widehat{f_{\tilde{y}}^{(1)}} = \frac{1}{2}\widehat{f_{y+\tilde{y}}^{(1)}} + \frac{1}{2}\widehat{f_{y-\tilde{y}}^{(1)}},$$
$$\widehat{f_y^{(1)}}\widehat{f_{\tilde{y}}^{(2)}} = \frac{1}{2}\widehat{f_{y+\tilde{y}}^{(2)}} + \frac{1}{2}\widehat{f_{\tilde{y}-y}^{(2)}},$$
$$\widehat{f_y^{(2)}}\widehat{f_{\tilde{y}}^{(2)}} = \frac{1}{2}\widehat{f_{y-\tilde{y}}^{(1)}} - \frac{1}{2}\widehat{f_{y+\tilde{y}}^{(1)}},$$

for any $y, \tilde{y} \in \Gamma^*$, that is $\mathscr{A}$ is a subalgebra of $C^\infty(T^n)$. Now note that $\mathscr{A}$ satisfies the assumptions of Theorem 2.1. Clearly (i) holds. On the other hand, let

$$x+\Gamma \neq \tilde{x}+\Gamma \text{ in } T^n, \text{ i.e. } x-\tilde{x} \notin \Gamma,$$

and assume $\widehat{f_y^{(1)}}(x+\Gamma) = \widehat{f_y^{(1)}}(\tilde{x}+\Gamma), \widehat{f_y^{(2)}}(x+\Gamma) = \widehat{f_y^{(2)}}(\tilde{x}+\Gamma)$ hold for any $y \in \Gamma^*$. Then $(x-\tilde{x})\cdot y \in \mathbb{Z}$ for any $y \in \Gamma^*$. It follows that $x-\tilde{x} \in (\Gamma^*)^* = \Gamma$, which is a contradiction, and $\mathscr{A}$ separates points of T^n. By Proposition 2.1 one gets the part of our statement concerning the eigenvalues of Δ. Finally, let us show that the system $\left\{1, \widehat{f_y^{(1)}}, \widehat{f_y^{(2)}} \middle| y \in \Gamma^* \setminus \{0\}\right\}$ is linearly independent over $\mathbb{R}$. Let $\widehat{f_y} : T^n \to \mathbb{C}$ be the smooth function induced by $f_y : \mathbb{R}^n \to \mathbb{C}$, $f_y(x) = \exp 2\pi i(x\cdot y)$, where "$\cdot$" denotes the canonical inner product on $\mathbb{R}^n$. Note that $\hat{f}_0 = 1$ and $\hat{f}_y \cdot \hat{f}_{y'} = \hat{f}_{y+y'}$ for every $y, y' \in \Gamma^*$. Since $\widehat{f_y^{(1)}} = \frac{1}{2}\widehat{f_y} + \frac{1}{2}\widehat{f_{-y}}$, $\widehat{f_y^{(2)}} = \frac{i}{2}\widehat{f_{-y}} - \frac{i}{2}\widehat{f_y}$, it is sufficient to prove that the system $S = \left\{\widehat{f_y^{(1)}} + i\widehat{f_y^{(2)}} = \widehat{f_y} \middle| y \in \Gamma^*\right\}$ is linearly independent over $\mathbb{C}$. By (2.2) one gets

$$\Delta\left(\widehat{f_y}\right) = \Delta\left(\widehat{f_y^{(1)}}\right) + i\Delta\left(\widehat{f_y^{(2)}}\right) = 4\pi\|y\|^2\widehat{f_y}. \tag{2.3}$$

First note that the subsystem consisting of one single function $\widehat{f_y}\ (y \in \Gamma^*)$ is linearly independent since $\widehat{f_y} \neq 0$ for any $y \in \Gamma^*$. Assume now that for given $k \geq 2$ this property is true for any choice of $k-1$ distinct functions of $\left\{\widehat{f_y} \middle| y \in \Gamma^*\right\}$. Should the k distinct functions $\widehat{f_{y_1}}, \ldots, \widehat{f_{y_k}}\ (y_j \in \Gamma^*,\ j = 1, \ldots, k)$ satisfy

$$\sum_{j=1}^{k} \alpha_j \widehat{f_{y_j}} = 0 \tag{2.4}$$

for a given choice of complex numbers $\alpha_1, \ldots, \alpha_k$, then multiplying by $\widehat{f_{-y_k}}$ on both sides of (2.4) and applying Δ, in virtue of (2.3), we would have

$$\sum_{j=1}^{k-1} \alpha_j \|y_j - y_k\|^2 \widehat{f_{y_j - y_k}} = 0.$$

The inductive assumption implies $\alpha_j = 0$ for $j = 1, \ldots, k-1$, so that (2.4) leads to $\alpha_k \hat{f}_{y_k} = 0$, that is $\alpha_k = 0$.

Q.E.D.

Remark 2.2. Proposition 2.3 states that $\lambda \in [0, +\infty)$ is an eigenvalue of the Laplace-Beltrami operator Δ on $\left(T^n, g_{can}/\Gamma\right)$ if and only if there exists $y \in \Gamma^*$ such that $\|y\|^2 = \frac{\lambda}{4\pi^2}$. The cardinal of the set $\left\{ y \in \Gamma^* \,\middle|\, \|y\|^2 = \frac{\lambda}{4\pi^2} \right\}$ is just the multiplicity of λ. The problem of determining the multiplicity of an eigenvalue $\lambda > 0$ of Δ reduces to a problem in number theory, which asks how many points of the dual lattice Γ^* to Γ lie on the $(n-1)$-sphere centered at the origin of radius $\sqrt{\lambda}/2\pi$ in $\mathbb{R}^n$.

Remark 2.3. Let $\left(K_{a,b}, g_{can}/\breve{\Gamma}_{a,b}\right)$ be the flat Klein bottle corresponding to the lattice $\Gamma_{a,b} = \mathbb{Z}(ae_1) + \mathbb{Z}(be_2)$, where (e_1, e_2) is the standard basis for $\mathbb{R}^2$ and $a > 0, b > 0$ (see Chapter 1, Example 2.18). By the definition of $\Gamma^*_{a,b}$, it follows that $\Gamma^*_{a,b} = \mathbb{Z}\left(\frac{1}{a}e_1\right) + \mathbb{Z}\left(\frac{1}{b}e_2\right)$. By Proposition 2.3 the eigenvalues of the Laplace-Beltrami operator associated to the flat torus $\left(\mathbb{R}^2/\Gamma_{a,b}, g_{can}/\Gamma_{a,b}\right)$ are given by $4\pi^2\left(\frac{k^2}{a^2} + \frac{l^2}{b^2}\right), k, l \in \mathbb{Z}$. Since the canonical projection of $\left(\mathbb{R}^2/\Gamma_{a,b}, g_{can}/\Gamma_{a,b}\right)$ onto $\left(K_{a,b}, g_{can}/\breve{\Gamma}_{a,b}\right)$ is a Riemannian covering, using Remark 2.1 one can show that the eigenvalues of the Laplace-Beltrami operator associated to $\left(K_{a,b}, g_{can}/\breve{\Gamma}_{a,b}\right)$ are given by $4\pi^2\left(\frac{k^2}{a^2} + \frac{l^2}{b^2}\right)$, with $k, l \in \mathbb{Z}$ satisfying the condition $l \neq 0$ if k is odd. Moreover, the eigenvalues of the form $4\pi^2 \frac{l^2}{b^2}, l \in \mathbb{Z}$, are of odd multiplicity, the other ones being of even multiplicity (for more details, see for example Berger, Gauduchon, Mazet [1971], pp. 151-153). In particular, if $\left(\mathbb{R}^2/\Gamma, g_{can}/\Gamma\right)$ [resp. $\left(K_{a,b}, g_{can}/\breve{\Gamma}_{a,b}\right)$] is a flat torus (resp. Klein bottle), then $\operatorname{Spec}\left(\mathbb{R}^2/\Gamma, g_{can}/\Gamma\right) \neq \operatorname{Spec}\left(K_{a,b}, g_{can}/\breve{\Gamma}_{a,b}\right)$. Indeed, let Δ (resp. Δ') be the Laplace-Beltrami operator associated to $\left(\mathbb{R}^2/\Gamma, g_{can}/\Gamma\right)$ [resp. $\left(K_{a,b}, g_{can}/\breve{\Gamma}_{a,b}\right)$]. The multiplicity of each eigenvalue $\lambda > 0$ of Δ is even, while Δ' has an infinite number of eigenvalues $\lambda' > 0$ of odd multiplicity.

B. The spectrum of the unit n-sphere

Let S^n be the unit sphere in $\mathbb{R}^{n+1}$, (S^n, g_{can}) the closed Riemannian manifold considered in Chapter 1, Example 2.5 and Δ its Laplace-Beltrami operator. Let us denote by r the distance of a point of $\mathbb{R}^{n+1}$ to the origin of $\mathbb{R}^{n+1}$ and by $\partial/\partial r$ the directional derivative with respect to the unit outer normal vector to S^n. Then for any $f \in C^\infty\left(\mathbb{R}^{n+1}\right)$ and its restriction $f\big|_{S^n}$ one has

$$\left(\Delta^{\mathbb{R}^{n+1}} f\right)\Big|_{S^n} = \Delta\left(f\big|_{S^n}\right) - \frac{\partial^2 f}{\partial r^2}\Big|_{S^n} - n\frac{\partial f}{\partial r}\Big|_{S^n}. \tag{2.5}$$

Indeed, choose an orthonormal basis $(e_1,\ldots,e_n)$ of the tangent space $T_xS^n = \mathbb{R}^n \times \{0\} \subset \mathbb{R}^{n+1}$ of S^n at $x \in S^n$, identify x with the normal unit vector to S^n at x and consider the coordinate system $\left(x^1,\ldots,x^n,r\right)$ with respect to the orthonormal basis $(e_1,\ldots,e_n,x)$. Let $\gamma_j(t) = (\cos t)x + (\sin t)e_j$ be the geodesic of (S^n, g_{can}) starting from x with the initial direction $e_j, j = 1,\ldots,n$.
Then one gets

$$\frac{d\left(f\circ\gamma_j\right)}{dt}(t) = \cos t\frac{\partial f}{\partial x^j}\left(\gamma_j(t)\right) - \sin t\frac{\partial f}{\partial r}\left(\gamma_j(t)\right)$$

and

$$\frac{d^2}{dt^2}\Big|_{t=0}\left(f\circ\gamma_j\right) = \frac{\partial^2 f}{\partial x^{j2}}(x) - \frac{\partial f}{\partial r}(x),$$

$j = 1,\ldots,n$. Hence

$$\left(\Delta\left(f\big|_{S^n}\right)\right)(x) = -\sum_{j=1}^{n}\frac{d^2}{dt^2}\Big|_{t=0}\left(f\circ\gamma_j\right) = -\sum_{j=1}^{n}\frac{\partial^2 f}{\partial x^{j2}}(x) + n\frac{\partial f}{\partial r}(x) \tag{2.6}$$

(see also Chapter 2, § 2, C) and (2.5) immediately follows from (2.6) (see Chapter 2, Example 2.1).

Let $\mathcal{P}_k = \left\{f \in \mathbb{R}\left[x^1,\ldots,x^{n+1}\right] \,\middle|\, f\left(\alpha x^1,\ldots,\alpha x^{n+1}\right) = \alpha^k f\left(x^1,\ldots,x^{n+1}\right) \text{ for any } \alpha \in \mathbb{R}\right\}$ be the real vector space of homogeneous polynomials of degree $k \in \mathbb{N}$ on $\mathbb{R}^{n+1}$ and $\mathcal{H}_k = \left\{f \in \mathcal{P}_k \,\middle|\, \Delta^{\mathbb{R}^{n+1}} f = 0\right\}$ the vector subspace of $\mathcal{P}_k$ consisting of harmonic homogeneous polynomials of degree k. Note that

$$\mathcal{P}_0 = \mathcal{H}_0 = \left\{\text{the constant functions on } \mathbb{R}^{n+1}\right\}$$

and

$$\mathcal{P}_1 = \mathcal{H}_1 = \left\{\text{the homogeneous polynomials of degree 1 on } \mathbb{R}^{n+1}\right\}.$$

In fact, for $\alpha = \left(\alpha^1,\ldots,\alpha^{n+1}\right) \in \mathbb{C}^{n+1}$, consider $u_\alpha(x) = \left(\alpha^1 x^1 + \cdots + \alpha^{n+1}x^{n+1}\right)^k$. A direct computation gives $\left(\Delta^{\mathbb{R}^{n+1}}\left(u_\alpha\right)\right)(x) = -k(k-1)\alpha^2\left(\alpha^1 x^1 + \cdots + \alpha^{n+1}x^{n+1}\right)^{k-2}$, where $\alpha^2 = \left(\alpha^1\right)^2 + \cdots + \left(\alpha^{n+1}\right)^2$. Hence $\Delta^{\mathbb{R}^{n+1}} u_\alpha = 0$ whenever $\alpha^2 = 0$. Moreover, the space of complex harmonic homogeneous polynomials of degree k is equal to the linear span of the set of polynomials of the form u_α with $\alpha^2 = 0$.

Proposition 2.4. For any $u \in \mathcal{H}_k$ and $k \in \mathbb{N}$, one has

$$\Delta\left(u\big|_{S^n}\right) = k(n+k-1)u\big|_{S^n},$$

that is $u\big|_{S^n}$ is an eigenfunction of Δ with eigenvalue $k(n+k-1)$.

Proof. For $f \in C^\infty(S^n)$, define a C^∞-function u on a neighbourhood of S^n in $\mathbb{R}^{n+1}$ by

$$u(y) = r^k(y) f\left(\frac{y}{\|y\|}\right).$$

Then $u\big|_{S^n} = f$, $\dfrac{\partial u}{\partial r} = kr^{k-1} f$, $\dfrac{\partial^2 u}{\partial r^2} = k(k-1)r^{k-2} f$, and by (2.5) it follows that

$$\Delta f = \left(\Delta^{\mathbb{R}^{n+1}}(u)\right)\Big|_{S^n} + k(n+k-1) f. \tag{2.7}$$

Therefore, if $u \in \mathcal{H}_k$, then $u(y) = r^k(y) f\left(\frac{y}{\|y\|}\right)$ for $f = u\big|_{S^n}$, and (2.7) leads to the above statement.

Q.E.D.

On each $\mathcal{P}_k$ one can introduce an inner product by

$$\langle P, Q\rangle = \int_{S^n} \tilde{P}\tilde{Q}\, d\mu_{g_{can}},$$

where $\tilde{f} = f\big|_{S^n}$ for $f \in \mathcal{P}_k$.

Proposition 2.5. With respect to the above inner product, we have the orthogonal decompositions

$$\mathcal{P}_{2k} = \mathcal{H}_{2k} \oplus r^2\mathcal{H}_{2k-2} \oplus \cdots \oplus r^{2k}\mathcal{H}_0,$$
$$\mathcal{P}_{2k+1} = \mathcal{H}_{2k+1} \oplus r^2\mathcal{H}_{2k-1} \oplus \cdots \oplus r^{2k}\mathcal{H}_1$$

for each $k \in \mathbb{N}$, where $r^2 = \left(x^1\right)^2 + \cdots + \left(x^{n+1}\right)^2$.

Proof. It suffices to prove that with respect to the above inner product we have the orthogonal decomposition

$$\mathcal{P}_{k+2} = \mathcal{H}_{k+2} \oplus r^2\mathcal{P}_k \qquad (2.8)_k$$

for each $k \in \mathbb{N}$. We proceed by induction on k. Let $\left(r^2\mathcal{P}_0\right)^\perp$ be the orthogonal complement of $r^2\mathcal{P}_0$ in $\mathcal{P}_2$ and $P \in \mathcal{H}_2$. By Proposition 2.4 and Green's formula (see Chapter 2, Corollary 2.2) one gets

$$\left\langle P, r^2\right\rangle = \int_{S^n} \tilde{P}\, d\mu_{g_{can}} = \frac{1}{2(n+1)}\int_{S^n} \Delta\tilde{P}\, d\mu_{g_{can}} = 0,$$

that is $\mathcal{H}_2 \subset \left(r^2\mathcal{P}_0\right)^{\perp}$.

On the other hand, if $P \in \left(r^2\mathcal{P}_0\right)^{\perp}$, then

$$\int_{S^n} \tilde{P}\, d\mu_{g_{can}} = 0. \tag{2.9}$$

From (2.7) one gets

$$\left(\Delta^{\mathbb{R}^{n+1}} P\right)\Big|_{S^n} = \Delta\tilde{P} - 2(n+1)\tilde{P},$$

so that (2.9) and Green's formula leads to

$$\int_{S^n} \widetilde{\Delta^{\mathbb{R}^{n+1}} P}\, d\mu_{g_{can}} = 0.$$

Since $\widetilde{\Delta^{\mathbb{R}^{n+1}} P}$ is constant, it must be 0. Therefore $\Delta^{\mathbb{R}^{n+1}} P = 0$, that is $P \in \mathcal{H}_2$. Hence $\left(r^2\mathcal{P}_0\right)^{\perp} \subset \mathcal{H}_2$ and

$$\mathcal{P}_2 = \left(r^2\mathcal{P}_0\right)^{\perp} \oplus r^2\mathcal{P}_0 = \mathcal{H}_2 \oplus r^2\mathcal{P}_0,$$

i.e. the equality $(2.8)_0$ is true. Similarly, one can show that the equality $(2.8)_1$ is true.

Now assume $(2.8)_l$ is true for $l = 0,1,\ldots,k-1$. Then

$$\mathcal{P}_k = \mathcal{H}_k \oplus r^2\mathcal{H}_{k-2} \oplus \cdots \oplus r^k\mathcal{H}_0 \text{ (or } r^{k-1}\mathcal{H}_1). \tag{2.10}$$

Let $\left(r^2\mathcal{P}_k\right)^{\perp}$ be the orthogonal complement of $r^2\mathcal{P}_k$ in $\mathcal{P}_{k+2}$ and $P \in \mathcal{H}_{k+2}$. For $Q \in \mathcal{H}_{k-2j}\ (k-2j = k,\ldots,1 \text{ or } 0)$, by Proposition 2.4 and Remark 2.2 in Chapter 2 one obtains

$$\begin{aligned}
\left\langle P, r^{2j+2}Q\right\rangle &= \int_{S^n} \tilde{P}\tilde{Q}\, d\mu_{g_{can}} = \frac{1}{(k+2)(n+k+1)}\int_{S^n} \Delta\tilde{P}\cdot\tilde{Q}\, d\mu_{g_{can}} \\
&= \frac{1}{(k+2)(n+k+1)}\int_{S^n} \tilde{P}\Delta\tilde{Q}\, d\mu_{g_{can}} = \frac{(k-2j)(n+k-2j-1)}{(k+2)(n+k+1)}\int_{S^n} \tilde{P}\tilde{Q}\, d\mu_{g_{can}} \\
&= \frac{(k-2j)(n+k-2j-1)}{(k+2)(n+k+1)}\left\langle P, r^{2j+2}Q\right\rangle.
\end{aligned}$$

Consequently, $\left\langle P, r^{2j+2}Q\right\rangle = 0$, which implies that $\mathcal{H}_{k+2} \subset \left(r^2\mathcal{P}_k\right)^{\perp}$. Conversely, assume $P \in \left(r^2\mathcal{P}_k\right)^{\perp} \subset \mathcal{P}_{k+2}$. By (2.7), Remark 2.2 in Chapter 2, Proposition 2.4 and the induction hypothesis (2.10) one gets

$$\left\langle \Delta^{\mathbb{R}^{n+1}} P, r^{2j} Q \right\rangle = \int_{S^n} \widetilde{\Delta^{\mathbb{R}^{n+1}} P} \cdot \tilde{Q}\, d\mu_{g_{can}} = \int_{S^n} \left[\Delta \tilde{P} - (k+2)(n+k+1) \tilde{P} \right] \tilde{Q}\, d\mu_{g_{can}}$$
$$= \int_{S^n} \Delta \tilde{P} \cdot \tilde{Q}\, d\mu_{g_{can}} - (k+2)(n+k+1) \left\langle P, r^{2j+2} Q \right\rangle$$
$$= \int_{S^n} \tilde{P} \Delta \tilde{Q}\, d\mu_{can} = (k-2j)(n+k-2j-1) \left\langle P, r^{2j+2} Q \right\rangle = 0,$$

for any $Q \in \mathscr{H}_{k-2j}\ (k-2j = k,\ldots,1 \text{ or } 0)$. Consequently, $\Delta^{\mathbb{R}^{n+1}} P \in \mathscr{P}_k$ is orthogonal to $\mathscr{P}_k$ by the induction hypothesis (2.10), and therefore $\Delta^{\mathbb{R}^{n+1}} P = 0$, that is $P \in \mathscr{H}_{k+2}$. Hence

$$\mathscr{P}_{k+2} = \left(r^2 \mathscr{P}_k \right)^{\perp} \oplus r^2 \mathscr{P}_k = \mathscr{H}_{k+2} \oplus r^2 \mathscr{P}_k,$$

i.e. $(2.8)_k$ is valid for any $k \in \mathbb{N}$.

Q.E.D.

The following statement improves Proposition 2.4.

Theorem 2.2. The eigenvalues of the Laplace-Beltrami operator Δ of (S^n, g_{can}) are given by $\lambda_k = k(n+k-1),\, k \in \mathbb{N}$, and the corresponding eigenspace $E(\Delta,\lambda_k)$ by $E(\Delta, \lambda_k) = \widetilde{\mathscr{H}}_k = \left\{ f|_{S^n} \,|\, f \in \mathscr{H}_k \right\}$. In particular, the multiplicity m_k of λ_k is equal to

$$m_k = \begin{cases} 1 & \text{if } k = 0, \\ \dfrac{(n+k-2)!}{(n-1)!k!}(n+2k-1) & \text{if } k \geq 1. \end{cases}$$

Proof. Since $r^2|_{S^n} = 1$, by Proposition 2.5, $\overset{\infty}{\underset{k=0}{\oplus}} \widetilde{\mathscr{H}}_k = \overset{\infty}{\underset{k=0}{\oplus}} \widetilde{\mathscr{P}}_k$. Since $\mathscr{P}_k \cdot \mathscr{P}_l \subset \mathscr{P}_{k+l}$, it follows that $\overset{\infty}{\underset{k=0}{\oplus}} \widetilde{\mathscr{P}}_k$ is a subalgebra of $C^\infty(S^n)$. Therefore $\mathscr{A} = \overset{\infty}{\underset{k=0}{\oplus}} \widetilde{\mathscr{H}}_k$ is a subalgebra of $C^\infty(S^n)$ that satisfies obviously the condition (i) of Theorem 2.1. As for condition (ii) of Theorem 2.1, note that for different points $x,y \in S^n$ one may choose a linear functional (for example, a coordinate function) $f \in \mathscr{H}_1 = \mathscr{P}_1$ such that $\tilde{f}(x) \neq \tilde{f}(y)$. Therefore, by Theorem 2.1, $\mathscr{A}$ is dense in $L^2(S^n)$ and Propositions 2.1 and 2.4 imply that all eigenvalues of Δ are exhausted by $k(n+k-1),\, k \in \mathbb{N}$, and moreover $E(\Delta, \lambda_k) = \widetilde{\mathscr{H}}_k$ for each $k \in \mathbb{N}$.

Concerning the multiplicities, first note that $\widetilde{\mathscr{H}}_k \cong \mathscr{H}_k$ and

$$\dim \mathscr{P}_k = \binom{n+k}{k} \tag{2.11}$$

for each $k \in \mathbb{N}$. In order to prove (2.11) we use the following equalities:

$$\dim \mathscr{P}_{n+1,0} = 1,\ \dim \mathscr{P}_{1,k} = 1,\ \ k \in \mathbb{N},$$

and

$$\mathscr{P}_{n+1,k} = \mathscr{P}_{n,k} \oplus x^{n+1} \mathscr{P}_{n+1,k-1},$$

where $\mathcal{P}_{n+1,k}$ denotes here the vector space of polynomials in the $x^1,\ldots,x^{n+1}$ variables which are homogeneous of degree k. The last equality implies

$$\dim \mathcal{P}_{n+1,k} = \dim \mathcal{P}_{n,k} + \dim \mathcal{P}_{n+1,k-1},$$

so that one gets (2.11) by induction. Therefore, by $(2.8)_{k-2}$ $(k \geq 2)$, one obtains

$$m_k = \dim \widetilde{\mathcal{H}_k} = \dim \mathcal{H}_k = \dim \mathcal{P}_k - \dim \mathcal{P}_{k-2} = \binom{n+k}{k} - \binom{n+k-2}{k-2}$$
$$= \frac{(n+k-2)(n+k-3)\cdots(n+1)n}{k!}(n+2k-1)$$

for $k \geq 2$. Concerning the cases $k=0$ and 1, note that

$$\dim \widetilde{\mathcal{H}_0} = \dim \mathcal{P}_0 = 1 \text{ and } \dim \widetilde{\mathcal{H}_1} = \dim \mathcal{P}_1 = n+1.$$

Q.E.D.

Since there are examples of closed Riemannian manifolds onto which the spheres cover or submerse with totally geodesic fibers, one can use Proposition 2.2 (see also Remark 2.1) in order to deduce their spectra. The simplest are the real projective spaces $(P^n(\mathbb{R}), g)$, where g is the canonical Riemannian metric on $P^n(\mathbb{R})$ (see Chapter 1, Example 2.15). Any $f \in C^\infty(S^n)$ that satisfies the property $f(x) = f(-x)$, for each $x \in S^n$, induces a unique function $\hat{f} \in C^\infty(P^n(\mathbb{R}))$ such that $\hat{f} \circ \pi = f$, where $\pi : S^n \to P^n(\mathbb{R})$ denotes the canonical projection. Consequently, the only eigenfunctions of Δ which descend from S^n to $P^n(\mathbb{R})$ are those whose degree of homogeneity is even. Thus one obtains

Corollary 2.1. The eigenvalues of the Laplace-Beltrami operator of $(P^n(\mathbb{R}), g)$ are given by $\lambda_k = 2k(n+2k-1), k \in \mathbb{N}$. The eigenspace corresponding to λ_k is given by $\{\hat{f} \in C^\infty(P^n(\mathbb{R})) \mid f \in \widetilde{\mathcal{H}_{2k}}\}$. In particular, the multiplicity m_k of λ_k is equal to

$$m_k = \begin{cases} 1 & \text{if } k=0, \\ \dfrac{(n+2k-2)!}{(n-1)!(2k)!}(n+4k-1) & \text{if } k \geq 1. \end{cases}$$

Using the Hopf fibration $\pi : S^{2n+1} \to P^n(\mathbb{C})$ (see Chapter 1, Example 2.14), one can see that the eigenfunctions of the Laplace-Beltrami operator associated to $P^n(\mathbb{C})$ are those eigenfunctions of the Laplace-Beltrami operator Δ of (S^{2n+1}, g_{can}) which are constant on the Hopf circles. These turn out to be the polynomials on $\mathbb{R}^{2n+2} = \mathbb{C}^{n+1}$ which are bihomogeneous of degree k in the complex coordinates z_j and in their conjugates $\bar{z}_j$. Therefore one gets that the eigenvalues are the "even" eigenvalues of Δ associated to (S^{2n+1}, g_{can}), i.e. the set $S = \{\lambda_k = 2k(2n+1+2k-1) = 4k(n+k) \mid k \in \mathbb{N}\}$. The multiplicity of λ_k is equal to

$$\begin{cases} 1 & \text{if } k=0, \\ n(n+2k)\left[\dfrac{n(n+1)\cdots(n+k-1)}{k!}\right]^2 & \text{if } k\geq 1 \end{cases}$$

(for more details, see Berger, Gauduchon and Mazet [1971]).

In a similar manner, one shows that the eigenvalues of the Laplace-Beltrami operator of $\left(P^n(\mathbb{H}), g\right)$ are given by $4k(2n+k+1), k \in \mathbb{N}$.

Similarly, the eigenfunctions of the Laplace-Beltrami operator of the lens space $\left(S^{2n+1}/G = L_{2n+1}(p;q_1,\ldots,q_{n+1}), g\right)$ (see Chapter 1, Example 2.16) lift to G-invariant eigenfunctions of the Laplace-Beltrami operator of the sphere $\left(S^{2n+1}, g_{can}\right)$. The eigenvalues of the Laplace-Beltrami operator of $\left(S^{2n+1}, g_{can}\right)$ are precisely the numbers $k(2n+k)$, $k \in \mathbb{N}$. Thus to determine the spectrum of the lens space $\left(L_{2n+1}(p;q_1,\ldots,q_{n+1}), g\right)$, one needs only to determine the dimension of the eigenspace corresponding to $k(2n+k)$, for each $k \in \mathbb{N}$. Consider the generating function

$$F_p(z;q_1,\ldots,q_{n+1}) = \sum_{k=0}^{\infty} m_k z^k,$$

where m_k is the multiplicity of $k(2n+k)$. Ikeda and Yamamoto [1979] showed that the generating function is rational:

$$F_p(z;q_1,\ldots,q_{n+1}) = \frac{1}{p}\sum_{l=1}^{p} \frac{1-z^2}{\prod_{j=1}^{n+1}\left(z-\gamma^{q_j l}\right)\left(z-\gamma^{-q_j l}\right)},$$

where γ is a primitive p-th root of unity.

Let us mention that one can calculate the spectra of compact Lie groups endowed with bi-invariant metrics by representation theory, and hence on symmetric spaces of compact type (see Cahn-Wolf [1976]). For example, one can show that the eigenvalues of the Laplace-Beltrami operator of the Cayley plane are given by $k(k+11), k \in \mathbb{N}$. Other examples for which the spectra can be explicitly computed are the Riemannian Heisenberg manifolds, that is Riemannian manifolds of the form $\left(H_n/\Gamma, g\right)$, where H_n is the $(2n+1)$-dimensional Heisenberg group, Γ is a uniform discrete subgroup, and g a Riemannian metric on H_n/Γ whose lift to H_n is left-invariant (see Chapter 1, Example 2.19). By Proposition 2.2 in Chapter 1 we may assume that g has the form

$$g = \begin{bmatrix} h & 0 \\ 0 & g_{2n+1} \end{bmatrix},$$

with h a positive-definite $2n \times 2n$ matrix and $g_{2n+1} > 0$. Let $\sum$ be the $2n \times 2n$ matrix $\begin{bmatrix} O & I_n \\ -I_n & O \end{bmatrix}$, where I_n is the $n \times n$ identity matrix. Since $h^{-1}\sum$ is similar to the skew-symmetric matrix $h^{-\frac{1}{2}}\sum h^{-\frac{1}{2}}$, it has pure imaginary eigenvalues. We denote them by $\pm\sqrt{-1}d_j^2, 1 \le j \le n$. For c a positive integer and $k = (k_1, \ldots, k_n)$ an n-tuple of non-negative integers, define

$$\lambda(c,k) = \frac{4\pi^2 c^2}{g_{2n+1}} + \sum_{j=1}^{n} 2\pi c d_j^2 (2k_j + 1).$$

Let $r = (r_1, \ldots, r_n) \in (\mathbb{Z}^+)^n$ be as in Example 2.19, Chapter 1. By $\sigma_2(r,g)$ we shall mean the collection of numbers λ which can be written in the form $\lambda(c,k)$, with the understanding that λ occurs $2c^n r_1 \ldots r_n$ times for each pair (c,k) such that $\lambda = \lambda(c,k)$.

Let us identify the subspace of matrices $\{A(x,y,0) | x, y \in \mathbb{R}^n\}$ of the Lie algebra of H_n (see the notations used in Example 2.19, Chapter 1) with $\mathbb{R}^{2n}$. Let $r = (r_1, \ldots, r_n) \in (\mathbb{Z}^+)^n$ and Γ_γ as in Example 2.19, Chapter 1, and define $\tilde{\Gamma}_r = \{A(x,y,0) | x \in r\mathbb{Z}^n, y \in \mathbb{Z}^n\}$. Then $\tilde{\Gamma}_r$ is a lattice in $\mathbb{R}^{2n}$ with lattice basis $(r_1 e_1, \ldots, r_n e_n, e_{n+1}, \ldots, e_{2n})$, where $(e_1, \ldots, e_{2n})$ is the standard basis for $\mathbb{R}^{2n}$.

The study of irreducible unitary representations of H_n leads to the following result.

Theorem 2.3. For g, r, Γ_γ and $\tilde{\Gamma}_r$ as before, the spectrum $\sigma(r,g)$ of $\left(H_n / \Gamma_r, g \right)$ is the join of the spectrum $\sigma_1(r,h)$ of the $2n$-dimensional flat torus $\left(\mathbb{R}^{2n} / \tilde{\Gamma}_r, h \right)$ and $\sigma_2(r,g)$ in the sense that the multiplicity of λ in $\sigma(r,g)$ is the sum of the multiplicities of λ in $\sigma_1(r,h)$ and $\sigma_2(r,g)$.

For a proof, see Gordon and Wilson [1986].

Also, let us mention that no techniques have been found for computing easily the explicit spectrum of a negatively curved Riemannian manifold, since the universal cover of such a manifold is non compact. However, a fair amount of numerical work has been carried out by number theorists (see, for example Hejhal [1983]).

3. CHARACTERIZING EIGENVALUES OF THE LAPLACE-BELTRAMI OPERATOR

Let (M,g) be a closed, n-dimensional smooth Riemannian manifold and $E(\Delta,\lambda)$ the space of eigenfunctions of the Laplace-Beltrami operator Δ with eigenvalue λ.

The real vector space $E(\Delta,\lambda)$ is finite dimensional and $E(\Delta,\lambda)$ is orthogonal to $E(\Delta,\mu)$ for $\lambda \neq \mu$ (see § 1).

Since

$$\langle \Delta f, f\rangle = \|df\|^2 = \|grad(f)\|^2$$

for any $f \in C^\infty(M)$ [see (2.23), Chapter 2], it follows that

$$\lambda = \frac{\|df\|^2}{\|f\|^2} \geq 0$$

for any $f \in E(\Delta, \lambda)\setminus\{0\}$, i.e. λ coincides with the Rayleigh-Ritz quotient

$$R(f) = \frac{\|df\|^2}{\|f\|^2}$$

of the eigenfunction $f \in E(\Delta, \lambda)\setminus\{0\}$ (cf. Rassias [1977], [1986] for a number of applications of the Rayleigh-Ritz quotient especially to the theory of minimal surfaces).

First, let us prove the following minimum characterization of eigenvalues of Δ.

Theorem 3.1. (Courant) Let

$$0 = \lambda_0 < \lambda_1 \leq \lambda_2 \leq \cdots$$

be the sequence of eigenvalues of Δ, each repeated according to its multiplicity, and $f_0, f_1, f_2, \ldots$ the corresponding complete orthonormal sequence of eigenfunctions of Δ. Then

$$\lambda_k = \inf_f \frac{\|df\|^2}{\|f\|^2},$$

where f ranges over the set A_k of all $f \in C^\infty(M)\setminus\{0\}$ such that f is orthogonal (in L^2-sense) to $f_0, f_1, \ldots, f_{k-1}$, $k \in \mathbb{N}^*$. Moreover, the equality

$$\lambda_k \|f\|^2 = \|df\|^2$$

holds if and only if $f \in E(\Delta, \lambda_k)$.

Proof. Let $\alpha_j = \langle f, f_j\rangle$, $j \in \mathbb{N}$ and fix $l \in \mathbb{N}^*$. Then one gets successively

$$0 \leq \left\| df - \sum_{j=0}^{l} \alpha_j df_j \right\|^2 = \|df\|^2 - 2\sum_{j=0}^{l} \alpha_j \langle df, df_j\rangle + \sum_{j=0}^{l} \alpha_j^2 \|df_j\|^2$$

$$= \|df\|^2 - 2\sum_{j=0}^{l} \alpha_j \langle f, \Delta f_j\rangle + \sum_{j=0}^{l} \alpha_j^2 \langle f_j, \Delta f_j\rangle$$

$$= \|df\|^2 - \sum_{j=0}^{l} \lambda_j \alpha_j^2,$$

that is

$$\|df\|^2 \geq \sum_{j=0}^{\infty} \lambda_j \alpha_j^2 . \tag{3.1}$$

If $f \in A_k$, then $\alpha_j = 0$ for $j = 1,\ldots,k-1$, so that

$$\sum_{j=0}^{\infty} \lambda_j \alpha_j^2 \geq \lambda_k \left(\sum_{j=0}^{\infty} \alpha_j^2 \right) = \lambda_k \|f\|^2. \tag{3.2}$$

Combining (3.1) and (3.2) one gets

$$\|df\|^2 \geq \lambda_k \|f\|^2 \text{ for any } f \in A_k.$$

Notice that the last inequality becomes equality for any $f \in E(\Delta, \lambda_k)$:

$$\|df\|^2 = \langle \Delta f, f \rangle = \lambda_k \|f\|^2.$$

Conversely, if

$$\|df\|^2 = \lambda_k \|f\|^2,$$

then $f \in E(\Delta, \lambda_k)$. Indeed, (3.2) is a strict inequality if at least one α_j is $\neq 0$, where the index j is such that $f_j \notin E(\Delta, \lambda_k)$. Consequently, if we have equality, it follows necessarily that in the Fourier series expansion of f one gets

$$f = \sum_{j, f_j \in E(\Delta, \lambda_k)} \alpha_j f_j,$$

that is $f \in E(\Delta, \lambda_k)$.

Q.E.D.

Remark 3.1. Theorem 3.1 remains also true in the case when M is with boundary.

Corollary 3.1. Let g, g' be two Riemannian metrics on the closed, n-dimensional smooth manifold M, $\lambda_k(g)$ [resp. $\lambda_k(g')$], $k \in \mathbb{N}$, be the k-th eigenvalue of the Laplace-Beltrami operator Δ_g (resp. $\Delta_{g'}$) corresponding to the Riemannian metric g (resp. g'). If

$$a^2 g' \leq g \leq b^2 g'$$

where $a, b \in \mathbb{R}_+^*$, $a < b$, then

$$\frac{a^n}{b^{n+2}} \lambda_k(g') \leq \lambda_k(g) \leq \frac{b^n}{a^{n+2}} \lambda_k(g')$$

for each $k \in \mathbb{N}$.

Proof. Since

$$a^n d\mu_{g'} \leq d\mu_g \leq b^n d\mu_{g'}$$

and

$$\frac{1}{b} |df|_{g'} \leq |df|_g \leq \frac{1}{a} |df|_{g'}$$

it follows that

$$\frac{a^n}{b^{n+2}}\frac{\|df\|_{g'}^2}{\|f\|_{g'}^2} \le \frac{\|df\|_g^2}{\|f\|_g^2} \le \frac{b^n}{a^{n+2}}\frac{\|df\|_{g'}^2}{\|f\|_{g'}^2},$$

where $\| \ \|_g$ (resp. $\| \ \|_{g'}$) denotes the canonical L^2-norms on the real vector spaces $C^\infty(M)$ and $A^1(M)$ induced by g (resp. g') (see Chapter 2, § 1, A, C). Then Theorem 3.1 leads us to the desired inequalities.

Q.E.D.

The above property tells us that if a sequence $g^{(j)}$ of Riemannian metrics on M converges to the Riemannian metric g in the C^0-topology, then $\lambda_k\left(g^{(j)}\right)$ converges to $\lambda_k(g)$ for each $k \in \mathbb{N}$.

Next we shall prove the following mini-max characterization of the k-th eigenvalue of Δ. Note that in this characterization are involved real vector subspaces which do not depend on the metric g.

Theorem 3.2. Let

$$0 = \lambda_0 < \lambda_1 \le \lambda_2 \le \cdots$$

be the sequence of the eigenvalues of Δ, each repeated according to its multiplicity. Then

$$\lambda_k = \inf_{V_{k+1}} \sup_{f \in V_{k+1}\setminus\{0\}} \left\{\frac{\|df\|^2}{\|f\|^2}\right\}, \qquad k \in \mathbb{N},$$

where V_{k+1} ranges over all (k+1)-dimensional linear subspaces of $C^\infty(M)$.

Proof. Let $f_0, f_1, f_2, \ldots$ be the corresponding complete orthonormal sequence for the completion $L^2(M)$ of $C^\infty(M)$ consisting of eigenfunctions of Δ. Let us denote

$$s(V_{k+1}) = \sup_{f \in V_{k+1}\setminus\{0\}} \left\{\frac{\|df\|^2}{\|f\|^2}\right\},$$

where V_{k+1} is a (k+1)-dimensional real linear subspace of $C^\infty(M)$. Let V_{k+1}^0 be the (k+1)-dimensional linear subspace of $C^\infty(M)$ generated by $f_0, f_1, \ldots, f_k$. Since $f_k \in V_{k+1}^0$ is an eigenfunction of Δ corresponding to the eigenvalue λ_k, it follows that

$$\lambda_k \|f_k\|^2 = \|df_k\|^2$$

(see Theorem 3.1), which implies that

$$s\left(V_{k+1}^0\right) \ge \lambda_k.$$

On the other hand, in virtue of Theorem 3.1 and of the fact that

$$0 < \lambda_1 \le \lambda_2 \le \cdots \le \lambda_k,$$

for any $f = \sum_{j=0}^{k} \alpha_j f_j \in V_{k+1}^0$ one gets

$$\|df\|^2 = \sum_{j=0}^{k} \alpha_j^2 \|df_k\|^2 = \sum_{j=0}^{k} \lambda_j \alpha_j^2 \|f_j\|^2 \le \lambda_k \|f\|^2,$$

so that

$$s\left(V_{k+1}^0\right) \le \lambda_k .$$

Consequently,

$$s\left(V_{k+1}^0\right) = \lambda_k ,$$

that is

$$\lambda_k \ge \inf_{V_{k+1}} \sup_{f \in V_{k+1} \setminus \{0\}} \left\{ \frac{\|df\|^2}{\|f\|^2} \right\}.$$

Now, let us assume that

$$\lambda_k > \inf_{V_{k+1}} \sup_{f \in V_{k+1} \setminus \{0\}} \left\{ \frac{\|df\|^2}{\|f\|^2} \right\}.$$

Then there exists a (k+1)-dimensional linear subspace V_{k+1}' of $C^\infty(M)$ such that

$$\lambda_k > s\left(V_{k+1}'\right).$$

But each $f = \sum_{j=0}^{\infty} \beta_j f_j \in V_{k+1}'$ satisfies the inequality

$$\|df\|^2 \le s\left(V_{k+1}'\right) \|f\|^2$$

and since

$$\|df\|^2 \ge \sum_{j=1}^{\infty} \lambda_j \beta_j^2$$

[see the inequality (3.1)] one obtains

$$\sum_{j=1}^{\infty} \lambda_j \beta_j^2 \le s\left(V_{k+1}'\right) \sum_{j=0}^{\infty} \beta_j^2 . \tag{3.3}$$

This inequality shows that there exists $j_0 \ge 1$ such that

$$\lambda_{j_0} \le s\left(V_{k+1}'\right).$$

Let us rewrite (3.3) under the form

$$\sum_{j, s\left(V_{k+1}'\right) < \lambda_j} \left[\lambda_j - s\left(V_{k+1}'\right)\right] \beta_j^2 \le \sum_{j, s\left(V_{k+1}'\right) \ge \lambda_j} \left[s\left(V_{k+1}'\right) - \lambda_j\right] \beta_j^2 . \tag{3.4}$$

Let $m = \max\left\{ j \middle| \lambda_j \le s\left(V_{k+1}'\right) \right\}$. If one defines the linear mapping

$$F : V'_{k+1} \to C^{\infty}(M) \text{ by } F(f) = \sum_{j=0}^{m} \langle f, f_j \rangle f_j, f \in V'_{k+1},$$

then $rank(F) < k+1$. Indeed,

$$\lambda_j \leq s(V'_{k+1}) < \lambda_k \text{ for each } j = 0, 1, \ldots, m,$$

which implies that

$$rank(F) \leq m+1 < k+1.$$

Consequently, the kernel rank of F is ≥ 1, so that there exists $a \in \ker F \setminus \{0\}$, that is $\langle a, f_j \rangle = 0$ for all j with the property

$$\lambda_j < s(V'_{k+1}).$$

On the other hand, applying (3.4) for $a \in V'_{k+1}$ it follows that each term of the left-hand side should be equal to zero. Hence $\beta_j = \langle a, f_j \rangle = 0$ for each j satisfying $\lambda_j \geq s(V'_{k+1})$, that is

$$a = \sum_{j=0}^{\infty} \langle a, f_j \rangle f_j = 0,$$

which contradicts $a \neq 0$.

Q.E.D.

4. GENERIC PROPERTIES OF THE RIEMANNIAN METRICS ON CLOSED SMOOTH MANIFOLDS

In this section we shall discuss two generic properties of the Riemannian metrics on closed smooth manifolds.

Let M be a connected closed smooth manifold of dimension n, $\mathcal{S}^2(M)$ the real vector space of all smooth symmetric tensor fields on M of type (2,0) and $\mathcal{M}$ the set of all smooth Riemannian metrics on M. First, let us introduce a complete distance d on $\mathcal{M}$ which gives the C^{∞}-topology on $\mathcal{M}$. In order to do this, we fix a finite covering $\{U_{\sigma}\}_{\sigma \in I}$ of M such that the closure of U_{σ} is contained in the open coordinate neighborhood V_{σ}. For $h \in \mathcal{S}^2(M)$, we denote by h_{ij} the components of h with respect to the coordinates $(x'_1, \ldots, x_n)$ on V_{σ}, $\sigma \in I$. For every $\sigma \in I$ and non-negative integer k, put

$$\|h\|_{\sigma,k} = \sup_{U_{\sigma}} \sum_{|\alpha| \leq k} \sum_{i,j=1}^{n} \left| \partial^{|\alpha|} h_{ij} \Big/ \partial x_1^{\alpha_1} \ldots \partial x_n^{\alpha_n} \right|,$$

where $\alpha = (\alpha_1, \ldots, \alpha_n)$ denotes an n-tuple of non-negative integers α_i and $|\alpha| = \alpha_1 + \cdots + \alpha_n$. Let us define a norm $\| \ \|$ on $\mathcal{S}^2(M)$ by

$$\|h\| = \sum_{k=0}^{\infty} 2^{-k} \|h\|_k \left(1 + \|h\|_k\right)^{-1}, \quad h \in \mathcal{S}^2(M),$$

where

$$\|h\|_k = \sum_{\sigma \in I} \|h\|_{\sigma,k},$$

and let us consider the corresponding distance d' on $\mathscr{S}^2(M)$ given by

$$d'(h_1,h_2) = \|h_2 - h_1\|, h_1, h_2 \in \mathscr{S}^2(M).$$

Now, let us define the distance d'' on $\mathcal{M}$ by

$$d''(g_1,g_2) = \sup_{x \in M} d''_x((g_1)_x,(g_2)_x), \quad g_1,g_2 \in \mathcal{M},$$

where

$$d''_x((g_1)_x,(g_2)_x) = \inf\left\{\delta > 0 \middle| e^{-\delta}(g_1)_x < (g_2)_x < e^{\delta}(g_1)_x\right\}$$

and, for $g', g'' \in \mathcal{M}, g'_x < g''_x$ means that the symmetric bilinear form $g''_x - g'_x : T_xM \times T_xM \to \mathbb{R}$ is positive definite. Then $d : \mathcal{M} \times \mathcal{M} \to [0,+\infty)$ given by $d = d' + d''$ is a distance on $\mathcal{M}$, the metric space $(\mathcal{M},d)$ is complete and the corresponding canonical topology on $\mathcal{M}$ is just the C^∞-topology (for more details, see Bando and Urakawa [1983]).

For $g \in \mathcal{M}$, let $\Delta_g : C^\infty(M) \to C^\infty(M)$ be the Laplace-Beltrami operator associated to (M,g) and

$$0 = \lambda_0(g) < \lambda_1(g) \le \lambda_2(g) \le \cdots \to \infty$$

the eigenvalues of Δ_g counted with their multiplicities. Regarding each eigenvalue $\lambda_k(g), k \in \mathbb{N}$, as a function of $g \in \mathcal{M}$, it depends continuously on g in $\mathcal{M}$ with respect to the C^∞-topology (see also Chapter 5, Proposition 3.2). More precisely, we have

Theorem 4.1. For each real number $\delta > 0$ and each $g, g' \in \mathcal{M}$, the inequality $d(g,g') < \delta$ implies

$$\exp(-(n+1)\delta) \le \frac{\lambda_k(g)}{\lambda_k(g')} \le \exp((n+1)\delta)$$

for each $k \in \mathbb{N}^*$. In particular each eigenvalue $\lambda_k(g)$ depends continuously on g in $\mathcal{M}$ with respect to the C^∞-topology.

Proof. For any fixed $\delta > 0$ and $g \in \mathcal{M}$, let $V_\delta(g)$ be the set $\left\{g' \in \mathcal{M} \middle| d''(g,g') < \delta\right\}$. Let $(x_1,\ldots,x_n)$ be a local coordinate system on an open subset U of M. For each $g' \in V_\delta(g)$, the symmetric positive definite bilinear forms defined by the matrices (g_{ij}) and (g'_{ij}) of the components of g and g' on U respectively with respect to this local coordinate system satisfy

$$\exp(-\delta)(g'_{ij}) < (g_{ij}) < \exp(\delta)(g'_{ij}). \tag{4.1}$$

Note that if A,B are real symmetric positive definite matrices of order n such that $A < B$, then $B^{-1} < A^{-1}$ and $\det(A) < \det(B)$. Consequently one gets

$$\exp(-\delta)(g'^{ij}) < (g^{ij}) < \exp(\delta)(g'^{ij})$$

and

$$\exp\left(-\frac{n}{2}\delta\right)\sqrt{\det(g'_{ij})} < \sqrt{\det(g_{ij})} < \exp\left(\frac{n}{2}\delta\right)\sqrt{\det(g'_{ij})},$$

where (g^{ij}) [resp. (g'^{ij})] is the inverse matrix of the matrix (g_{ij}) [resp. (g'_{ij})]. Hence for each $f \in C^\infty(M)$ with the support contained in U one obtains successively

$$\begin{aligned}\|f\|_g^2 &= \int_M f^2(x)\,d\mu_g(x) = \int_M f^2(x)\sqrt{\det(g_{ij})}\,dx_1\ldots dx_n \\ &\le \exp\left(\frac{n}{2}\delta\right)\int_M f^2(x)\sqrt{\det(g'_{ij})}\,dx_1\ldots dx_n \\ &= \exp\left(\frac{n}{2}\delta\right)\int_M f^2(x)\,d\mu_{g'}(x) = \exp\left(\frac{n}{2}\delta\right)\|f\|_{g'}^2\end{aligned}$$

and similarly

$$\|f\|_g^2 \ge \exp\left(-\frac{n}{2}\delta\right)\|f\|_{g'}^2.$$

Hence

$$\exp\left(-\frac{n}{2}\delta\right)\|f\|_{g'}^2 \le \|f\|_g^2 \le \exp\left(\frac{n}{2}\delta\right)\|f\|_{g'}^2 \tag{4.2}$$

for any $f \in C^\infty(M)$ with $\operatorname{supp}(f) \subset U$.

On the other hand, for each $\omega \in A^1(M)$ with the support contained in U one obtains successively

$$\begin{aligned}\|\omega\|_g^2 &= \int_M \sum_{i,j=1}^n g^{ij}\omega_i\omega_j\,d\mu_g = \int_M \sum_{i,j=1}^n g^{ij}\omega_i\omega_j\sqrt{\det(g_{ij})}\,dx_1\ldots dx_n \\ &\le \exp(\delta)\exp\left(\frac{n}{2}\delta\right)\int_M \sum_{i,j=1}^n g'^{ij}\omega_i\omega_j\sqrt{\det(g'_{ij})}\,dx_1\ldots dx_n \\ &= \exp\left(\left(\frac{n}{2}+1\right)\delta\right)\int_M \sum_{i,j=1}^n g'^{ij}\omega_i\omega_j\,d\mu_{g'} = \exp\left(\left(\frac{n}{2}+1\right)\delta\right)\|\omega\|_{g'}^2,\end{aligned}$$

where $\omega_1,\ldots,\omega_n$ are the components of the 1-form ω with respect to the above local coordinate system. Similarly,

$$\|\omega\|_g^2 \ge \exp\left(-\left(\frac{n}{2}+1\right)\delta\right)\|\omega\|_{g'}^2$$

and consequently

$$\exp\left(-\left(\frac{n}{2}+1\right)\delta\right)\|\omega\|_{g'}^2 \le \|\omega\|_g^2 \le \exp\left(\left(\frac{n}{2}+1\right)\delta\right)\|\omega\|_{g'}^2 \tag{4.3}$$

for each $\omega \in A^1(M)$ with $\operatorname{supp}(\omega) \subset U$. But the inequalities (4.2) and (4.3) are valid – via a smooth partition of unity on M – for every $f \in C^\infty(M)$ and $\omega \in A^1(M)$, so that the inequalities

$$\exp(-(n+1)\delta)\frac{\|df\|_{g'}^2}{\|f\|_{g'}^2} \le \frac{\|df\|_g^2}{\|f\|_g^2} \le \exp((n+1)\delta)\frac{\|df\|_{g'}^2}{\|f\|_{g'}^2}$$

are valid for every $f \in C^\infty(M)\setminus\{0\}$. Therefore, by Theorem 3.2 one gets

$$\exp(-(n+1)\delta)\lambda_k(g') \le \lambda_k(g) \le \exp((n+1)\delta)\lambda_k(g') \tag{4.4}$$

for each $k \in \mathbb{N}^*$. In particular, the inequalities (4.4) imply the continuity of the real functions $\lambda_k, k \in \mathbb{N}^*$: $g' \in V_\delta(g)$ implies

$$|\lambda_k(g') - \lambda_k(g)| \le [\exp((n+1)\delta) - 1]\lambda_k(g) \tag{4.5}$$

for each $k \in \mathbb{N}^*$.

Q.E.D.

The proof of Theorem 4.1 shows that $d(g, g') < \delta$ implies that

$$\exp(-(n+1)\delta) \le \frac{\lambda_k(g)}{\lambda_k(g')} \le \exp((n+1)\delta)$$

for each $k \in \mathbb{N}^*$, that is if the Riemannian metrics g and g' are close to each other with respect to the distance d, then the ratio $\lambda_k(g)/\lambda_k(g')$ is close to 1 uniformly in $k \in \mathbb{N}^*$. This leads immediately to the following consequence.

Corollary 4.1. For each $k \in \mathbb{N}^*$, the multiplicity

$$m_k(g) = \#\{j \,|\, \lambda_j(g) = \lambda_k(g)\}$$

of the eigenvalue $\lambda_k(g)$ depends upper-semicontinuously on $g \in \mathcal{M}$: for each $g \in \mathcal{M}$ and $k \in \mathbb{N}^*$, there exists $\delta > 0$ such that $d(g, g') < \delta$ implies $m_k(g') \le m_k(g)$.

Remark 4.1. For each $g \in \mathcal{M}$, let

$$0 = \lambda_0(g) < \lambda_1(g) < \lambda_2(g) < \cdots \to +\infty$$

be the sequence of the distinct eigenvalues of Δ_g and $m_k(g)$ be the multiplicity of the eigenvalue $\lambda_k(g)$. It is known that if $n = \dim M \ge 3$, then m_k is unbounded on the space of smooth Riemannian metrics of volume 1 on the closed manifold M for each fixed $k \in \mathbb{N}^*$ (for example, see Colin de Verdière [1987]). Furthermore, for compact rank-one symmetric spaces [(S^n, g_{can}), $(P^n(\mathbb{R}), g_{can})$, $(P^n(\mathbb{C}), g_{can})$, $(P^n(\mathbb{H}), g_{can})$ and the Cayley projective plane $\left(P^2(\mathbb{C}a), g_{can}\right)$] one has the asymptotic property $m_k(g) = ak^{n-1} + O\left(k^{n-2}\right)$ for certain $a > 0$. Zelditch [1996] proves that if (M, g) is an n-dimensional Riemannian manifold for which $m_k(g)$ satisfies this asymptotic relation for any $k \in \mathbb{N}^*$, then the geodesic flow associated to (M, g), i.e. the Hamilton flow of the norm function, is periodic. Such a metric g is called Zoll metric. The spectrum of the Laplace-Beltrami operator associated to a Zoll metric is a union of eigenvalue clusters which can be described quite precisely.

For each $k \in \mathbb{N}^*$, let $\mathcal{S}_k$ be the set of all Riemannian metrics $g \in \mathcal{M}$ for which the first k eigenvalues of Δ_g are simple, that is they have multiplicity one, i.e.

$$\mathcal{S}_k = \left\{ g \in \mathcal{M} \middle| 0 < \lambda_1(g) < \cdots < \lambda_k(g) \right\}.$$

Corollary 4.2. For each $k \in \mathbb{N}^*$, $\mathcal{S}_k$ is open in $(\mathcal{M}, d)$.

Proof. First note that $\mathcal{S}_k \neq \varnothing$ (see Remark 1.6), so it is sufficient to show that for each $g \in \mathcal{S}_k$ there exists $\delta > 0$ such that $V_\delta(g) \subset \mathcal{S}_k$. Let

$$\delta' = \min\left\{ \lambda_2(g) - \lambda_1(g), \ldots, \lambda_k(g) - \lambda_{k-1}(g) \right\} > 0$$

and let us choose $\delta > 0$ so small that

$$\delta' \left[2\lambda_k(g) \right]^{-1} > \exp\left((n+1)\delta \right) - 1.$$

By the inequalities (4.5), for each $g' \in V_\delta(g)$ and $j = 0, 1, \ldots, k-1$, one obtains

$$\begin{aligned} \delta' &\leq \lambda_{j+1}(g) - \lambda_j(g) \leq \left|\lambda_{j+1}(g) - \lambda_{j+1}(g')\right| + \left|\lambda_{j+1}(g') - \lambda_j(g')\right| + \left|\lambda_j(g') - \lambda_j(g)\right| \\ &\leq \left[\exp\left((n+1)\delta - 1\right)\right]\left[\lambda_{j+1}(g) + \lambda_j(g)\right] + \left|\lambda_{j+1}(g') - \lambda_j(g')\right| \\ &\leq 2\left[\exp\left((n+1)\delta - 1\right)\right]\lambda_k(g) + \left|\lambda_{j+1}(g') - \lambda_j(g')\right|, \end{aligned}$$

so that

$$0 < \delta' - 2\left[\exp\left((n+1)\delta\right) - 1\right]\lambda_k(g) \leq \left|\lambda_{j+1}(g') - \lambda_j(g')\right|$$

for $j - 0, 1, \ldots, k-1$, which implies that $g' \in \mathcal{S}_k$ and therefore $V_\delta(g) \subset \mathcal{S}_k$.

Q.E.D.

Next, following Bando and Urakawa [1983], we shall prove that the set of all $g \in \mathcal{M}$ all of whose eigenvalues of Δ_g have multiplicity one is a residual set in $(\mathcal{M}, d)$. Before doing this one needs some preparatory material concerning the splitting of the eigenvalues under certain deformations of $g \in \mathcal{M}$. First let us fix some notations.

Let $C^\infty(M;\mathbb{C})$ be the space of all complex valued smooth functions on M. For a fixed Riemannian metric g on M, let $\langle\ ,\ \rangle_g$ be the inner product on $C^\infty(M;\mathbb{C})$ defined by

$$\langle f_1, f_2 \rangle_g = \int_M f_1(x)\overline{f_2(x)}\,d\mu_g, \quad f_1, f_2 \in C^\infty(M;\mathbb{C}),$$

where $d\mu_g$ is the canonical measure associated to (M,g) (see Chapter 2, §1). For every non-negative integer s, let $H^s(M)$ be the Sobolev space on M, which is the completion of $C^\infty(M;\mathbb{C})$ with respect to the following inner product $\langle\ ,\ \rangle_s$:

$$\langle f_1, f_2\rangle_s = \left\langle \left(I+\Delta_g\right)^s f_1, f_2\right\rangle_g, \quad f_1, f_2 \in C^\infty(M;\mathbb{C}),$$

where I is the identity operator and $\left(I+\Delta_g\right)^s$ the power of order s of the operator $I+\Delta_g$ (see also Chapter 2, §2, C). Put $\|f\|_s = \sqrt{\langle f, f\rangle_s}$ for $f \in H^s(M)$.

Now let us recall the definitions of the real analytic family of vectors and of the real analytic family of bounded operators. Let X,Y be complex Banach spaces and D a domain in $\mathbb{R}$.

Definition 4.1. A family of vectors x_t in X, $t \in D$, is said to be real analytic if it can be expanded as a convergent power series, i.e. for an arbitrary fixed $t_0 \in D$, there exist a sequence x_k, $k \in \mathbb{N}$, in X such that

$$x_t = \sum_{k=0}^{\infty} \left(t-t_0\right)^k x_k$$

for every $t \in D, |t-t_0| < \varepsilon$, where the series converges in the strong topology of X.

Definition 4.2. A family of bounded operators $A_t : X \to Y, t \in D$, is said to be real analytic if it can be expanded as a convergent power series of bounded operators, i.e. for an arbitrary fixed $t_0 \in D$, there exist a sequence of bounded operators $T_k : X \to Y, k \in \mathbb{N}$, such that

$$A_t = \sum_{k=0}^{\infty} \left(t-t_0\right)^k T_k$$

for every $t \in D, |t-t_0| < \varepsilon$, where the series converges in the uniform topology.

The following lemma gives an example of real analytic family of bounded operators.

Lemma 4.1. Let L_t, $|t| < \varepsilon$, be a family of differential operators of order m which can be expressed locally as

$$L_t = \sum_{|\alpha| \le m} a_\alpha(x,t) D_x^\alpha,$$

where $D_x^\alpha = \partial^{|\alpha|} \big/ \partial x_1^{\alpha_1} \cdots \partial x_n^{\alpha_n}$, $\alpha = (\alpha_1, \ldots, \alpha_n)$ being an n-tuple of non-negative integers, $|\alpha| = \alpha_1 + \cdots + \alpha_n$, and $a_\alpha(x,t)$ is real analytic in t, $|t| < \varepsilon$, where x belongs to an open subset of M on which local coordinates are defined. Then the family of bounded operators $L_t : H^m(M) \to H^0(M)$ is real analytic.

Proof. By assumption, $a_\alpha(x,t)$ can be expressed as $a_\alpha(x,t) = \sum_{k=0}^{\infty} a_{\alpha,k}(x) t^k$, where $a_{\alpha,k}(x)$ satisfy the inequalities

$$\left|a_{\alpha,k}(x)\right| \le C r^k \quad \text{for all } \alpha, |\alpha| \le m, k \in \mathbb{N} \text{ and } x,$$

where the constants $C>0$ and $r>0$ do not depend on α, k and x. Using a partition of unity on M, define the sequence of differential operators $L_k, k\in\mathbb{N}$, of order m which can be expressed locally as

$$L_k=\sum_{|\alpha|\le m} a_{\alpha,k}(x)D_x^\alpha .$$

Since L_k satisfy the inequalities

$$\|L_k(f)\|_0\le m^n C' r^k\|f\|_m,\ f\in H^m(M),$$

for a certain constant $C'>0$, they are bounded operators of $H^m(M)$ into $H^0(M)$ and the series $\sum_{k=0}^{\infty} t^k L_k$ converges to L_t in the uniform topology.

Q.E.D.

The next statement gives some properties of certain real analytic family of bounded operators, which will be usefull later.

Lemma 4.2. Let D be a small bounded domain in $\mathbb{R}$ containing the origin and $s_1>s_0$ non-negative integers. Let $A_t: H^{s_1}(M)\to H^{s_0}(M), t\in D$, be a real analytic family of bounded operators such that each $A_t, t\in D$, satisfies $\langle A_t(f_1),f_2\rangle_{s_0}=\langle f_1,A_t(f_2)\rangle_{s_0}$ for all $f_1,f_2\in H^{s_1}(M)$, and there exists a constant $C>0$ such that $\langle A_0(f),f\rangle_{s_0}\ge C\langle f,f\rangle_{s_0}$ for all $f\in H^{s_1}(M)$, that is A_0 is bounded below. Let λ be an eigenvalue of A_0. Then the kernel of $A_0-\lambda I$ is finite dimensional and there exists a subdomain D' of D containing the origin, $l=\dim\ker(A_0-\lambda I)$ real analytic families of vectors $f_1(t),\ldots,f_l(t)$ in $H^{s_1}(M)$ and l real analytic real valued functions $\lambda_1(t),\ldots,\lambda_l(t),\ t\in D'$, such that

$$A_t f_j(t)=\lambda_j(t) f_j(t),\ j=1,\ldots l,\ t\in D', \tag{4.6}$$

$$\langle f_j(t),f_k(t)\rangle_{s_0}=\delta_{jk},\ j,k=1,\ldots l,\ t\in D' \tag{4.7}$$

and

$$\lambda_j(0)=\lambda,\ j=1,\ldots l. \tag{4.8}$$

For the proof of Lemma 4.2 and some comments of it, see Bando and Urakawa [1983].

Now let $g_t, |t|<\varepsilon$, be a one-parameter family of Riemannian metrics on M depending real analytically on the parameter t. In the following, we denote merely by Δ_t [resp. $\langle\ ,\ \rangle_t$] the Laplace-Beltrami operator [resp. the inner product $\langle\ ,\ \rangle_{g_t}$ on $C^\infty(M;\mathbb{C})$] associated to g_t. Then one has:

Theorem 4.2. Let $g_t, |t|<\varepsilon$, be a one-parameter family of Riemannian metrics on M depending real analytically on t. For any eigenvalue λ of Δ_0 with multiplicity l, there exist $\varepsilon'>0, \varepsilon'<\varepsilon$, and l families $f_1(t),\ldots,f_l(t)\in C^\infty(M;\mathbb{C})$, which are real analytic in $H^0(M)=L^2(M)$ in the sense of Definition 4.1 and l real analytic (with respect to the parameter t) real valued functions $\lambda_1(t),\dots,\lambda_l(t)$ such that

$$\Delta_t f_j(t) = \lambda_j(t) f_j(t) \text{ for } j = 1,\dots,l \text{ and for } t \text{ with } |t| < \varepsilon', \tag{4.9}$$

$$\langle f_j(t), f_k(t)\rangle_t = \delta_{jk} \text{ for } j,k = 1,\dots,l \text{ and for } t \text{ with } |t| < \varepsilon', \tag{4.10}$$

and

$$\lambda_j(0) = \lambda \text{ for } j = 1,\dots,l. \tag{4.11}$$

Proof. For a function f on M and for t with $|t| < \varepsilon$, put

$$(U_t f)(x) = \sqrt[4]{\det(g_{ij}(x,0)) \Big/ \det(g_{ij}(x,t))}\, f(x), \quad x \in M,$$

where $(g_{ij}(\ ,t)), |t| < \varepsilon$, are the components of g_t with respect to the local coordinate system $(x^1,\dots,x^n)$ around x. Then $U_t : H^s(M) \to H^s(M), |t| < \varepsilon$, is a real analytic family of bounded operators in the sense of Definition 4.2 for every non-negative integer s. By definition of U_t, it is an isometry of the Hilbert space $(H^0(M), \langle\ ,\ \rangle_t)$ into the Hilbert space $(H^0(M), \langle\ ,\ \rangle_0)$. Since the Laplace-Beltrami operator $\Delta_t, |t| < \varepsilon$, is a self-adjoint operator on $H^0(M)$ with respect to the inner product $\langle\ ,\ \rangle_t$, the operator $\tilde{\Delta}_t = U_t \circ \Delta_t \circ U_t^{-1}$ is self-adjoint with respect to the inner product $\langle\ ,\ \rangle_0$. Moreover, by Lemma 4.1, $A_t = \tilde{\Delta}_t + I : H^2(M) \to H^0(M), |t| < \varepsilon$, is a real analytic family of bounded operators in the sense of Definition 4.2 and satisfies the hypotheses of Lemma 4.2. Therefore by Lemma 4.2 there exist a $\varepsilon' > 0, \varepsilon' < \varepsilon$, and l families $\tilde{f}_1(t),\dots,\tilde{f}_l(t)$, which are real analytic in $H^2(M)$ in the sense of Definition 4.1, and l real analytic (with respect to the parameter t) real valued functions $1+\lambda_1(t),\dots,1+\lambda_l(t)$ satisfying the conditions corresponding to (4.6), (4.7) and (4.8). Then the vectors $f_j(t), j = 1,\dots,l$, in $H^0(M)$ defined by $f_j(t) = U_t^{-1}\tilde{f}_j(t)$ satisfy $\Delta_t f_j(t) = \lambda_j(t) f_j(t), |t| < \varepsilon'$, in the sense of distributions and the condition (4.10). By hypoellipticity of $\Delta_t, f_1(t),\dots,f_l(t)$ belong to $C^\infty(M;\mathbb{C})$ and satisfy (4.9).

Q.E.D.

In the following, let us consider a particular deformation $g(t)$ of $g \in \mathcal{M}$ given by

$$g(t) = (1 + t\sigma)g, \text{ where } \sigma \in C^\infty(M). \tag{4.12}$$

For small enough $\varepsilon(\sigma) > 0$, one has $g(t) \in \mathcal{M}$ for all t with $|t| < \varepsilon(\sigma)$.

Now let λ be a non-zero eigenvalue of Δ_g with multiplicity l. Applying Theorem 4.2 to $g(t)$, one obtains the existence of $\lambda_j(t) \in \mathbb{R}$ and $f_j(t) \in C^\infty(M)$, $j = 1,\dots,l$, that depend real analytically on t, $|t| < \varepsilon(\sigma)$, and satisfying the conditions:

$$\Delta_{g(t)} f_j(t) = \lambda_j(t) f_j(t) \text{ for each } j = 1,\dots,l \text{ and for } t \text{ with } |t| < \varepsilon(\sigma), \tag{4.13}$$

$$\left\langle f_j(t), f_k(t)\right\rangle_{g(t)} = \delta_{jk} \text{ for each } j,k=1,\ldots,l \text{ and for } t \text{ with } |t|<\varepsilon(\sigma), \tag{4.14}$$

and

$$\lambda_j(0)=\lambda \text{ for each } j=1,\ldots,l. \tag{4.15}$$

Since $\lambda_j(t)$ and $f_j(t)$, $j=1,\ldots,l$, depend real analytically on t, it follows that

$$\lambda_j(t)=\lambda+t\alpha_j+t^2\beta_j(t) \text{ for } |t|<\varepsilon(\sigma),\ j=1,\ldots,l, \tag{4.16}$$

where $\alpha_1,\ldots,\alpha_l$ are real constants, $\beta_1(t),\ldots,\beta_l(t)$ are real analytic (in t) real valued functions, and

$$\left\langle f_j(t),u\right\rangle_g,\ j=1,\ldots,l, \text{ are real analytic functions in } t, |t|<\varepsilon(\sigma), \tag{4.17}$$

for every $u\in C^\infty(M)$. Then one has the following

Lemma 4.3. Let λ be a non-zero eigenvalue of Δ_g with multiplicity l. For $\sigma\in C^\infty(M)$, let $g(t)$ be the deformation of $g\in\mathcal{M}$ given by (4.12). Let $f_1(t),\ldots,f_l(t)\in C^\infty(M)$ as above satisfying (4.13)-(4.15) and $\alpha_1,\ldots,\alpha_l$ the real constants given by (4.16). Then one has

$$\left\langle\left[\left(1-\frac{n}{2}\right)grad_g(\sigma)-\lambda\sigma\right]f_j,f_k\right\rangle_g=\alpha_j\delta_{jk},\ 1\le j,k\le l,$$

where $f_j=f_j(0)$, $j=1,\ldots,l$.

Proof. Let us apply the equality (2.21) of Chapter 2 to $g(t)=a(t)g$ with $a(t)=1+t\sigma>0$ for $|t|<\varepsilon(\sigma)$. Consequently, (4.13) leads to

$$a(t)\Delta_g f_j(t)+\left(1-\frac{n}{2}\right)grad_g(a)f_j(t)-a^2(t)\lambda_j(t)f_j(t)=0,\ j=1,\ldots,l, |t|<\varepsilon(\sigma),$$

which imply

$$\left\langle a(t)\Delta_g f_j(t)+\left(1-\frac{n}{2}\right)t\ grad_g(\sigma)f_j(t)-a^2(t)\lambda_j(t)f_j(t),u\right\rangle_g=0,\ j=1,\ldots,l, |t|<\varepsilon(\sigma),$$

for every $u\in C^\infty(M)$. Differentiating both sides of the above equality at $t=0$, one obtains by (4.16) and (4.17)

$$\left\langle\left[\left(1-\frac{n}{2}\right)grad_g(\sigma)-\lambda\sigma-\alpha_j\right]f_j+(\Delta_g-\lambda I)u_j,u\right\rangle_g=0,\ \ j=1,\ldots,l,$$

for every $u\in C^\infty(M)$, where $u_j=f_j'(0)$. In particular, for an eigenfunction $u\in E(\Delta_g,\lambda)$, one gets

$$\left\langle \left[\left(1-\frac{n}{2}\right) grad_g(\sigma) - \lambda\sigma - \alpha_j \right](f_j), u \right\rangle_g = -\langle (\Delta_g - \lambda I)(u_j), u \rangle_g$$
$$= -\langle u_j, (\Delta_g - \lambda I)(u) \rangle_g = 0, \ j = 1, \ldots, l,$$

which implies the desired equality.

Q.E.D.

Lemma 4.4. Assume $n = \dim M \geq 2$. With the above notations, there exists a real smooth function $\sigma \in C^\infty(M)$ such that at least two of $\alpha_1, \ldots, \alpha_l$ in (4.16) are distinct.

Proof. Let $P : C^\infty(M) \to E(\Delta_g, \lambda)$ be the orthogonal projection, where $E(\Delta_g, \lambda)$ is the eigenspace of Δ_g corresponding to the eigenvalue λ. For $\sigma \in C^\infty(M)$, define the operator $G_\sigma : E(\Delta_g, \lambda) \to E(\Delta_g, \lambda)$ by

$$G_\sigma(f) = P\left[\left(1-\frac{n}{2}\right) grad_g(\sigma)(f) - \lambda\sigma f\right], \ f \in E(\Delta_g, \lambda).$$

If $(f_1, \ldots, f_l)$ is the orthonormal basis of $E(\Delta_g, \lambda)$ with respect to $\langle\ ,\ \rangle_g$ considered in Lemma 4.3, then

$$\langle G_\sigma(f_j), f_k \rangle_g = \left\langle \left(1-\frac{n}{2}\right) grad_g(\sigma)(f_j) - \lambda\sigma f_j, f_k \right\rangle_g = \alpha_j \delta_{jk}, 1 \leq j, k \leq l,$$

by Lemma 4.3. Thus the matrix of the operator G_σ with respect to the orthonormal basis $(f_1, \ldots, f_l)$ is a diagonal one whose diagonal entries are $\alpha_1, \ldots, \alpha_l$.

Assume that $\alpha_1 = \cdots = \alpha_l$. Then the matrix of G_σ with respect to this basis can be expressed as a constant multiple of I_l (= the identity matrix of order l), and hence the same property of G_σ holds with respect to any basis of $E(\Delta_g, \lambda)$. Therefore, in order to prove Lemma 4.4, one has only to find $\sigma \in C^\infty(M)$ such that $\langle G_\sigma(f_1), f_2 \rangle_g \neq 0$.

Note that for $\sigma \in C^\infty(M)$, by Corollary 2.4 (ii) of Chapter 2, one gets

$$\begin{aligned} \langle G_\sigma(f_1), f_2 \rangle_g &= \left\langle P\left[\left(1-\frac{n}{2}\right) grad_g(\sigma)(f_1) - \lambda\sigma f_1\right], f_2 \right\rangle_g \\ &= \left\langle \left(1-\frac{n}{2}\right) grad_g(\sigma)(f_1) - \lambda\sigma f_1, f_2 \right\rangle_g \qquad (4.18) \\ &= \left\langle \sigma, \left(1-\frac{n}{2}\right) \delta(f_2 df_1) - \lambda f_1 f_2 \right\rangle_g . \end{aligned}$$

Case 1. $\left(1-\frac{n}{2}\right) \delta(f_2 df_1) - \lambda f_1 f_2 \neq 0$.

In this case, putting $\sigma = \left(1-\frac{n}{2}\right) \delta(f_2 df_1) - \lambda f_1 f_2$, one obtains

$$\langle G_\sigma(f_1), f_2 \rangle_g \neq 0.$$

Case 2. $\left(1-\frac{n}{2}\right)\delta(f_2 df_1) - \lambda f_1 f_2 = 0$.

In this case, one gets

$$f_1 f_2 = 0. \tag{4.19}$$

Indeed, this is obvious if $n = 2$. Assuming $n > 2$, one has successively

$$\begin{aligned}\left[\left(1-\frac{n}{2}\right)\Delta_g - 2\lambda I\right](f_1 f_2) &= \left(1-\frac{n}{2}\right)\delta d(f_1 f_2) - 2\lambda f_1 f_2 \\ &= \left(1-\frac{n}{2}\right)\delta(f_1 df_2 + f_2 df_1) - 2\lambda f_1 f_2 \\ &= \left[\left(1-\frac{n}{2}\right)\delta(f_1 df_2) - \lambda f_1 f_2\right] + \left[\left(1-\frac{n}{2}\right)\delta(f_2 df_1) - \lambda f_1 f_2\right] = 0,\end{aligned}$$

by Corollary 2.4 (i) of Chapter 2 and the hypothesis. Since $2-n<0$, if $f_1 f_2 \neq 0$, then Δ_g would have a negative eigenvalue, which is a contradiction (see Chapter 2, Remark 2.2). Therefore, (4.19) is valid.

Let us take

$$u_1 = \frac{1}{2}(f_1 + f_2), u_2 = \frac{1}{2}(f_1 - f_2), u_3 = f_3, \ldots, u_l = f_l,$$

as an orthonormal basis of $E(\Delta_g, \lambda)$ with respect to $\langle\ ,\ \rangle_g$. Put $\sigma = \left(1-\frac{n}{2}\right)\delta(u_2 du_1) - \lambda u_1 u_2$. Then by (4.18) one has

$$\langle G_\sigma(u_1), u_2 \rangle_g = \langle \sigma, \sigma \rangle_g = \int_M \sigma^2 d\mu_g.$$

So we have only to prove $\sigma \neq 0$. Otherwise, by Corollary 2.4 (i) of Chapter 2, one gets successively

$$\begin{aligned}0 = 4\sigma &= \left(1-\frac{n}{2}\right)\delta\big((f_1 - f_2)d(f_1 + f_2)\big) - \lambda(f_1 + f_2)(f_1 - f_2) \\ &= \left(1-\frac{n}{2}\right)\big(\delta(f_1 df_1) - \delta(f_2 df_2)\big) - \lambda\left(f_1^2 - f_2^2\right) \\ &= \left[\frac{1}{4}(2-n)\delta d - \lambda I\right]\left(f_1^2 - f_2^2\right) \\ &= \left[\frac{1}{4}(2-n)\Delta_g - \lambda I\right]\left(f_1^2 - f_2^2\right).\end{aligned}$$

Since $2-n \leq 0$, it follows that $f_1^2 - f_2^2 = 0$. Therefore, by (4.19) one obtains

$$0 = \int_M \left(f_1^2 - f_2^2\right)^2 d\mu_g = \int_M \left(f_1^4 - 2f_1^2 f_2^2 + f_2^4\right) d\mu_g = \int_M \left(f_1^4 + f_2^4\right) d\mu_g,$$

which is a contradiction. Consequently, $\sigma \neq 0$.

Q.E.D.

Let $\mathcal{S}_k$ be the set of all elements $g \in \mathcal{M}$ for which the first k eigenvalues have multiplicity one, that is

$$\mathcal{S}_k = \{g \in \mathcal{M} : 0 < \lambda_1(g) < \cdots < \lambda_{k-1}(g) < \lambda_k(g)\},$$

for each $k \in \mathbb{N}^*$. Then one has

$$\mathcal{M} = \mathcal{S}_1 \supset \mathcal{S}_2 \supset \cdots \supset \mathcal{S}_k \supset \cdots$$

and the following result is true.

Theorem 4.3. Let M be a closed connected smooth manifold of dimension $n \geq 2$. Then each $\mathcal{S}_{k+1}$, $k \in \mathbb{N}^*$, is dense in $\mathcal{S}_k$ with respect to the topology induced by $(\mathcal{M},d)$.

Proof. It is sufficient to show that for each $g \in \mathcal{S}_k$, there exists an element $g' \in \mathcal{S}_{k+1}$ which is arbitrarily close to g. Let $g \in \mathcal{S}_k$, that is $0 = \lambda_0(g) < \lambda_1(g) < \cdots < \lambda_k(g)$. Assume that the k-th eigenvalue $\lambda_k(g)$ has multiplicity l, i.e.

$$\lambda_k(g) = \cdots = \lambda_{k+l-1}(g) = \lambda$$

and

$$0 = \lambda_0(g) < \lambda_1(g) < \cdots < \lambda_{k-1}(g) < \lambda < \lambda_{k+l}(g) \leq \cdots.$$

Let us consider a deformation $g(t) \in \mathcal{M}$ of g of the type (4.12) with $|t| < \varepsilon(\sigma)$ and let $\lambda_j(t)$, $j = 1,\ldots,l$, be such eigenvalues of $\Delta_{g(t)}$ as in (4.16). By Theorem 4.1 there exists a positive number $\varepsilon'(\sigma) \leq \varepsilon(\sigma)$ such that

$$0 = \lambda_0(g(t)) < \lambda_1(g(t)) < \cdots < \lambda_{k-1}(g(t)) < \lambda_j(t) < \lambda_{k+l}(g(t)) \leq \cdots,$$

for each $|t| < \varepsilon'(\sigma)$ and $j = 1,\ldots,l$. On the other hand, by assumption and Lemma 4.4 one can choose $\sigma \in C^\infty(M)$ such that at least two of the $\alpha_1,\ldots,\alpha_l$ in (4.16) are distinct. Let $\alpha_i \neq \alpha_j$ where $1 \leq i, j \leq l$. For this $\sigma \in C^\infty(M)$, one can choose a positive number $\varepsilon''(\sigma) \leq \varepsilon'(\sigma)$ such that $\lambda_i(t) \neq \lambda_j(t)$ for $0 < |t| < \varepsilon''(\sigma)$. Therefore all the first k eigenvalues of $\Delta_{g(t)}$, $|t| < \varepsilon''(\sigma)$, have multiplicity one and the k-th eigenvalue $\lambda_k(g(t))$ has multiplicity at most $l-1$. Repeating this process, one can choose $g' \in \mathcal{S}_{k+1}$ as close to g as one wants.

Q.E.D.

Before stating the main result of this section, let us recall some definitions and a result on Baire spaces. A subset S of a topological space X is **residual** if S is a countable intersection of open dense subsets of X. A topological space X is called a **Baire space** if any residual subset of X is dense in X. It is well-known that a complete metric space (X,d) is a Baire space and a residual subset of such a space is a subset of the second category.

Corollary 4.3. (**Uhlenbeck** [1976]) Let M be a smooth manifold as in Theorem 4.3 and $d : \mathcal{M} \times \mathcal{M} \to [0,+\infty)$ the complete distance on the set $\mathcal{M}$ of all smooth Riemannian metrics on $\mathcal{M}$ defined above. Then the set

$$\mathscr{S} = \{g \in \mathcal{M} \mid 0 = \lambda_0(g) < \lambda_1(g) < \cdots < \lambda_k(g) < \cdots\}$$

of all $g \in \mathcal{M}$ with the property that the eigenvalues of the Laplace-Beltrami operator Δ_g have multiplicity one is a residual set in $(\mathcal{M},d)$, i.e. $\mathscr{S}$ is a countable intersection of open dense subsets of $(\mathcal{M},d)$. In particular, $\mathscr{S}$ is a subset of the second category and dense in $(\mathcal{M},d)$.

Proof. Indeed, with the above notations, one has $\mathscr{S} = \bigcap_{k=1}^{\infty} \mathscr{S}_k$, where $\mathscr{S}_1 = \mathcal{M}$, so that the statement is an immediate consequence of Corollary 4.2 and Theorem 4.3.

Q.E.D.

Finally, let us prove the genericity of the Riemannian metrics with finite isometry groups.

Lemma 4.5. With the above notations, if M is a closed, connected, smooth manifold, and $g \in \mathscr{S}$, then the group of all isometries of (M,g) is finite.

Proof. For $g \in \mathcal{M}$, let us denote the eigenvalues of Δ_g as follows.

$$0 = \lambda_0(g) = \lambda_{j_0}(g) < \lambda_1(g) = \cdots = \lambda_{j_1}(g) < \lambda_{j_1+1}(g) = \cdots = \lambda_{j_2}(g) < \cdots.$$

Let $E(\Delta_g, \lambda_{j_l}(g))$ be the eigenspace of Δ_g corresponding to the eigenvalue $\lambda_{j_l}(g), l \in \mathbb{N}$. Notice that

$$\dim E(\Delta_g, \lambda_{j_l}(g)) = \begin{cases} 1 & \text{for } l = 0, \\ j_1 & \text{for } l = 1, \\ j_l - j_{l-1} & \text{for } l \in \mathbb{N} \setminus \{0,1\}. \end{cases}$$

Let $\{f_k \in C^\infty(M) \mid k \in \mathbb{N}\}$ be a complete orthonormal system of eigenfunctions of Δ_g in $L^2(M) = H^0(M)$ (see Corollary 1.4). Take a large natural number k so that the mapping

$$F : M \to \mathbb{R}^N, F(x) = (f_0(x), f_1(x), \ldots, f_{N-1}(x)), x \in M,$$

where $N = 1 + j_k$, is injective. Let G be the compact Lie group of all isometries of (M,g). Each $\Phi \in G$, which is a smooth diffeomorphism of M onto M, induces a linear mapping $\Phi^* : C^\infty(M) \to C^\infty(M)$ given by $[\Phi^*(f)](x) = f(\Phi^{-1}(x))$, $x \in M, f \in C^\infty(M)$ (see Chapter 1, §1). Moreover,

$$\langle \Phi^*(f_1), \Phi^*(f_2) \rangle_g = \langle f_1, f_2 \rangle_g$$

for all $f_1, f_2 \in C^\infty(M)$ (see Chapter 2, §1,A) and

$$(\Phi_1 \circ \Phi_2)^* = \Phi_1^* \circ \Phi_2^*$$

for all $\Phi_1, \Phi_2 \in G$. Since $\Delta_g \circ \Phi^* = \Phi^* \circ \Delta_g$ for each $\Phi \in G$ [see the equality (2.22) of Chapter 2], it follows that Φ^* maps each eigenspace $E(\Delta_g, \lambda_{j_l}(g))$, $l \in \mathbb{N}$, onto itself. Thus one obtains a Lie group morphism $\varphi : G \to O(E), \varphi(\Phi) = \Phi^*|_E, \Phi \in G$, where

$O(E)$ is the orthogonal group of the Euclidean vector space $\left(E = \sum_{l=0}^{k} E\left(\Delta_g, \lambda_{j_l}(g)\right), \langle\, , \rangle_g\right)$.

Note that φ is injective. Indeed, if $\Phi_1, \Phi_2 \in G$ are such that $\Phi_1^* = \Phi_2^*$, then for all $x \in M$,

$$F\left(\Phi_1^{-1}(x)\right) = \left(\Phi_1^*(f_0), \dots, \Phi_1^*(f_{N-1})\right)(x) = \left(\Phi_2^*(f_0), \dots, \Phi_2^*(f_{N-1})\right)(x) = F\left(\Phi_2^{-1}(x)\right).$$

Since F is injective, it follows that $\Phi_1^{-1}(x) = \Phi_2^{-1}(x)$ for all $x \in M$, that is $\Phi_1 = \Phi_2$.

Now, if $g \in \mathscr{S}$, then j_l is equal to l for each $l \in \mathbb{N}$. Moreover, each eigenspace $E_l = E\left(\Delta_g, \lambda_l\right)$, $l \in \mathbb{N}$, is of dimension one so that $O\left(E_l\right) = \left\{1_{E_l}, -1_{E_l}\right\}$ for all $l \in \mathbb{N}$, which implies that the Lie subgroup $\varphi(G)$ of $O(E)$ is contained in a finite group. Hence $\varphi(G)$ is finite and since φ is injective, G itself is finite.

Q.E.D.

Combining Lemma 4.5 with Corollary 4.3 one obtains the following result.

Corollary 4.4. (Ebin [1968]). Let M be a closed, connected smooth manifold of dimension $n \geq 2$. Then the set $\left\{g \in \mathcal{M} \,\middle|\, \text{the isometry group of } (M, g) \text{ is finite}\right\}$ contains a residual subset of $(\mathcal{M},d)$. In particular, this set is open and dense in $\mathcal{M}$.

Consequently, for most Riemannian metrics on a closed, connected smooth manifold of dimension ≥ 2, the corresponding isometry groups are finite.

Note that the assumption $n \geq 2$ in Corollary 4.4 is essential: for any Riemannian metric g on S^1, the isometry group of (S^1,g) is just the orthogonal group $O(2)$.

Remark 4.2. The previous property namely that the generic spectrum is simple, that is all eigenvalues have multiplicity one, has been generalized for certain finite Riemannian coverings (see Chapter 1, Example 2.8) by Zelditch [1990]. More precisely, let $\Phi: M \to M_0$ be a finite normal covering projection, with covering transformation group G (see Chapter 1, Remark 2.2), i.e. $\left(M, \Phi, M_0\right)$ is a principal G-bundle over M_0 with M closed, where G is a finite group, $\mathcal{M}_0$ the set of smooth Riemannian metrics on M_0, $\Phi^*\mathcal{M}_0 \subset \mathcal{M}$ be their pullbacks to M and let $\mathcal{M}_G$ be the G-invariant Riemannian metrics on M. The normality of Φ implies $\mathcal{M}_G = \Phi^*\mathcal{M}_0$. If M and M_0 are equipped with Riemannian metrics g and g_0 respectively such that $\Phi^*\left(g_0\right) = g$, then $\Phi:(M, g) \to \left(M_0, g_0\right)$ is a Riemannian covering (see Chapter 1, Example 2.8). When $\Phi:(M, g) \to \left(M_0, g_0\right)$ is a normal Riemannian covering with covering transformation group G, it is obvious that the real eigenspaces $E\left(\Delta_g, \lambda\right)$ are orthogonal representations of G (see also Chapter 4, §3). In general they are reducible. Zelditch [1990] proves that the set

$$\left\{g_0 \in \mathcal{M}_0 \,\middle|\, \text{for any } k \in \mathbb{N}, \text{ the real eigenspace } E\left(\Delta_{\Phi^*(g_0)}, \lambda_k\right) \text{ is an irreducible, real } G\text{-module}\right\}$$

is residual in $\mathcal{M}_0$ and hence dense in $\mathcal{M}_0$, under the assumption that $\dim M \geq \max\left\{\deg \sigma \,\middle|\, \sigma \in \hat{G}_0\right\}$, where $\hat{G}_0$ denotes the set of equivalence classes of real orthogonal irreducibles of G.

5. ESTIMATES OF THE EIGENVALUES THROUGH GEOMETRIC DATA

In § 1 it was pointed out that the geometry determines the spectrum. There are only few examples of compact Riemannian manifolds (M,g) for which the spectrum is explicitly known (see § 2). A natural question in spectral geometry is to understand the ways in which the spectrum of the Laplace-Beltrami operator (or other natural self-adjoint elliptic differential operators) associated to a compact Riemannian manifold (M,g) influences the geometry and topology of M (inverse spectral theory). One way in which one may extract geometric information from spectral data is given by geometric bounds. Unlike other methods described in next chapters, this method consists of a collection of techniques for estimating various geometric features of a compact Riemannian manifold (M,g) in terms of spectral data, usually $\dim M$, $\mathrm{Vol}(M,g)$ and the eigenvalues of Δ, together with geometric data, such as the diameter $D(M,g)$ and curvature bounds. In particular, one basic question in spectral geometry is to find upper and lower bounds of eigenvalues of Δ in terms of $\dim M$, $\mathrm{Vol}(M,g)$ and geometric data.

In the following, we consider the case of compact Riemannian manifolds. Let us start to discuss lower bounds for the first non-zero eigenvalue of the Laplace-Beltrami operator Δ. For bounded domains in $\mathbb{R}^n$ estimating the first eigenvalue is a problem with a long history. Among others, C. Faber, E. Krahn, W. Hayman, L.E. Payne, G. Polyà, G. Szegö, H. Weinberger had made contributions to this problem. For closed Riemannian manifolds with Ricci curvature bounded from below, the first estimate for the lower bound of λ_1 is the following result.

Theorem 5.1 (**Lichnerowicz** [1958]). Let (M,g) be a closed Riemannian manifold of dimension $n \geq 2$ and c_2R its Ricci tensor field. Assume that for some given constant $k \in (0,+\infty)$,

$$-c_2R(X,X) \geq kg(X,X) \tag{5.1}$$

for all $X \in \mathfrak{X}(M)$. Then

$$\lambda_1 \geq \frac{n}{n-1}k.$$

Proof. If $f \neq 0$ is an eigenfunction of Δ corresponding to λ_1, i.e. $\Delta f = \lambda_1 f$, then

$$-\frac{1}{2}\Delta\left(|df|^2\right) = |H_f|^2 - \lambda_1|df|^2 - c_2R\big(grad(f), grad(f)\big), \tag{5.2}$$

where H_f is the Hessian form of f (see Chapter 2, Proposition 2.4).

Now let us note that

$$\Delta f = -trace\left(H_f\right) \tag{5.3}$$

for any $f \in C^\infty(M)$ (see Chapter 2, § 2, C).

If one integrates over M the equality (5.2) and one takes into account that

$$\int_M \Delta\left(|df|^2\right)d\mu_g = 0$$

(see Chapter 2, Corollary 2.2), one gets

$$\|H_f\|^2 - \lambda_1 \|df\|^2 - \int_M (c_2 R)(grad(f), grad(f))\, d\mu_g = 0 .$$

Then by (5.1) one deduces that

$$\|H_f\|^2 - \lambda_1 \|df\|^2 + k\|df\|^2 \le 0 .$$

Since

$$\|\Delta f\|^2 = \lambda_1 \langle \Delta f, f \rangle = \lambda_1 \|grad(f)\|^2 = \lambda_1 \|df\|^2$$

(see Chapter 2, Proposition 2.3), it follows that

$$\|H_f\|^2 - \|\Delta f\|^2 + \frac{k}{\lambda_1}\|\Delta f\|^2 \le 0 .$$

Now using (5.3) and the Cauchy-Schwarz inequality we have

$$(\Delta f)^2 = \left[\sum_{i=1}^{n} H_f(X_i, X_i)\right]^2 \le n|H_f|^2 ,$$

where $(X_1, \ldots, X_n)$ is a local orthonormal frame on (M,g), and consequently

$$\|\Delta f\|^2 \le n\|H_f\|^2 .$$

Therefore one can conclude that

$$\lambda_1 \ge \frac{nk}{n-1} ,$$

as required.

Q.E.D.

If (M,g) is isometric to (S^n, g_{can}), $n \ge 2$, then its sectional curvature σ is equal to 1. Hence $\tau = -n(n-1)$, so that $-c_2 R = (n-1)g$. Consequently, in this case one can consider $k = n-1$ in Theorem 5.1 and $\lambda_1 = n = \frac{nk}{n-1}$ (see also Theorem 2.2). Obata [1962] proved that the converse statement is also true:

Theorem 5.2 Let (M,g) be a closed Riemannian manifold of dimension $n \ge 2$ satisfying the assumption (5.1). If the equality $\lambda_1 = \frac{nk}{n-1}$ holds, then (M,g) is isometric to the n-sphere of constant sectional curvature $\frac{k}{n-1}$.

For an elegant proof of this statement based on the generalized Toponogov theorem, see Chavel [1984], p. 84.

Based on some lower bounds given by Cheeger for the first non-zero eigenvalue λ_1 in terms of his isoperimetric constants, Yau gave some estimates for λ_1 in terms of geometric data such as the diameter, volume and lower bound for $-\rho$, where ρ denotes the Ricci curvature, which are easier to compute. From 1979 on, Li and Yau developed the method of getting estimates on λ_1 via the gradient estimate on the first eigenfunction. For example, using this method they proved that if (*M*,*g*) is a closed Riemannian manifold satisfying $\rho \le 0$, then $\lambda_1 \ge \frac{\pi^2}{2D^2(M,g)}$, where *D*(*M*,*g*) is the diameter of (*M*,*g*) (for the proof of this result, see Schoen and Yau [1994]). Using a similar technique, Zhong and Yang [1984] improved this result, namely they proved the following beautiful estimate.

Theorem 5.3 Let (*M*,*g*) be a closed Riemannian manifold such that $c_2R(X,X) \le 0$ for all $X \in \mathfrak{X}(M)$. Then

$$\lambda_1 \ge \frac{\pi^2}{D^2(M,g)}.$$

For a proof of this theorem, see Schoen and Yau [1994]. One should note that the above estimate is optimal under the assumption $\rho \le 0$.

Yang [1999] recently generalized Theorem 5.3, namely for closed Riemannian manifolds (M,g) with $-c_2R(X,X) \ge kg(X,X)$ for all $X \in \mathfrak{X}(M)$ and for some constant $k \in [0,+\infty)$, he proves the lower bound $\lambda_1 \ge \frac{1}{4}k + \frac{\pi^2}{D^2(M,g)}$. Notice that for closed Riemannian manifolds with small diameter, this last estimate is better than Lichnerowicz's estimate for λ_1 given in Theorem 5.1.

Li and Yau have been able to weaken the assumption concerning c_2R in Theorem 5.3, and the result they having get is the following one.

Theorem 5.4 (Li-Yau) Let (M,g) be a closed connected Riemannian manifold of dimension $n \ge 2$. Assume that for some given constant $k \in [0,+\infty)$,

$$c_2R(X,X) \le (n-1)kg(X,X) \tag{5.4}$$

for all $X \in \mathfrak{X}(M)$. Then

$$\lambda_1 \ge \frac{\exp\left(-1-\sqrt{1+4(n-1)^2 kD^2(M,g)}\right)}{(n-1)D^2(M,g)},$$

where *D*(*M*,*g*) denotes the diameter of (*M*,*g*).

Proof. Let *f* be an eigenfunction of Δ corresponding to λ_1. Since

$$\int_M f d\mu_g = \frac{1}{\lambda_1}\int_M \Delta f d\mu_g = 0,$$

one can assume $\sup f = 1$. For any $\beta > 1$, let us consider the real smooth function

$$F = \frac{|grad(f)|^2}{(\beta - f)^2},$$

where $|grad(f)|^2 = g(grad(f), grad(f))$. Assume *F* takes its maximum at $x_0 \in M$. Then $(grad(F))_{x_0} = 0$ and $(\Delta F)(x_0) \ge 0$. Indeed, let $(e_1,\dots,e_n)$ be an orthonormal basis of $T_{x_0}M$ and γ_j the geodesic with $\gamma_j(0)=x_0$, $\dot{\gamma}_j(0)=e_j$, $j=1,\dots,n$. Since $F\circ\gamma_j$ has a maximum at $t=0$, by elementary calculus one gets

$$dF(e_j)=\frac{d}{dt}\bigg|_{t=0}(F\circ\gamma_j)(t)=0,\ \frac{d^2}{dt^2}\bigg|_{t=0}(F\circ\gamma_j)(t)\leq 0$$

for every $j=1,\dots,n$. Consequently, the equality

$$(\Delta F)(x_0)=-\sum_{j=1}^{n}\frac{d^2}{dt^2}\bigg|_{t=0}(F\circ\gamma_j)(t)$$

(see Chapter 2, § 2) implies that $(\Delta F)(x_0)\geq 0$.

Since $|grad(f)|^2=(\beta-f)^2F$, it follows that

$$\Delta\left(|grad(f)|^2\right)=(\beta-f)^2\Delta F-2g\left(grad\left((\beta-f)^2\right),grad(F)\right)+F\Delta\left((\beta-f)^2\right)$$

[see the property (2.11) in Chapter 2]. Therefore,

$$\left(\Delta\left(|grad(f)|^2\right)\right)(x_0)-F(x_0)\left(\Delta(\beta-f)^2\right)(x_0)\geq 0. \tag{5.5}$$

On the other hand, note that

$$\begin{aligned}\Delta(\beta-f)^2&=2(\beta-f)\Delta(\beta-f)-2g\left(grad(\beta-f),grad(\beta-f)\right)\\&=-2(\beta-f)\Delta(f)-2|grad(f)|^2=-2\lambda_1 f(\beta-f)-2|grad(f)|^2,\end{aligned}$$

so that by (5.2) and (5.4), (5.5) leads to

$$\begin{aligned}&|H_f|^2(x_0)-\lambda_1|grad(f)|^2(x_0)-(n-1)k|grad(f)|^2(x_0)\\&-F(x_0)\left[\lambda_1 f(x_0)(\beta-f)(x_0)+|grad(f)|^2(x_0)\right]\leq 0.\end{aligned}\tag{5.6}$$

One may choose a normal coordinate system $(x^1,\dots,x^n)$ around $x_0\in M$ such that

$$\frac{\partial f}{\partial x^1}(x_0)=|grad(f)|(x_0),\ \frac{\partial f}{\partial x^j}(x_0)=0 \text{ for } j\geq 2.$$

In virtue of the equality

$$|grad(f)|^2=\sum_{k,l}g^{kl}\frac{\partial f}{\partial x^k}\frac{\partial f}{\partial x^l}$$

and of the Proposition 3.4 of Chapter 1, in this coordinate system one has

$$|grad(f)|^2(x_0)=\sum_{j=1}^{n}\left(\frac{\partial f}{\partial x^j}(x_0)\right)^2,\ |H_f|^2(x_0)=\sum_{j,k=1}^{n}\left(\frac{\partial^2 f}{\partial x^j\partial x^k}(x_0)\right)^2.$$

Then

$$\frac{\partial}{\partial x^j}\left(|grad(f)|\right)=\frac{1}{2|grad(f)|}\left(\sum_{k,l=1}^{n}\frac{\partial g^{kl}}{\partial x^j}\frac{\partial f}{\partial x^k}\frac{\partial f}{\partial x^l}+2\sum_{k,l=1}^{n}g^{kl}\frac{\partial f}{\partial x^k}\frac{\partial^2 f}{\partial x^l\partial x^j}\right),$$

so that

$$\left[\frac{\partial}{\partial x^j}\left(|grad(f)|\right)\right](x_0)=\frac{\partial^2 f}{\partial x^1\partial x^j}(x_0).$$

Since

$$\frac{\partial F}{\partial x^j}=\frac{2|grad(f)|}{(\beta-f)^3}\left[(\beta-f)\frac{\partial}{\partial x^j}\left(|grad(f)|\right)+|grad(f)|\frac{\partial f}{\partial x^j}\right],$$

from $\left(grad(F)\right)_{x_0}=0$, one gets consequently

$$\begin{cases}\dfrac{\partial^2 f}{\partial x^{1^2}}(x_0)=-\dfrac{|grad(f)|^2(x_0)}{\beta-f(x_0)},\\[2mm] \dfrac{\partial^2 f}{\partial x^1\partial x^j}(x_0)=0 \ \text{ for } j\geq 2.\end{cases}\tag{5.7}$$

Using the Cauchy-Schwarz inequality, one obtains

$$\begin{aligned}\sum_{j,k=2}^{n}\left(\frac{\partial^2 f}{\partial x^j\partial x^k}(x_0)\right)^2 &\geq\sum_{j=2}^{n}\left(\frac{\partial^2 f}{\partial x^{j^2}}(x_0)\right)^2\geq\frac{1}{n-1}\left(\sum_{j=2}^{n}\frac{\partial^2 f}{\partial x^{j^2}}(x_0)\right)^2\\ &=\frac{1}{n-1}\left[(\Delta f)(x_0)+\frac{\partial^2 f}{\partial x^{1^2}}(x_0)\right]^2=\frac{1}{n-1}\left(\lambda_1 f(x_0)+\frac{\partial^2 f}{\partial x^{1^2}}(x_0)\right)^2\\ &=\frac{1}{n-1}\left[\lambda_1^2 f^2(x_0)+2\lambda_1 f(x_0)\frac{\partial^2 f}{\partial x^{1^2}}(x_0)+\left(\frac{\partial^2 f}{\partial x^{1^2}}(x_0)\right)^2\right]\\ &\geq\frac{1}{2(n-1)}\left(\frac{\partial^2 f}{\partial x^{1^2}}(x_0)\right)^2-\frac{1}{n-1}\lambda_1^2 f^2(x_0),\end{aligned}$$

that is

$$\sum_{j,k=2}^{n}\left(\frac{\partial^2 f}{\partial x^j\partial x^k}(x_0)\right)^2\geq\frac{1}{2(n-1)}\left(\frac{\partial^2 f}{\partial x^{1^2}}(x_0)\right)^2-\frac{1}{n-1}\lambda_1^2 f^2(x_0).\tag{5.8}$$

On the other hand, note that

$$\frac{f(x_0)}{\beta-f(x_0)}\leq\frac{1}{\beta-f(x_0)}\leq\frac{1}{\beta-1}.\tag{5.9}$$

If one takes into account (5.7), (5.8) and (5.9), (5.6) leads to

$$F^2(x_0)-2(n-1)\left[\frac{\lambda_1\beta}{\beta-1}+(n-1)k\right]F(x_0)-2\frac{\lambda_1^2}{(\beta-1)^2}\leq 0.$$

Hence

$$F(x_0) \le \alpha \le 4(n-1)\left[\frac{\lambda_1\beta}{\beta-1}+(n-1)k\right],$$

where α is the positive root of the equation in the real variable t:

$$t^2-2(n-1)\left[\frac{\lambda_1\beta}{\beta-1}+(n-1)k\right]t-\frac{2\lambda_1^2}{(\beta-1)^2}=0,$$

and consequently

$$\frac{|grad(f)|}{\beta-f} \le \left[4(n-1)\left(\frac{\lambda_1\beta}{\beta-1}+(n-1)k\right)\right]^{\frac{1}{2}}. \tag{5.9}$$

Let $x_1, x_2 \in M$ be such that $f(x_1)=0$, $f(x_2)=1$ and let γ be a minimal geodesic joining x_1 and x_2. It follows that

$$\ln\frac{\beta}{\beta-1}=\ln\frac{\beta-f(x_1)}{\beta-f(x_2)}=\left|\int_\gamma \frac{d\ln((\beta-f)\circ\gamma)(s)}{ds}ds\right| \le \int_\gamma \frac{|grad(f)|}{\beta-f}ds,$$

where ln denotes the natural real logarithmic function. So, combining with (5.9), one gets

$$\ln\frac{\beta}{\beta-1} \le \left[4(n-1)\left(\frac{\lambda_1\beta}{\beta-1}+(n-1)k\right)\right]^{\frac{1}{2}} D(M,g),$$

that is

$$\lambda_1 \ge \frac{\beta-1}{\beta}\left[\frac{1}{4(n-1)D^2(M,g)}\left(\ln\frac{\beta}{\beta-1}\right)^2-(n-1)k\right] \tag{5.10}$$

for any $\beta>1$.

Let us maximize the right-hand side of (5.10) as a smooth function in the real variable $\beta>1$. The derivative of this function has the expression

$$\frac{1}{4(n-1)D^2(M,g)}\left[\left(\ln\frac{\beta}{\beta-1}\right)^2-2\ln\frac{\beta}{\beta-1}-4(n-1)^2kD^2(M,g)\right],$$

and hence the following equation for the maximum point $\beta_0>1$:

$$\ln\frac{\beta_0}{\beta_0-1}=1+\sqrt{1+4(n-1)^2kD^2(M,g)},$$

which leads us immediately to the desired inequality.

Q.E.D.

The results of Yang [1999] and Li-Yau (see Theorem 5.4) display the interesting general principle that the scaling invariant $D^2(M,g)\lambda_1$ can be estimated from below in terms of the dimension n and the scaling invariant $D^2(M,g)k$ of the Riemannian manifold. For Riemannian manifolds (M,g) with large diameter $D(M,g)$, the Li-Yau estimate given in Theorem 5.4 shows that $D^2(M,g)\lambda_1$ has a lower bound which decays exponentially in $D(M,g)$ with decay rate $2(n-1)\sqrt{k}$. Kalka, Mann, Yang and Zinger [1997] recently showed that the optimal decay rate is given by $\frac{1}{2}(n-1)\sqrt{k}$. More precisely, for a closed connected, n-dimensional Riemannian manifold $(M,g), n\geq 2$, satisfying (5.4) for some constant $k\in[0,+\infty)$, they proved that

$$\lambda_1 \geq \frac{48}{(3\pi+1)(n+1)D^2(M,g)}\left\{1+\left[1+\frac{1}{4}(n-1)^2 kD^2(M,g)\right]^{\frac{1}{2}}\right\}$$
$$\bullet\exp\left(-\left\{1+\left[1+\frac{1}{4}(n-1)^2 kD^2(M,g)\right]^{\frac{1}{2}}\right\}\right).$$

Now let us discuss lower estimates on higher eigenvalues. First, let us recall the Dirichlet eigenvalue problem for a bounded domain Ω in $\mathbb{R}^n$:

$$\begin{cases}\Delta f=\lambda f,\\ f|_{\partial\Omega}=0.\end{cases}$$

Using the heat kernel and a Tauberian theorem, H. Weyl [1912] proved the asymptotic formula

$$\mu_j \sim c_n\left(\frac{j}{Vol(\Omega)}\right)^{\frac{2}{n}} \qquad \text{as } j\to\infty$$

(see also Chapter 4, §1), where

$$c_n=(2\pi)^2\Big/\left(\frac{Vol(S^{n-1},g_{can})}{n}\right)^{\frac{2}{n}}.$$

Based on this formula, Polyà [1961] conjectured that

$$\mu_j\geq c_n\left(\frac{j}{Vol(\Omega)}\right)^{\frac{2}{n}}$$

for each $j\in\mathbb{N}^*$.

In the case $n=2$, Polyà proved that the conjecture holds for some special planar domains. Lieb [1980] proved that there exists a constant $\tilde{c}_n<c_n$ such that

$$\mu_j\geq \tilde{c}_n\left(\frac{j}{\mathrm{Vol}(\Omega)}\right)^{\frac{2}{n}}$$

for each $j\in\mathbb{N}^*$.

Following Schoen-Yau [1994], we now present Cheng-Li's method of using the properties of the heat kernel to estimate μ_j. More precisely, the following result is true:

Theorem 5.5 Let (M,g) be an n-dimensional connected smooth Riemannian manifold, $n \geq 3$, and K a precompact open subset of M with smooth boundary as in §1. Then for any $j \in \mathbb{N}^*$, one has

$$\mu_j = \mu_j(K) \geq \frac{c}{e}\left[\frac{j}{\mathrm{Vol}(M,g)}\right]^{\frac{2}{n}}, \tag{5.11}$$

where $c > 0$ is a constant depending only on n and on the Sobolev constant of $\overline{K}$ (see Chapter 2, §2, C). In particular, the same estimates hold for any domain K in $\mathbb{R}^n$ $(n \geq 3)$ with compact closure and nonempty smooth boundary.

Proof. Let $E \overset{not}{=} E_K : K \times K \times (0,+\infty) \to \mathbb{R}$ be the heat kernel considered in §1. Then

$$E(x,y,t) = \int_K E(x,z,s)E(z,y,t-s)\,d\mu_g(z)$$

for all $(x,y) \in K \times K$ and $0 < s < t$, so that in particular

$$E(x,x,2t) = \int_K (E(x,z,t))^2\,d\mu_g(z).$$

Fix $x \in K$. Since Green's formula is still valid for precompact open subsets K of M, that is

$$\int_K f_1 \Delta f_2\,d\mu_g = \int_K g(\mathrm{grad}(f_1), \mathrm{grad}(f_2))\,d\mu_g,$$

where f_1 and f_2 are real smooth functions in K such that one of them vanishes in a neighborhood of the boundary ∂K, one gets

$$\begin{aligned}
\frac{\partial}{\partial t}\int_K (E(x,z,t))^2\,d\mu_g(z) &= 2\int_K E(x,z,t)\frac{\partial E}{\partial t}(x,z,t)\,d\mu_g(z) \\
&= -2\int_K E(x,z,t)\Delta E(x,z,t)\,d\mu_g(z) \\
&= -2\int_K g_z((grad\ E)(x,z,t),(grad\ E)(x,z,t))\,d\mu_g(z) \\
&\leq -2c\left(\int_K (E(x,z,t))^{\frac{2n}{n-2}}\,d\mu_g(z)\right)^{\frac{n-2}{n}}.
\end{aligned} \tag{5.12}$$

The last inequality is a consequence of the Sobolev inequality (see Chapter 2, §2, C). By Hölder's inequality, namely take $p = \dfrac{n+2}{n-2}$ and $q = \dfrac{n+2}{4}$ in the standard notation, one gets

$$\begin{aligned}&\int_K (E(x,z,t))^2 \, d\mu_g(z) = \int_K (E(x,z,t))^{\frac{2n}{n+2}} (E(x,z,t))^{\frac{4}{n+2}} \, d\mu_g(z) \\ &\leq \left(\int_K (E(x,z,t))^{\frac{2n}{n-2}} \, d\mu_g(z) \right)^{\frac{n-2}{n+2}} \cdot \left(\int_K E(x,z,t) \, d\mu_g(z) \right)^{\frac{4}{n+2}} \\ &\leq \left(\int_K (E(x,z,t))^{\frac{2n}{n-2}} \, d\mu_g(z) \right)^{\frac{n-2}{n+2}} .\end{aligned} \tag{5.13}$$

(for the last inequality see §1). Consequently, (5.13) leads to

$$\left(\int_K (E(x,z,t))^{\frac{2n}{n-2}} \, d\mu_g(z) \right)^{\frac{n-2}{n}} \geq \left(\int_K (E(x,z,t))^2 \, d\mu_g(z) \right)^{\frac{n+2}{n}} .$$

(5.12) and the last inequality implies that

$$\frac{\partial}{\partial t} \int_K (E(x,z,t))^2 \, d\mu_g(z) \leq -2c \left(\int_K (E(x,z,t))^2 \, d\mu_g(z) \right)^{\frac{n+2}{n}} ,$$

that is the differential inequality

$$\frac{\partial}{\partial t} E(x,x,2t) \leq -2c \left(E(x,x,2t) \right)^{\frac{n+2}{n}} ,$$

or equivalently

$$\frac{\partial}{\partial t} \left[E(x,x,2t) \right]^{-\frac{2}{n}} \geq \frac{4c}{n} \tag{5.14}$$

for each $x \in K$ and $t > 0$. Since $\lim_{t \searrow 0} E(x,x,t) = +\infty$, by integrating (5.14) it follows that

$$\left[E(x,x,2t) \right]^{-\frac{2}{n}} \geq \frac{4ct}{n}$$

for each $x \in K$ and $t > 0$, that is

$$E(x,x,2t) \leq \left(\frac{4ct}{n} \right)^{-\frac{n}{2}} \tag{5.15}$$

for each $x \in K$ and $t > 0$.

Since

$$\int_K E(x,x,2t) \, d\mu_g(x) = \sum_{k=1}^{\infty} e^{-2\mu_k t}$$

for each $t > 0$ and $0 \leq \mu_1 \leq \mu_2 \leq \cdots \leq \mu_j$, (5.15) implies that

$$j e^{-2\mu_j t} \leq \sum_{k=1}^{\infty} e^{-2\mu_k t} \leq \left(\frac{4c}{n} t \right)^{-\frac{n}{2}} \operatorname{Vol}(K,g)$$

for each $t>0$. Choose t such that $2\mu_j t=\frac{n}{2}$. Then the inequality

$$je^{-2\mu_j t}\leq\left(\frac{4c}{n}t\right)^{-\frac{n}{2}}\operatorname{Vol}(K,g)$$

implies that

$$je^{-\frac{n}{2}}\leq\left(\frac{c}{\mu_j}\right)^{-\frac{n}{2}}\operatorname{Vol}(K,g),$$

that is (5.11).

Q.E.D.

In general, it is easier to give an upper bound for λ_1 than a lower bound. The basic tool in order to obtain upper bounds is the following comparison theorem given by S.Y. Cheng [1975] for the first eigenvalue λ_1.

Theorem 5.6 Let (M,g) be an n-dimensional complete Riemannian manifold $(n\geq 2)$ such that $c_2R\leq-(n-1)kg$, $B(x_0,r)$ the geodesic ball in (M,g) with center x_0 and radius r, $V_n(k,r)$ a ball of radius r in the n-dimensional simply connected space of constant sectional curvature k and $\mu_1(B(x_0,r))$ [resp. $\mu_1(V_n(k,r))$] the first eigenvalue corresponding to Dirichlet boundary condition on $B(x_0,r)$ [resp. $V_n(k,r)$]. Then $\mu_1(B(x_0,r))\leq\mu_1(V_n(k,r))$. In particular, if (M,g) is a closed Riemannian manifold, then

$$\lambda_j\leq\mu_1\left(V_n\left(k,\frac{D(M,g)}{2j}\right)\right)$$

for each $j=1,2,\ldots$.

Proof. (Schoen-Yau [1994]). Let E_1 (resp. E_2) be the heat kernel of $B(x_0,r)$ [resp. $V_n(k,r)$]. From the invariance of E_2 under isometry it follows that $E_2(x,y,t)$ is only a function of $d(x,y)$ (=distance from x to y) and t, that is $E_2(x,y,t)=E_2(d(x,y),t)$. E_2 can be thought of as a function on $B(x_0,r)$ in an obvious way. Fixing an arbitrary point $x\in M$ and a $r>0$, the Cheeger-Yau [1981] comparison theorem for heat kernels gives

$$E_2(d(x,y),t)\leq E_1(x,y,t)\text{ for all }(x,y,t)\in B(x_0,r)\times B(x_0,r)\times(0,+\infty).$$

In particular

$$E_2(0,t)\leq E_1(x,x,t)\text{ for all }(x,t)\in B(x_0,r)\times(0,+\infty).$$

Now

$$E_2(0,t)=\sum_j e^{-\tilde{\mu}_j t}\tilde{f}_j^2(0),$$

$$E_1(x,x,t)=\sum_j e^{-\mu_j t}f_j^2(x),$$

where $\tilde{\mu}_j = \mu_j\left(V_n(k,r)\right)$, $\mu_j = \mu_j\left(B(x_0,r)\right)$ and $\tilde{f}_j, f_j$, $j \in \mathbb{N}^*$, are the corresponding eigenfunctions. Therefore,

$$e^{-\tilde{\mu}_1 t}\left[\tilde{f}_1^2(0)+e^{-(\tilde{\mu}_2-\tilde{\mu}_1)t}\tilde{f}_2^2(0)+\cdots\right] \le e^{-\mu_1 t}\left[f_1^2(x)+e^{-(\mu_2-\mu_1)t}f_2^2(x)+\cdots\right],$$

that is

$$e^{(\mu_1-\tilde{\mu}_1)t}\left[\tilde{f}_1^2(0)+e^{-(\tilde{\mu}_2-\tilde{\mu}_1)t}\tilde{f}_2^2(0)+\cdots\right] \le f_1^2(x)+e^{-(\mu_2-\mu_1)t}f_2^2(x)+\cdots.$$

Note that $f_1^2(x)>0$, $\tilde{f}_1^2(x)>0$ since f_1 (resp. $\tilde{f}_1$) does not change sign, and $\mu_j > \mu_1$ (resp. $\tilde{\mu}_j > \tilde{\mu}_1$) for each $j \ge 2$. Letting $t \to \infty$ in the above inequality, one gets

$$\mu_1 - \tilde{\mu}_1 \le 0,$$

that is the desired inequality.

In order to prove the last statement, let $j \ge 1$ and let us choose $x_1, \dots, x_{j+1} \in M$ such that $B\left(x_k, \dfrac{D(M,g)}{2j}\right)$, $k = 1,\dots, j+1$, are pairwise disjoint. Let $f^{(k)}$ be the first eigenfunction on $B\left(x_k, \dfrac{D(M,g)}{2j}\right)$ with Dirichlet boundary condition, $k = 1,\dots, j+1$. Integrating the identity

$$f^{(k)}\Delta f^{(k)} = \mu_1\left(B_{x_k}, \frac{D(M,g)}{2j}\right)\left(f^{(k)}\right)^2$$

and using Green's formula for compact Riemannian manifolds with boundary (see Sakai [1996]) as well as $f^{(k)}\Big|_{\partial B\left(x_k, \frac{D(M,g)}{2j}\right)} = 0$, one obtains

$$\int_{B\left(x_k, \frac{D(M,g)}{2j}\right)} \left|df^{(k)}\right|^2 = \mu_1\left(B\left(x_k, \frac{D(M,g)}{2j}\right)\right)\int_{B\left(x_k, \frac{D(M,g)}{2j}\right)} \left(f^{(k)}\right)^2,$$

so that by the first part of the theorem,

$$\int_{B\left(x_k, \frac{D(M,g)}{2j}\right)} \left|df^{(k)}\right|^2 \le \mu_1\left(V_n\left(k, \frac{D(M,g)}{2j}\right)\right)\int_{B\left(x_k, \frac{D(M,g)}{2j}\right)} \left(f^{(k)}\right)^2.$$

Extend $f^{(k)}$ to be zero outside $B\left(x_k, \dfrac{D(M,g)}{2j}\right)$. Let $f_0, f_1, \dots$ be the eigenfunctions on M corresponding to $0 = \lambda_0 < \lambda_1 \le \cdots \le \lambda_j \le \cdots$. There exist real constants $\alpha_1, \dots, \alpha_{j+1}$ such that $\sum_{k=1}^{j+1} \alpha_k f^{(k)} \ne 0$ and $\sum_{k=1}^{j+1} \alpha_k f^{(k)} \perp \left\{f_0, f_1, \dots, f_{j+1}\right\}$ in the $L^2(M)$-sense. By Theorem 3.1,

$$\lambda_j \int_M \left(\sum_{k=1}^{j+1} \alpha_k f^{(k)} \right)^2 d\mu_g \le \int_M \left| \sum_{k=1}^{j+1} \alpha_k df^{(k)} \right|^2 d\mu_g = \sum_{k=1}^{j+1} \alpha_k^2 \int_M \left| df^{(k)} \right|^2 d\mu_g$$

$$\le \mu_1 \left(V_n \left(k, \frac{D(M,g)}{2j} \right) \right) \sum_{k=1}^{j+1} \alpha_k^2 \int_M \left(f^{(k)} \right)^2 d\mu_g$$

$$= \mu_1 \left(V_n \left(k, \frac{D(M,g)}{2j} \right) \right) \int_M \left(\sum_{k=1}^{j+1} \alpha_k f^{(k)} \right)^2,$$

which implies the desired inequality.

Q.E.D.

By estimating the first eigenvalue μ_1 of geodesic balls in space forms, S.Y. Cheng obtained the following upper bounds for λ_j under the specified curvature conditions on (M,g):

(i) If $c_2 R \le 0$, then $\lambda_j \le \dfrac{2j^2 n(n+4)}{D^2(M,g)}$ for each $j = 1, 2, \ldots$;

(ii) If $c_2 R \le -(n-1)g$, then $\lambda_1 \le \dfrac{n\pi^2}{D^2(M,g)}$;

(iii) If $c_2 R \le (n-1)kg \ (k > 0)$, then $\lambda_1 \le \dfrac{(n-1)^2}{4} k + \dfrac{c_n}{D^2(M,g)}$,

for some constant $c_n > 0$ depending only on n.

Now let us point out a nice consequence of the previous results.

Corollary 5.1. Let (M,g) be an n-dimensional connected, complete Riemannian manifold without boundary, $n \ge 2$, such that $\sigma \ge b > 0$ and $D(M,g) = \dfrac{\pi}{\sqrt{b}}$. Then (M,g) is isometric to $\left(S^n\left(\dfrac{1}{\sqrt{b}}\right), g_{can} \right)$, where $S^n\left(\dfrac{1}{\sqrt{b}}\right)$ denotes the n-sphere of radius $\dfrac{1}{\sqrt{b}}$.

Proof. Indeed, the condition $D(M,g) = \dfrac{\pi}{\sqrt{b}}$ implies that M is closed and $\sigma \ge b$ implies that $c_2 R \le -(n-1)bg$. By Theorem 5.1 it follows that $\lambda_1 \ge nb$. On the other hand, if $x_1, x_2 \in M$ are such that $d(x_1, x_2) = D(M,g)$, then any geodesic starting from x_1 passes through x_2. By Theorem 5.6 one gets

$$\lambda_1 \le \mu_1 \left(V_n \left(b, \frac{1}{2} D(M,g) \right) \right) = \mu_1 \left(V_n \left(b, \frac{\pi}{2\sqrt{b}} \right) \right) = nb.$$

Therefore $\lambda_1 = nb$ and Theorem 5.2 shows that (M,g) is isometric to $\left(S^n\left(\dfrac{1}{\sqrt{b}}\right), g_{can} \right)$.

Q.E.D.

Next, let us discuss upper bounds of λ_1 in terms of geometric data for closed Riemannian manifolds of dimension two. First, let us recall some basic facts about conformal maps.

Definition 5.1. Let (M,g), (N,h) be Riemannian manifolds having the same

dimension. A smooth bijective map $F : M \to N$ is called **conformal** if, for every $x \in M$, the differential $dF_x : T_x M \to T_{F(x)} N$ is a similarity, that is
$h_{F(x)}\left(dF_x(u), dF_x(u)\right) = a(x) g_x(u,u)$ for all $u \in T_x M$, where $a : M \to (0,+\infty)$ is smooth, that is $F^*(h) = ag$.

The composition of two conformal maps is conformal. In particular, the set of conformal maps from (M,g) to (M,g), denoted by Conf(M,g), has a natural group structure. For example, one can show that $\mathrm{Conf}\left(\mathbb{R}^n, g_{can}\right)$ is just the group of similarities of $\left(\mathbb{R}^n, g_{can}\right)$, that is the elements of the affine group of $\mathbb{R}^n$ that preserve ratios between distances. On the other hand, for every $n \geq 2$, $\mathrm{Conf}\left(S^n, g_{can}\right)$ coincides with the Möbius group $M\left(S^n\right)$ of S^n [and also with $M\left(\widehat{\mathbb{R}^n}\right)$] (see Chapter 1, §2 and Berger [1987] for details). Consequently, the group $\mathrm{Conf}\left(S^n, g_{can}\right)$ can be described via the canonical extension of stereographic projection.

In particular, $F_{x,\lambda} = \hat{\Psi}_x^{-1} \circ \hat{H}_\lambda \circ \hat{\Psi}_x$ is a conformal map of (S^n, g_{can}) for any vector homothety $H_\lambda : \mathbb{R}^n \to \mathbb{R}^n$ of ratio $\lambda \in (0,+\infty)$, where $\widehat{\Psi_x} : S^n \to \widehat{\mathbb{R}^n}$ denotes the canonical extension of the stereographic projection $\Psi_x : S^n \setminus \{x\} \to \mathbb{R}^n$, with x as the pole, and $\widehat{H_\lambda} : \widehat{\mathbb{R}^n} \to \widehat{\mathbb{R}^n}$ the canonical extension of H_λ (see Chapter 1, §2). If $\lambda > 1$, the iterates $F_{x,\lambda}^k \left(k \in \mathbb{N}\right)$ make every point in $S^n \setminus \{-x\}$ converge to x. Notice that as $\lambda \to \infty$, $F_{x,\lambda}(y) \to x$ for all $y \in S^n \setminus \{-x\}$.

$\mathrm{Conf}\left(S^n, g_{can}\right)$ contains a subset $\mathrm{Conf}_0\left(S^n, g_{can}\right)$ which is homeomorphic to the open unit $(n+1)$-ball $B^{n+1} = \left\{x \in \mathbb{R}^{n+1} \mid \|x\| < 1\right\}$ in the following way. Given $a \in B^{n+1}$, define $G_a(x) \in \mathbb{R}^{n+1}$ by

$$G_a(x) = \begin{cases} x & \text{if } a = 0 \text{ for } x \in \mathbb{R}^{n+1}, \\ \dfrac{1}{1 - 2\langle a, x\rangle + \|a\|^2 \|x\|^2}\left[\left(1 - \|a\|^2\right)x - \left(1 - 2\langle a,x\rangle + \|x\|^2\right)a\right] & \text{if } a \neq 0 \text{ for } x \in \mathbb{R}^{n+1} \setminus \left\{\dfrac{a}{\|a\|^2}\right\}, \end{cases}$$

where $\langle\ ,\ \rangle$ (resp. $\|\ \|$) denotes the canonical inner product (resp. norm) on $\mathbb{R}^{n+1}$. For $a \in B^{n+1} \setminus \{0\}$, note that

$$1 - 2\langle a, x\rangle + \|a\|^2 \|x\|^2 = \|a\|^2 \left\| x - \frac{a}{\|a\|^2} \right\|^2 \geq 0$$

for every $x \in \mathbb{R}^{n+1}$, where $\left\| \dfrac{a}{\|a\|^2} \right\| > 1$, and $G_a = G_b \left(a, b \in B^{n+1}\right)$ implies $a = b$. A direct computation gives

$$1-\|G_a(x)\|^2=\frac{\left(1-\|a\|^2\right)\left(1-\|x\|^2\right)}{1-2\langle a,x\rangle+\|a\|^2\|x\|^2}$$

for every $x\in\mathbb{R}^{n+1}\setminus\left\{\frac{a}{\|a\|^2}\right\}$, which shows that G_a induces mappings

$$G_a:S^n\to S^n,\ G_a:B^{n+1}\to B^{n+1},$$

provided $a\in B^{n+1}$. Moreover, for a unit vector $x\in\mathbb{R}^{n+1}$ normal to S^n, we have

$$(dG_a)_x(u)=\frac{1-\|a\|^2}{\left(1-2\langle a,x\rangle+\|a\|^2\right)^2}\left[\left(1-2\langle a,x\rangle+\|a\|^2\right)u+2\langle a,u\rangle(x-a)\right],$$

where $u\in T_xS^n\subset\mathbb{R}^{n+1}$ is a tangent vector, so that

$$\langle (dG_a)_x(u),(dG_a)_x(v)\rangle=\frac{\left(1-\|a\|^2\right)^2}{\left(1-2\langle a,x\rangle+\|a\|^2\right)^2}\langle u,v\rangle$$

for every $u,v\in T_xS^n\subset\mathbb{R}^{n+1}$. Consequently, $G_a:S^n\to S^n$ is a conformal mapping for every $a\in B^{n+1}$. On the other hand the definition of G_a implies that as a approaches $x_0\in S^n$, then $G_a(x)$ converges to $-\frac{\|x-x_0\|^2}{\|x-x_0\|^2}x_0=-x_0$ for every $x\in S^n\setminus\{x_0\}$.

Now, let us prove the following result.

Theorem 5.7. (Hersch [1970]). For any Riemannian metric g on S^2,

$$\lambda_1(g)\le\frac{8\pi}{Area\left(S^2,g\right)},$$

or equivalently

$$\lambda_1(g)Area\left(S^2,g\right)\le\lambda_1(g_{can})Area\left(S^2,g_{can}\right)$$

(see also Theorem 2.2).

Proof. By Theorem 3.1,

$$\lambda_1(g)=\inf_f\frac{\int_{S^2}|grad_g(f)|_g^2\,d\mu_g}{\int_{S^2}f^2d\mu_g},$$

where $|grad_g(f)|_g^2=g\left(grad_g(f),grad_g(f)\right)$ and f ranges over the set

$$A_1=\left\{f\in C^\infty\left(S^2\right)\setminus\{0\}\,\middle|\,\int_{S^2}f\,d\mu_g=0\right\}.$$

Let $x^1,x^2,x^3:S^2\to\mathbb{R}$ be the coordinate functions on $\mathbb{R}^3$ restricted to

$S^2 = \{x \in \mathbb{R}^3 \mid \|x\| = 1\}$. By Theorem 2.2 they are eigenfunctions of the Laplace-Beltrami operator Δ of (S^2, g_{can}) corresponding to the eigenvalue $\lambda_1(g_{can}) = 2$, forming a basis for the eigenspace $E(\Delta,2)$.

On the other hand, the Uniformization Theorem says that for any Riemannian metric g on S^2, always there exists a conformal diffeomorphism $F:(S^2, g) \to (S^2, g_{can})$, i.e. there exists a positive function $a \in C^\infty(S^2)$ such that $F^*(g) = a g_{can}$. Consequently, $F:\left(S^2, \frac{1}{a} g\right) \to (S^2, g_{can})$ is an isometry, so that by the formulae (2.22) and (2.21) (in dimension two) of Chapter 2, one gets

$$a\Delta_g \circ F^* = F^* \circ \Delta_{g_{can}},$$

where Δ_g (resp. $\Delta_{g_{can}}$) is the Laplace-Beltrami operator corresponding to g (resp. g_{can}). Therefore, by Proposition 2.3 (Green's formula) of Chapter 2, one obtains

$$\begin{aligned}\int_{S^2} g\left(grad_g(f \circ F), grad_g(f \circ F)\right) d\mu_g &= \int_{S^2} (f \circ F)\Delta_g(f \circ F)\, d\mu_g \\ &= \int_{S^2} (f \circ F)\left(\Delta_{g_{can}}(f) \circ F\right)\frac{1}{a}\, d\mu_g \\ &= \int_{S^2} f\Delta_{g_{can}}(f)\, d\mu_{g_{can}},\end{aligned}$$

for every $f \in C^\infty(S^2)$. In particular,

$$\int_{S^2} \left|\mathrm{grad}\left(x^i \circ F\right)\right|_g^2 d\mu_g = \int_{S^2} x^i \Delta_{g_{can}}\left(x^i\right) d\mu_{g_{can}} = 2\int_{S^2} \left(x^i\right)^2 d\mu_{g_{can}},$$

$i = 1,2,3$. The last equality is a consequence of a previous remark.

Now, let us compute $I_1 = \int_{S^2} \left(x^1\right)^2 d\mu_{g_{can}}$, using the spherical coordinates $(\cos\theta\cos\varphi, \cos\theta\sin\varphi, \sin\theta)$ on S^2, which implies $g'_{11} = 1,\ g'_{12} = g'_{21} = 0,\ g'_{22} = \cos^2\theta$. One obtains

$$I_1 = \int_0^{2\pi} \cos^2\theta|\cos\theta|\,d\theta \int_{-\frac{\pi}{2}}^{\frac{\pi}{2}} \cos^2\varphi\, d\varphi = \frac{1}{2}\left(\int_0^{\frac{\pi}{2}} \cos^3\theta\, d\theta - \int_{\frac{\pi}{2}}^{\frac{3\pi}{2}} \cos^3\theta\, d\theta + \int_{\frac{3\pi}{2}}^{2\pi} \cos^3\theta\, d\theta\right)\int_{-\frac{\pi}{2}}^{\frac{\pi}{2}} (1+\cos 2\varphi)\, d\varphi$$

$$= \frac{\pi}{2}\left[\int_0^{\frac{\pi}{2}}\left(\frac{3}{4}\cos\theta + \frac{1}{4}\cos 3\theta\right)d\theta - \int_{\frac{\pi}{2}}^{\frac{3\pi}{2}}\left(\frac{3}{4}\cos\theta + \frac{1}{4}\cos 3\theta\right)d\theta + \int_{\frac{3\pi}{2}}^{2\pi}\left(\frac{3}{4}\cos\theta + \frac{1}{4}\cos 3\theta\right)d\theta\right] = \frac{4\pi}{3}.$$

In the same way,

$$\int_{S^2} (x^2)^2 \, d\mu_{g_{can}} = \int_{S^2} (x^3)^2 \, d\mu_{g_{can}} = \frac{4\pi}{3}.$$

Hence

$$\int_{S^2} g\left(grad_g\left(x^i \circ F\right), grad_g\left(x^i \circ F\right)\right) d\mu_g = \frac{8\pi}{3} \tag{5.16}$$

for every $i = 1,2,3$. On the other hand, note that

$$Area\left(S^2, g\right) = \int_{S^2} 1 \cdot d\mu_g = \sum_{i=1}^{3} \int_{S^2} \left(x^i \circ F\right)^2 d\mu_g . \tag{5.17}$$

Therefore, at least for one i_0, $1 \le i_0 \le 3$,

$$\int_{S^2} \left(x^{i_0} \circ F\right)^2 d\mu_g \ge \frac{1}{3} Area\left(S^2, g\right), \tag{5.18}$$

because otherwise one contradicts (5.17). Thus, if one can choose F such that $\int_{S^2} \left(x^{i_0} \circ F\right) d\mu_g = 0$, by Theorem 3.1, (5.16) and (5.18) one obtains

$$\lambda_1(g) \le \frac{\int_{S^2} \left|grad_g\left(x^{i_0} \circ F\right)\right|_g^2 d\mu_g}{\int_{S^2} \left(x^{i_0} \circ F\right)^2 d\mu_g} \le \frac{8\pi}{Area\left(S^2, g\right)},$$

as desired.

The problem is reduced now to find a conformal map $F : \left(S^2, g\right) \to \left(S^2, g_{can}\right)$ such that $\int_{S^2} \left(x^{i_0} \circ F\right) d\mu_g = 0$. Fix a conformal map $F_0 : \left(S^2, g\right) \to \left(S^2, g_{can}\right)$ and let $Conf_0\left(S^2, g_{can}\right)$ be the subset of the conformal group $Conf\left(S^2, g_{can}\right)$ as above. Then $G_a \circ F_0 : \left(S^2, g\right) \to \left(S^2, g_{can}\right)$ is conformal for any $G_a \in Conf_0\left(S^2, g_{can}\right), a \in B^3$. Let us define $H : B^3 \to B^3$ by

$$H(a) = -\frac{1}{Area\left(S^2, g\right)} \left(\int_{S^2} x^1 \circ G_a \circ F_0 d\mu_g, \int_{S^2} x^2 \circ G_a \circ F_0 d\mu_g, \int_{S^2} x^3 \circ G_a \circ F_0 d\mu_g \right),$$

$a \in B^3$. Recall that as $a \in B^3$ approaches $x_0 = \left(x_0^1, x_0^2, x_0^3\right) \in S^2$, then $G_a(x)$ converges to $-x_0$ for every $x \in S^2 \setminus \{x_0\}$. Therefore, $\int_{S^2} x^i \circ G_a \circ F_0 d\mu_g$ converges to $-x_0^i Area\left(S^2, g\right), i = 1,2,3$, as a approaches $\left(x_0^1, x_0^2, x_0^3\right) \in S^2$, that is $H(a)$ converges to x_0 as a approaches $x_0 \in S^2$, and consequently H can be extended to $\bar{H} : D^3 \to D^3$ such that its restriction and corestriction to S^2 is the identity, where $D^3 = \left\{x \in \mathbb{R}^3 \mid \|x\| \le 1\right\}$. The map $\bar{H}$ is surjective. Indeed, let us assume that H is not surjective, namely the existence of a point $y_0 \in B^3$ such that $\bar{H}(x) \neq y_0$ for every $x \in D^3$. Then $r: D^3 \to S^2$ defined by

$$r(x) = y_0 + \frac{\langle y_0, y_0 - \bar{H}(x)\rangle + \sqrt{\langle y_0, \bar{H}(x) - y_0\rangle^2 + \left(1 - \|y_0\|^2\right)\|\bar{H}(x) - y_0\|^2}}{\|\bar{H}(x) - y_0\|^2}\left(\bar{H}(x) - y_0\right)$$

$$= \text{the unique point of } S^2 \text{ on the ray from } y_0 \text{ to } \bar{H}(x),$$

$x \in D^3$, has the property $r(x) = x$ for every $x \in S^2$, that is S^2 is a retract of D^3, in contradiction to the Brouwer fixed-point theorem. Therefore, there exists a point $a \in B^3$ such that $H(a) = 0 \in \mathbb{R}^3$, i.e.

$$\int_{S^2} x^i \circ G_a \circ F_0 d\mu_g = 0, i = 1, 2, 3.$$

Q.E.D.

Remark 5.1. Urakawa [1979] pointed out that Hersch's result cannot be generalized directly to higher dimensions, that is the inequality

$$\lambda_1(g)\left(Vol(M, g)\right)^{\frac{2}{n}} \leq c(n),$$

where $n = \dim M$, is not true in general. The bound should depend on some other geometric quantities (see also Chapter 8 as well as the above Cheng's estimates). On the other hand, Hersch's theorem was generalized to closed surfaces M_k^2 of genus $k \geq 1$ by Yang and Yau [1980]: For any Riemannian metric g on M_k^2, $\lambda_1(g) Area\left(M_k^2, g\right) \leq 8\pi(1+k)$. One method to prove this result is similar to Hersch's, by using branched conformal maps $F: M_k^2 \to S^2$. The existence of such a F comes from the Riemann-Roch theorem, i.e. on M_k^2 there exists a nonconstant meromorphic function having only one zero with multiplicity $\leq k+1$. Therefore, one can find a conformal branched covering $F: M_k^2 \to S^2$ with degree $\leq k+1$. In studying this type of problem, Li and Yau [1982] developed the important notion of conformal volume, which is closely related to λ_1 and to Willmore conjecture. Moreover, by using this concept, the Yang-Yau theorem has an even simpler proof (see Schoen and Yau [1994] for details).

Remark 5.2 If one scales the smooth Riemannian metric g on S^2 so that $\text{Area}\left(S^2, g\right) = 4\pi$, then Hersch's theorem becomes $\lambda_1(g) \leq 2$. For Riemannian metrics invariant under the standard S^1-action on $S^2 \subset \mathbb{R}^3$ given by rotation around the vertical axis (fixing the North and South poles), one can restrict the Laplace-Beltrami operator to the subspace of S^1-invariant functions and consider its spectrum there. The corresponding eigenvalues are called invariant eigenvalues and let us denote by

$$0 = \lambda_0(g) < \lambda_1(g) < \lambda_2(g) < \cdots$$

these invariant eigenvalues corresponding to the S^1-invariant Riemannian metric g. Using ideas from symplectic geometry, Abreu and Freitas [1999] showed that within the class of smooth S^1-invariant Riemannian metrics g on S^2 with total area 4π, the first invariant eigenvalue $\lambda_1(g)$ can be any number strictly between zero and $+\infty$. On the other hand, they show that within the class of smooth S^1-invariant metrics g on S^2 with total area 4π and such that (S^2, g) is isometric to a closed surface of revolution in $\mathbb{R}^3$, one has

$$\lambda_j(g) < \frac{\xi_j^2}{2},\ j \in \mathbb{N}^*,$$

where ξ_j is the $\left(\frac{(j+1)}{2}\right)$-th positive zero of the Bessel function J_0 if j is odd, and the $\left(\frac{j}{2}\right)$-th positive zero of J_0' if j is even. Moreover, these bounds are optimal. In particular,

$$\lambda_1(g) < \frac{\xi_1^2}{2} \approx 2.89.$$

There are also many interesting results concerning estimates for lower and upper bounds of the eigenvalues corresponding to the Dirichlet problem for bounded domains in $\mathbb{R}^n$ (for example, see Schoen and Yau [1994]). Next we shall present, following Hile and Protter [1980], an estimate of gaps among such higher eigenvalues.

Let us start with the following upper bound given by Payne, Polyà and Weinberger [1956]: Let M be a bounded domain in $\mathbb{R}^2$ with smooth boundary and μ_k be its k-th eigenvalue corresponding to the Dirichlet problem. Then these eigenvalues satisfy the inequality

$$\frac{1}{\mu_{k+1} - \mu_k} \sum_{j=1}^{k} \mu_j \geq \frac{k}{2} \tag{5.19}$$

for all $k = 1, 2, \ldots$.

In fact, the simple (and weaker) inequality

$$\mu_{k+1} \leq 3\mu_k \tag{5.20}$$

is obtained by replacing all the μ_j by μ_k in the left hand side of (5.20). One of the most interesting features of (5.19) and (5.20) is that the inequalities are independent of the size or shape of the domain, that is they are in some sense universal inequalities.

The Payne, Polyà and Weinberger result has a natural extension to bounded domains in $\mathbb{R}^n$, namely

$$\frac{1}{\mu_{k+1} - \mu_k} \sum_{j=1}^{k} \mu_j \geq \frac{nk}{4},\ k = 1, 2, \ldots.$$

Now, let M be such a domain (=connected open set) in $\mathbb{R}^n$, $n \geq 2$,

$$0 < \mu_1 \leq \mu_2 \leq \mu_3 \leq \cdots \to \infty$$

be $\mathbf{Spec}_D(\boldsymbol{M}, \boldsymbol{g}_{can})$ corresponding to the Dirichlet boundary problem and $f_1, f_2, f_3, \ldots$ an associated orthonormal system of real eigenfunctions of Δ, that is

$$\int_M f_i f_j \, d\mu_{g_{can}} = \delta_{ij},\ i, j \in \mathbb{N}^*.$$

Next we shall prove a stronger version of the previous inequality, namely the Hile-Protter inequality, that is

$$\sum_{j=1}^{k}\frac{\mu_j}{\mu_{k+1}-\mu_j}\geq\frac{nk}{4},\quad k=1,2,\ldots.$$

Before doing this, let us recall the classical Rayleigh-Ritz principle, which gives a simple way to bound eigenvalues corresponding to a Dirichlet boundary problem. It reads as follows:

$$\mu_{k+1}=\inf\left\{\frac{\int_M|\operatorname{grad}(f)|^2\,d\mu_{g_{can}}}{\int_M f^2 d\mu_{g_{can}}}\,\middle|\, f\in H_0^1(M)\setminus\{0\},\int_M ff_j\,d\mu_{g_{can}}=0, j=1,\ldots,k\right\},$$

where $H_0^1(M)$ denotes the completion of $C_0^\infty(M)$ with respect to the norm $\|\ \|$ defined by

$$\|f\|^2=\int_M f^2 d\mu_{g_{can}}+\int_M|\operatorname{grad}(f)|^2\,d\mu_{g_{can}},\ f\in C_0^\infty(M).$$

Theorem 5.8. (Hile and Protter [1980]). If t_0 is the unique root on $(\mu_m,+\infty)$ of the equation

$$\sum_{j=1}^{m}\frac{\mu_j}{t-\mu_j}=\frac{mn}{4},\tag{5.21}$$

then $t_0\geq\mu_{m+1}$.

This theorem will be a consequence of some auxiliary results.

We first prove the following

Lemma 5.1. The equation (5.21) has a unique root on the interval $(\mu_m,+\infty)$.

Proof. Use the Rolle sequence. Let

$$f(t)=\sum_{j=1}^{m}\frac{\mu_j}{t-\mu_j}-\frac{mn}{4}.$$

Then

$$f'(t)=-\sum_{j=1}^{m}\frac{\mu_j}{(t-\mu_j)^2}<0,$$

$$\lim_{\substack{t\to\mu_m\\ t>\mu_m}}f(t)=\frac{\mu_1}{\mu_m-\mu_1}+\cdots+\frac{\mu_{m-1}}{\mu_m-\mu_{m-1}}+\lim_{\substack{t\to\mu_m\\ t>\mu_m}}\frac{\mu_m}{t-\mu_m}=+\infty,$$

$$\lim_{t\to\infty}f(t)=-\frac{mn}{4}<0.$$

Therefore there exists a unique root of the equation (5.21) on the interval $(\mu_m,+\infty)$, as required.

Q.E.D.

Lemma 5.2. For any $i, 1 \le i \le m$, let u_i be the trial function given by

$$u_i = x_1 f_i - \sum_{j=1}^{m} a_{ij} f_j ,$$

where the constants a_{ij} are given by

$$a_{ij} = \int_M x_1 f_i f_j d\mu_{g_{can}}, \; i, j = 1, \ldots, m .$$

Then

(i) Each trial function u_i is orthogonal to $f_1, \ldots, f_m$;

(ii) $\Delta u_i = x_1 \mu_i f_i - 2\dfrac{\partial f_i}{\partial x_1} - \sum_{j=1}^{m} a_{ij} \mu_j f_j, \; i = 1, \ldots, m$;

(iii) $\int_M u_i \Delta u_i d\mu_{g_{can}} = \mu_i \int_M x_1 f_i u_i d\mu_{g_{can}} - 2 \int_M u_i \dfrac{\partial f_i}{\partial x_1} d\mu_{g_{can}}, \; i = 1, \ldots, m$;

(iv) $\int_M u_i^2 d\mu_{g_{can}} = \int_M x_1 f_i u_i d\mu_{g_{can}}, \; i = 1, \ldots, m$.

Proof. (i) For each $i, k = 1, \ldots, m$, one can write successively

$$\begin{aligned}
\langle u_i, f_k \rangle &= \int_M u_i f_k d\mu_{g_{can}} = \int_M x_1 f_i f_k d\mu_{g_{can}} - \int_M \sum_{j=1}^{m} a_{ij} f_j f_k d\mu_{g_{can}} \\
&= \int_M x_1 f_i f_k d\mu_{g_{can}} - \sum_{j=1}^{m} a_{ij} \delta_{jk} = \int_M x_1 f_i f_k d\mu_{g_{can}} - \sum_{j=1}^{m} \left(\int_M x_1 f_i f_j d\mu_{g_{can}} \right) \delta_{jk} \\
&= \int_M x_1 f_i f_k d\mu_{g_{can}} - \int_M x_1 f_i f_k d\mu_{g_{can}} = 0.
\end{aligned}$$

(ii) Since

$$\Delta f_i = \mu_i f_i, \; i = 1, \ldots, m ,$$

it follows that

$$\begin{aligned}
\Delta u_i &= \Delta(x_1 f_i) - \sum_{j=1}^{m} a_{ij} \Delta f_j = -\left[\frac{\partial^2}{\partial x_1^2}(x_1 f_i) + \cdots + \frac{\partial^2}{\partial x_n^2}(x_1 f_i) \right] - \sum_{j=1}^{m} a_{ij} \mu_j f_j \\
&= -2\frac{\partial f_i}{\partial x_1} - x_1 \left(\frac{\partial^2 f_i}{\partial x_1^2} + \cdots + \frac{\partial^2 f_i}{\partial x_n^2} \right) - \sum_{j=1}^{m} a_{ij} \mu_j f_j \\
&= x_1 \Delta f_i - 2\frac{\partial f_i}{\partial x_1} - \sum_{j=1}^{m} a_{ij} \mu_j f_j = x_1 \mu_i f_i - 2\frac{\partial f_i}{\partial x_1} - \sum_{j=1}^{m} a_{ij} \mu_j f_j ,
\end{aligned}$$

for each $i = 1, \ldots, m$.

(iii) Let us multiply the equality

$$\Delta u_i = x_1 \mu_i f_i - 2\frac{\partial f_i}{\partial x_1} - \sum_{j=1}^{m} a_{ij} \mu_j f_j$$

by u_i and let us integrate over M. One gets

$$\begin{aligned}\int_M u_i \Delta u_i d\mu_{g_{can}} &= \mu_i \int_M x_1 f_i u_i d\mu_{g_{can}} - 2\int_M u_i \frac{\partial f_i}{\partial x_1} d\mu_{g_{can}} - \sum_{j=1}^m a_{ij}\mu_j \int_M f_j u_i d\mu_{g_{can}} \\ &= \mu_i \int_M x_1 f_i u_i d\mu_{g_{can}} - 2\int_M u_i \frac{\partial f_i}{\partial x_1} d\mu_{g_{can}},\end{aligned}$$

for each $i = 1,\ldots,m$, where the last equality is a direct consequence of the property (i).

(iv) The definition of u_i implies

$$u_i^2 = x_1 f_i u_i - \sum_{j=1}^m a_{ij} f_j u_i, \quad i = 1,\ldots,m,$$

so that if we integrate over M, one obtains

$$\int_M u_i^2 d\mu_{g_{can}} = \int_M x_1 f_i u_i d\mu_{g_{can}} - \sum_{j=1}^m a_{ij} \int_M f_j u_i d\mu_{g_{can}} = \int_M x_1 f_i u_i d\mu_{g_{can}},$$

for each $i = 1,\ldots,m$, where the last equality is again a direct consequence of the property (i).

Q.E.D.

Lemma 5.3. For each $m \in \mathbb{N}$ and $i = 1,\ldots,m$, one has

$$\mu_{m+1} \le \frac{\int_M u_i \Delta u_i d\mu_{g_{can}}}{\int_M u_i^2 d\mu_{g_{can}}}.$$

Proof. Using the mini-max characterization of the (m+1)-th eigenvalue μ_{m+1} of Δ, one has

$$\mu_{m+1} = \inf_f \frac{\int_M |\mathrm{grad}(f)|^2 d\mu_{g_{can}}}{\int_M f^2 d\mu_{g_{can}}},$$

where f ranges over the set

$$A_{n+1} = \left\{ f \in C^\infty(M)\setminus\{0\} \,\middle|\, \int_M f f_j d\mu_{g_{can}} = 0, j = 1,\ldots,m, \text{ and } f|_{\partial M} = 0 \right\}.$$

Clearly, $u_i|_{\partial M} = 0, i = 1,\ldots,m$. Then by Lemma 5.2(i) one gets

$$\mu_{m+1} \le \frac{\int_M |grad(u_i)|^2 d\mu_{g_{can}}}{\int_M u_i^2 d\mu_{g_{can}}},$$

so that Green's formula implies the desired inequality.

Q.E.D.

Using now Lemmas 5.3 and 5.2(iii),(iv) one obtains successively

$$\mu_{m+1}\int_M u_i^2 d\mu_{g_{can}} \le \int_M u_i \Delta u_i d\mu_{g_{can}} = \mu_i \int_M x_1 f_i u_i d\mu_{g_{can}} - 2\int_M u_i \frac{\partial f_i}{\partial x_1} d\mu_{g_{can}}$$
$$= \mu_i \int_M u_i^2 d\mu_{g_{can}} - 2\int_M u_i \frac{\partial f_i}{\partial x_1} d\mu_{g_{can}}$$

for each $i = 1,\ldots,m$. Therefore,

$$\mu_{m+1}\sum_{i=1}^m \int_M u_i^2 d\mu_{g_{can}} \le \sum_{i=1}^m \mu_i \int_M u_i^2 d\mu_{g_{can}} - 2\sum_{i=1}^m \int_M u_i \frac{\partial f_i}{\partial x_1} d\mu_{g_{can}}. \tag{5.22}$$

Lemma 5.4. The following equality holds:

$$-2\sum_{i=1}^m \int_M u_i \frac{\partial f_i}{\partial x_1} d\mu_{g_{can}} = m.$$

Proof. Indeed, using Green's formula and the symmetry of a_{ij} one can write successively

$$-2\sum_{i=1}^m \int_M u_i \frac{\partial f_i}{\partial x_1} d\mu_{g_{can}} = -2\sum_{i=1}^m \int_M g_{can}\left(grad(x_1), grad(f_i)\right) u_i d\mu_{g_{can}}$$
$$= -2\sum_{i=1}^m \int_M g_{can}\left(x_1 grad(x_1), f_i grad(f_i)\right) d\mu_{g_{can}} + 2\sum_{i,j=1}^m a_{ij}\int_M g_{can}\left(f_j grad(f_i), grad(x_1)\right) d\mu_{g_{can}}$$
$$= \sum_{i=1}^m\left[-\frac{1}{2}\int_M g_{can}\left(grad(x_1^2), grad(f_i^2)\right) d\mu_{g_{can}}\right] + \sum_{i,j=1}^m a_{ij}\int_M g_{can}\left(grad(f_i f_j), grad(x_1)\right) d\mu_{g_{can}}$$
$$= \sum_{i=1}^m\left[-\frac{1}{2}\int_M f_i^2 \Delta(x_1^2) d\mu_{g_{can}}\right] + \sum_{i,j=1}^m a_{ij}\int_M (f_i f_j)\Delta(x_1) d\mu_{g_{can}} = \sum_{i=1}^m \int_M f_i^2 d\mu_{g_{can}} = m.$$

Q.E.D.

Consequently, (5.22) and Lemma 5.4 imply

$$\mu_{m+1}\sum_{i=1}^m \int_M u_i^2 d\mu_{g_{can}} \le \sum_{i=1}^m \mu_i \int_M u_i^2 d\mu_{g_{can}} + m. \tag{5.23}$$

On the other hand, note that Lemma 5.4 implies

$$-2(1+\beta)\sum_{i=1}^m \int_M u_i \frac{\partial f_i}{\partial x_1} d\mu_{g_{can}} = (\beta+1)m,$$

where β is a positive real constant to be determined later, so that

$$m = -2(1+\beta)\sum_{i=1}^m \int_M u_i \frac{\partial f_i}{\partial x_1} d\mu_{g_{can}} - \beta m.$$

If one inserts this value of m in (5.23) one obtains

$$\mu_{m+1}\sum_{i=1}^m \int_M u_i^2 d\mu_{g_{can}} \le \sum_{i=1}^m \mu_i \int_M u_i^2 d\mu_{g_{can}} - 2(1+\beta)\sum_{i=1}^m \int u_i \frac{\partial f_i}{\partial x_1} d\mu_{g_{can}} - m\beta. \tag{5.24}$$

Lemma 5.5. Let $\alpha_1,\ldots,\alpha_m \in (0,+\infty)$ be constants. Then

$$-2(1+\beta)\int_M u_i \frac{\partial f_i}{\partial x_1} d\mu_{g_{can}} \le \alpha_i \int_M u_i^2 d\mu_{g_{can}} + \frac{(1+\beta)^2}{\alpha_i} \int_M \left(\frac{\partial f_i}{\partial x_1}\right)^2 d\mu_{g_{can}} \tag{5.25}$$

for each $i=1,\ldots,m$.

Proof. Indeed, notice that

$$\left(\sqrt{\alpha_i}u_i + \frac{1+\beta}{\sqrt{\alpha_i}}\frac{\partial f_i}{\partial x_1}\right)^2 \ge 0$$

for each $i=1,\ldots,m$ and β, or equivalently

$$\alpha_i u_i^2 + \frac{(1+\beta)^2}{\alpha_i}\left(\frac{\partial f_i}{\partial x_1}\right)^2 + 2(1+\beta)u_i\frac{\partial f_i}{\partial x_1} \ge 0$$

for each $i=1,\ldots,m$ and β. Hence

$$-2(1+\beta)u_i\frac{\partial f_i}{\partial x_1} \le \alpha_i u_i^2 + \frac{(1+\beta)^2}{\alpha_i}\left(\frac{\partial f_i}{\partial x_1}\right)^2$$

for each $i=1,\ldots,m$, so that integrating this over M one gets the desired inequalities.

Q.E.D.

Inserting now (5.25) in (5.24) one can conclude that

$$\begin{aligned}\mu_{m+1}\sum_{i=1}^m \int_M u_i^2 d\mu_{g_{can}} \le \sum_{i=1}^m \mu_i \int_M u_i^2 d\mu_{g_{can}} + \sum_{i=1}^m \alpha_i \int_M u_i^2 d\mu_{g_{can}} \\ + \sum_{i=1}^m \frac{(1+\beta)^2}{\alpha_i} \int_M \left(\frac{\partial f_i}{\partial x_1}\right)^2 d\mu_{g_{can}} - m\beta.\end{aligned} \tag{5.26}$$

Let us set $\alpha_m = \alpha$ and choose $\alpha_1,\ldots,\alpha_{m-1}$ so that

$$\alpha_i = \alpha + \mu_m - \mu_i, \quad i=1,\ldots,m-1.$$

Let us define

$$S_1 = \sum_{i=1}^m \int_M u_i^2 d\mu_{g_{can}}$$

and let us put (5.26) in the form

$$\mu_{m+1}S_1 \le (\mu_m+\alpha)S_1 + (1+\beta)^2\sum_{i=1}^m \frac{1}{\alpha_i}\int_M \left(\frac{\partial f_i}{\partial x_1}\right)^2 d\mu_{g_{can}} - m\beta.$$

In place of the trial functions u_i one may choose the functions

$$u_{ik} = x_k u_i - \sum_{j=1}^m a_{ijk} u_j ,$$

$i=1,\dots,m$, $k=1,\dots,n$, where

$$a_{ijk}=\int_M x_k f_i f_j d\mu_{g_{can}},\ a_{ij1}=a_{ij}.$$

Clearly, $u_{i1}=u_i$, $i=1,\dots,m$. The above analysis may be performed for $k=2,\dots,n$, and denoting

$$S_k=\sum_{i=1}^m\int_M u_{ik}^2 d\mu_{g_{can}},\ k=1,\dots,n,$$

one gets the following n inequalities:

$$\mu_{m+1}S_k\le(\mu_m+\alpha)S_k+(1+\beta)^2\sum_{i=1}^m\frac{1}{\alpha_i}\int_M\left(\frac{\partial f_i}{\partial x_k}\right)^2 d\mu_{g_{can}}-m\beta, \tag{5.27}$$

$k=1,\dots,n$. Setting

$$S=\sum_{k=1}^n S_k$$

and adding all inequalities (5.27) one obtains

$$\mu_{m+1}S\le(\mu_m+\alpha)S+(1+\beta)^2\sum_{i=1}^m\frac{1}{\alpha_i}\int_M|grad(f_i)|^2 d\mu_{g_{can}}-mn\beta,$$

or equivalently

$$\mu_{m+1}S\le(\mu_m+\alpha)S+(1+\beta)^2\sum_{i=1}^m\frac{\mu_i}{\alpha_i}-mn\beta. \tag{5.28}$$

The selection of α such that

$$(1+\beta)^2\sum_{i=1}^m\frac{\mu_i}{\alpha_i}-mn\beta\le 0$$

i.e. so that

$$\sum_{i=1}^m\frac{\mu_i}{\alpha_i}\le\frac{\beta}{(1+\beta)^2}mn \tag{5.29}$$

yields an inequality for the α_i in terms of β. The right hand side of (5.29) is maximized when $\beta=1$ and thus the condition on α becomes

$$\sum_{i=1}^m\frac{\mu_i}{\alpha_i}\le\frac{1}{4}mn$$

or equivalently

$$\sum_{i=1}^m\frac{\mu_i}{\alpha+\mu_m-\mu_i}\le\frac{1}{4}mn. \tag{5.30}$$

The inequality (5.28) can now be written as

$$\mu_{m+1}S \le (\mu_m+\alpha)S+(1+\beta)^2\sum_{i=1}^{m}\frac{\mu_i}{\alpha_i}-mn\beta<(\mu_m+\alpha)S,$$

because

$$(1+\beta)^2\sum_{i=1}^{m}\frac{\mu_i}{\alpha_i}-mn\beta<0.$$

Since $S>0$, one gets

$$\mu_{m+1}\le\mu_m+\alpha. \tag{5.31}$$

Let us observe now that if one takes $t=\alpha+\mu_m$, then (5.30) becomes

$$\sum_{i=1}^{m}\frac{\mu_i}{t-\mu_i}-\frac{1}{4}mn\le 0.$$

Therefore, (5.31) implies

$$\mu_{m+1}\le t_0,$$

which finishes the proof of Theorem 5.8.

Using now Theorem 5.8 one can obtain some surprising inequalities.

Corollary 5.2. Under the above hypotheses on M, one has

$$\sum_{i=1}^{m}\frac{\mu_i}{\mu_{m+1}-\mu_i}\ge\frac{mn}{4} \tag{5.32}$$

for $m=1,2,\dots$.

Proof. We have the following distribution of eigenvalues of Δ (see Figure 5.1). It follows that

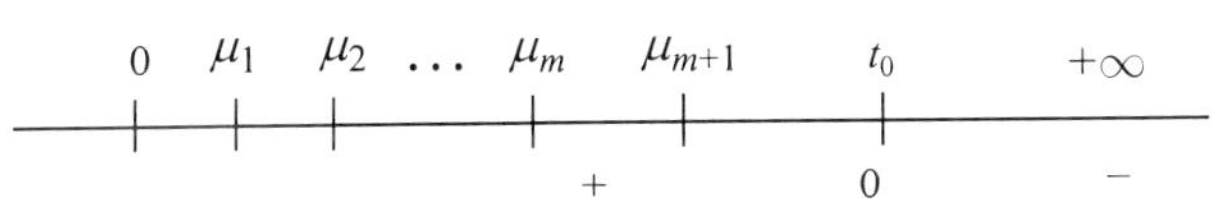

Figure 5.1: Universal inequalities, where $\sum_{i=1}^{m}\frac{\mu_i}{t_0-\mu_i}=\frac{mn}{4}$.

$$f(\mu_{m+1})\ge 0,$$

where

$$f(t)=\sum_{i=1}^{m}\frac{\mu_i}{t-\mu_i}-\frac{mn}{4}$$

(see the proof of Lemma 5.1), that is the desired inequality.

Q.E.D.

As we have already noticed in dimension $n=2$, it is an amazing fact that the inequality in Corollary 5.2 does not depend of the size or shape of the bounded domain M for Dirichlet boundary conditions.

Remark 5.3. From Corollary 5.2 one can also see that the weaker inequality of Payne, Pólya and Weinberger

$$\frac{1}{\mu_{m+1}-\mu_m}\sum_{i=1}^{m}\mu_i \geq \frac{mn}{4} \tag{5.33}$$

for $m = 1,2,\ldots$ holds for each bounded domain $M \subset \mathbb{R}^n$. Indeed, if one replaces μ_i appearing in the denominator of the left-hand side of (5.32) by μ_m one obtains (5.33).

Let us notice now that using the last inequality one can write successively

$$\mu_{m+1}-\mu_m \leq \frac{4}{mn}\sum_{i=1}^{m}\mu_i \leq \frac{4}{mn}\sum_{i=1}^{m}\mu_m = \frac{4}{n}\mu_m$$

and therefore

$$\mu_{m+1} \leq \frac{n+4}{n}\mu_m \tag{5.34}$$

for $m = 1,2,\ldots$.

Hence one can conclude that no sequence of positive real numbers with the property, say, that one element in the sequence is more than $(n+4)/n$ times its immediate predecessor, can be in $\boldsymbol{Spec}_D(\boldsymbol{M}, \boldsymbol{g}_{can})$ of a bounded domain M in $\mathbb{R}^n$ with smooth boundary.

Thus the inequalities (5.34) impose restrictions of another type as the multiplicity in order that the result of Colin de Verdière mentioned in Remark 1.5 to be true for bounded domains of $\mathbb{R}^n$ in the setting of the Dirichlet eigenvalue problem. On the other hand, one can show that the terms of any finite sequence $\{0 = \nu_0 < \nu_1 \leq \cdots \leq \nu_k\}$ (resp. $\{0 = \nu_0 < \nu_1 < \cdots < \nu_k\}$) are the first $k+1$ eigenvalues corresponding to the Neumann eigenvalue problem on a bounded domain of $\mathbb{R}^n$, $n \geq 3$ (resp. $n = 2$) with smooth boundary (for details, see Colin de Verdière [1987]).

Remark 5.4 Hile-Protter's inequality (5.32) has been extended by H.C. Yang [1991] as follows:

$$\sum_{i=1}^{m}(\mu_{m+1}-\mu_i)\left(\mu_{m+1}-\frac{n+4}{n}\mu_i\right) \leq 0 \text{ for } m = 1,2,\ldots. \tag{5.35}$$

Noting that the left-hand side of (5.35) is quadratic with respect to μ_{m+1}, one can write it as

$$m\mu_{m+1}^2 - \frac{2(n+2)}{n}\left(\sum_{i=1}^{m}\mu_i\right)\mu_{m+1} + \frac{n+4}{4}\sum_{i=1}^{m}\mu_i^2 \leq 0 \tag{5.36}$$

and derive the upper bound

$$\lambda_{m+1} \leq [\text{larger root of the left-hand side of (5.36)}]$$

$$= \frac{1}{m}\left\{\frac{n+2}{n}\sum_{i=1}^{m}\mu_i + \left[\left(\frac{n+2}{n}\right)^2\left(\sum_{i=1}^{m}\mu_i\right)^2 - m\frac{n+4}{n}\sum_{i=1}^{m}\mu_i^2\right]^{\frac{1}{2}}\right\} \tag{5.37}$$

for $m=1,2,\ldots$. This is the best general upper bound yet derived by any methods for that matter. Using the Cauchy-Schwarz inequality to replace $m\sum_{i=1}^{m}\mu_i^2$ by $\left(\sum_{i=1}^{m}\mu_i\right)^2$ on the right-hand side of (5.37), one gets

$$\mu_{m+1} \leq \frac{n+4}{mn}\sum_{i=1}^{m}\mu_i \text{ for } m=1,2,\ldots. \tag{5.38}$$

Note that (5.38) implies (5.34), since the average $\frac{1}{m}\sum_{i=1}^{m}\mu_i$, occuring on the right-hand side of (5.38), is clearly less than μ_m (and is, in fact, strictly less for all $m \geq 2$ since $\mu_1 < \mu_k$ for all $k \geq 2$). For a proof of (5.35) as well as for further comments and observations concerning the inequalities (5.32), (5.33), (5.38) and (5.35), we refer to the very nice paper of Ashbaugh [1999].

Concerning a lower bound for $\mu_2 - \mu_1$, let us mention the following result of Singer, Wong, Yau, S.-T. and Yau, S.S.-T. [1985].

Theorem 5.9. Let M be a bounded convex domain in $\mathbb{R}^n$ with smooth boundary, μ_1 (resp. μ_2) be the first (resp. second) Dirichlet eigenvalue of Δ. Then

$$\mu_2 - \mu_1 \geq \frac{\pi^2}{4D^2\left(M, g_{can}\right)},$$

where $D\left(M, g_{can}\right)$ is the diameter of $\left(M, g_{can}\right)$.

For a proof of this result, see Schoen and Yau [1994].

REFERENCES

ABREU, M. and FREITAS, P., *On the invariant spectrum of S^1-invariant metrics on S^2*, preprint September 1999.

ASHBAUGH, M.S., *Isoperimetric and universal inequalities for eigenvalues*, in "Spectral Theory and Geometry" (edited by B. Davies and Y. Safarov), ICMS Instructional Conference Edinburgh, 30 March-9 April 1998, pp. 95-139, London Mathematical Society, Lecture Note Series 273, Cambridge University Press, 1999.

ALBERT, J.H., *Generic properties of eigenfunctions of elliptic partial differential equations*, Trans. Am. Math. Soc. 238(1978), 341-354.

BANDO, S. and URAKAWA, H., *Generic properties of the eigenvalues of the Laplacian for compact Riemannian manifolds*, Tôhoku Math. Journal 35(1983), 155-172.

BÉRARD BERGERY, L. and BOURGUIGNON, J.P., *Laplacians and Riemannian submersions with totally geodesic fibers*, Illinois J. Math. 26(1982), 181-200.

BÉRARD, P. and GALLOT, S., *Inégalités isopérimétriques pour l'equation de la chaleur et applications à l'estimation de quelques invariants géométri-*

ques, in Séminaire Goulaouic-Meyer-Schwarz 1983-84, exposé no. XV, École Polytechnique, Palaiseau, 1984.

BERGER, M., *Geometry II*, Universitext, Springer-Verlag, Berlin-Heidelberg, 1987.

BERGER, M., GAUDUCHON, P. and MAZET, E., *Le Spectre d'une Variété Riemannienne*, Lecture Notes in Math., Vol. 194, Springer-Verlag, Berlin, 1971.

BESSON, G., *A Kato type inequality for Riemannian submersions with totally geodesic fibers*, Ann. Glob. Analysis and Geometry, 4(1986), 273-289.

BESSON, G. and BORDONI, M., *On the spectrum of Riemannian submersions with totally geodesic fibers*, Rend. Mat. Acc. Lincei, s. 9, v. 1(1990), 335-340.

BORDONI, M., *Bounding from below the spectrum of a Riemannian covering*, Proc. Int. Conf. Diff. Geom. Appl., Dubrovnik (1988), 83-90.

BORDONI, M., *An estimate for finite sums of eigenvalues of fiber spaces*, C.R. Acad. Sci. Paris, t. 315, s. I(1992), 1079-1083.

BORDONI, M., *Comparing heat operators through local isometries or fibrations*, Bull. Soc. Math. France, 128(2000), 151-178.

BUSER, P., *Geometry and Spectra of Compact Riemann Surfaces*, Progress in Mathematics, Vol. 106, Birkhäuser, Boston, 1992.

CAHN, R. and WOLF, J., *Zeta functions and their asymptotic expansions for compact symmetric spaces of rank one*, Comment. Math. Helvetici, 51(1976), 1-21.

CHAVEL, I., *Eigenvalues in Riemannian Geometry*, Academic Press, New York, 1984.

CHEEGER, J., *A lower bound for the smallest eigenvalue of the Laplacian*, "Problems in Analysis", pp. 195-199, Princeton Univ. Press, Princeton, New Jersey, 1970.

CHEEGER, J. and YAU, S. –T., *A lower bound for the heat kernel*, Commun. Pure Appl. Math. 34(1981), 465-480.

CHENG, S.Y., *Eigenvalue comparison theorems and its geometric applications*, Math. Z., 143(1975), 289-297.

CHENG, S.Y. and LI, P., *Heat kernel estimates and lower bounds of eigenvalues*, Comment. Math. Helv. 56(1981), 327-338.

COLIN de VERDIÈRE, Y., *Spectres de variétés riemanniennes et spectres de graphes*, Proceedings of ICM, 1986, Berkeley, California, USA, Vol. 1, 522-530.

COLIN de VERDIÈRE, Y., *Construction de laplaciens dont une partie finie du spectre est donnée*, Ann. Sci. Éc. Norm. Sup. Paris, 20(1987), 599-615.

CRAIOVEANU, M. and PUTA, M., *Introducere in Geometria Spectrală*, Editura Academiei Române, Bucuresti, 1988.

DAVIES, E.B., *Heat kernels and spectral theory*, Cambridge University Press, Cambridge-New York-New Rochelle, 1989.

DETURCK, D.M., *Audible and inaudible geometric properties*, Rendiconti Seminari Facolta Scienze Università Cagliari, Supplemento al Vol. 58(1988), 1-26.

DONNELLY, H., *Asymptotic expansions for the compact quotients of proper discontinuous group actions*, Ill. J. Math. 23(1979), 485-496.

EBIN, D., *The manifold of Riemannian metrics*, Proc. Sympos. in Pure Math. AMS, XV (1968), 11-40.

GILKEY, P.B., *The Index Theorem and the Heat Equation*, Publish or Perish Lecture Notes 4, Boston, Massachusetts, 1974.

GILKEY, P.B. and PARK, J.H., *Riemannian submersions which preserve the eigenforms of the Laplacian*, Illinois J. Math. 40(1996), 194-201.

GORDON, C.S. and WILSON, E.N., *The spectrum of the Laplacian on Riemannian Heisenberg manifolds*, Mich. Math. J. 33(1986), 253-271.

GRIGOR'YAN, A., *Estimates of heat kernels on Riemannian manifolds*, in "Spectral Theory and Geometry" (edited by B. Davies and Y. Safarov), ICMS Instructional Conference, Edinburgh, 30 March - 9 April 1998, pp. 140-225, London Mathematical Society, Lecture Note Series 273, Cambridge University Press, 1999.

HEJHAL, D.A., *The Selberg trace formula for* $PSL(2,\mathbb{R})$, *Vol. II*, Lecture Notes in Math., Vol. 1001, Springer-Verlag, Berlin and New York, 1983.

HERMANN, R., *A sufficient condition that a mapping of Riemannian manifolds be a fibre bundle*, Proc. Amer. Math. Soc. 11(1960), 236-242.

HERSCH, J., *Quatre propriétés isopérimetriques de membranes sphériques*, C.R. Acad. Sci., Paris 270(1970), 1645-1648.

HESS, H., SCHRADER, R. and UHLENBROCK, D.A., *Kato's inequality and the spectral distribution of Laplacians on compact Riemannian manifolds*, J. Differential Geometry 15(1980), 27-38.

HILE, G.N. and PROTTER, M.H., *Inequalities for eigenvalues of the Laplacian*, Indiana Univ. Math. Journal, vol. 29, no. 4(1980), 523-538.

HILE, G.N. and XU, Z., *Inequalities for sums of reciprocals of eigenvalues*, J. Math. Anal. Appl. 180(1993), 412-430.

IKEDA, A. and YAMAMOTO, Y., *On the spectra of 3-dimensional lens spaces*, Osaka J. Math., 16(1979), 447-469.

KAC, M., *Can one hear the shape of a drum?*, Amer. Math. Monthly 73(1966), 1-23.

KALKA, M., MANN, E., YANG, D. and ZINGER, A., *The exponential decay rate of the lower bound for the first eigenvalue of compact manifolds*, Internat. J. Math. 8(1997), no. 3, 345-355.

Chapter 4
ISOSPECTRAL CLOSED RIEMANNIAN MANIFOLDS

1. ASYMPTOTIC EXPANSION FOR THE TRACE OF THE HEAT KERNEL AND CONSEQUENCES

Since it is difficult to study $\mathrm{Spec}(M,g)$ directly, instead one introduces certain functions of the eigenvalues of Δ, which can be used to extract more geometric information. Some useful such functions having applications to spectral geometry are the heat coefficients. These can be made obvious by the method of heat asymptotics, which is discussed below.

Let (M,g) be a closed, connected, n-dimensional smooth Riemannian manifold with distance function $d: M\times M \to [0,+\infty)$ (see Chapter 1, §2). There exists $\varepsilon>0$ [for example, take $\varepsilon>0$ less or equal then the injectivity radius of (M,g)] such that for all $x\in M$, the exponential map $\exp_x$ takes the open ball $B_\varepsilon(0_x)$ of radius ε of T_xM centered at 0_x diffeomorphically onto an open set $O_{x,\varepsilon}$ in M containing x (see Chapter 1, §5). For $y\in O_{x,\varepsilon}$, the length of the geodesic joining x with y is strictly less than ε. Let us consider the neighborhood U_ε of the diagonal in $M\times M$ defined by
$U_\varepsilon=\{(x,y)\in M\times M \mid y\in O_{x,\varepsilon}\}$.

In Chapter 3, we have considered the existence of the heat kernel on closed Riemannian manifolds (see Definition 1.1 and Corollary 1.1 of that Chapter). In order to prove this existence, one constructs a parametrix $H_k: M\times M\times(0,+\infty)\to\mathbb{R}$ for the heat operator $\Delta+\frac{\partial}{\partial t}$, where k is a natural number $>\frac{n}{2}+2$, that is H_k is a smooth function such that

(i) $\left(\Delta+\frac{\partial}{\partial t}\right)H_k: M\times M\times[0,+\infty)\to\mathbb{R}$ is C^2-smooth, where Δ is the Laplace-Beltrami operator associated to (M,g) acting on the second variable, and

(ii) $\lim_{t\searrow 0}\int_M H_k(x,y,t)f(y)\,d\mu_g(y)=f(x)$ for any $x\in M$ and for all continuous functions $f: M\to\mathbb{R}$.

In fact one shows that H_k can be defined as follows

$$H_k(x,y,t)=\begin{cases}(4\pi t)^{-\frac{n}{2}}\eta(x,y)e^{-\frac{d^2(x,y)}{4t}}\sum_{j=0}^{k}u_j(x,x)t^j, & (x,y,t)\in U_\varepsilon\times(0,+\infty),\\ 0 & , \ (x,y,t)\in(M\times M\setminus U_\varepsilon)\times(0,+\infty),\end{cases}$$

where $(u_j)_{j\in\mathbb{N}}$ is a sequence of suitable chosen C^∞-smooth real functions defined on the open neighborhood U_ε of the diagonal in $M\times M$ and $\eta\in C^\infty(M\times M)$ is a bump function with $\eta(x,y)\in[0,1]$ for any $(x,y)\in M\times M$, $\eta|_{M\times M\setminus U_\varepsilon}=0$ and $\eta|_{U_{\varepsilon/2}}=1$. In particular, $u_0(x,x)=1$ for any $x\in M$.

Then one proves that the real function $E: M\times M\times(0,+\infty)\to\mathbb{R}$ defined by

$$E = H_k - Q_k * H_k$$

where Q_k is the series $\sum_{j=1}^{\infty}(-1)^{j+!}K_k^{*j}$, $Q_k * H_k$ denotes the convolution of the functions Q_k and H_k, that is

$$(Q_k * H_k)(x,y,t)=\int_0^t d\tau\int_M Q_k(x,z,\tau)H_k(z,y,t-\tau)d\mu_g(z), (x,y,t)\in M\times M\times[0,+\infty),$$

and K_k^{*j} denotes the j-fold convolution product $K_k*\cdots*K_k$, $K_k^{*1}=K_k=\left(\Delta+\frac{\partial}{\partial t}\right)H_k$, is just the heat kernel associated to (M,g) (if $k>\frac{n}{2}+2$) and by the uniqueness of the heat kernel must be independent of k. Concerning the series Q_k, one can check that $\left|K_k^{*j}(x,y,t)\right|$ is bounded by a constant times

$$\frac{(At)^{j-1}t^{k-\left(\frac{n}{2}\right)}}{\Gamma\left(k-\frac{n}{2}+j-2\right)} \tag{1.1}$$

for some constant A, where Γ denotes the gamma Euler function. Thus the ratio test shows that $\sum_{j=1}^{\infty}\left|K_k^{*j}\right|$ converges, what implies that $\sum_{j=1}^{\infty}(-1)^{j+1}K_k^{*j}$ converges to a C^2-smooth function on $M\times M\times[0,+\infty)$ if $k>\frac{n}{2}+2$ (for more details, see Berger, Gauduchon, Mazet [1971], Chavel [1984], Craioveanu and Puta [1988] or Rosenberg [1997]).

As one might expect on physical grounds, the dominant factor in $E(x,x,t)$ for small time is the local geometry of (M,g) at $x\in M$. This is made precise next.

The estimate (1.1) also implies that

$$|Q_k|\le Ct^{k-\frac{n}{2}}$$

for some constant $C>0$ and for all t in some finite interval, so that

$$(4\pi t)^{\frac{n}{2}}|Q_k * H_k|\le Ct^{k+1}.$$

As

$$(4\pi t)^{\frac{n}{2}}H_k(x,x,t)=\sum_{j=0}^{k}u_j(x,x)t^j,$$

it follows the property of the heat kernel formulated in Proposition 1.1.

Before stating this Proposition, let us recall a definition.

Definition 1.1 The (formal) series $\sum_{k=k_0}^{\infty} b_k t^k$, where $k_0 \in \mathbb{N}$, $b_k \in \mathbb{R}$ for any $k \geq k_0$, is called **an asymptotic expansion as** $t \searrow 0$ for the real function $A:(0,+\infty) \to \mathbb{R}$ if

$$\lim_{t \searrow 0} \frac{A(t) - \sum_{k=k_0}^{N} b_k t^k}{t^N} = 0$$

for each natural number $N \geq k_0$.

Note that if the function $A:(0,+\infty) \to \mathbb{R}$ admits an asymptotic expansion as $t \searrow 0$, it is unique. Indeed, let $\sum_{k=k_0}^{\infty} b_k t^k$, $\sum_{k=k_0'}^{\infty} b_k' t^k$ be asymptotic expansions as $t \searrow 0$ for $A:(0,+\infty) \to \mathbb{R}$ and assume that $k_0 > k_0'$. Then

$$\lim_{t \searrow 0} \frac{\sum_{k=k_0}^{N} (b_k - b_k') t^k - \sum_{k=k_0'}^{k_0 - 1} b_k' t^k}{t^N} = \lim_{t \searrow 0} \frac{A(t) - \sum_{k=k_0'}^{N} b_k' t^k - \left[A(t) - \sum_{k=k_0}^{N} b_k t^k \right]}{t^N} = 0$$

for each $N \geq k_0$. So for $N = k_0$ one gets

$$\lim_{t \searrow 0} \left[-\sum_{k=k_0'}^{k_0 - 1} b_k' t^{k-k_0} + \left(b_{k_0} - b_{k_0}' \right) \right] = 0,$$

that is $b_k' = 0$ for each $k < k_0$ and $b_{k_0} = b_{k_0}'$, and by an induction argument one obtains $b_k = b_k'$ for all $k \geq k_0$.

The real numbers $b_k, k \geq k_0$, in Definition 1.1 are called **the coefficients of the asymptotic expansion** and in this case we shall write

$$A(t) \sim \sum_{k=k_0}^{\infty} b_k t^k \text{ as } t \searrow 0.$$

Proposition 1.1. The real function $E(x,x,\,):(0,+\infty) \to \mathbb{R}$ corresponding to the heat kernel E on (M,g) has the following asymptotic expansion as $t \searrow 0$:

$$E(x,x,t) \sim (4\pi t)^{-\frac{n}{2}} \sum_{j=0}^{\infty} u_j(x,x) t^j,$$

that is

$$\lim_{t \searrow 0} \frac{E(x,x,t) - (4\pi t)^{-\frac{n}{2}} \sum_{j=0}^{k} u_j(x,x) t^j}{t^k} = 0$$

for all $k \in \mathbb{N}$ and any $x \in M$.

Moreover, using a normal coordinate system $(x^1, \ldots, x^n)$ at x (see Chapter 1, §5), it can be shown that

$$u_1(x,x) = -\frac{1}{6}\tau(x), \tag{1.2}$$

$$u_2(x,x) = \frac{1}{360}\left[2|R_x|^2 - 2|(c_2R)_x|^2 + 5\tau^2(x) - 12(\Delta\tau)(x)\right] \tag{1.3}$$

(see Berger, Gauduchon, Mazet [1971] or Craioveanu and Puta [1988]),

$$\begin{aligned} u_3(x,x) = \frac{1}{7!}\Big[& -18\tau_{;jjkk}(x) + 17\tau_{;k}(x)\tau_{;k}(x) - 2(c_2R)_{jk;l}(x)(c_2R)_{jk;l}(x) \\ & -4(c_2R)_{jk;l}(x)(c_2R)_{jl;k}(x) + 9R_{ijkl;m}(x)R_{ijkl;m}(x) \\ & +28\tau(x)\tau_{;kk}(x) - 8(c_2R)_{jk}(x)(c_2R)_{jk;ll}(x) + 24(c_2R)_{jk}(x)(c_2R)_{jl;kl}(x) \\ & +12R_{ijkl}(x)R_{ijkl;mm}(x) - \frac{35}{9}\tau^3(x) + \frac{14}{3}\tau(x)|(c_2R)_x|^2 - \frac{14}{3}\tau(x)|R_x|^2 \\ & +\frac{208}{9}(c_2R)_{jk}(x)(c_2R)_{kl}(x)(c_2R)_{lj}(x) + \frac{64}{3}(c_2R)_{ij}(x)(c_2R)_{kl}(x)R_{ikjl}(x) \\ & +\frac{16}{3}(c_2R)_{jk}(x)R_{jilm}(x)R_{kilm}(x) + \frac{44}{9}R_{ijkl}(x)R_{klmp}(x)R_{mpij}(x) \\ & +\frac{80}{9}R_{ijkl}(x)R_{imkp}(x)R_{jmlp}(x)\Big] \end{aligned} \tag{1.4}$$

(see Gilkey [1975], [1995]), where $\tau = (c_2R)_{jj}$ is the scalar curvature of (M,g), R (resp. c_2R) the curvature (resp. Ricci) tensor field of the Levi-Civita connection associated to (M,g), $R_{ijkl} = g\left(R\left(\frac{\partial}{\partial x^k}, \frac{\partial}{\partial x^l}\right)\frac{\partial}{\partial x^j}, \frac{\partial}{\partial x^i}\right)$, $(c_2R)_{jk} = R_{ijki}, i,j,k,l = 1,\ldots,n$, the notation ";" denotes covariant differentiation and in (1.4) and other formulas one sums over repeated indices, $|\ |$ the corresponding pointwise norm defined in Chapter 2, §1, C, and more generally

$$u_j(x,x),\ j \geq 4,$$

is an $O(n)$-invariant polynomial in the components of the curvature tensor field R_x at x and of its covariant derivatives $\left(\overset{\circ}{\nabla} R\right)_x, \left(\overset{\circ}{\nabla}{}^2 R\right)_x, \ldots, \left(\overset{\circ}{\nabla}{}^{2j-4} R\right)_x$ at x. The coefficients of these polynomials do not depend on the closed Riemannian manifold (M,g), but only on the dimension n of the manifold M.

If one takes into account Corollary 1.3 of Chapter 3, Proposition 1.1 and the formulas (1.2)-(1.4) lead us to the following asymptotic expansion for the trace of the heat kernel.

Corollary 1.1. Let (M,g) be as above and

$$0 = \lambda_0 < \lambda_1 \leq \lambda_2 \leq \cdots \tag{1.5}$$

the spectrum of (M,g). Then

$$Z_{(M,g)}(t) \overset{def}{=} \sum_{j=0}^{\infty} e^{-\lambda_j t} \sim \sum_{k=0}^{\infty} a_k t^{\frac{k-n}{2}} \text{ as } t \searrow 0,$$

where

$$a_{2k} = (4\pi)^{-\frac{n}{2}} \int_M u_k(x,x)\, d\mu_g(x) \text{ and } a_{2k+1} = 0, k \in \mathbb{N}$$

(see also Chapter 5, Theorem 4.1). In particular,

$$a_0 = (4\pi)^{-\frac{n}{2}} \operatorname{Vol}(M,g),$$

$$a_2 = -\frac{(4\pi)^{-\frac{n}{2}}}{6} \int_M \tau(x)\, d\mu_g(x),$$

$$a_4 = \frac{(4\pi)^{-\frac{n}{2}}}{360} \int_M \left[2|R_x|^2 - 2|(c_2R)_x|^2 + 5\tau^2(x)\right] d\mu_g(x),$$

$$a_6 = \frac{(4\pi)^{-\frac{n}{2}}}{7!} \int_M \left[-\frac{7}{9}\left|\overset{\circ}{\nabla} R\right|^2 - \frac{26}{9}\left|\overset{\circ}{\nabla}(c_2R)\right|^2 - \frac{142}{9}\left|\overset{\circ}{\nabla}\tau\right|^2 - \frac{35}{9}\tau^3 + \frac{14}{3}\tau|c_2R|^2 \right.$$
$$-\frac{14}{3}\tau|R|^2 + 4(c_2R)_{ij}(c_2R)_{jk}(c_2R)_{ki} + \frac{20}{9}(c_2R)_{ij}(c_2R)_{kl}R_{ikjl} + \frac{8}{9}(c_2R)_{ij}R_{iklm}R_{jklm}$$
$$\left. + \frac{8}{3}R_{ijkl}R_{ijmp}R_{mpkl} \right] d\mu_g$$

(see also Chapter 5, Theorem 4.2), where we use the metric tensor field g to identify the tangent and cotangent bundles of M and we sum over repeated indices.

Definition 1.2. Two closed Riemannian manifolds (M_1, g_1) and (M_2, g_2) are called **isospectral** (or more precisely 0-isospectral, see also Chapter 5, Definition 4.1) if $\mathbf{Spec}(M_1, g_1) = \mathbf{Spec}(M_2, g_2)$.

The next statement gives the first example of how the spectrum of a closed, connected Riemannian manifold determines certain geometric or topological data.

Corollary 1.2. Let (M', g') and (M'', g'') be isospectral, closed, connected Riemannian manifolds. Then the following assertions are true:

(i) $n' \overset{not}{=} \dim M' = \dim M'' \overset{not}{=} n''$ and $a'_{2k} = a''_{2k}$ for each $k \in \mathbb{N}$. In particular, $\operatorname{Vol}(M', g') = \operatorname{Vol}(M'', g'')$ and $\int_{M'} \tau' d\mu_{g'} = \int_{M''} \tau'' d\mu_{g''}$.

(ii) If $n' = n'' = 2$, then $\chi(M') = \chi(M'')$. Thus, if (M', g') and (M'', g'') are orientable (or nonorientable) surfaces, then M' is diffeomorphic to M''. In particular, the genus of a closed, connected, orientable (or nonorientable) surface is spectrally determined.

Proof. (i) It follows from the hypothesis and Corollary 1.1 that

$$\lim_{t \searrow 0} \frac{\sum_{k=0}^{N} a'_{2k} t^{k-\frac{n'}{2}} - \sum_{k=0}^{N} a''_{2k} t^{k-\frac{n''}{2}}}{t^N} = 0 \tag{1.6}$$

for any natural number N.

For $N = 0$ one gets

$$\lim_{t \searrow 0} \left(a'_0 t^{-\frac{n'}{2}} - a''_0 t^{-\frac{n''}{2}} \right) = 0 .$$

Assuming $n' > n''$, this equality implies a contradiction:

$$0 = \lim_{t \searrow 0} \frac{1}{t^{\frac{n''}{2}}} \left(\frac{a'_0}{t^{\frac{n'-n''}{2}}} - a''_0 \right) = +\infty .$$

Similarly, supposing $n' < n''$, the same equality implies again a contradiction. Hence $n' = n''$ and the same equality also leads to $a'_0 = a''_0$, i.e. $\mathrm{Vol}(M', g') = \mathrm{Vol}(M'', g'')$.

Thus, if one denotes $n' = n'' = n$, (1.6) gives

$$\lim_{t \searrow 0} \frac{\sum_{k=0}^{N} \left(a'_{2k} - a''_{2k} \right) t^{k-\frac{n}{2}}}{t^N} = 0$$

for any natural number N. So for $N = 1$ one gets

$$\lim_{t \searrow 0} \left(a'_2 - a''_2 \right) t^{-\frac{n}{2}} = 0 ,$$

which implies $a'_2 = a''_2$, i.e.

$$\int_{M'} \tau' d\mu_{g'} = \int_{M''} \tau'' d\mu_{g''} .$$

By induction one obtains $a'_{2k} = a''_{2k}$ for all $k \in \mathbb{N}$.

(ii) It follows from the Gauss-Bonnet theorem and (i) that

$$-4\pi\chi(M') = \int_{M'} \tau' d\mu_{g'} = \int_{M''} \tau'' d\mu_{g''} = -4\pi\chi(M''),$$

i.e. $\chi(M') = \chi(M'')$. Since two closed oriented (or nonoriented) smooth surfaces with the same Euler-Poincaré characteristic are diffeomorphic, the last assertion follows immediately.

Q.E.D.

The next statement, namely **Weyl's asymptotic formula**, describes the asymptotic behaviour of the eigenvalues of the Laplace-Beltrami operator on a closed Riemannian manifold.

Corollary 1.3. Let (M, g) be a closed, connected, n-dimensional Riemannian manifold and (1.5) the spectrum of (M,g). Then

$$\lim_{k\to\infty}\frac{\lambda_k^{\frac{n}{2}}}{k}=\frac{(4\pi)^{\frac{n}{2}}\Gamma\left(\frac{n}{2}+1\right)}{\mathrm{Vol}(M,g)},\tag{1.7}$$

where $\Gamma:(0,+\infty)\to\mathbb{R}$, $\Gamma(x)=\int_0^{+\infty}e^{-t}t^{x-1}dt$, $x\in(0,+\infty)$, is the gamma Euler function.

Indeed, let

$$N(\lambda)=\#\{\lambda_k\,|\,\lambda_k\le\lambda\}$$

be the counting function. Note that $N(\lambda_k)\sim k$ as $k\to+\infty$, so (1.7) is equivalent to

$$\lim_{\lambda\to+\infty}\frac{\lambda^{\frac{n}{2}}}{N(\lambda)}=\frac{(4\pi)^{\frac{n}{2}}\Gamma\left(\frac{n}{2}+1\right)}{\mathrm{Vol}(M,g)}.\tag{1.8}$$

Let μ be the positive locally finite Borel measure on $[0,+\infty)$ given by $\mu([0,\lambda])=N(\lambda)$. Then one has

$$Z_{(M,g)}(t)=\int_0^{+\infty}e^{-\lambda t}d\mu(\lambda).$$

On the other hand, it follows from Corollary 1.1 that

$$\lim_{t\searrow 0}\left[Z_{(M,g)}(t)-(4\pi t)^{-\frac{n}{2}}\mathrm{Vol}(M,g)\right]=0.\tag{1.9}$$

That (1.8) follows from (1.9) is a particular case of a result known as **Karamata's Tauberian theorem**: If μ is a positive locally finite Borel measure on $[0,+\infty)$, $\alpha\in(0,+\infty)$ and A real constants, then

$$\lim_{t\searrow 0}\left[\int_0^{+\infty}e^{-\lambda t}d\mu(\lambda)-\frac{A}{t^{\alpha}}\right]=0$$

implies

$$\lim_{\lambda\to+\infty}\left[\int_0^{\lambda}d\mu(x)-\frac{A}{\Gamma(\alpha+1)}\lambda^{\alpha}\right]=0.$$

For a neat proof of this theorem, see Taylor [1996].

Remark 1.1. If $N(\lambda)$ denotes the counting function of the eigenvalues of the Laplace-Beltrami operator on a bounded, open subset M of $\mathbb{R}^n$, with Dirichlet boundary condition, then (1.8) holds (for details, see Taylor [1996]).

Next let us show that spaces of constant sectional curvature in dimension ≤ 5 and of strictly positive constant sectional curvature in dimension 6 are characterized by their spectrum.

Proposition 1.2. Let $(M',g'),(M'',g'')$ be isospectral, closed Riemannian manifolds. Then $\dim M' = \dim M'' \overset{not}{=} n$ (see Corollary 1.2) and

(i) if (M',g') has constant sectional curvature c, so does (M'',g'') for $n=2,3,4,5$.

(ii) if (M',g') is conformally flat and has constant scalar curvarute c, so does (M'',g'') for $n=6$.

Proof. The hypothesis $\mathbf{Spec}(M',g')=\mathbf{Spec}(M'',g'')$ gives

$$a_0' = a_0'', \text{ i.e. } \int_{M'} d\mu_{g'} = \int_{M''} d\mu_{g''}, \tag{1.10}$$

$$a_2' = a_2'', \text{ i.e. } \int_{M'} \tau' d\mu_{g'} = \int_{M''} \tau'' d\mu_{g''}, \tag{1.11}$$

$$\begin{aligned} a_4' = a_4'', \text{ i.e. } &\int_{M'} \left(2|R'|'^2 - 2|c_2R'|'^2 + 5\tau'^2\right) d\mu_{g'} \\ &= \int_{M''} \left(|R''|''^2 - 2|c_2R''|''^2 + 5\tau''^2\right) d\mu_{g''}. \end{aligned} \tag{1.12}$$

(i) If (M',g') has constant sectional curvature c, then $\tau' = -n(n-1)c$, that is τ' is constant. The Cauchy-Schwarz inequality, (1.11), the fact that τ' is constant and (1.10) gives

$$\begin{aligned} \left(\int_{M''} d\mu_{g''}\right)\left(\int_{M''} \tau''^2 d\mu_{g''}\right) &\geq \left(\int_{M''} \tau'' d\mu_{g''}\right)^2 = \left(\int_{M'} \tau' d\mu_{g'}\right)^2 = \tau'^2\left(\int_{M'} d\mu_{g'}\right)^2 \\ &= \tau'^2\left(\int_{M'} d\mu_{g'}\right)\cdot\left(\int_{M''} d\mu_{g''}\right) = \int_{M'} \tau'^2 d\mu_{g'} \int_{M''} d\mu_{g''}, \end{aligned}$$

that is

$$\int_{M''} \tau''^2 d\mu_{g''} \geq \int_{M'} \tau'^2 d\mu_{g'}. \tag{1.13}$$

If one shows that τ'' is also constant, then (1.11) and (1.10) implies that $\tau' = \tau'' = -n(n-1)c$, that is if (M'',g'') has constant sectional curvature, then it is also equal to c.

First assume $n=2$. Since $|R'|'^2 = 2|c_2R'|'^2 = \tau'^2, |R''|''^2 = 2|c_2R''|''^2 = \tau''^2$ (see Chapter 1, Example 4.6), (1.12) leads to

$$\int_{M'} \tau'^2 d\mu_{g'} = \int_{M''} \tau''^2 d\mu_{g''}. \tag{1.14}$$

Consequently, (1.11), the Cauchy-Schwarz inequality, (1.10) and (1.14) implies that

$$\left(\int\limits_{M'} \tau' d\mu_{g'}\right)^2 = \left(\int\limits_{M''} \tau'' d\mu_{g''}\right)^2 \le \left(\int\limits_{M''} \tau''^2 d\mu_{g''}\right)\left(\int\limits_{M''} d\mu_{g''}\right) = \left(\int\limits_{M'} \tau'^2 d\mu_{g'}\right)\left(\int\limits_{M'} d\mu_{g'}\right). \tag{1.15}$$

Since τ' is constant (namely equal to $-2c$), one has

$$\left(\int\limits_{M'} \tau' d\mu_{g'}\right)^2 = \int\limits_{M'} \tau'^2 d\mu_{g'} \int\limits_{M'} d\mu_{g'},$$

so it follows from (1.15) that

$$\left(\int\limits_{M''} \tau'' d\mu_{g''}\right)^2 = \int\limits_{M''} \tau''^2 d\mu_{g''} \int\limits_{M''} d\mu_{g''}.$$

Consequently, τ'' is constant [and hence (M'', g'') has constant sectional curvature], so by a previous remark $\tau' = \tau'' = -2c$, that is (M'', g'') has constant sectional curvature $-\frac{1}{2}\tau'' = c$.

Next assume $n = 3$. Since $|R'|'^2 = |c_2 R'|'^2 = \frac{1}{3}\tau'^2$ (see Chapter 1, Propositions 4.6 and 4.7), $|R''|''^2 - 4|c_2 R''|''^2 + \tau''^2 = 0$ (see Chapter 1, Example 4.7), (1.12) leads to

$$\int\limits_{M'} 5\tau'^2 d\mu_{g'} = \int\limits_{M''} \left[6\left(|c_2 R''|''^2 - \frac{1}{3}\tau''^2\right) + 5\tau''^2\right] d\mu_{g''}. \tag{1.16}$$

On the other hand, the inequality $\tau''^2 \le 3|c_2 R''|''^2$ (see Chapter 1, Proposition 4.6) and (1.16) implies that

$$\int\limits_{M''} \tau''^2 d\mu_{g''} \le \int\limits_{M'} \tau'^2 d\mu_{g'},$$

so using (1.13) one has

$$\int\limits_{M'} \tau'^2 d\mu_{g'} = \int\limits_{M''} \tau''^2 d\mu_{g''}. \tag{1.17}$$

Since τ' is constant (namely equal to $-6c$), it follows from (1.11), (1.17) and (1.10) that

$$\left(\int\limits_{M''} \tau'' d\mu_{g''}\right)^2 = \left(\int\limits_{M'} \tau' d\mu_{g'}\right)^2 = \left(\int\limits_{M'} \tau'^2 d\mu_{g'}\right)\left(\int\limits_{M'} d\mu_{g'}\right) = \left(\int\limits_{M''} \tau''^2 d\mu_{g''}\right)\left(\int\limits_{M''} d\mu_{g''}\right),$$

that is τ'' is constant. Hence by a previous remark $\tau' = \tau'' = -6c$. Now use (1.16) and (1.17) to get

$$\left|c_2 R''\right|''^2 - \frac{1}{3}\tau''^2 = 0,$$

so (M'', g'') is Einsteinian (see Chapter 1, Proposition 4.6), i.e. (M'', g'') has constant sectional curvature $-\frac{1}{6}\tau'' = c$ (see Chapter 1, Corollary 4.2). This finishes the proof for the case $n = 3$.

Finally, use relation (4.15) of Chapter 1 and (1.12) to obtain

$$\begin{aligned}&\int_{M'}\left\{2\left|C'\right|'^2 + \frac{2(6-n)}{n-2}\left(\left|c_2 R'\right|'^2 - \frac{1}{n}\tau'^2\right) + \left[\frac{2(6-n)}{n(n-2)} + \frac{5n(n-3)+6}{(n-1)(n-2)}\right]\tau'^2\right\}d\mu_{g'}\\ &= \int_{M''}\left\{2\left|C''\right|''^2 + \frac{2(6-n)}{n-2}\left(\left|c_2 R''\right|''^2 - \frac{1}{n}\tau''^2\right) + \left[\frac{2(6-n)}{n(n-2)} + \frac{5n(n-3)+6}{(n-1)(n-2)}\right]\tau''^2\right\}d\mu_{g''}.\end{aligned} \tag{1.18}$$

Since (M', g') has constant sectional curvature c, it follows that $C' = 0$ and Einsteinian (see Chapter 1, Corollary 4.2), that is $C' = 0$ and $\left|c_2 R'\right|'^2 = \frac{1}{n}\tau'^2$ (see Chapter 1, Proposition 4.6).

Now assume $4 \le n \le 5$ and use this last remark as well as (1.13). Then (1.18) leads to

$$\begin{aligned}&\int_{M''}\left[2\left|C''\right|''^2 + \frac{2(6-n)}{n-2}\left(\left|c_2 R''\right|''^2 - \frac{1}{n}\tau''^2\right)\right]d\mu_{g''}\\ &\quad = \frac{5n^3 - 17n^2 + 20n - 12}{n(n-1)(n-2)}\left[\int_{M'}\tau'^2 d\mu_{g'} - \int_{M''}\tau''^2 d\mu_{g''}\right] \le 0.\end{aligned}$$

Note that $\frac{6-n}{n-2} > 0$ for $n = 4, 5$, so by Proposition 4.6 of Chapter 1 one obtains

$$\int_{M''}\left[2\left|C''\right|''^2 + \frac{2(6-n)}{n-2}\left(\left|c_2 R''\right|''^2 - \frac{1}{n}\tau''^2\right)\right]d\mu_{g''} = 0$$

and

$$\int_{M'}\tau'^2 d\mu_{g'} = \int_{M''}\tau''^2 d\mu_{g''}. \tag{1.19}$$

Therefore, $C'' = 0$ and $\left|c_2 R''\right|''^2 - \frac{1}{n}\tau''^2 = 0$, that is (M'', g'') has constant sectional curvature (see again Chapter 1, Proposition 4.6 and Corollary 4.2). Since τ' is constant, using (1.11), (1.19) and (1.10), as in the case $n = 3$, one gets

$$\left(\int_{M''}\tau'' d\mu_{g''}\right)^2 = \left(\int_{M''}\tau''^2 d\mu_{g''}\right)\left(\int_{M''}d\mu_{g''}\right),$$

so τ'' is constant, hence $\tau' = \tau'' = -n(n-1)c$ and (M'', g'') has constant sectional curvature c.

(ii) Assume $n=6$. In this case (1.18) leads to

$$\int_{M'}\left(2\left|C'\right|'^2+\frac{24}{5}\tau'^2\right)d\mu_{g'}=\int_{M''}\left(2\left|C''\right|''^2+\frac{24}{5}\tau''^2\right)d\mu_{g''}. \tag{1.20}$$

Since $C'=0$, it follows from (1.20) that

$$\int_{M'}\frac{24}{5}\tau'^2 d\mu_{g'}=\int_{M''}\left(2\left|C''\right|''^2+\frac{24}{5}\tau''^2\right)d\mu_{g''},$$

so by (1.13), which is true because τ' is constant, one gets $C''=0$, that is (M'',g'') is conformally flat and

$$\int_{M'}\tau'^2 d\mu_{g'}=\int_{M''}\tau''^2 d\mu_{g''}. \tag{1.21}$$

Since τ' is constant, using (1.11), (1.21) and (1.10), as in the case $n=3$, one gets

$$\left(\int_{M''}\tau'' d\mu_{g''}\right)^2=\int_{M''}\tau''^2 d\mu_{g''}\int_{M''}d\mu_{g''},$$

so τ'' is constant, and hence $\tau'=\tau''$. Consequently, (M'',g'') has the same constant scalar curvature as (M',g').

Q.E.D.

Proposition 1.3. (**Tanno** [1973]). Let (M',g'), (M'',g'') be isospectral 6-dimensional closed Riemannian manifolds. If (M',g') has constant sectional curvature $c>0$, so does (M'',g'').

Proof. We keep the previous notations. As before we adopt the Einstein convention and sum over repeated indices. We use the Riemannian metric tensor to identify the tangent and cotangent bundles of M''. Let $(x''^1,\dots,x''^6)$ be a normal coordinate system at $x_0''\in M''$. If one requires that $g_{ij}''(x_0'')=\delta_{ij}$, $i,j=1,\dots,6$, then this choice of coordinates is unique up to the action of the orthogonal group $O(n)$.

Since (M',g') has constant sectional curvature $c>0$, it follows that its scalar curvature $\tau'=-30c<0$ is constant and $C'=0$ (see Chapter 1, Corollary 4.2), so by Proposition 1.2 (ii) one has $C''=0$ and $\tau''=\tau'<0$.

The relation (4.15) of Chapter 1 gives

$$\tau''\left|R''\right|''^2=\tau''\left|c_2R''\right|''^2-\frac{1}{10}\tau''^3, \tag{1.22}$$

and the formula for C_{ijkl}'' (see Chapter 1, §4) leads in this case to

$$R_{ijkl}''=\frac{1}{4}\left(\delta_{il}R_{jk}''-\delta_{ik}R_{jl}''+\delta_{jk}R_{il}''-\delta_{jl}R_{ik}''\right)-\frac{\tau''}{20}\left(\delta_{jk}\delta_{il}-\delta_{jl}\delta_{ik}\right), \tag{1.23}$$

where $R''_{ijkl} = g''\left(R''\left(\frac{\partial}{\partial x''^k}, \frac{\partial}{\partial x''^l}\right)\frac{\partial}{\partial x''^j}, \frac{\partial}{\partial x''^i}\right)$, $R''_{jk} = \left(c_2 R''\right)_{jk} = R''_{ijki}$, $\tau'' = R''_{jj}$,
$i,j,k,l = 1,\ldots,6$.
Using (1.23) one gets successively

$$\left|\mathring{\nabla}'' R''\right|''^2 = \left|\mathring{\nabla}''\left(c_2 R''\right)\right|''^2, \tag{1.24}$$

$$R''_{ij}R''_{kl}R''_{ikjl} = \frac{1}{2}R''_{ij}R''_{jk}R''_{ki} - \frac{11}{20}\tau''\left|c_2 R''\right|''^2 + \frac{1}{20}\tau''^3, \tag{1.25}$$

$$R''_{ij}R''_{iklm}R''_{jklm} = \frac{1}{4}R''_{ij}R''_{jk}R''_{ki} + \frac{7}{40}\tau''\left|c_2 R''\right|''^2 - \frac{1}{40}\tau''^3, \tag{1.26}$$

$$\begin{aligned} R''_{ijkl}R''_{ijmp}R''_{mpkl} &= -R''_{ij}R''_{iklm}R''_{jklm} + \frac{1}{10}\tau''\left|R''\right|''^2 = \left[\text{using (1.23) and (1.26)}\right] \\ &= -\frac{1}{4}R''_{ij}R''_{jk}R''_{ki} - \frac{3}{40}\tau''\left|c_2 R''\right|''^2 + \frac{3}{200}\tau''^3, \end{aligned} \tag{1.27}$$

where $\mathring{\nabla}''$ denotes the covariant differential operator associated to the Levi-Civita connection ∇'' (see Chapter 1, Definition 3.3).

Now we shall compute $\int_{M''} R''_{ij}R''_{jk}R''_{ki}\,d\mu_{g''}$ in terms of $\left|\mathring{\nabla}''\left(c_2 R''\right)\right|''^2$, $\left|c_2 R''\right|''^2$ and τ''.

First note that from second Bianchi's identity

$$R''_{ijkl;m} + R''_{ijlm;k} + R''_{ijmk;l} = 0,$$

where ";" denotes covariant differentiation [see Chapter 1, Proposition 4.1 (iv)] one gets

$$R''_{jk;l} - R''_{jl;k} + R''_{ijlk;i} = 0 \tag{1.28}$$

for each $j,k,l = 1,2,\ldots,6$. If one takes into account that $\tau''_{;m} = 0$ for any $m = 1,2,\ldots,6$ and that $R''_{jk} = R''_{ijki}$, one obtains

$$R''_{jk;j} = 0 \tag{1.29}$$

for each $k = 1,2,\ldots,6$. It follows from (1.28), (1.23) and (1.29) that

$$R''_{jk;l} - R''_{jl;k} + \frac{1}{4}R''_{jk;l} - \frac{1}{4}R''_{jl;k} = 0,$$

or equivalently

$$R''_{jk;l} = R''_{jl;k}, \tag{1.30}$$

for each $j,k,l=1,2,\ldots,6$. Now note that

$$-\frac{1}{2}\Delta''\left(\left|c_2R''\right|''^2\right)=-\frac{1}{2}\Delta''\left(R''_{ij}R''_{ij}\right)=\frac{1}{2}\nabla''_k\nabla''_k\left(R''_{ij}R''_{ij}\right)$$
$$=R''_{ij}R''_{ij;kk}+R''_{ij;k}R''_{ij;k}=R''_{ij}R''_{ij;kk}+\left|\overset{\circ}{\nabla}''\left(c_2R''\right)\right|''^2, \tag{1.31}$$

where $\nabla''_k=\nabla''_{\frac{\partial}{\partial x''^k}}$, $k=1,2,\ldots,6$. On the other hand, the commutation law

$$\left(\overset{\circ}{\nabla}''^2\left(c_2R''\right)\right)\left(X'',Y'',Z''_1,Z''_2\right)-\left(\overset{\circ}{\nabla}''^2\left(c_2R''\right)\right)\left(Y'',X'',Z''_1,Z''_2\right)$$
$$=-c_2R''\left(R''\left(X'',Y''\right)Z''_1,Z''_2\right)-c_2R''\left(Z''_1,R''\left(X'',Y''\right)Z''_2\right)$$

for any $X'',Y'',Z''_1,Z''_2\in\mathfrak{X}\left(M''\right)$ (see Chapter 1, §4) gives in local coordinates the so-called Ricci identity, that is

$$R''_{ij;kl}-R''_{ij;lk}=-R''_{im}R''_{mjkl}-R''_{jm}R''_{mikl},$$

so in virtue of (1.30) and (1.29) one obtains

$$R''_{ij;kk}=R''_{ik;jk}=R''_{ik;kj}-R''_{li}R''_{lkjk}-R''_{lk}R''_{lijk}=R''_{il}R''_{jl}-R''_{lk}R''_{lijk}. \tag{1.32}$$

Consequently, (1.31) and (1.32) gives

$$\left|\overset{\circ}{\nabla}''\left(c_2R''\right)\right|''^2=-R''_{ij}R''_{jl}R''_{li}+R''_{ij}R''_{lk}R''_{lijk}-\frac{1}{2}\Delta''\left(\left|c_2R''\right|''^2\right). \tag{1.33}$$

Again, using the formula (1.23) this immediately implies

$$R''_{ij}R''_{lk}R''_{lijk}=\frac{1}{4}R''_{ij}R''_{lk}\left(\delta_{lk}R''_{ij}-\delta_{lj}R''_{ik}+\delta_{ij}R''_{lk}-\delta_{ik}R''_{lj}\right)-\frac{\tau''}{20}R''_{ij}R''_{lk}\left(\delta_{ij}\delta_{lk}-\delta_{ik}\delta_{lj}\right)$$
$$=-\frac{1}{2}R''_{ij}R''_{jk}R''_{ki}+\frac{11}{20}\tau''\left|c_2R''\right|''^2-\frac{\tau''^3}{20}.$$

Thus, it follows from (1.33) that

$$\left|\overset{\circ}{\nabla}''\left(c_2R''\right)\right|^2=-\frac{3}{2}R''_{ij}R''_{jk}R''_{ki}+\frac{11}{20}\tau''\left|c_2R''\right|''^2-\frac{\tau''^3}{20}-\frac{1}{2}\Delta''\left(\left|c_2R''\right|''^2\right),$$

or equivalently

$$R''_{ij}R''_{jk}R''_{ki}=-\frac{2}{3}\left|\overset{\circ}{\nabla}''\left(c_2R''\right)\right|''^2+\frac{11}{30}\tau''\left|c_2R''\right|''^2-\frac{\tau''^3}{30}-\frac{1}{3}\Delta''\left(\left|c_2R''\right|''^2\right),$$

so integrating over M'' one gets

$$\int_{M''} R''_{ij} R''_{jk} R''_{ki} d\mu_{g''} = \int_{M''} \left[-\frac{2}{3} \left| \overset{\circ}{\nabla}{}'' \left(c_2 R'' \right) \right|''^2 + \frac{11}{30} \tau'' \left| c_2 R'' \right|''^2 - \frac{\tau''^3}{30} \right] d\mu_{g''} . \quad (1.34)$$

Finally, taking into account the expression of a''_6 given in Corollary 1.1 and substituting there the corresponding terms that occur in the formulas (1.22), (1.24)-(1.27) and (1.34) one obtains

$$\begin{aligned} a''_6 &= \frac{(4\pi)^{-3}}{7!} \int_{M''} \left[-\frac{61}{9} \left| \overset{\circ}{\nabla}{}'' \left(c_2 R'' \right) \right|''^2 + \frac{4}{9} \tau'' \left| c_2 R'' \right|''^2 - \frac{2328}{675} \tau''^3 \right] d\mu_{g''} \\ &= \frac{(4\pi)^{-3}}{7!} \int_{M''} \left[-\frac{61}{9} \left| \overset{\circ}{\nabla}{}'' \left(c_2 R'' \right) \right|''^2 + \frac{4}{9} \tau'' \left(\left| c_2 R'' \right|''^2 - \frac{1}{6} \tau''^2 \right) - \frac{2278}{675} \tau''^3 \right] d\mu_{g''} . \end{aligned} \quad (1.35)$$

Since (M', g') has constant sectional curvature, it is Einsteinian (see Chapter 1, Corollary 4.2), so $\left| c_2 R' \right|'^2 = \frac{1}{6} \tau'^2$ (see Chapter 1, Proposition 4.6). Hence $c_2 R' = \frac{\tau'}{6} g'$ and this immediately implies $\overset{\circ}{\nabla}{}' \left(c_2 R' \right) = 0$ (see Chapter 1, Proposition 3.2). Consequently,

$$a'_6 = \frac{(4\pi)^{-3}}{7!} \left[\int_{M'} -\frac{2278}{675} \tau'^3 d\mu_{g''} \right] .$$

Because (M', g') and (M'', g'') are isospectral, $a'_6 = a''_6$, so $\tau' = \tau'' < 0$ and (1.35) leads to $\overset{\circ}{\nabla}{}'' \left(c_2 R'' \right) = 0$ and $\left| c_2 R'' \right|''^2 = \frac{\tau''^2}{6}$, that is (M'', g'') is also Einsteinian. Since $C'' = 0, (M'', g'')$ has constant sectional curvature (see Chapter 1, Corollary 4.2) $-\frac{\tau''}{30} = -\frac{\tau'}{30} > 0$.

Q.E.D.

Corollary 1.4. (S^n, g_{can}) and $(P^n(\mathbb{R}), g_{can})$ are completely characterized by their spectrum for $n \le 6$, that is if the closed, connected Riemannian manifold (M, g) is isospectral to (S^n, g_{can}) [resp. $(P^n(\mathbb{R}), g_{can})$], then it is isometric to (S^n, g_{can}) [resp. $(P^n(\mathbb{R}), g_{can})$] for $n \le 6$.

Proof. It follows from the hypothesis that if $n = 1$, then M is diffeomorphic to S^1, so there exists a unique Riemannian metric g' on S^1 such that (M, g) is isometric to (S^1, g'). In particular, $\mathbf{Spec}(S^1, g') = \mathbf{Spec}(S^1, g_{can})$ therefore $\text{Length}(S^1, g') = \text{Lenght}(S^1, g_{can})$ and this immediately implies that (S^1, g') is isomorphic to (S^1, g_{can}) (see also Remark 2.1 of this Chapter).

Suppose now $2 \le n \le 6$ and let (M, g) be a closed, connected Riemannian manifold isospectral to (S^n, g_{can}) [resp. $(P^n(\mathbb{R}), g_{can})$], M' the universal covering space of M, g' the Riemannian metric on M' such that the corresponding covering projection $\pi : (M', g') \to (M, g)$ is a Riemannian covering (see Chapter 1, Example 2.8) and s the number of sheets of π. It follows from Proposition 1.2 (i) and Proposition 1.3 that (M, g) has constant sectional curvature 1, and so does (M', g') (see Chapter 1, Remark 4.1 and Example 4.3). Because M' is simply connected, it follows from Berger's rigidity theorem (see Sakai [1996], p. 207), that (M', g') is isometric to (S^n, g_{can}) (see also Wolf [1967], Theorem 2.5.1). Thus, if (M, g) is isospectral to (S^n, g_{can}) [resp. $(P^n(\mathbb{R}), g_{can})$], then $\mathrm{Vol}(M, g) = \mathrm{Vol}(S^n, g_{can})$ [resp. $= \mathrm{Vol}(P^n(\mathbb{R}), g_{can})$], therefore

$$\mathrm{Vol}(S^n, g_{can}) = \mathrm{Vol}(M', g') = s\,\mathrm{Vol}(M, g)$$

$$= s\,\mathrm{Vol}(S^n, g_{can}) \text{ [resp. } = \frac{s}{2}\mathrm{Vol}(S^n, g_{can})]$$

(see also Chapter 2, Remark 1.2 and Example 1.1). Hence $s = 1$ (resp. 2), that is (M', g') is isometric to (M, g) [resp. $\pi : (M', g') = (S^n, g_{can}) \to (M, g)$ is a Riemannian covering with two sheets], or equivalently, (S^n, g_{can}) is isometric to (M, g) [resp. (M, g) is isometric to $(P^n(\mathbb{R}), g_{can})$].

Q.E.D.

Remark 1.2. Let (M, g) be an n-dimensional closed Riemannian manifold such that g is sufficiently close to a Riemannian metric of constant sectional curvature in the sense explained by Tanno [1980], p. 274, and $\mathbf{Spec}(M, g) = \mathbf{Spec}(S^n, g_{can})$. Using the heat coefficients a_0, a_2, a_4, a_6 of the asymptotic expansion by Minakshisundaram-Pleijel, Tanno [1980] has proven that, under these assumptions, (M, g) is isometric to (S^n, g_{can}). In particular, any isospectral one-parameter deformation of g_{can} on S^n is trivial.

Remark 1.3. Using heat asymptotics, Brüning and Lesch [1996] have shown that from the spectrum of an algebraic curve, equipped via projective embedding with the Fubini-Study metric, one can tell whether the curve has singularities, other than multiple points.

While the heat coefficients are a very useful tool for obtaining information from the spectrum, they nonetheless have severe limitations. For example, any two closed Riemannian manifolds with a common Riemannian cover and with the same volume, e.g. two Riemann surfaces of a given volume, will have identical heat coefficients.

Remark 1.4. In particular, it follows from Corollary 1.4 that the standard Riemannian metric g_{can} on S^n for $n \le 6$ is spectrally determined, that is if g is any smooth $(= C^\infty)$ Riemannian metric on S^n satisfying $\mathbf{Spec}(S^n, g) = \mathbf{Spec}(S^n, g_{can})$, then (S^n, g) is isometric to (S^n, g_{can}). In fact g_{can} is the only Riemannian metric on S^n $(n \le 6)$ presently known to have this property. Let $\mathcal{M}^*$ be the class of rotationally invariant analytic metrics on S^2 which are of "simple type" and satisfy some generic non-degeneracy conditions (see Zelditch [1998]). In particular, they satisfy the generalized "simple length spectrum" condition that the length functional on the loop space is a clean Bott-Morse function which takes on distinct values on distinct components of its critical set (up to orientation). Zelditch[1998] succeeded to show that if (S^2, g_1), (S^2, g_2), g_1, $g_2 \in \mathcal{M}^*$, are isospectral, then (S^2, g_1) is isometric to (S^2, g_2).

Remark 1.5. Studying the Laplace-Beltrami operators associated to Zoll metrics with the property that for each such an operator there is precisely one distinct eigenvalue in each of the eigenvalue clusters (see also Chapter 3, Remark 4.1), Zelditch [1996] has been able to show that if g is a Riemannian metric on S^2 carrying an effective S^1-action by isometries, and $m_k(g) = m_k(g_{can}) = 2k+1$ for each $k \in \mathbb{N}^*$, then $g = g_{can}$, where $m_k(g)$ [resp. $m_k(g_{can})$] denotes the multiplicity of the eigenvalue $\lambda_k(g)$ [resp. $\lambda_k(g_{can})$]. This result is consistent with a problem raised by Yau [1992], Problem 41: Let (M, g) be a closed surface and assume that $m_k(g) = m_k(S^2, g_{can}) = 2k+1$ for any $k \in \mathbb{N}^*$. Then is $(M, g) \cong (S^2, g_{can})$? Also, he proves that if g is a Riemannian metric on $P^2(\mathbb{R})$ for which $m_k(g) = m_k(g_{can})$, then $g = g_{can}$, i.e. $(P^2(\mathbb{R}), g_{can})$ is determined by the multiplicities of the eigenvalues of $\Delta_{g_{can}}$.

Remark 1.6. Another way in which one may extract geometric information from spectral data is given by the asymptotics of the trace of the wave kernel. The fundamental solution (wave kernel) F of the wave equation problem associated to a closed Riemannian manifold (M, g)

$$\left(\frac{\partial^2}{\partial t^2} + \Delta_x\right) F(x,y,t) = 0,\ F(x,y,0) = \delta_y,\ \frac{\partial F}{\partial t}(x,y,0) = 0, \text{ for all } y \in M,$$

where Δ_x denotes the Laplace-Beltrami operator corresponding to g, acting on the first variable, and δ_y is the Dirac delta distribution at y, is a distribution on $M \times M \times \mathbb{R}$ that can be represented formally by

$$F(x,y,t) = \sum_{k=0}^{\infty} \cos\left(\sqrt{\lambda_k} t\right) f_k(x) f_k(y),$$

where $0 = \lambda_0 < \lambda_1 \le \lambda_2 \le \cdots$ denotes $\mathbf{Spec}(M, g)$, and $f_0, f_1, \ldots$ the corresponding complete orthonormal system of eigenfunctions of Δ. Consequently, the trace of the wave kernel is given formally by

$$trace(Q_t) = \int_M F(x,x,t)\, d\mu_g(x) = \sum_{k=0}^{\infty} \cos\left(\sqrt{\lambda_k} t\right).$$

In general, F and $trace(Q_t)$ are not smooth functions. It is precisely their lack of smoothness which provides new and useful information. Chazarain [1974], Duistermaat and Guillemin [1975] proved that the singular support (see Hörmander [1964]) of the spectral distribution given by $trace(Q_t)$ is contained in the set $\mathscr{L} = \{\pm L | L$ is the length of a closed geodesic on $(M,g)\} \cup \{0\}$. In other words, $trace(Q_t)$ is C^∞ outside $\mathscr{L}$. More precisely, assume that the Riemannian metric g is such that the geodesics of (M,g) are isolated and the linear Poincaré map around each geodesic is non-degenerate. If $L > 0$ is the length of a closed geodesic of (M,g), let $T \in C_0^\infty(\mathbb{R})$ be identically one near $t = L$ and assume that the support of T is contained in a sufficiently small neighborhood of L. Then there is an asymptotic expansion as $\tau \to \infty$ of the Fourier transform

$$\mathscr{F}(\tau) = \left\langle trace(Q_t), T(t)e^{-\sqrt{-1}t\tau} \right\rangle = \int_{-\infty}^{+\infty} trace(Q_t)T(t)e^{-\sqrt{-1}t\tau}dt \sim e^{-\sqrt{-1}\tau L}\sum_{k=0}^{\infty}\beta_k\tau^{-k}$$

(see Duistermaat and Guillemin [1975], Donnelly [1978]). Moreover, if there is a single isolated geodesic γ of length L, then

$$\beta_0 = \frac{l}{2}\left(\sqrt{-1}\right)^{-\sigma}\left|\det\left(I - P_\gamma\right)\right|^{-\frac{1}{2}},$$

where l is the primitive period of γ, σ is its Morse index, and P_γ the Poincaré map around γ. The Poincaré map P_γ is the linear symplectic map on the complex vector space $J_\gamma^\perp \otimes \mathbb{C}$ defined by $(P_\gamma Y)(t) = Y(t + L_\gamma)$, where L_γ is the length of γ and $J_\gamma^\perp \otimes \mathbb{C}$ denotes the space of complex normal Jacobi vector fields along γ, a symplectic vector space of complex dimension $2(n-1)$ with respect to the symplectic form $\omega(X,Y) = g\left(X, \frac{D}{ds}Y\right) - g\left(\frac{D}{ds}X, Y\right)$. If there are several geodesics of length L, then one merely sums the formulas corresponding to each individual geodesic. From this one can see that a given closed geodesic γ makes a non-zero contribution to the singular support of $trace(Q_t)$ if it is non-degenerate in the sense of Morse theory, and if there are no other geodesics of the same length as γ, or more generally if all the geodesics of the same length as γ have Morse index which do not differ with the index of γ by $2(\mathrm{mod}\,4)$. When g satisfies the above conditions, then the singular support of $trace(Q_t)$ is precisely equal to the set of lengths of closed geodesics along with the additive inverses of those lengths. In this case, the lengths of all the closed geodesics of (M,g) are determined by $\mathbf{Spec}(\boldsymbol{M},\boldsymbol{g})$. In particular this happens when (M,g) has negative sectional curvature, since all the geodesics of (M,g) have Morse index 0, and so make a non-trivial contribution to the singular support of $trace(Q_t)$. Along these lines, let us mention that there is a new approach to the problem of determining the compact Riemannian manifold (M,g) knowing $\mathbf{Spec}(\boldsymbol{M},\boldsymbol{g})$ (see also §3 of this Chapter). The strategy, originated by Guillemin [1993], [1996] and developed by Zelditch in a series of papers (see the references in Zelditch [1999]) consists of three steps. First, having an isolated non-degenerate closed geodesic γ

on (M,g), the wave group of (M,g) [that is the unitary group $U(t)=e^{\sqrt{-1}t\sqrt{\Delta}}$, where Δ is the Laplace-Beltrami operator associated to (M,g)] is put into a (microlocal) normal form around γ. Secondly, assuming γ is the only one with the given length, the normal form is determined by $\mathbf{Spec}(M,g)$. Finally, the metric g is found from the normal form. As for the first two steps, impressive results have been obtained by Zelditch and Guillemin, the strategy as a whole has been realised until now for some examples only, and it is the last step which seems to be the bottleneck. Let us mention that an application of the last step has been already pointed out in Remark 1.4.

2. ISOSPECTRAL FLAT TORI

Although it is rare that one can compute the spectrum of a closed Riemannian manifold explicitly, various special pairs of such manifolds have been shown to be isospectral by a direct comparison of their spectra. Let us illustrate this for the case of flat tori.

Recall that the spectrum of the n-dimensional flat torus $\left(T^n=\mathbb{R}^n/\Gamma, g_{can}/\Gamma\right)$, where Γ is a lattice in $\mathbb{R}^n$ of rank n (see Chapter 1, Example 2.17), is known explicitly, namely $4\pi^2$ times the set of squares of distances from the origin of $\mathbb{R}^n$ to the points of the dual lattice $\Gamma^*=\left\{\gamma^*\in\mathbb{R}^n \,\middle|\, \langle\gamma^*,\gamma\rangle\in\mathbb{Z} \text{ for any } \gamma\in\Gamma\right\}$ to Γ, counted with their multiplicities (see Chapter 3, Proposition 2.3). Therefore, the partition function $Z_{\left(T^n, g_{can}/\Gamma\right)}$ can be written as

$$Z_{\left(T^n, g_{can}/\Gamma\right)}(t)=\sum_{y\in\Gamma^*} e^{-4\pi^2\|y\|^2 t}.$$

If one makes the substitution $s=e^{-4\pi^2 t}$, then one gets a power series what is usually called the **theta series** of the lattice $\Gamma^*\subset\mathbb{R}^n : \Theta_{\Gamma^*}(s)=\sum_{y\in\Gamma^*} s^{\|y\|^2}$. In particular, if the lattices Γ_1 and Γ_2 in $\mathbb{R}^n$ are congruent, that is there exists an orthogonal automorphism $\Phi:\mathbb{R}^n\to\mathbb{R}^n$ such that $\Phi(\Gamma_1)=\Gamma_2$, then $\Theta_{\Gamma_1^*}=\Theta_{\Gamma_2^*}$. More generally, the flat tori $\left(\mathbb{R}^n/\Gamma_1, g_{can}/\Gamma_1\right)$ and $\left(\mathbb{R}^n/\Gamma_2, g_{can}/\Gamma_2\right)$ are isospectral if and only if $\Theta_{\Gamma_1^*}=\Theta_{\Gamma_2^*}$ (see Chapter 3, Proposition 1.1), that is Γ_1^* and Γ_2^* have the same theta series. This last remark suggests the following definition.

Definition 2.1. Two non-congruent lattices in $\mathbb{R}^n$ with the same theta series are called **isospectral**.

We shall see below that isospectral lattices in $\mathbb{R}^n$ exist if and only if n is at least 4.

An important result concerning flat tori is the Poisson formula (see Chavel [1984], Chapter XII, §3), which connects the spectrum of the flat torus $\left(\mathbb{R}^n/\Gamma, g_{can}/\Gamma\right)$ to the length spectrum of $\left(\mathbb{R}^n/\Gamma, g_{can}/\Gamma\right)$: for any $t\in(0,+\infty)$,

$$\sum_{\gamma^*\in\Gamma^*} e^{-4\pi^2\|\gamma^*\|^2 t} = (4\pi t)^{-\frac{n}{2}} \operatorname{Vol}\left(\mathbb{R}^n/\Gamma, g_{can}/\Gamma\right) \sum_{\gamma\in\Gamma} e^{-\frac{\|\gamma\|^2}{4t}} .$$

This formula has two important consequences:

(i) Two isospectral flat tori have the same volume (see also §1 of this Chapter);

(ii) $\left(\mathbb{R}^n/\Gamma_1, g_{can}/\Gamma_1\right)$ and $\left(\mathbb{R}^n/\Gamma_2, g_{can}/\Gamma_2\right)$ are isospectral if and only if they have the same length spectrum, that is

$$\{\|\gamma_1\| \,|\, \gamma_1\in\Gamma_1\} = \{\|\gamma_2\| \,|\, \gamma_2\in\Gamma_2\} \text{ (with multiplicities).}$$

Before showing the existence of isospectral lattices in $\mathbb{R}^n$, $n\ge 4$, let us discuss the connection between lattices and quadratic forms and prove the fact that any lattice in $\mathbb{R}^2$ is determined up to a congruence by its theta series (see also Watson [1979], [1980] and Pesce [1992]).

Let $\Gamma = A\mathbb{Z}^n$, $A\in GL(n,\mathbb{R})$, be a lattice. An element $\gamma\in\Gamma$ can be put in the form $\gamma = Ax$, where $x\in\mathbb{Z}^n$, and $\|\gamma\|^2 = \langle x, {}^tAAx\rangle$. To each element $A\in GL(n,\mathbb{R})$ one can associate the quadratic form whose matrix with respect to the standard basis of $\mathbb{R}^n$ is ${}^tAA$. Consequently one has a canonical mapping of $GL(n,\mathbb{R})$ into the set $S_n^+(\mathbb{R})$ of positive definite symmetric matrices, namely $A\mapsto {}^tAA$. This mapping induces a bijection of $O(n,\mathbb{R})\backslash GL(n,\mathbb{R})$ onto $S_n^+(\mathbb{R})$. In what follows all quadratic forms are positive definite and we shall identify a quadratic form (resp. linear operator) on $\mathbb{R}^n$ with its matrix with respect to the canonical basis of $\mathbb{R}^n$. Now let us assume that $\Gamma = A\mathbb{Z}^n = A'\mathbb{Z}^n$, where $A' = AB$ with $B\in GL(n,\mathbb{Z})$. Then one has ${}^tA'A' = {}^tB({}^tAA)B$. Hence, if the quadratic form q (resp. q') has the matrix ${}^tAA$ (resp. ${}^tA'A'$), then there exists $B\in GL(n,\mathbb{Z})$ such that $q'(x) = q(Bx)$ for all $x\in\mathbb{R}^n$. Let us define on $S_n^+(\mathbb{R})$ the following equivalence relation: $q\sim q'$ if and only if there exists $B\in GL(n,\mathbb{Z})$ such that $q'(x) = q(Bx)$ for all $x\in\mathbb{R}^n$. Let us denote by $[q]$ the equivalence class of q and by $S_n^+(\mathbb{R})/GL(n,\mathbb{Z})$ the set of all these equivalence classes. Let us note that one has a canonical bijection of $O(n,\mathbb{R})\backslash\left(GL(n,\mathbb{R})/GL(n,\mathbb{Z})\right)$ onto $S_n^+(\mathbb{R})/GL(n,\mathbb{Z})$, that is there is a canonical bijection between the set of isometry classes of flat tori and the set $S_n^+(\mathbb{R})/GL(n,\mathbb{Z})$ of equivalence classes of quadratic forms. Indeed, there is a canonical mapping π of $GL(n,\mathbb{R})/GL(n,\mathbb{Z})$ onto $S_n^+(\mathbb{R})/GL(n,\mathbb{Z})$ given by $\pi(\Gamma) = [q]$, where $q(x) = \langle Ax, Ax\rangle$ for any $x\in\mathbb{R}^n$, $\Gamma = A\mathbb{Z}^n$, $A\in GL(n,\mathbb{R})$. Moreover, $\pi(\Gamma) = \pi(\Gamma_1)$ if and only if there exists $B\in GL(n,\mathbb{Z})$ such that $\langle A'x, A'x\rangle = \langle A_1x, A_1x\rangle$ for any $x\in\mathbb{R}^n$, where $A' = AB$, $\Gamma_1 = A_1\mathbb{Z}^n$, $A_1\in GL(n,\mathbb{R})$. This implies that $A_1(A')^{-1}\in O(n,\mathbb{R})$ and notice that $\Gamma_1 = \left[A_1 (A')^{-1}\right]\Gamma$.

For $q \in S_n^+(\mathbb{R})$, let us define $\operatorname{Spec}(q) = \{q(x) | x \in \mathbb{Z}^n\}$. Note that if $\Gamma = A\mathbb{Z}^n$, where $A \in GL(n,\mathbb{R})$ and $q \in S_n^+(\mathbb{R})$ is given by $q(x) = \langle x, {}^tAAx\rangle = \|Ax\|^2$ for all $x \in \mathbb{R}^n$, then $\operatorname{Spec}(q)$ is just the length spectrum of $\left(\mathbb{R}^n/\Gamma, g_{can}/\Gamma\right)$. Consequently, by the above remark concerning the length spectra of two isospectral flat tori, it is obvious that two n-dimensional flat tori represented respectively by $[q_1]$ and $[q_2]$ are isospectral if and only if $\operatorname{Spec}(q_1) = \operatorname{Spec}(q_2)$. Note that if $[q_1] = [q_2]$, then $\operatorname{Spec}(q_1) = \operatorname{Spec}(q_2)$.

Let us recall the concept of **Minkowski's reduction**, one of the principal concepts of reduction which appear in the geometry of numbers.

Definition 2.2. Let $q \in S_n^+(\mathbb{R})$ and $(f_1,\ldots,f_n)$ a basis for $\mathbb{Z}^n$. The quadratic form q is said to be **reduced** (in the sense of **Minkowski**) with respect to the basis $(f_1,\ldots,f_n)$ if

(i) $q(f_1) = \inf\{q(x) | x \in \mathbb{Z}^n \setminus \{0\}\}$;

(ii) $q(f_j) \le q(x)$ for any $x \in \mathbb{Z}^n$ such that $(f_1,\ldots,f_{j-1},x)$ can be extended to a basis for $\mathbb{Z}^n$;

(iii) For any $j \in \{1,\ldots,n-1\}$, $\breve{q}(f_j, f_{j+1}) \ge 0$, where $\breve{q}$ is the symmetric bilinear form associated to q.

If $q \in S_n^+(\mathbb{R})$ is reduced with respect to the basis $(f_1,\ldots,f_n)$ for $\mathbb{Z}^n$, then $q(f_1) \le q(f_2) \le \cdots \le q(f_n)$.

Before showing the existence of a basis $(f_1,\ldots,f_n)$ for $\mathbb{Z}^n$ such that $q \in S_n^+(\mathbb{R})$ is reduced with respect to $(f_1,\ldots,f_n)$, let us recall the following lemma.

Lemma 2.1. Let $(f_1,\ldots,f_n)$ be a basis for Γ and $f = \sum_{j=1}^{n} \lambda^j f_j \in \Gamma$. The system $(f_1,\ldots,f_{m-1},f)$, $m < n$, can be extended to a basis for Γ if and only if the greatest common divisor (denoted by g.c.d.) of $\lambda^m, \lambda^{m+1},\ldots,\lambda^n$ is equal to 1.

For a proof of this result, see Cassels [1959], p. 14.

Lemma 2.2. If $q \in S_n^+(\mathbb{R})$, there exists a basis $(f_1,\ldots,f_n)$ for $\mathbb{Z}^n$ such that q is reduced with respect to $(f_1,\ldots,f_n)$.

Proof. We follow Cassels [1959]. First note that there exists a real constant $C > 0$ such that $q(x) \ge C\|x\|^2$ for any $x \in \mathbb{R}^n$. Indeed, q attains its infimum $C > 0$ on the (compact) unit sphere $\|x\| = 1$, so that the above inequality follows immediately by homogeneity. In particular, there are only a finite number of vectors $x \in \mathbb{Z}^n$ such that $f(x)$ is less than any given number. Let $f_1 \in \mathbb{Z}^n \setminus \{0\}$ for which $q(f_1) = \inf\{q(x) | x \in \mathbb{Z}^n \setminus \{0\}\}$. Such f_1 exist, and there are only a finite number of them. If $f_1 = \lambda x$, $x \in \mathbb{Z}^n$, where $\lambda > 1$ is an integer, then

$$0 < q(x) = \frac{1}{\lambda^2} q(f_1) < q(f_1),$$

contrary to the way of choice of f_1. By Lemma 2.1 one may extend f_1 to a basis $f_1, e'_2, \dots, e'_n$ for $\mathbb{Z}^n$. Let now choose f_j, $j = 2, \dots, n$, in succession. Assume that $f_1, \dots, f_{j-1}$ have already been chosen and are extensible to a basis $f_1, \dots, f_{j-1}, e'_j, \dots, e'_n$ for $\mathbb{Z}^n$. Then f_j is one of the finite number of vectors with the property that $f_1, \dots, f_j$ can be extended to a basis for $\mathbb{Z}^n$ and for which $q(f_j)$ is as small as possible. Such f_j exist but are finite in number. In this way one gets finally a basis $(f_1, \dots, f_n)$ for $\mathbb{Z}^n$ that satisfies the conditions (i) and (ii) of Definition 2.2. In order to fulfil also the condition (iii) notice that it can happen that some of the vectors f_j should be replaced by $-f_j$ in the above obtained basis.

Q.E.D.

Assume that the quadratic form q is reduced with respect to the basis $(f_1, \dots, f_n)$ for $\mathbb{Z}^n$. There exists $A \in GL(n, \mathbb{Z})$ such that $Ae_j = f_j$ for $j \in \{1, \dots, n\}$, where $(e_1, \dots, e_n)$ is the canonical basis for $\mathbb{R}^n$. If one considers $q' = q \circ A$, then $[q] = [q']$ and q' is reduced with respect to the basis $(e_1, \dots, e_n)$. Thus in each equivalence class $[q] \in S_n^+(\mathbb{R}) / GL(n, \mathbb{Z})$ there is an element reduced with respect to the standard basis for $\mathbb{R}^n$.

Remark 2.1. Assume $n = 1$ and let $q, q' \in S_1^+(\mathbb{R})$. If $\mathrm{Spec}(q) = \mathrm{Spec}(q')$, then $[q] = [q']$, that is if two 1-dimensional flat tori (i.e. circles with their canonical metric) are isospectral, then they are isometric. Indeed, without loss of generality, one can assume that q and q' are reduced with respect to the canonical basis $e_1 = 1$ for $\mathbb{R}$. Since $\mathrm{Spec}(q) = \mathrm{Spec}(q')$, by Definition 2.2 it follows that $q(e_1) = q'(e_1)$. But this immediately implies $q = q'$.

Now assume that q is reduced with respect to the standard basis $(e_1, \dots, e_n)$ for $\mathbb{Z}^n$. By definition

$$q(e_1) = \inf \{\mathrm{Spec}(q) \setminus \{0\}\}$$

is determined by $\mathrm{Spec}(q)$. The same happens with $q(e_2)$. More precisely,

Lemma 2.3. If the quadratic form $q \in S_n^+(\mathbb{R})$ is reduced with respect to the standard basis $(e_1, \dots, e_n)$ for $\mathbb{Z}^n$, then

$$q(e_2) = \inf \{\mathrm{Spec}(q) \setminus \{q(\alpha e_1) | \alpha \in \mathbb{Z}\}\}.$$

Proof. Choose $u = \sum_{j=1}^{n} u^j e_j \in \mathbb{Z}^n$ such that

$$q(u) = \inf \{\mathrm{Spec}(q) \setminus \{q(\alpha e_1) | \alpha \in \mathbb{Z}\}\}.$$

Since $e_2 \notin \mathbb{Z}e_1$, it follows that $q(e_2) \geq q(u)$. If one verifies that (e_1, u) can be extended to a basis for $\mathbb{Z}^n$, by Definition 2.2 one gets $q(u) \geq q(e_2)$, what proves the Lemma.

Now let us write $u = u^1 e_1 + dv$, where the g.c.d. d of $(u^2, \ldots, u^n)$ is ≥ 1 and v is a primitive lattice point in $\mathbb{Z}^{n-1} = \mathbb{Z}e_2 \oplus \cdots \oplus \mathbb{Z}e_n$, that is not of the shape $\lambda v'$, $v' \in \mathbb{Z}^{n-1}$ for an integer $\lambda > 1$.

By Lemma 2.2 it is sufficient to show that $d = 1$. Changing eventually u with $-u$ and v with $-v$, one can suppose that $u^1 \geq 0$. Let $\|x\| = \sqrt{q(x)}$ and assume that $d \geq 2$. Let us define $n^0 \in \{0, \ldots, u^1\}$ such that

$$\|n^0 e_1 - w\| = \min\left\{\|ne_1 - w\| \,\middle|\, 0 \leq n \leq u^1\right\},$$

where

$$w = \left(1 - \frac{1}{d}\right) u^1 e_1 .$$

Since

$$\|n^0 e_1 - w\| \leq \|u^1 e_1 - w\| = \frac{1}{d}\|e_1\| \leq \frac{1}{d}\|u\| \leq \frac{1}{2}\|u\|$$

and

$$\left\|w - \left(u^1 e_1 + v\right)\right\| = \frac{1}{d}\|u\| \leq \frac{1}{2}\|u\|$$

it follows that

$$\left\|n^0 e_1 - \left(u^1 e_1 + v\right)\right\| < \|n^0 e_1 - w\| + \left\|w - \left(u^1 e_1 + v\right)\right\| \leq \frac{1}{2}\|u\| + \frac{1}{2}\|u\| = \|u\|,$$

which contradicts the way of choice of u. Note that the vectors $n^0 e_1 - w$ and $w - \left(u^1 e_1 + v\right)$ cannot belong to the same linear line in $\mathbb{R}^n$, so that the first inequality is strict. Hence $d = 1$ and the proof is finished.

Q.E.D.

Theorem 2.1. Assume $n = 2$ and let $q, q' \in S_2^+(\mathbb{R})$. If $\operatorname{Spec}(q) = \operatorname{Spec}(q')$, then $[q] = [q']$, that is if two 2-dimensional flat tori are isospectral, then they are isometric.

Proof. It is sufficient to prove that if two quadratic forms q and q' reduced with respect to the canonical basis (e_1, e_2) for $\mathbb{R}^2$ are such that $\operatorname{Spec}(q) = \operatorname{Spec}(q')$, then $q = q'$. Let q and q' be two such quadratic forms. Since $\operatorname{Spec}(q) = \operatorname{Spec}(q')$, by Definition 2.2 it follows that $q(e_1) = q'(e_1)$. On the other hand, Lemma 2.3 implies that $q(e_2) = q'(e_2)$. Recall that if $q \in S_n^+(\mathbb{R})$ is given by $q(x) = \langle Ax, Ax \rangle$, $x \in \mathbb{R}^n$, where $A \in GL(n, \mathbb{R})$, then $\det(q) = \left[\operatorname{Vol}\left(\mathbb{R}^n / A\mathbb{Z}^n, g_{can} / A\mathbb{Z}^n\right)\right]^2$ (see also Chapter 2, Example 1.2). Consequently, $\operatorname{Spec}(q) = \operatorname{Spec}(q')$ and Poisson's formula implies that $\det(q) = \det(q')$, that is

$$q(e_1)q(e_2)-\left[\breve{q}(e_1,e_2)\right]^2=q'(e_1)q'(e_2)-\left[\breve{q}'(e_1,e_2)\right]^2.$$

But $\breve{q}(e_1,e_2)\geq 0$ and $\breve{q}'(e_1,e_2)\geq 0$ by Definition 2.2, so that the last equality implies that $\breve{q}(e_1,e_2)=\breve{q}'(e_1,e_2)$, that is $q=q'$.

Q.E.D.

Remark 2.2. Quite recently, Schiemann [1997] has shown that any lattice in $\mathbb{R}^3$ is also determined up to a congruence by its theta series. Hence any two isospectral 3-dimensional flat tori are isometric.

Before showing what happens in higher dimensions, let us discuss some properties of the theta series of lattices or of translates of lattices.

The theta series of lattices (or translates of lattices) can be related to the classical theta functions of Jacobi (see Whittaker and Watson [1969], Chapter 21 or Conway and Sloane [1999]). For example, the theta series of the lattice $\mathbb{Z}\subset\mathbb{R}$ is

$$\Theta_{\mathbb{Z}}(s)=\sum_{k=-\infty}^{+\infty}s^{k^2}=1+2s+2s^4+2s^9+\cdots,$$

which is the same as the Jacobi theta function $\theta_3(0,s)\overset{\text{not}}{=}\theta_3(s)$ (see Whittaker and Watson [1969], p. 464). The translate $\frac{1}{2}+\mathbb{Z}=\left\{\cdots,-\frac{3}{2},-\frac{1}{2},\frac{1}{2},\frac{3}{2},\cdots\right\}$ of the lattice $\mathbb{Z}$ in $\mathbb{R}$ has the theta series

$$\Theta_{\frac{1}{2}+\mathbb{Z}}(s)=\sum_{k=-\infty}^{+\infty}s^{\left(k+\frac{1}{2}\right)^2}=2s^{\frac{1}{4}}+2s^{\frac{9}{4}}+2s^{\frac{25}{4}}+\cdots,$$

which is the same as the Jacobi theta function $\theta_2(0,s)\overset{\text{not}}{=}\theta_2(s)$.

Two properties of the theta series enable us to calculate the theta series of certain higher-dimensional lattices in terms of Jacobi theta functions.

Proposition 2.1. (i) Let $\Sigma_k\subset\mathbb{R}^{n_k}$ be a lattice or a translate of a lattice, $k=1,2$ and $n=n_1+n_2$. Identify $\mathbb{R}^{n_1}$ (resp. $\mathbb{R}^{n_2}$) with the vector subspace $\mathbb{R}^{n_1}\times\{0\}$ (resp. $\{0\}\times\mathbb{R}^{n_2}$) of $\mathbb{R}^n$ and Σ_1 (resp. Σ_2) with its image in $\mathbb{R}^n$ with respect to these identifications. If $\Sigma=\Sigma_1\overset{\perp}{+}\Sigma_2$ is the orthogonal sum in $\mathbb{R}^n$ of Σ_1 and Σ_2, then

$$\Theta_\Sigma=\Theta_{\Sigma_1}\cdot\Theta_{\Sigma_2}.$$

(ii) If $\Sigma=\Sigma'\overset{\circ}{\cup}\Sigma''$ is the disjoint union of two lattices (or translates of lattices) in $\mathbb{R}^n$, then

$$\Theta_\Sigma=\Theta_{\Sigma'}+\Theta_{\Sigma''}.$$

Proof. (i) Indeed, one has successively

$$\Theta_\Sigma(s)=\sum_{y\in\Sigma}s^{\|y\|^2}=\sum_{\substack{y_k\in\Sigma_k\\k=1,2}}s^{\|y_1\|^2+\|y_2\|^2}=\sum_{\substack{y_k\in\Sigma_k\\k=1,2}}s^{\|y_1\|^2}\cdot s^{\|y_2\|^2}=\Theta_{\Sigma_1}(s)\cdot\Theta_{\Sigma_2}(s).$$

The statement contained in (ii) is obvious.

Q.E.D.

Now let us consider another Jacobi function, namely $\theta_4(0,s) \overset{\text{not}}{=} \theta_4(s)$, where

$$\theta_4(s) = \sum_{k=-\infty}^{+\infty} (-s)^{k^2} = 1 - 2s + 2s^4 - 2s^9 + \cdots .$$

Let us express the theta series of Σ in terms of the functions $\theta_2, \theta_3, \theta_4$ derived from the Jacobi theta functions in the following cases:

$$\Sigma = \mathbb{Z}^n,$$

$$\Sigma = C_n = \left\{ \left(x^1, \ldots, x^n\right) \in \mathbb{Z}^n \middle| \sum_{k=1}^{n} x^k \in 2\mathbb{Z} \right\},$$

$$\Sigma = D_n = \left\{ \left(x^1, \ldots, x^n\right) \in \mathbb{Z}^n \middle| \sum_{k=1}^{n} x^k \in 2\mathbb{Z} + 1 \right\},$$

$$\Sigma = E_n = C_n \mathring{\cup} \left(u_n + C_n\right), u_n = \left(\frac{1}{2}, \ldots, \frac{1}{2}\right) \in \mathbb{R}^n$$

where $n \in \mathbb{N}^*$. Before doing this, let us make the following remarks.

Remark 2.3. Let $(e_1, \ldots, e_n)$ be the standard basis for $\mathbb{R}^n$.

(i) C_n is a lattice in $\mathbb{R}^n$, namely $C_n = \mathbb{Z}f_1 + \mathbb{Z}f_2 + \cdots + \mathbb{Z}f_{n-1} + \mathbb{Z}f_n$, where $f_1 = e_1 - e_n, f_2 = e_2 - e_n, \ldots, f_{n-1} = e_{n-1} - e_n, f_n = e_{n-1} + e_n$;

(ii) D_n is a translate of the lattice C_n;

(iii) If n is even, then E_n is a lattice in $\mathbb{R}^n$. For instance,

$$E_2 = \mathbb{Z}\left(\frac{1}{2}, \frac{1}{2}\right) + \mathbb{Z}(1,-1), E_4 = \mathbb{Z}g_1 + \mathbb{Z}g_2 + \mathbb{Z}g_3 + \mathbb{Z}g_4, E_8 = \mathbb{Z}u_8 + \mathbb{Z}f_2 + \cdots + \mathbb{Z}f_8,$$

where $g_1 = \left(\frac{1}{2}, \frac{1}{2}, \frac{1}{2}, \frac{1}{2}\right)$, $g_2 = \left(\frac{1}{2}, -\frac{1}{2}, -\frac{1}{2}, \frac{1}{2}\right)$, $g_3 = \left(\frac{1}{2}, \frac{1}{2}, -\frac{1}{2}, -\frac{1}{2}\right)$,

$g_4 = \left(\frac{1}{2}, -\frac{1}{2}, \frac{1}{2}, -\frac{1}{2}\right)$, $u_8 = \left(\frac{1}{2}, \ldots, \frac{1}{2}\right) \in \mathbb{R}^8$, $f_2 = (0,1,0,0,0,0,0,-1)$,

$f_3 = (0,0,1,0,0,0,0,-1)$, $f_4 = (0,0,0,1,0,0,0,-1)$, $f_5 = (0,0,0,0,1,0,0,-1)$,

$f_6 = (0,0,0,0,0,1,0,-1)$, $f_7 = (0,0,0,0,0,0,1,-1)$, $f_8 = (0,0,0,0,0,0,1,1)$.

Note that for odd n, E_n is not a lattice, because the vector $2u_n = (1,1,\ldots,1)$ is not in E_n.

(iv) If $n \in 8\mathbb{N}^*$, then $E_n = E_n^*$ (for a proof of this remark, see Berger, Gauduchon, Mazet [1971], p. 156). E_n is called the n-dimensional diamond packing because the points of E_3 correspond to the positions of carbon atoms in a diamond.

Proposition 2.2. The theta series of the above considered lattices or translates of lattices can be expressed in terms of the functions $\theta_2, \theta_3, \theta_4$ as follows:

$$\Theta_{\mathbb{Z}^n} = \theta_3^n, \tag{2.1}$$

$$\Theta_{C_n} = \frac{1}{2}\left(\theta_3^n + \theta_4^n\right), \tag{2.2}$$

$$\Theta_{D_n} = \frac{1}{2}\left(\theta_3^n - \theta_4^n\right), \tag{2.3}$$

$$\Theta_{E_n} = \frac{1}{2}\left(\theta_2^n + \theta_3^n + \theta_4^n\right). \tag{2.4}$$

Proof. Let $(e_1, \ldots, e_n)$ be the standard basis for $\mathbb{R}^n$. Since

$$\mathbb{Z}^n = \Sigma_1 \overset{\perp}{+} \cdots \overset{\perp}{+} \Sigma_n,$$

where $\Sigma_k = \mathbb{Z}e_k$, $k = 1, \ldots, n$, by Proposition 2.1 (i) one gets $\Theta_{\mathbb{Z}^n} = (\Theta_{\mathbb{Z}})^n = \theta_3^n$.

We shall prove (2.2) and (2.3) by induction on n. First note that $C_1 = 2\mathbb{Z}$,so that

$$\Theta_{C_1}(s) = \sum_{k=-\infty}^{+\infty} s^{(2k)^2} = 1 + 2s^4 + 2s^{16} + \cdots = \frac{1}{2}\left[\theta_3(s) + \theta_4(s)\right].$$

Since $\mathbb{Z} = C_1 \overset{\circ}{\cup} D_1$, Proposition 2.1 (ii) implies that

$$\Theta_{D_1} = \Theta_{\mathbb{Z}} - \Theta_{C_1} = \frac{1}{2}\left(\theta_3 - \theta_4\right).$$

Assume that (2.2) and (2.3) are true for $n \in \mathbb{N}^*$ and let us prove them for $n+1$. Note that $(x^1, \ldots, x^n, x^{n+1}) \in C_{n+1}$ if and only if $(x^1, \ldots, x^n) \in C_n$ and $x^{n+1} \in C_1$ or $(x^1, \ldots, x^n) \in D_n$ and $x^{n+1} \in D_1$. Consequently,

$$C_{n+1} = \left(C_n \overset{\perp}{+} C_1\right) \overset{\circ}{\cup} \left(D_n \overset{\perp}{+} D_1\right),$$

so that by Proposition 2.2 one gets

$$\Theta_{C_{n+1}} = \Theta_{C_n}\Theta_{C_1} + \Theta_{D_n}\Theta_{D_1} = \frac{1}{4}\left[\left(\theta_3^n + \theta_4^n\right)\left(\theta_3 + \theta_4\right) + \left(\theta_3^n - \theta_4^n\right)\left(\theta_3 - \theta_4\right)\right] = \frac{1}{2}\left(\theta_3^{n+1} + \theta_4^{n+1}\right).$$

Since $\mathbb{Z}^{n+1} = C_{n+1} \overset{\circ}{\cup} D_{n+1}$, Proposition 2.2 (ii) implies that

$$\Theta_{D_{n+1}} = \Theta_{\mathbb{Z}^{n+1}} - \Theta_{C_{n+1}} = \frac{1}{2}\left(\theta_3^{n+1} - \theta_4^{n+1}\right).$$

Similarly, using the equalities

$$\Theta_{u_1+C_1} = \Theta_{u_1+D_1} = \frac{1}{2}\theta_2,\; u_{n+1} + C_{n+1} = \left[\left(u_n + C_n\right) \overset{\perp}{+} \left(u_1 + C_1\right)\right] \overset{\circ}{\cup} \left[\left(u_n + D_n\right) \overset{\perp}{+} \left(u_1 + D_1\right)\right]$$

and

$$u_{n+1}+D_{n+1}=\left[(u_n+C_n)\stackrel{\perp}{+}(u_1+D_1)\right]\stackrel{\circ}{\cup}\left[(u_n+D_n)\stackrel{\perp}{+}(u_1+C_1)\right]$$

one obtains

$$\Theta_{u_n+C_n}=\Theta_{u_n+D_n}=\frac{1}{2}\theta_2^n,$$

what implies via (2.2) the desired equality for Θ_{E_n}.

Q.E.D.

Proposition 2.3. The functions $\theta_2,\theta_3,\theta_4$ satisfy the following identities:

$$\theta_2^4+\theta_4^4=\theta_3^4, \tag{2.5}$$

$$\theta_3^{12}+\theta_4^{12}=\frac{1}{2}\left(\theta_3^4+\theta_4^4\right)\left(\theta_2^8+\theta_3^8+\theta_4^8\right), \tag{2.6}$$

$$\theta_2^{16}+\theta_3^{16}+\theta_4^{16}=\frac{1}{2}\left(\theta_2^8+\theta_3^8+\theta_4^8\right)^2. \tag{2.7}$$

Proof. Let us keep the notations used in Remark 2.3. Since (g_1,g_2,g_3,g_4) is an orthonormal basis for $\mathbb{R}^4$, there exists a unique orthogonal automorphism $\Phi:\mathbb{R}^4\to\mathbb{R}^4$ such that $\Phi(g_k)=e_k$, $k=1,2,3,4$, that is the lattices E_4 and $\mathbb{Z}^4$ in $\mathbb{R}^4$ are congruent. Hence

$$\Theta_{E_4}=\Theta_{\mathbb{Z}^4}$$

and (2.5) is an immediate consequence of (2.4) and (2.1). A straightforward computation shows that the identities (2.6) and (2.7) are direct consequences of (2.5).

Q.E.D.

Now let us describe two classical examples in dimensions 12 and 16 of non-isometric isospectral flat tori.

Theorem 2.2. Let C_4,C_{12},E_8,E_{16} be the previously defined lattices [see also Remark 2.3 (i), (iii)], $\Gamma_1=E_8\stackrel{\perp}{+}E_8$, $\Gamma_2=E_{16}\subset\mathbb{R}^{16}$ and $\Gamma_3=\left(E_8\stackrel{\perp}{+}C_4\right)^*$, $\Gamma_4=(C_{12})^*\subset\mathbb{R}^{12}$.

Then the following is true:

(i) The 16-dimensional flat tori $\left(\mathbb{R}^{16}/\Gamma_1, g_{can}/\Gamma_1\right)$ and $\left(\mathbb{R}^{16}/\Gamma_2, g_{can}/\Gamma_2\right)$ are isospectral and non-isometric;

(ii) The 12-dimensional flat tori $\left(\mathbb{R}^{12}/\Gamma_3, g_{can}/\Gamma_3\right)$ and $\left(\mathbb{R}^{12}/\Gamma_4, g_{can}/\Gamma_4\right)$ are isospectral and non-isometric.

Proof. (i) First note that $E_n=E_n^*$ for any $n\in 8\mathbb{N}^*$ [see Remark 2.3 (iv)]. By Proposition 2.1 (i) and the identities (2.4) and (2.7) one gets

$$\Theta_{\Gamma_1}=\left(\Theta_{E_8}\right)^2=\frac{1}{4}\left(\theta_2^8+\theta_3^8+\theta_4^8\right)^2=\frac{1}{2}\left(\theta_2^{16}+\theta_3^{16}+\theta_4^{16}\right)=\Theta_{E_{16}}=\Theta_{\Gamma_2}.$$

This implies that the given tori are isospectral. In order to show that these tori are non-isometric, let us suppose the contrary, namely that they should be isometric, that is there exists an orthogonal automorphism $\Phi:\mathbb{R}^{16}\to\mathbb{R}^{16}$ such that $\Phi(\Gamma_1)=\Gamma_2$. Let $(u_8, f_2,\ldots,f_8)$ be the basis for E_8 considered in Remark 2.3 (iii) and $B=((u_8,0),(f_2,0),\ldots,(f_8,0),(0,u_8),(0,f_2),\ldots,(0,f_8))$ the associated basis for Γ_1. Each vector of B has the length $\sqrt{2}$. On the other hand, $\Phi(B)$ is a basis for Γ_2, hence in particular a system of generators for Γ_2. The lattice Γ_2 contains either vectors of the form $\sum_{k=1}^{16} x^k e_k$ or vectors of the form $\sum_{k=1}^{16}\left(x^k+\frac{1}{2}\right)e_k$, $\sum_{k=1}^{16} x^k\in 2\mathbb{Z}$, where $(e_1,\ldots,e_{16})$ is the standard basis for $\mathbb{R}^{16}$. But any system of generators for Γ_2 must contain at least one vector of the second form, whose length is ≥ 2:

$$\left\|\sum_{k=1}^{16}\left(x^k+\frac{1}{2}\right)e_k\right\|=\sqrt{\sum_{k=1}^{16}\left(x^k+\frac{1}{2}\right)^2}\geq\sqrt{16\times\frac{1}{4}}=2,$$

a contradiction.

(ii) The isospectrality of the above defined 12-dimensional flat tori is an immediate consequence of Proposition 2.1 (i) and of the equalities (2.2), (2.4) and (2.6).

Now assume, like in the proof of (i), the existence of an orthogonal automorphism $\Phi:\mathbb{R}^{12}\to\mathbb{R}^{12}$ such that $\Phi(\Gamma_3^*)=\Gamma_4^*$, that is $\Gamma_3\equiv\Gamma_4$. First notice that the vectors in C_{12} of length $\sqrt{2}$ are just those of the form $\pm e_i\pm e_j$ with $i\neq j$, where $B=(e_1,e_2,\ldots,e_{12})$ is the standard basis for $\mathbb{R}^{12}$. The value of the modulus of the cosine of the angle between two vectors of this form $\pm e_i\pm e_j$ and $\pm e_k\pm e_l\ (i\neq j, k\neq l)$ is

(i) 1 if and only if $\pm e_i\pm e_j$ is collinear with $\pm e_k\pm e_l$;

(ii) $\frac{1}{2}$ if and only if $\{i,j\}\cap\{k,l\}$ contains just one element;

(iii) 0 if and only if the indices i,j,k,l are mutually distinct, or $\{i,j\}=\{k,l\}$ and the vectors $\pm e_i\pm e_j$ and $\pm e_k\pm e_l$ are not collinear.

Now, let us consider the basis

$$B_3=\{f_1=(u_8,0), f_2=e_2-e_8,\ldots,f_7=e_7-e_8, f_8=e_7+e_8,$$
$$f_9=e_9-e_{12}, f_{10}=e_{10}-e_{12}, f_{11}=e_{11}-e_{12}, f_{12}=e_{11}+e_{12}\}$$

for $\Gamma_3=E_8\overset{\perp}{+}C_4$, where $u_8=\left(\frac{1}{2},\ldots,\frac{1}{2}\right)\in\mathbb{R}^8$ [see also Remark 2.3 (iii)], and let us denote $h_j=\Phi(f_j), j=1,\ldots,12$. Each vector of B_3 has the length $\sqrt{2}$ and $B_4=\{h_j\,|\,j=1,\ldots,12\}$ should be a basis for C_{12} formed by vectors of length $\sqrt{2}$. Let us write a part of the matrix corresponding to the operation that associates to two vectors h_j,h_k in the assumed basis B_4 the modulus of the cosine of the angle between h_j and h_k:

	h_1	h_2	h_3	h_4	h_5	h_6	h_7	h_8
h_1	1	0	0	0	0	0	0	$\frac{1}{2}$
h_7	0	$\frac{1}{2}$	$\frac{1}{2}$	$\frac{1}{2}$	$\frac{1}{2}$	$\frac{1}{2}$	1	0
h_8	$\frac{1}{2}$	$\frac{1}{2}$	$\frac{1}{2}$	$\frac{1}{2}$	$\frac{1}{2}$	$\frac{1}{2}$	0	1

Since $\left|\cos\sphericalangle(h_1,h_8)\right|=\frac{1}{2}$, the above mentioned decompositions of the vectors h_1 and h_8 have in common a vector from B. Changing the enumeration of the vectors in B and eventually their sign, one can suppose that $h_1=e_1+e_2, h_8=e_1+e_3$. For $h_7=\pm e_i\pm e_j$ one has $\cos\sphericalangle(h_7,h_1)=0$ and $\cos\sphericalangle(h_7,h_8)=0$. We cannot have $\{i,j\}=\{1,2\}$, since otherwise $\left|\cos\sphericalangle(h_7,h_8)\right|=\frac{1}{2}$, a contradiction. Similarly, we cannot have $\{i,j\}=\{1,3\}$, since otherwise $\left|\cos\sphericalangle(h_7,h_1)\right|=\frac{1}{2}$, again a contradiction. Consequently, the indices $i,j,1,2,3$ are mutually distinct and one can assume that $h_7=e_4+e_5$.

For $j=2,\ldots,6$ and $h_j=h_j^1+h_j^2, h_j^1\in\{\pm e_k\}, h_j^2\in\{\pm e_l\}$, one has $\left|\cos\sphericalangle(h_j,h_7)\right|=\frac{1}{2}$, hence one of the vectors h_j^1 or h_j^2 should be in the set $\{\pm e_4,\pm e_5\}$. Similarly, from the equality $\left|\cos\sphericalangle(h_j,h_8)\right|=\frac{1}{2}$, it follows that the other one among h_j^1 and h_j^2 should be in $\{\pm e_1,\pm e_3\}$. This shows that $\{h_1,\ldots,h_8\}\subset\operatorname{Span}\{e_1,e_2,e_3,e_4,e_5\}$, that is $h_1,\ldots,h_8$ are linearly dependent vectors, contradicting the fact that B_4 is a basis.

Q.E.D.

Remark 2.4. The lattice-theoretic result made obvious in Theorem 2.2 (i) was first noted by Witt [1941] and applied by Milnor [1964] to the isospectrality problem. Kneser [1967] subsequently observed that C_{12} and $\left(E_8\overset{\perp}{+}C_4\right)$ are noncongruent lattices in $\mathbb{R}^{12}$ with the same theta series. Still later, Kitaoka [1977] gave examples in $\mathbb{R}^8$.

Now, let us describe a simple pair (depending on four parameters) of lattices in $\mathbb{R}^4$ with the same theta series, which was found by Conway and Sloane [1992]. Their construction provides some geometric understanding of some computer-generated examples (see, for instance, Schiemann [1990]) and pairs of isospectral lattices in $\mathbb{R}^4$ in the sense of Definition 2.1.

Let f_1,f_2,f_3,f_4 be four mutually orthogonal vectors in $\mathbb{R}^4$ of the following lengths, namely

$$\|f_1\|=\sqrt{\frac{\alpha}{12}},\|f_2\|=\sqrt{\frac{\beta}{12}},\|f_3\|=\sqrt{\frac{\gamma}{12}},\|f_4\|=\sqrt{\frac{\delta}{12}},$$

where $\alpha,\beta,\gamma,\delta>0$ and let $\left[x^1,x^2,x^3,x^4\right]$ denote the vector $x^1f_1+x^2f_2+x^3f_3+x^4f_4$. Define $L^+=L^+(\alpha,\beta,\gamma,\delta)$ to be the lattice in $\mathbb{R}^4$ spanned by $v_1^+=[+3,-1,-1,-1]$, $v_2^+=[+1,+3,+1,-1]$, $v_3^+=[+1,-1,+3,+1]$, $v_4^+=[+1,+1,-1,+3]$.

Similarly, let us define the lattice $L^- = L^-(\alpha,\beta,\gamma,\delta)$ replacing $+3$ by -3: it is spanned by the vectors: $v_1^- = [-3,-1,-1,-1]$, $v_2^- = [+1,-3,+1,-1]$, $v_3^- = [+1,-1,-3,+1]$, $v_4^- = [+1,+1,-1,-3]$.

Theorem 2.3. (Conway and Sloane [1992]). The theta series corresponding to the lattices $L^+(\alpha,\beta,\gamma,\delta)$ and $L^-(\alpha,\beta,\gamma,\delta)$ are identical. In particular, $L^+(1,7,13,49)$ and $L^-(1,7,13,49)$ are isospectral lattices in $\mathbb{R}^4$ in the sense of Definition 2.1.

Sketch of proof. For $\sigma = +$ or $-$, it is easy to see that $L^\sigma(\alpha,\beta,\gamma,\delta)$ has a sublattice $M^\sigma = M^\sigma(\alpha,\beta,\gamma,\delta)$ spanned by the 8 vectors $[\pm3,\pm3,\pm3,\pm3]$ in which the product of the signs is σ. Indeed, one has

$$
\begin{aligned}
v_1^\pm + v_2^\pm - v_3^\pm &= [\pm3,\pm3,\mp3,-3],\\
v_2^\pm + v_3^\pm + v_4^\pm &= [+3,\pm3,\pm3,\pm3],\\
v_1^\pm + v_3^\pm - v_4^\pm &= [\pm3,-3,\pm3,\mp3],\\
v_1^\pm - v_2^\pm + v_4^\pm &= [\pm3,\mp3,-3,\pm3]
\end{aligned}
$$

(the signs are in correspondence). Changing the sign of any coordinate of a vector of $M^\sigma(\alpha,\beta,\gamma,\delta)$ yields a vector of $M^{-\sigma}(\alpha,\beta,\gamma,\delta)$, that is M^+ is congruent to M^-. Representatives for the nontrivial cosets of M^σ in L^σ are $\pm v_1^\sigma, \pm v_2^\sigma, \pm v_3^\sigma, \pm v_4^\sigma$. Indeed, these vectors belong to different cosets because the difference of any two of them has at least one coordinate equal to $+2$ or -2. Moreover, $v_k^+ + M^+$ (resp. $-v_k^+ + M^+$) is congruent to $+v_k^- + M^-$ (resp. $-v_k^- + M^-$). One can combine these two remarks into a simple rule that gives a one-to-one length-preserving correspondence between the vectors $[x^1,x^2,x^3,x^4]$ of L^+ and those of L^- by changing the sign of the first coordinate of $[x^1,x^2,x^3,x^4]$ that is divisible by 3. Finally, one can be easily verified that $L^+(1,7,13,19)$ and $L^-(1,7,13,19)$ are not congruent.

Q.E.D.

Remark 2.5. (i) $L^\sigma(\alpha,\beta,\gamma,\delta) \equiv L^\sigma(\alpha,\delta,\beta,\gamma) \equiv L^{-\sigma}(\beta,\alpha,\gamma,\delta)$. Hence if any two of $\alpha,\beta,\gamma,\delta$ are equal, then $L^+(\alpha,\beta,\gamma,\delta) \equiv L^-(\alpha,\beta,\gamma,\delta)$.

(ii) One can also verify for other small values of $\alpha,\beta,\gamma,\delta$ (1,7,13,43;1,7,25,31,etc) that the corresponding lattices L^+ and L^- are not congruent.

(iii) Conway and Sloane [1992] have conjectured that $L^+(\alpha,\beta,\gamma,\delta)$ and $L^-(\alpha,\beta,\gamma,\delta)$ are non-congruent whenever $\alpha < \beta < \gamma < \delta$.

Corollary 2.1. Let $\Gamma_5 = L^+(1,7,13,19)$ and $\Gamma_6 = L^-(1,7,13,19)$ be the previously defined lattices in $\mathbb{R}^4$. Then the 4-dimensional flat tori $\left(\mathbb{R}^4/\Gamma_5, g_{can}/\Gamma_5\right)$ and $\left(\mathbb{R}^4/\Gamma_6, g_{can}/\Gamma_6\right)$ are isospectral and non-isometric.

The following result shows that the existence of a pair of non-isometric, isospectral *n*-

dimensional flat tori implies also the existence of such a pair in any dimension > n.

Proposition 2.4. Let $\left(\mathbb{R}^n/\Gamma', g_{can}/\Gamma'\right)$ and $\left(\mathbb{R}^n/\Gamma'', g_{can}/\Gamma''\right)$ be two non-isometric, isospectral, n-dimensional flat tori. Then for each $k \geq 1$, there exists a lattice Γ in $\mathbb{R}^k$ such that the $(n+k)$-dimensional flat tori $\left(\mathbb{R}^{n+k}/\Gamma_1, g_{can}/\Gamma_1\right)$ and $\left(\mathbb{R}^{n+k}/\Gamma_2, g_{can}/\Gamma_2\right)$ are isospectral and non-isometric, where we have identified $\mathbb{R}^n$ (resp. $\mathbb{R}^k$) with the vector subspace $\mathbb{R}^n \times \{0\}$ (resp. $\{0\} \times \mathbb{R}^k$) of $\mathbb{R}^{n+k}, \Gamma', \Gamma''$ (resp. Γ) with its images (resp. image) in $\mathbb{R}^{n+k}$ with respect to these identifications, and $\Gamma_1 = \Gamma' \overset{\perp}{+} \Gamma$ (resp. $\Gamma_2 = \Gamma'' \overset{\perp}{+} \Gamma$), denotes the lattice in $\mathbb{R}^{n+k}$ defined by the corresponding orthogonal sum.

Proof. It suffices to suppose that $k = 1$. Choose a real number $\alpha_0 \in (0, +\infty)$ which is strictly smaller as the length of any vector of $\Gamma' \setminus \{0\}$ or $\Gamma'' \setminus \{0\}$. Let $f_{n+1} = (0, \ldots, 0, \alpha_0) \in \mathbb{R}^{n+1}$, $\Gamma = \mathbb{Z} f_{n+1}$ and $\Gamma_1 = \Gamma' \overset{\perp}{+} \Gamma$, $\Gamma_2 = \Gamma'' \overset{\perp}{+} \Gamma$. Proposition 2.1 (i) shows that $\Theta_{\Gamma_1} = \Theta_{\Gamma_2}$ and consequence (ii) of Poisson's formula implies that the flat tori $\left(\mathbb{R}^{n+1}/\Gamma_1, g_{can}/\Gamma_1\right)$ and $\left(\mathbb{R}^{n+1}/\Gamma_2, g_{can}/\Gamma_2\right)$ are isospectral. In order to show that these tori are non-isometric, let us suppose the contrary, namely that they should be isometric, that is there exists an orthogonal automorphism $\Phi : \mathbb{R}^{n+1} \to \mathbb{R}^{n+1}$ such that $\Phi(\Gamma_1) = \Gamma_2$. Then one gets necessarily $\Phi(f_{n+1}) = \pm f_{n+1}$, $\Phi(\mathbb{R}^n) = \mathbb{R}^n$ and $\Phi(\Gamma') = \Gamma''$. It follows therefore that $\Gamma' \equiv \Gamma''$, which is in contradiction with our assumption.

Q.E.D.

Starting from the above examples, Proposition 2.4 shows that it is possible to construct non-isometric, isospectral, n-dimensional flat tori for any $n \geq 4$.

Remark 2.6. Wolpert [1978] proved that almost all flat tori are uniquely determined by their spectra. The exceptions lie in a proper real subvariety of the moduli space of all flat tori. Moreover, Kneser (unpublished) and Wolpert [1978] have established that there exists only a finite number of isospectral flat tori which are two by two isometric. Pesce [1992] succeeded to give in any dimension an explicit bound, depending on the geometry, for the number of isometry classes of flat tori isospectral to a given flat torus.

While additional examples exist of spectrally rigid Riemannian manifolds (see also §§1 and 3), the known examples are very sparse. The scarcity of examples is most likely due to our very limited techniques for obtaining rigidity results (see also the comments in Remark 1.5 of this Chapter). It may well be the case that generic closed Riemannian manifolds are uniquely determined by their spectra.

Remark 2.7. Using explicit computation of the spectra, Ikeda [1983], [1989] constructed many examples of isospectral space forms, i.e. quotients S^n/Γ by finite groups Γ of orthogonal transformations of $\mathbb{R}^{n+1}$ acting without fixed points, including lens spaces (see Chapter 1, Example 2.16 and Example 3.5 in this Chapter). Using also explicit computation of the spectra, Szabó [1999] succeeded to construct pairs of isospectral closed Riemannian manifolds with different local geometry. For these Riemannian manifolds the isometry groups have different dimension. Among these examples are a pair of isospectral closed Riemannian manifolds, one of which is homogeneous and the other not.

3. SUNADA'S THEOREM AND PESCE'S APPROACH TO ISOSPECTRALITY

Recall that two closed Riemannian manifolds (M_1, g_1) and (M_2, g_2) are said to be **isospectral** if $\operatorname{Spec}(M_1, g_1) = \operatorname{Spec}(M_2, g_2)$ (see also Chapter 3, Definition 1.2).

The question, often posed as "*Can you hear the shape of a drum*", asks whether isospectral Riemannian manifolds are necessarily isometric. The answer is well-known to be negative. The first counterexample was a pair of isospectral tori given by Milnor [1964] (see §2). Until 1979, however, the only other examples discovered were a few additional pairs of tori or twisted products with tori. The past two decades has been a flood of examples of pairs of isospectral, non-isometric Riemannian manifolds as well as fairly general techniques for constructing examples (see e.g. Ejiri [1979], Vignéras [1980], Urakawa [1982], Ikeda [1983], [1989], [1997], Sunada [1985], Bérard [1989], [1992], [1993], DeTurck, Gordon [1989], Spatzier [1989], Zelditch [1990], Buser [1992], Dotti Miatello, R.J. Miatello [1992], Reid [1992], Gordon [1992], [1993], [1994], Gordon, Webb, Wolpert [1992], Buser, Conway, Doyle, Semmler [1994], Gordon, Webb [1994], Pesce [1994], [1996], Gornet [1996], [2000], Gordon, Gornet [1997], Gordon, Wilson [1997], Miatello, Rossetti [1999], Schüth [1999], Bergeron [2000], Gordon, Szabo [2000]). Isospectral, non-isometric Riemann surfaces have been constructed in every genus greater than three (see Buser [1986], Brooks, Tse [1987] and Brooks, Gornet, Gustafson [1998]). For a very recent expository article on isospectral Riemannian manifolds, see Gordon [2000a].

In contrast, some closed Riemannian manifolds such as the canonical sphere (S^n, g_{can}) and the real projective space $(P^n(\mathbb{R}), g_{can}), n \leq 6$ (see §1), an 3-dimensional spherical space form $\left(S^3/G, g_{can}\right)$, where G is a finite group of fixed point free isometries of (S^3, g_{can}) (in particular, an 3-dimensional lens space), an homogeneous 5-dimensional lens space, a two-dimensional flat torus (see §2) are uniquely determined up to isometry by their spectrum on functions.

For most known examples, a pair of isospectral closed Riemannian manifolds has a common Riemannian covering, i.e. the pair is of the form $\left(M_1 = M/\Gamma_1, g_1\right)$ and $\left(M_2 = M/\Gamma_2, g_2\right)$, where (M, g) is a (not necessarily closed) Riemannian manifold and Γ_1, Γ_2 are discrete groups of isometries acting freely on M. Most of these examples can be treated by representation theoretic techniques. As in the case of flat tori, the Laplace-Beltrami operator of the quotient Riemannian manifold M_j may be viewed as the Laplace-Beltrami operator of the Riemannian covering restricted to smooth functions which are Γ_j-invariant, $j = 1,2$ (see Chapter 3, Remark 2.1).

In order to state and prove the main results of this section, let us fix some notation and recall some preliminaries on representation theory, following Warner [1972], Donnelly [1978] and Pesce [1996].

Let G be a locally compact, separable, topological group with the property that the right and the left Haar measures for G coincide and $(H, \langle\ ,\ \rangle)$ a separable complex Hilbert space. By a **representation of *G* in *H*** we shall mean a homomorphism ρ of G into the group $U(H)$ of unitary operators of *H*, that is $\rho(xy) = \rho(x) \circ \rho(y)$ for all $x, y \in G$, $\langle \rho(x)u, \rho(x)v \rangle = \langle u, v \rangle$ for all $x \in G$ and $u, v \in H$, $\rho(x): H \to H$ is onto for all $x \in G$, and the mapping $G \times H \ni (x, u) \mapsto \rho(x)u \in H$ is continuous. If ρ is a finite dimensional representation of *G* in *H*, that is the dimension of *H* is finite, the function $\chi_\rho : G \to \mathbb{C}, \chi_\rho(x) = \mathrm{Trace}(\rho(x)), x \in G$, is called the **character** of ρ. In this case, if $(e_1, \ldots, e_N)$ is an orthonormal basis in *H*, then a **unitary operator** $\rho(x): H \to H$ is represented by a unitary matrix $[D_{ij}(x)]$:

$$(\rho(x))(e_j) = \sum_{i=1}^{N} D_{ij}(x) e_i, \; j = 1, \ldots, N,$$

or equivalently

$$D_{ij}(x) = \langle (\rho(x))(e_j), e_i \rangle, i, j = 1, \ldots, N.$$

Moreover, each matrix element $D_{ij} : G \to \mathbb{C}$ is a continuous function and

$$D_{ij}(xy) = \sum_{k=1}^{N} D_{ik}(x) D_{kj}(y), \; i, j = 1, \ldots, N,$$

for all $x, y \in G$. Note that

$$\chi_\rho(x) = \sum_{i=1}^{N} \langle (\rho(x))(e_i), e_i \rangle = \sum_{i=1}^{N} D_{ii}(x) \tag{3.1}$$

and

$$\begin{aligned} \chi_\rho(x^{-1}) &= \mathrm{Trace}(\rho(x^{-1})) = \mathrm{Trace}(\rho(x))^* \\ &= \sum_{i=1}^{N} \overline{D_{ii}(x)} = \overline{\chi_\rho(x)} = \chi_{\bar{\rho}}(x) \end{aligned} \tag{3.2}$$

for all $x \in G$.

A representation ρ of *G* in *H* is said to be **irreducible** if the only two closed invariant subspaces of $\rho(x): H \to H$ are the null-space $\{0\}$ and the whole space *H* for every $x \in G$. The following statement gives a criterion of irreducibility: a representation ρ of *G* in *H* is irreducible if and only if the only operators commuting with all the $\rho(x)$ are scalar multiples of the identity. One can show that if $\rho : G \to U(H)$ is a finite dimensional irreducible representation, then its matrix elements D_{ij} satisfy the orthogonality relations

$$N \int_G D_{ij}(x) \overline{D_{kl}(x)} dx = \delta_{ik} \delta_{jl}, \; i, j, k, l = 1, \ldots, N,$$

where *dx* is the Haar measure on *G* and *N* the dimension of *H*. In particular, for such a representation one gets

$$\int_G \chi_\rho(x)\overline{\chi_\rho(x)}dx = \int_G \left(\sum_{i=1}^N D_{ii}(x)\right)\left(\sum_{j=1}^N \overline{D_{jj}(x)}\right)dx$$
$$= \sum_{k=1}^N \int_G D_{kk}(x)\overline{D_{kk}(x)}dx = 1 \tag{3.3}$$

and

$$\chi_\rho(y^{-1}xy) = \text{Trace}\left(\rho(y^{-1}xy)\right) = \text{Trace}\left(\rho(y^{-1})\rho(x)\rho(y)\right)$$
$$= \text{Trace}\left(\rho(x)\right) = \chi_\rho(x) \tag{3.4}$$

for all $x \in G$, so that χ_ρ is constant on the conjugacy classes of G.

If ρ is a representation of G in H and H' is a closed vector subspace of H invariant with respect to $\rho(x)$ for all $x \in G$, then $\rho' : G \to U(H')$, $\rho'(x) = {}_{H'}|\rho(x)|_{H'}$, $x \in G$, defines a representation of G in H', which is called the **subrepresentation** of ρ associated to H'.

If $\rho_1, \rho_2, \ldots$ are representations of G in the Hilbert spaces $H_1, H_2, \ldots$, then one defines their direct sum $\rho = \sum_j \rho_j$ as follows. Let H be the (Hilbert space) direct sum of the H_j. Given $u = (u_1, u_2, \ldots)$ in H, that is $\sum_j \|u_j\|^2 < \infty$, $x \in G$, put $[\rho(x)](u) = (\rho_1(x)u_1, \rho_2(x)u_2, \ldots)$. Then ρ is a representation of G in H called the direct sum of the ρ_j. Note that a finite dimensional representation of any topological group is a direct sum of irreducible representations.

Two representations ρ_1, ρ_2 of G in the Hilbert spaces H_1 and H_2 are said **unitarily equivalent** if there exists a unitary isomorphism $T : H_1 \to H_2$ such that $T \circ \rho_1(x) = \rho_2(x) \circ T$ for every $x \in G$. We shall write in this case $\rho_1 \cong \rho_2$. This operation "$\cong$" is an equivalence relation, so that it partitions off the set of all representations of G into disjoint classes of unitarily equivalent representations. Let $\hat{G}$ be the set of all such equivalence classes of irreducible representations of G.

Let ρ_1, ρ_2 be finite dimensional irreducible representations of G. Note that if ρ_1, ρ_2 are unitarily equivalent, then $\chi_{\rho_1} = \chi_{\rho_2}$.

Let ρ be a representation of G in H. One says that ρ is **discretely decomposable** if ρ is unitarily equivalent to a direct sum $\rho_1 + \rho_2 + \cdots$, where the ρ_j are irreducible. For example, every representation of a compact Lie group G in H is discretely decomposable. If the representation $\rho = \sum_j \rho_j$ of G in H is discretely decomposable and σ is an irreducible subrepresentation of ρ, then the finite number of indices j for which σ is unitarily equivalent to ρ_j is called **the multiplicity of σ in the decomposition of ρ.**

From now on, let us assume that G is a compact Lie group. Let $R(G)$ be the free abelian group with basis $\hat{G}$, that is the Grothendieck group of G. The group $R(G)$ has a natural symmetric bilinear form $\langle\ ,\ \rangle_G$ relative to which the basis $\hat{G}$ is orthonormal. If

$\sigma \in \hat{G}$ and ρ is a representation of G, then $\langle \sigma, \rho \rangle_G$ is just the multiplicity of the irreducible representation σ as a constituent of ρ. For any $\rho \in R(G)$ we have

$$\rho = \sum_{\sigma \in \hat{G}} \langle \sigma, \rho \rangle_G \sigma .$$

On the other hand, under this assumption on G, every irreducible representation ρ of G in H is finite dimensional, that is the dimension of H is finite. Moreover, every representation ρ of G in H is a direct sum $\sum_j \rho_j$ of irreducible finite dimensional subrepresentations ρ_j of ρ.

If $\rho = \sum_{j=1}^{k} \rho_j$ is a finite dimensional representation of G, then by (3.1) one has

$$\chi_\rho(x) = \sum_{j=1}^{k} \langle \rho_j, \rho \rangle_G \chi_{\rho_j}(x).$$

Using (3.3) one gets

$$\langle \rho_j, \rho \rangle_G = \int_G \chi_\rho(x) \overline{\chi_{\rho_j}(x)} dx, \quad j = 1, \ldots, k,$$

and

$$\sum_{j=1}^{k} \langle \rho_j, \rho \rangle_G^2 = \int_G \chi_\rho(x) \overline{\chi_\rho(x)} dx .$$

Note that the representation ρ is irreducible if and only if $\int_G \chi_\rho(x) \overline{\chi_\rho(x)} = 1$. In particular, if G is finite, the number of irreducible representations of G is equal to the number of conjugacy classes of G and the multiplicity $\langle \rho_j, \rho \rangle_G$ is given by

$$\langle \rho_j, \rho \rangle_G = \frac{1}{|G|} \sum_{x \in G} \chi_{\rho_j}(x) \overline{\chi_\rho(x)}. \tag{3.5}$$

Now let us consider a closed subgroup G_1 of G and let us point out some relations between the representations of G_1 and those of G.

If $\tau : G \to U(H)$ is a representation of G in H, then $\tau|_{G_1}$ defines a representation of G_1 in H. Let us denote by $\mathrm{Res}_{G_1}^{G}(\tau)$ this representation of G_1. Conversely, if $\sigma : G_1 \to U(H)$ is a representation of G_1 in H, let us denote by V_0 the complex vector space of all continuous functions $\varphi : G \to H$ such that $\varphi(x_1 x) = \sigma(x_1)\varphi(x)$ for all $x_1 \in G_1$ and $x \in G$. Putting

$$(\varphi, \psi) = \int_G \langle \varphi(x), \psi(x) \rangle dx,$$

where dx denotes a Haar measure on G, one gets an Hermitian product on V_0 and let V be the completion of V_0 with respect to this Hermitian product. The mapping

$$\pi : G \to U(V), [(\pi(x))(\varphi)](y) = \varphi(yx), x, y \in G, \varphi \in V,$$

defines a representation of G in the complex Hilbert space V. It is denoted by $\mathrm{Ind}_{G_1}^{G}(\sigma)$ and it is called the **induced representation** of G from σ. In case σ is the trivial representation, that is $\sigma(x_1) = 1_H$ for all $x_1 \in G_1$, which we shall denote by 1_{G_1}, then $V = L^2\left(G/G_1, \mu\right)$, where G/G_1 is the homogeneous space of the right G_1-cosets, μ any quasi-invariant measure on it, and the induced representation $\mathrm{Ind}_{G_1}^{G}(1_{G_1})$ of G from 1_{G_1} is called the **quasi-regular representation of *G***. For example, if the discrete subgroup G_1 of G (which can be an arbitrary Lie group) is cocompact, i.e. G/G_1 is compact, then G is necessarily unimodular, that is the Haar measure on G, uniquely defined up to rescaling, is bi-invariant. The Haar measure induces a measure μ on G/G_1. The right translations $R_x : G/G_1 \to G/G_1$, $R_x(G_1 y) = G_1(yx)$, $x, y \in G$, are volume preserving. Letting $L^2\left(G/G_1, \mu\right)$ be the space of measurable functions on G/G_1, which are square integrable with respect to μ, the quasi-regular representation $\pi = \mathrm{Ind}_{G_1}^{G}(1_{G_1})$ of G is given by

$$\pi(x)\varphi = \varphi \circ R_x$$

for $x \in G$ and $\varphi \in L^2\left(G/G_1, \mu\right)$. Note that $\pi(x)$ is a unitary automorphism of $L^2\left(G/G_1, \mu\right)$, that is π is a unitary representation of G. In particular, if G is finite, then μ is, up to rescaling, just the counting measure of G/G_1, $L^2\left(G/G_1, \mu\right)$ is a finite-dimensional vector space, and

$$\chi_{\mathrm{Ind}_{G_1}^{G}(1_{G_1})}(x) = \frac{\#\left\{y \in G \mid yxy^{-1} \in G_1\right\}}{\#(G_1)}, \tag{3.6}$$

where $x \in G$ and $\#(S)$ denotes the number of elements in the finite set S.

In what follows denote by 1_G the trivial one-dimensional representation of G. Weil's formulation of the classical Frobenius reciprocity rule states (see Warner [1972], p. 430):

Theorem 3.1 Let G be a compact Lie group and G_1 a closed subgroup of G. If $\sigma \in \hat{G}_1$ and $\tau \in \hat{G}$, then the number of times that the induced representation $\mathrm{Ind}_{G_1}^{G}(\sigma)$ contains τ as a discrete direct summand is equal to the number of times that τ restricted to G_1 contains σ as a discrete direct summand.

An equivalent formulation of the Frobenius reciprocity rule is as follows: if $\sigma \in \hat{G}_1$ and $\tau \in \hat{G}$, where G and G_1 are as above, then

$$\left\langle \sigma, \mathrm{Res}_{G_1}^G(\tau) \right\rangle_{G_1} = \left\langle \tau, \mathrm{Ind}_{G_1}^G(\sigma) \right\rangle_G, \tag{3.7}$$

that is $\mathrm{Res}_{G_1}^G$ and $\mathrm{Ind}_{G_1}^G$ are adjoint with respect to $\langle\,,\rangle_{G_1}$ and $\langle\,,\rangle_G$.

Now let K be a closed subgroup of the compact Lie group G, and let ρ be a (unitary) representation of G in H. Let $H^K = \left\{ v \in H \,\|\, \rho(x) \,](v) = v \text{ for each } x \in K \right\}$ be the subspace of the vectors of H fixed by every element of K, and let H_K denote the smallest closed G-invariant subspace of H containing H^K. Since the subspace H_K is G-invariant, it follows that one gets a subrepresentation ρ_K of ρ associated to H_K.

Note that H_K is the adherence of the vector space generated by $\left\{ H^{xKx^{-1}} \,|\, x \in G \right\}$.

Remark 3.1 Assume that the (discrete) direct sum decomposition of ρ into irreducibles is given by

$$\rho = \sum_{\sigma \in \hat{G}} m_\sigma \sigma,$$

where $m_\sigma = \langle \sigma, \rho \rangle_G$ is the multiplicity of σ as a constituent of ρ. Then the irreducible decomposition of ρ_k is given by

$$\rho_K = \sum_{\sigma \in \hat{G}_K} m_\sigma \sigma,$$

where $\hat{G}_K$ denotes the subset of $\hat{G}$ consisting of irreducible representations σ of G in H such that the trivial representation of K occurs as a constituent of $\mathrm{Res}_K^G(\rho)$, i.e. such that $\left\langle 1_K, \mathrm{Res}_K^G(\rho) \right\rangle_K \neq 0$, where 1_K denotes the trivial one-dimensional representation of K. Indeed H_K is generated as a G-space by the irreducible subrepresentations σ or ρ containing non-trivial K-fixed vectors.

If ρ_1 and ρ_2 are two unitarily equivalent representations of G in H, then $(\rho_1)_K$ and $(\rho_2)_K$ are also unitarily equivalent. The converse statement is not true, and we shall see that such examples appear in the isospectrality problems.

Definition 3.1. (Pesce [1996]) Two representations ρ_1, ρ_2 of G in H are ***K*-equivalent** if $(\rho_1)_K$ and $(\rho_2)_K$ are unitarily equivalent representations of G.

By Remark 3.1 this means that all irreducible representations of G in H which contain K-fixed vectors occur with the same multiplicities in ρ_1 and ρ_2.

For example, assume that G is discrete, K is a finite subgroup of G and ρ_1, ρ_2 are two finite dimensional representations of G in H. Then ρ_1 and ρ_2 are K-equivalent if and only if

$$\sum_{k \in K} \chi_1(kx) = \sum_{k \in K} \chi_2(kx)$$

for any $x \in G$ (for the proof of a more general result, see Pesce [1996]).

Also, let us note the following ereditary property concerning the relative equivalence of representations:

Proposition 3.1. Let L and K be closed subgroups of the compact Lie group G with the following property: for each $x \in G$ there exists $y \in L$ such that $(xKx^{-1}) \cap L \supseteq y(K \cap L)y^{-1}$. If σ_1 and σ_2 are two representations of L in H_1 and respectively H_2 which are $(K \cap L)$-equivalent, then the induced representations $\mathrm{Ind}_L^G(\sigma_1)$ and $\mathrm{Ind}_L^G(\sigma_2)$ of G from σ_1 and respectively from σ_2 are K-equivalent.

For the proof of this statement, see Pesce [1996].

Next, let us recall some results concerning Lie group actions on smooth manifolds. Let G be a compact Lie group acting smoothly and effectively on the compact, connected smooth manifold M. Let $G_x = \{g \in G \mid g \cdot x = x\}$ be the isotropy group at x. One can show that there exists a compact subgroup K of G, which is called the **principal isotropy group** of the action of G, having the following two properties:

(i) K is conjugate to a subgroup of G_x for each $x \in M$, and

(ii) there is a dense open subset $M_{(K)}$ in M such that its image $\bar{M}_{(K)}$ in the orbit space $\bar{M} = M/G$ is connected and if $x \in M_{(K)}$, then K and G_x are conjugated (see Bredon [1972], pp. 179, 309).

Example 3.1 Let (M, g) be a closed Riemannian manifold and assume that G is a compact Lie group acting isometrically and transitively on M. Then $M = G/K$ for some closed subgroup K of G. It is obvious that K can be taken any of the isotropy groups of the action of G on M and K is the principal isotropy group of the action of G on M. In particular, if $M = S^n$, then $G = O(n+1)$ and $K = O(n)$.

Example 3.2 Let (M, g) be a closed Riemannian manifold and assume that G is a finite group acting isometrically and effectively on M. Since the fixed point set F_ϕ of any $\phi \in G \setminus \{1_M\}$ is a closed subset of M with empty interior, by Baire's theorem it follows that $\bigcup_{\phi \in G \setminus \{1_M\}} F_\phi$ has empty interior, so that it is different from M. Consequently, in this case the principal isotropy group of the action G on M is trivial.

Now let M be any smooth G-manifold and let us discuss the decomposition of the G-representation space $L^2(M)$ into irreducibles. If G is a compact Lie group and M a transitive G-space, that is $M = G/K$ for some closed subgroup K of G, then the representation of G in $L^2(M)$ may be decomposed into a direct sum of finite dimensional representations of G. Moreover, Theorem 3.1 leads us in this case to the following consequence.

Corollary 3.1 If ρ is any irreducible representation of G, then the number of times that ρ is contained in the representation of G in $L^2(M)$ is equal to the number of times that ρ restricted to K contains the trivial representation.

Now let M be any smooth G-manifold. Since $M \setminus M_{(K)}$ has measure zero, it follows that $L^2(M) = L^2(M_{(K)})$. On the other hand, it is well-known that $M_{(K)}$ is a fiber bundle over $\bar{M}_{(K)}$ with fiber G/K (see Bredon [1972], p. 88). By removing a set of measure zero in $\bar{M}_{(K)}$ one may assume that this is a trivial bundle and consequently

$L^2(M) = L^2(\bar{M}) \times L^2(G/K)$, where the orbit space $\bar{M}$ is regarded as a trivial G-space. Thus, one may deduce

Theorem 3.2 The decomposition of $L^2(M)$ into G-irreducibles contains precisely those finite dimensional irreducible representations ρ which appear in $L^2(G/K)$. If the orbit space $\bar{M}$ has dimension greater than one, then each G-irreducible appears an infinite number of times.

Assume now that the closed smooth G-manifold M is endowed with a G-invariant Riemannian metric g. Let Δ be the Laplace-Beltrami operator associated to (M,g), $\lambda \in [0,+\infty)$ an eigenvalue of Δ and $E(\Delta,\lambda)$ the corresponding eigenspace. Then one has the following orthogonal decomposition

$$L^2(M) = \sum_{\lambda} E(\Delta,\lambda)$$

[see Chapter 3, Remark 1.2 (i)]. Since any $\phi \in G$ is an isometry of (M,g), ϕ commutes with Δ (see Chapter 2, §1) and preserves each of the eigenspaces $E(\Delta,\lambda)$. Thus, if ρ is any irreducible representation of G in $L^2(M)$, there exists an eigenvalue λ of Δ such that $\rho \subset E(\Delta,\lambda)$. It is well-known that the eigenspaces $E(\Delta,\lambda)$ are irreducible G-spaces if M is a symmetric space of compact type of rank 1. But in general the eigenspaces $E(\Delta,\lambda)$ need not be irreducible G-spaces. For example, if G is any compact Lie group, then the number of times that ρ is contained in the representation of G in $L^2(M)$ is equal to $\dim(\rho)$ by Corollary 3.1. However, the Laplace-Beltrami operator associated to the bi-invariant Riemannian metric on G may be identified with the Casimir operator of $L^2(G)$ (see Cahn, Wolf [1976], p. 3). Thus in this case each copy of ρ appears in the same eigenspace. It is well-known that $\dim(\rho) > 1$ for some ρ if G is compact semi-simple (Jacobson [1962], p. 257).

The following consequence of the previous remarks contrasts with Corollary 4.3 of Chapter 3 that for a generic Riemannian metric on a closed smooth manifold the eigenspaces of the corresponding Laplace-Beltrami operator are one-dimensional.

Corollary 3.2 (Donnelly [1978]) Let (M,g) be a closed Riemannian manifold and assume that a compact semi-simple Lie group acts isometrically and effectively on M. Then the Laplace-Beltrami operator associated to (M,g) has eigenspaces of arbitrarily large dimension.

Proof. If K is the principal isotropy group of the action of G on M, then $L^2(M)$ contains each of the infinite number of distinct irreducible representations $\rho \subset L^2(G/K)$. In particular $L^2(M)$ contains representations ρ of arbitrarily large dimension (Jacobson [1962], p. 257). On the other hand, each $\rho \subset E(\Delta,\lambda)$ for some eigenvalue λ of Δ, which implies that $\dim(\rho) \le \dim E(\Delta,\lambda)$.

Q.E.D.

The classical technique for constructing isospectral Riemannian manifolds based on the representation theory of Lie groups is as follows (see also Gordon [1992]). Let G be a Lie group. A Riemannian metric g on G is said to be **left-invariant** if the left translations $L_x : G \to G, L_x(y) = xy, x \in G$, are isometries of (G,g) for all $x \in G$. In this case, the Riemannian measure (see Chapter 2, §1) is a Haar measure. Note that a left-invariant metric is detemined by a choice of an orthonormal basis of the Lie algebra of G. Let $\Gamma \backslash G$ be the space obtained by quoting out the left translations of G by Γ, that is we mod out by Γ on the left. Note that the group G acts by right translations on $\Gamma \backslash G$. If Γ is any cocompact discrete subgroup of G, i.e. $\Gamma \backslash G$ is compact, then g induces a Riemannian metric on $\Gamma \backslash G$, denoted g', in such a way that the natural projection $\pi : (G,g) \to (\Gamma \backslash G, g')$ is a Riemannian covering. The elements of G define local isometries of $(\Gamma \backslash G, g')$ via the locally defined left translation action of G on $\Gamma \backslash G$.

Definition 3.2. Two cocompact discrete subgroups Γ_1 and Γ_2 of G are **representation equivalent** if the quasi-regular representations π_1 and π_2 of G in $L^2(\Gamma_1 \backslash G, \mu_1)$ and $L^2(\Gamma_2 \backslash G, \mu_2)$ respectively, defined as above, are unitarily equivalent, i.e. if there exists a unitary isomorphism $T : L^2(\Gamma_1 \backslash G, \mu_1) \to L^2(\Gamma_2 \backslash G, \mu_2)$ (called **intertwining operator**) such that

$$T \circ (\pi_1(x)) \circ T^{-1} = \pi_2(x)$$

for all $x \in G$.

Note that for two flat tori $\mathbb{R}^n/\Gamma_1, \mathbb{R}^n/\Gamma_2$ this condition holds only if the tori are isometric. In case of a Lie group G endowed with a left-invariant Riemannian metric g, the condition for the cocompact discrete subgroups Γ_1 and Γ_2 of G to be representation equivalent is a sufficient condition in order that $\mathbf{Spec}(\boldsymbol{\Gamma_1 \backslash G, g_1}) = \mathbf{Spec}(\boldsymbol{\Gamma_2 \backslash G, g_2})$. Indeed, the Laplace-Beltrami operator Δ associated to the left-invariant metric g is given by $\Delta = -\sum_{k=1}^{n} E_k^2$, where $(E_1, \ldots, E_n)$ is an orthonormal basis of the Lie algebra of G relative to g. Recall that a left-invariant vector field X on G acts on $C^\infty(G)$ as

$$(Xf)(x) = \frac{d}{dt}\Big|_{t=0} \left(f\left(x \exp(tX)\right)\right),$$

where exp is the Lie group exponential map. The space $L^2(\Gamma_j \backslash G)$ is identified with the space $\{f \in L^2_{loc}(G) : f \circ L_x = f \text{ for all } x \in \Gamma_j\}$. On the other hand, the Lie algebra of G may be identified with the space of left-invariant vector fields on G (see also Chapter 1, Example 4.5). Such vector fields may also be viewed as vector fields on $\Gamma_j \backslash G, j = 1,2$. Let R^j denote the right action of G on $\Gamma_j \backslash G$. Since

$$X(f_j) = \frac{d}{dt}\Big|_{t=0} \left(f_j \circ R^j_{\exp(tX)}\right) = \frac{d}{dt}\Big|_{t=0} \left[\pi_j(\exp(tX)) f_j\right]$$

for $f_j \in L^2(\Gamma_j \backslash G), j = 1,2$, the extension of Δ to $L^2(\Gamma_j \backslash G)$ is given by

$$\Delta = -\sum_{k=1}^{n} \left[\left(R^{j} \right)_{*} \left(E_{k} \right) \right]^{2}.$$

Consequently, the Laplace-Beltrami operator associated to $\left(\Gamma_j \backslash G, g_j\right)$ acts through π_j. Under the assumption on Γ_1 and Γ_2, the intertwining operator T between π_1 and π_2, given as in Definition 3.2, intertwines the Laplace-Beltrami operators Δ_1, Δ_2 associated to $(\Gamma_1 \backslash G, g_1)$ and $(\Gamma_2 \backslash G, g_2)$ respectively, that is Δ_1 and Δ_2, viewed as defined on $L^2(\Gamma_1 \backslash G)$ and $L^2(\Gamma_2 \backslash G)$ respectively, are unitarily equivalent, hence isospectral.

Example 3.3. Vignéras [1980] found a pair of representation equivalent cocompact, discrete subgroups Γ_1 and Γ_2 of $G = SL(2,\mathbb{R})$. Thus $(M_1 = \Gamma_1 \backslash G, g_1)$ and $(M_2 = \Gamma_2 \backslash G, g_2)$, where g_j is the Riemannian metric on $\Gamma_j \backslash G$ induced by any left-invariant Riemannian metric on G, are isospectral. Note that the hyperbolic plane $\left(H^2, h\right)$ (see Chapter 1, Example 2.3) may be identified with $G/SO(2)$ equipped with a suitable left-invariant Riemannian metric. The quotients N_1 and N_2 of H^2 by the discrete groups of isometries Γ_1 and Γ_2 respectively are compact Riemann surfaces. Moreover, Vignéras [1980] remarked that these Riemann surfaces N_1 and N_2 are also isospectral. One proof of this result is based on the following remark. Given isospectral closed Riemannian manifolds (M_1, g_1), (M_2, g_2) and Riemannian submersions $\pi_j : (M_j, g_j) \to (N_j, h_j)$, $j = 1,2$, with totally geodesic fibers, one can in special circumstances use the isospectrality of (M_1, g_1) and (M_2, g_2) to prove the isospectrality of (N_1, h_1) and (N_2, h_2). In case under discussion, the projection from M_j onto the Riemann surface N_j is a Riemannian submersion with totally geodesic fibers. Moreover, it is a principal bundle with fiber $SO(2)$. The spectrum of $\left(N_j, h_j\right)$ is the part of the spectrum of $\left(M_j, g_j\right)$ corresponding to functions invariant under the right action of the subgroup $SO(2)$ of $SL(2,\mathbb{R})$. The intertwining operator corresponding to the right actions of $G = SL(2,\mathbb{R})$ on $\Gamma_1 \backslash G$ and $\Gamma_2 \backslash G$ intertwines the Laplace-Beltrami operators associated to (M_1, g_1) and (M_2, g_2). Moreover, this intertwining operator carries $SO(2)$-invariant functions to $SO(2)$-invariant functions, so that the Riemann surfaces (N_1, h_1) and (N_2, h_2) are isospectral. Vignéras [1980] similarly constructed the first known examples of isospectral Riemannian 3-manifolds with different fundamental groups.

Sunada [1985] introduced a systematic method for constructing pairs of isospectral Riemannian manifolds $\left(M/\Gamma_1, g_1\right)$ and $\left(M/\Gamma_2, g_2\right)$ with a common Riemannian cover (M, g). His method was based on an interpretation of isospectrality in terms of finite groups. Many interesting examples have been constructed using this method (in fact, most of the authors mentioned before, who have constructed isospectral but non-isometric closed Riemannian manifolds, used this method or a variation of it – see also Remark

3.7), and some examples constructed prior to Sunada are in retrospect best understood via Sunada's theorem, which can be stated in representation theoretic terms as follows (see Pesce [1994]):

Theorem 3.3. If the finite groups Γ_1 and Γ_2 of fixed point free isometries (except the identity) of the closed Riemannian manifold (M,g) are "representation equivalent" subgroups of a larger closed subgroup G of the (compact Lie) isometry group of (M,g), i.e. if the quasi-regular representations of G in $L^2(\Gamma_1\backslash G)$ and $L^2(\Gamma_2\backslash G)$ are equivalent (see the definitions above), then $\left(M/\Gamma_1, g_1\right)$ and $\left(M/\Gamma_2, g_2\right)$ are isospectral, where g_j is the Riemannian metric on M/Γ_j induced by g (see Chapter 1, Theorem 2.3).

In Sunada's original formulation the group G was finite (in this case, the principal isotropy group K of the action of G on M is trivial) and the condition concerning the subgroups Γ_1 and Γ_2 was the following one: for any conjugacy class $[x]_G = \{yxy^{-1} \mid y \in G\}$ of G, one has

$$\#\left(\Gamma_1 \cap [x]_G\right) = \#\left(\Gamma_2 \cap [x]_G\right), \tag{3.8}$$

that is each G-conjugacy class $[x]_G$ intersects Γ_1 and Γ_2 in the same number of elements. In particular, (3.8) implies $\#(\Gamma_1) = \#(\Gamma_2)$. Taking into account our previous notation, (3.8) is equivalent to the following condition: the quasi-regular representations $\operatorname{Ind}_{\Gamma_1}^G(1_{\Gamma_1})$ and $\operatorname{Ind}_{\Gamma_2}^G(1_{\Gamma_2})$ are equivalent. Indeed, two finite dimensional representations are equivalent if and only if they have the same characters. Since

$$\#\{y \in G \mid yxy^{-1} \in \Gamma_j\} = \#(C_G(x)) \cdot \#\left(\Gamma_j \cap [x]_G\right),\ j = 1, 2,$$

where $C_G(x)$ denotes the centralizer of x in G, it follows that the condition (3.8) is equivalent to: for all $x \in G$,

$$\#\{y \in G \mid yxy^{-1} \in \Gamma_1\} = \#\{y \in G \mid yxy^{-1} \in \Gamma_2\}.$$

Thus the statement follows immediately from the formula (3.6). It seems that Mackey [1964] was the first to consider this equivalence condition for the induced representations (see Brooks [1999]). Mackey was motivated by the question of when one can deduce the "classical mechanics" of a group action from its "quantum mechanics", in the following sense: given a group G acting on measure spaces X_1 and X_2 by measure-preserving transformations, assume that $L^2(X_1)$ is G-equivalent to $L^2(X_2)$. Does it follow that there is a measurable G-equivalence from X_1 onto X_2? He remarks that any triple (G, Γ_1, Γ_2) of finite groups satisfying the above equivalence condition, where Γ_1 is not conjugate to Γ_2 in G will provide a counterexample since a G-equivalence $G/\Gamma_1 \to G/\Gamma_2$ is precisely a conjugation of Γ_1 to Γ_2 in G. For a very nice survey paper which summarizes the "state of the art" of the classical Sunada method until 1998 and reviews the historical context of this method, emphasizing in particular the connections to number theory, see Brooks [1999].

Let us give Sunada's original proof, simplified slightly by Buser [1992]. Since the Laplace-Beltrami operator Δ associated to (M,g) commutes with isometries of (M,g), the group G acts naturally on the eigenspaces of Δ. The eigenfunctions of the Laplace-Beltrami operator Δ_j associated to $\left(M/\Gamma_j, g_j\right)$ can be identified with those eigenfunctions of Δ which are Γ_j-invariant, $j=1,2$ (see Chapter 3, Remark 2.1), i.e. which are fixed with respect to the action of Γ_j. Recall that $\lambda \in [0,+\infty)$ is an eigenvalue of Δ_j if and only if there exists a Γ_j-invariant function $f \in C^\infty(M)\setminus\{0\}$ such that $\Delta f = \lambda f$. To prove Sunada's theorem, one needs only compare the dimensions of the subspaces of Γ_1-invariant functions and Γ_2-invariant functions in each eigenspace of Δ. Let E be an eigenspace of Δ. Then E is a finite dimensional real vector space. Asssume that G acts on E by linear transformations, i.e. there exists a homomorphism from G into $GL(E)$. Denote the action of $\phi \in G$ on $e \in E$ by $\phi(e)$. Let E_j be the vector subspace of all vectors from E fixed by Γ_j and define the mapping $P_j : E \to E$ by

$$P_j(e) = \frac{1}{|\Gamma_j|}\sum_{\phi\in\Gamma_j}\phi(e),\ e\in E,\ j=1,2.$$

Note that $P_j(e)=e$ for all $e\in E_j$ and $\phi\circ P_j = P_j$ [that is $P_j(e)\in E_j$ for all $e \in E$] for any $j=1,2$, i.e. P_j is the projection of E onto E_j parallel to $\ker P_j$. Consequently

$$\dim(E_j) = trace(P_j) = \frac{1}{|\Gamma_j|}\sum_{\phi\in\Gamma_j} trace(\phi).$$

Since conjugate elements of G have the same trace as linear operators on E, the condition (3.8) implies that $\dim(E_1)=\dim(E_2)$, so

$$\text{Spec}\left(M/\Gamma_1, g_1\right) = \text{Spec}\left(M/\Gamma_2, g_2\right).$$

Q.E.D.

Remark 3.2. The proof of Sunada's theorem uses from the point of view of the Laplace-Beltrami operator only the fact that this operator commutes with the isometries. It follows that pairs of closed Riemannian manifolds constructed in such a way are actually strongly isospectral, i.e. they are isospectral for any natural self-adjoint elliptic operator, such as the Hodge-de Rham operators acting on forms (see Chapter 2, §2), the Lichnérowicz Laplacians acting on tensor fields, etc. Sunada's original proof of his theorem was an elegant application of the trace formula for finite Riemannian coverings (see Chapter 3, §1). A slight variant of it was given by Brooks [1988]. This proof of the Sunada theorem shows that his method applies to any self-adjoint operator which is natural with respect to finite coverings and which is of trace class. This remark allows to construct isospectral potentials (see Brooks [1999]).

Remark 3.3. Note that the quotient Riemannian manifolds $\left(M_1 = M/\Gamma_1, g_1\right)$ and

$\left(M_2 = M/\Gamma_2, g_2\right)$ will be isometric precisely when Γ_1 and Γ_2 are conjugate subgroups of the full isometry group $\mathrm{Iso}(M,g)$ (see Chapter 1, §2). In many cases, one does not know the full isometry group of (M,g), only some subgroup G containing Γ_1 and Γ_2. In this case, even after making sure that Γ_1 and Γ_2 are not conjugate in G, one must still check, generally by ad hoc methods, whether the Riemannian manifolds (M_1,g_1) and (M_2,g_2) are isometric. Consequently, if one intends to construct by using Sunada's theorem non-isometric isospectral Riemannian manifolds, it is necessary to choose a triple such that (3.8) is satisfied and Γ_1, Γ_2 are not conjugate in G. It seems that Gassmann [1926] was the first to study such triples related to a problem concerning number fields. Gassmann did obtain examples and many other examples have been obtained since (see Gerst [1970], Komatsu [1976], Brooks [1988] and Bérard [1989]). Zelditch [1990] was able to construct examples of pairs of isospectral Riemannian manifolds of this type whose eigenvalue spectrum is simple. This is especially noteworthy, since one would initially expect that Sunada's method would tend to produce examples whose spectrum had a high degree of multiplicity.

Example 3.4. Let $G = SL(3,\mathbb{Z}_2)$, Γ_1 the set of all matrices in G of the form

$$\begin{pmatrix} 1 & * & * \\ 0 & * & * \\ 0 & * & * \end{pmatrix}$$

and let Γ_2 be the transpose of Γ_1. The subgroups Γ_1 and Γ_2 are not conjugate in G. However, each element of Γ_1 is conjugate to its transpose in Γ_2, so Γ_1 and Γ_2 satisfy (3.8). The group G has order 168 and Γ_1, Γ_2 are of order 7. For other examples of triples (G,Γ_1,Γ_2), where G is a finite group and Γ_1, Γ_2 are subgroups of G satisfying the condition (3.8), see Buser, Conway, Doyle, Semmler [1994] and Gordon [2000a].

One way to find a Riemannian manifold (M,g) on which G acts by isometries is to construct first a manifold $\widehat{M}$ whose fundamental group maps onto G. Let G_0 be the kernel of this surjection, and set $M = \widetilde{M}/G_0$, where $\widetilde{M}$ is the universal covering of $\widehat{M}$. Then M is a regular covering of $\widehat{M}$ with covering group G. One can give on M a G-invariant Riemannian metric g. Buser [1986] used the subgroups Γ_1, Γ_2 of $SL(3,\mathbb{Z}_2)$ given in Example 3.3 to construct isospectral flat surfaces.

Sunada's method and that given above account for most of the known examples of isospectral closed Riemannian manifolds.

Pesce [1996] succeeded to give a major strengthening of Sunada's theorem and gave several converse statements. Besides allowing for the construction of new examples of isospectral Riemannian manifolds, Pesce's generalization of Sunada's theorem allows some previously known examples of isospectral, but not strongly isospectral, Riemannian manifolds – e.g. isospectral lens spaces constructed by Ikeda [1989] – to be viewed in this context.

Before stating Pesce's results, let us recall the picture in which we shall work. Let

(M,g) be a closed Riemannian manifold, G a closed subgroup of the isometry group of (M,g) and Γ_1,Γ_2 two finite subgroups of G acting freely on M. Then the group G is compact and the Riemannian metric g induces a unique Riemannian metric g_j on the orbit closed manifold M/Γ_j such that the canonical projection of (M,g) onto $\left(M/\Gamma_j, g_j\right)$ is a Riemannian covering, $j=1,2$ (see Chapter 1, Theorem 2.3).

The next result due to Pesce [1996] gives an algebraic condition in order that the closed Riemannian manifolds $\left(M/\Gamma_1, g_1\right)$ and $\left(M/\Gamma_2, g_2\right)$ should be isospectral. Note that the Riemannian manifolds $\left(M/\Gamma_1, g_1\right)$ and $\left(M/\Gamma_2, g_2\right)$ are locally isometric, since each one is locally isometric to (M,g).

Theorem 3.4 (Pesce [1996]) Let (M,g) be a closed Riemannian manifold, G a compact (Lie) group of isometries of (M,g), Γ_1,Γ_2 finite subgroups of G acting freely on M and K the principal isotropy group of the action of G on M. If the induced representations $\mathrm{Ind}_{\Gamma_1}^G(1_{\Gamma_1})$ and $\mathrm{Ind}_{\Gamma_2}^G(1_{\Gamma_2})$ are K-equivalent, then the closed Riemannian manifolds $\left(M/\Gamma_1, g_1\right)$ and $\left(M/\Gamma_2, g_2\right)$ are isospectral.

Proof. Let Δ (resp. Δ_j) be the Laplace-Beltrami operator associated to (M,g) [resp. $\left(M/\Gamma_j, g_j\right)$] and $E(\Delta,\lambda)$ the complex eigenspace of Δ corresponding to the eigenvalue $\lambda\in[0,+\infty)$ of Δ. Hence the multiplicity of λ in $\mathrm{Spec}(M,g)$ is equal to $\dim_{\mathbb{C}}(E(\Delta,\lambda))$. If G_1 is a closed subgroup of the isometry group of (M,g), then one gets a representation of G_1 in $E(\Delta,\lambda)$ by setting

$$\left[\pi_\lambda^{G_1}(\phi)\right](f)=f\circ\phi^{-1}$$

for each $\phi\in G_1$ and $f\in E(\Delta,\lambda)$. Viewing functions on M/Γ_j as Γ_j-invariant functions on M, it is obvious that the set of eigenvalues of Δ includes the set of eigenvalues of $\Delta_j, j=1,2$ (see Chapter 3, Proposition 2.2). Now G acts on the space $L^2(M,\mathbb{C})$ of complex-valued square-integrable functions on M, and for each eigenvalue λ of Δ, the eigenspace $E(\Delta,\lambda)$ is G-invariant. The multiplicity of λ as an eigenvalue of Δ_j, that is $\dim_{\mathbb{C}}(E(\Delta_j,\lambda))$, is just the dimension $\dim_{\mathbb{C}}(E(\Delta,\lambda))^{\Gamma_j}$, where $(E(\Delta,\lambda))^{\Gamma_j}$ denotes the complex vector space of Γ_j-invariant eigenfunctions of Δ corresponding to λ (see Chapter 3, Remark 2.1). In other words, the multiplicity of λ as an eigenvalue of Δ_j is just the multiplicity of the trivial representation of Γ_j in $\pi_\lambda^{\Gamma_j}=\mathrm{Res}_{\Gamma_j}^G(\pi_\lambda^G)$:

$$\dim_{\mathbb{C}}(E(\Delta_j,\lambda))=\left\langle 1_{\Gamma_j},\mathrm{Res}_{\Gamma_j}^G(\pi_\lambda^G)\right\rangle_{\Gamma_j}.$$

Consequently, $\left(M/\Gamma_1, g_1\right)$ and $\left(M/\Gamma_2, g_2\right)$ are isospectral if and only if

$$\left\langle 1_{\Gamma_1}, \mathrm{Res}^G_{\Gamma_1}\left(\pi^G_\lambda\right)\right\rangle_{\Gamma_1} = \left\langle 1_{\Gamma_2}, \mathrm{Res}^G_{\Gamma_2}\left(\pi^G_\lambda\right)\right\rangle_{\Gamma_2} \tag{3.9}$$

for any eigenvalue λ of Δ. Let Γ denote Γ_1 or Γ_2. One gets successively

$$\begin{aligned}\left\langle 1_\Gamma, \mathrm{Res}^G_\Gamma\left(\pi^G_\lambda\right)\right\rangle_\Gamma &= \left\langle 1_\Gamma, \mathrm{Res}^G_\Gamma\left[\sum_{\sigma\in\hat{G}}\left\langle\sigma,\pi^G_\lambda\right\rangle_G \sigma\right]\right\rangle_\Gamma = \sum_{\sigma\in\hat{G}}\left\langle\sigma,\pi^G_\lambda\right\rangle_G \left\langle 1_\Gamma, \mathrm{Res}^G_\Gamma(\sigma)\right\rangle_\Gamma \\ &= \sum_{\sigma\in\hat{G}}\left\langle\sigma,\pi^G_\lambda\right\rangle_G \left\langle \mathrm{Ind}^G_\Gamma(1_\Gamma),\sigma\right\rangle_G,\end{aligned}$$

where the last equality is a direct consequence of the Frobenius reciprocity theorem [see also (3.7)]. Consequently, the isospectrality of $\left(M/\Gamma_1, g_1\right)$ and $\left(M/\Gamma_2, g_2\right)$ is equivalent to the following condition:

For any eigenvalue λ of Δ,

$$\sum_{\sigma\in\hat{G}}\left\langle\sigma,\pi^G_\lambda\right\rangle_G\left[\left\langle \mathrm{Ind}^G_{\Gamma_1}(1_{\Gamma_1}),\sigma\right\rangle_G - \left\langle \mathrm{Ind}^G_{\Gamma_2}(1_{\Gamma_2}),\sigma\right\rangle_G\right] = 0. \tag{3.10}$$

Notice that an irreducible representation σ of G is a subrepresentation of one of the representations π^G_λ if and only if it belongs to $\hat{G}_K$: if ρ is the representation of G in $L^2(M,\mathbb{C})$, then $\rho = \mathrm{Ind}^G_K(1_K)$, so that $\langle\sigma,\rho\rangle_G = \left\langle \mathrm{Res}^G_K(\sigma), 1_K\right\rangle_K$. Hence (3.10) is equivalent to:

For any eigenvalue λ of Δ,

$$\sum_{\sigma\in\hat{G}_K}\left\langle\sigma,\pi^G_\lambda\right\rangle_G\left[\left\langle \mathrm{Ind}^G_{\Gamma_1}(1_{\Gamma_1}),\sigma\right\rangle_G - \left\langle \mathrm{Ind}^G_{\Gamma_2}(1_{\Gamma_2}),\sigma\right\rangle_G\right] = 0. \tag{3.11}$$

But $\mathrm{Ind}^G_{\Gamma_1}(1_{\Gamma_1})$ and $\mathrm{Ind}^G_{\Gamma_2}(1_{\Gamma_2})$ are K-equivalent, by hypothesis, so that as we noticed before

$$\left\langle \mathrm{Ind}^G_{\Gamma_1}(1_{\Gamma_1}),\sigma\right\rangle_G = \left\langle \mathrm{Ind}^G_{\Gamma_2}(1_{\Gamma_2}),\sigma\right\rangle_G$$

for any $\sigma\in\hat{G}_K$. This implies that the condition (3.11) is satisfied, that is $\left(M/\Gamma_1, g_1\right)$ and $\left(M/\Gamma_2, g_2\right)$ are isospectral.

Q.E.D.

Remark 3.4 DeTurck and Gordon [1989] have strengthened Sunada's theorem to permit G to be a Lie group. Pesce's insight [1996] that the induced representations need only be K-equivalent affords much greater flexibility, and allows an understanding of isospectral Riemannian manifolds such as those of Ikeda [1983] (see also Example 3.5), which formerly did not appear to fit into Sunada's framework, as well as the construction of new examples of isospectral Riemannian manifolds (see Theorem 3.5 and Pesce [1996]). On the other hand, let us notice that the examples of isospectral closed Riemannian manifolds which do not enter within the framework of Sunada's theorem are rare. More precisely, let M be a closed smooth manifold, G a finite group acting freely on M

and $\mathcal{M}$ the set of all smooth Riemannian metrics on M endowed with the C^k-topology, $k \geq 5$. Let $\mathcal{M}^G$ be the set of G-invariant Riemannian metrics on M and S^G the set of Riemannian metrics $g \in \mathcal{M}^G$ which are characterized by the following property: If Γ_1 and Γ_2 are two subgroups of G such that the Riemannian manifolds $\left(M/\Gamma_1, g_1\right)$ and $\left(M/\Gamma_2, g_2\right)$ are isospectral, where g_j is as above $(j=1,2)$, then the induced representations $\mathrm{Ind}_{\Gamma_1}^G(1_{\Gamma_1})$ and $\mathrm{Ind}_{\Gamma_2}^G(1_{\Gamma_2})$ are equivalent. Under these assumptions and using the relationships between the spectrum of the Laplace-Beltrami operator and the length spectrum (that is the collection of lengths of closed geodesics counted with multiplicities) established by Colin de Verdière [1973], Pesce [1997] has shown that S^G is an open dense set in $\mathcal{M}^G$. For example, if G is of order pq, where p and q are two prime numbers, then $S^G = \mathcal{M}^G$. Indeed, if Γ_1 and Γ_2 are two subgroups of G such that the closed Riemannian manifolds $\left(M/\Gamma_1, g_1\right)$ and $\left(M/\Gamma_2, g_2\right)$ are isospectral, then Γ_1 and Γ_2 have the same cardinal since these Riemannian quotients have the same volume (see §1), and consequently, by Sylow's theorem, Γ_1 and Γ_2 are conjugated in G, hence $\left(M/\Gamma_1, g_1\right)$ and $\left(M/\Gamma_2, g_2\right)$ are isometric.

Remark 3.5 If (M,g) is a Riemannian manifold with boundary, then either Dirichlet or Neumann boundary conditions can be imposed.

Remark 3.6 Using the transplantation method of Bérard [1992] (which allows one explicitly to "transplant" eigenfunctions on one Riemannian manifold to eigenfunctions for the same eigenvalue on the second Riemannian manifold) or Selberg's trace formula, Pesce [1996] has proved a slightly more general result: Let (M,g) be a complete Riemannian manifold, G a Lie subgroup of the isometry group of $(M,g), \Gamma_1, \Gamma_2$ discrete subgroups of G such that the orbit spaces M/Γ_1 and M/Γ_2 are closed manifolds (that is Γ_j acts freely and properly on $M, j=1,2$, see Chapter 1, Theorem 2.3). Let K be the principal isotropy group of the action of G on M, and assume that the induced representations $\mathrm{Ind}_{\Gamma_1}^G(1_{\Gamma_1})$ and $\mathrm{Ind}_{\Gamma_2}^G(1_{\Gamma_2})$ are K-equivalent. Then (M_1, g_1) and (M_2, g_2) are isospectral. One can even relax the assumption that Γ_1 and Γ_2 act freely on M. In this case, the conclusion is that M/Γ_1 and M/Γ_2 are isospectral as orbifolds, as in Gordon, Webb and Wolpert [1992]. Representation-theoretic proofs of Sunada's theorem were given by Bérard [1992] and Zelditch [1992]. Bérard's proof allows the extension to orbifolds and furnishes an explicit combinatorial "transplantation" of eigenfunctions on M/Γ_1 to eigenfunctions on M/Γ_2.

Remark 3.7. Riemannian nilmanifolds and solvmanifolds now account for all known examples of isospectral closed Riemannian manifolds that do not fall under a Sunada-like setting. A Riemannian **nilmanifold** (resp. **solvmanifold**) is a closed Riemannian manifold of the form $(\Gamma \backslash G, g)$, where G is a connected and simply connected nilpotent Lie group (resp. solvable Lie group such that the exponential map from the Lie algebra of G onto G is a diffeomorphism), Γ is a discrete subgroup of G such that $\Gamma \backslash G$ is compact, and g is a left invariant metric on G, which descends to a Riemannian metric on $\Gamma \backslash G$, also denoted by g (for details concerning nilpotent Lie groups and discrete co-compact subgroups, see Corwin and Greenleaf [1990]). A Riemannian nilmanifold $(\Gamma \backslash G, g)$ is called a ***k*-step nilmanifold** if the Lie algebra of G is k-step nilpotent, that is all brackets of at least $k+1$ elements of the Lie algebra of G are 0 but not all brackets of order k are. Every Riemannian nilmanifold is a solvmanifold.

Remark 3.8. In all the considered examples (see also §2), the isospectral, closed Riemannian manifolds arise as different quotients (by discrete groups of isometries) of a common Riemannian covering manifold. In particular, such quotients can neither differ in their local geometry nor be simply connected. Gordon [1993] constructed the first pairs of isospectral closed Riemannian manifolds which are not locally isometric. The Riemannian manifolds are of the form $(\Gamma_j \backslash G_j, g_j)$, $j = 1, 2$, where each G_j is a simply connected nilpotent Lie group, Γ_j is a co-compact discrete subgroup, and g_j is a Riemannian metric whose lift to G_j is left-invariant. Szabó had constructed the first such examples in the case of Riemannian manifolds with boundary even earlier. He constructed Riemannian metrics on the product of a ball by a torus (the boundary of this manifold is the product of a sphere by the same torus) such that the corresponding Riemannian manifolds were both Dirichlet and Neumann isospectral but locally non-isometric (see Szabó [1999]). Moreover, in a revised version of this paper, it is shown that the restriction of these metrics to the boundary of his examples are isospectral with different local geometries. Among the later examples there are pairs of isospectral closed Riemannian manifolds one of which is homogeneous and the other is locally inhomogeneous. This is the first proof of the fact that one can not hear the local homogeneity property of closed Riemannian manifolds. Szabó's examples were proven isospectral first by explicit computation of the spectra and later by construction of an intertwining operator between the Laplace-Beltrami operators. Gordon and Wilson [1997] were able to construct even continuous families of isospectral Riemannian metrics with different local geometries in the case of manifolds with boundary. More precisely, in their examples the underlying manifolds are products of a ball with a torus. Gordon, Gornet, Schüth, Webb and Wilson [1998] showed that also the boundaries of these Riemannian manifolds are isospectral, which led to the first examples of isospectral deformations with changing local geometry in the case of closed manifolds, namely on products of a sphere with a torus. Szabó [1999] independently gave finite families of isospectral manifolds which represent special cases of these examples. Schüth [1999] had constructed the first examples of simply connected, closed Riemannian manifolds which are isospectral, but not locally isometric. She even gave continuous families of such Riemannian manifolds. More precisely, she showed that there exist continuous isospectral families of Riemannian metrics on $S^n \times S^3 \times S^3$ which are not locally isometric for every $n \geq 4$. This answers negatively the long-standing question of whether isospectral simply-connected manifolds must be isometric (for the construction of several new classes of isospectral Riemannian manifolds with different local geometries, see Schüth [2001]). All these examples (see also Remark 2.7) as well as all known isospectral Riemannian manifolds with different local geometry (see Gordon, Wilson [1997], Gordon, Gornet, Schüth, Webb, Wilson [1998]) can be explained by applying the following theorem for constructing such manifolds: Let T be a torus of dimension greater than one. Suppose that the Riemannian manifolds (M_1, g_1) and (M_2, g_2) are principal T-bundles and that the fibers, with the induced Riemannian metrics, are totally geodesic flat tori. For each subtorus S of T of codimension at most one, assume that the quotient manifolds M_1/S and M_2/S, with the induced Riemannian metrics, are isospectral. Then (M_1, g_1) and (M_2, g_2) are isospectral. For a proof of this result, see Gordon [2000a]. Most recently, using a generalization of this method (the torus fibers are not totally geodesic), Gordon and Szabó [2000] gave examples of pairs of isospectral closed Riemannian manifolds, one of which has constant scalar curvature and the other variable scalar curvature. These examples show that from the spectrum on functions alone one cannot tell that the Riemannian manifold has constant scalar curvature [see also Chapter 5, Theorem 4.3 (iii)]. This method is further generalized by Gordon [2000b]. In this paper, she constructs non-trivial continuous isospectral deformations of Riemannian metrics on the ball and on the sphere in $\mathbb{R}^n$ for every $n \geq 9$. The metrics on the sphere can be chosen arbitrarily close to the round metric; in particular, they can be chosen to be positively curved. Recently, Schüth [2001a] was able to construct isospectral pairs of Riemannian metrics on the five (resp. six)-dimensional sphere (resp. ball) and thus decreasing the minimal dimensions of Gordon's examples by three. Independently, Szabó [2000a,b] obtained pairs of isospectral Riemannian metrics on balls and spheres using a different method ("anticommutator technique") involving explicit construction of an intertwining operator by a careful analysis of the function spaces. This new method inherits no "tricks" from the previous investigations performed on this field, since all the tools applied on the product of a ball by a torus type manifolds so far break down on the ball-type domains and on their sphere-type boundaries. The lowest dimensional examples obtained by Szabó are however on the 10-sphere and the 11-ball. The most striking examples are constructed on the geodesic spheres of certain harmonic spaces. Szabó showed that geodesic balls of the same radius in different such harmonic spaces are both Dirichlet and Neumann isospectral. Moreover, the geodesic spheres are also isospectral. These examples provide isospectral pairs of Riemannian metrics constructed on a sphere such that one of them is a homogeneous and the other is a locally inhomogeneous metric. They arose from the construction by Damek and Ricci [1992] of simply connected homogeneous harmonic manifolds of negative curvature which are not symmetric.

Now let us turn to a special case of a converse to Pesce's theorem.

Theorem 3.5 (Pesce [1996]) Let (M, g) be a closed Riemannian manifold, G the isometry group of (M, g), K the principal isotropy group of the action of G on M and Γ_1, Γ_2 discrete subgroups of G such that the quotients M/Γ_1 and M/Γ_2 are closed manifolds. If all the real eigenspaces of the Laplace-Beltrami operator associated to (M, g) are absolutely irreducible and $\left(M/\Gamma_1, g_1\right), \left(M/\Gamma_2, g_2\right)$ are isospectral, then the induced representations $\mathrm{Ind}_{\Gamma_1}^G(1_{\Gamma_1})$ and $\mathrm{Ind}_{\Gamma_2}^G(1_{\Gamma_2})$ are K-equivalent.

Proof. We keep the previous notations. First recall that if τ is a real irreducible representation, then for its complexification $\tau^{\mathbb{C}}$ three possibilities can occur:

(i) $\tau^{\mathbb{C}}$ is irreducible, that is τ is of real type;

(ii) $\tau^{\mathbb{C}} = \rho \oplus \rho^*$, where ρ is irreducible and ρ^* is the contragredient representation, that is τ is of complex type;

(iii) $\tau^{\mathbb{C}} = \rho \oplus \rho$, where ρ is irreducible, that is τ is of quaternionic type.

On the other hand, the proof of Theorem 3.4 shows that the isospectrality of $\left(M/\Gamma_1, g_1 \right)$ and $\left(M/\Gamma_2, g_2 \right)$ is equivalent to the condition (3.11). If $\rho \in \hat{G}_K$, there is an eigenvalue λ of Δ such that ρ is a subrepresentation of π_λ^G. If the representation $\left(\pi_\lambda^G\right)^{\mathbb{R}}$ is of real type, then $\pi_\lambda^G = \rho$ and from (3.11) one obtains

$$\left\langle \rho, \operatorname{Ind}_{\Gamma_1}^G \left(1_{\Gamma_1}\right)\right\rangle_G = \left\langle \rho, \operatorname{Ind}_{\Gamma_2}^G \left(1_{\Gamma_2}\right)\right\rangle_G.$$

If the real representation $\left(\pi_\lambda^G\right)^{\mathbb{R}}$ is of complex type, then $\pi_\lambda^G = \rho \oplus \rho^*$ and from (3.11) one gets

$$\begin{aligned} &\left\langle \rho, \operatorname{Ind}_{\Gamma_1}^G \left(1_{\Gamma_1}\right)\right\rangle_G \left\langle \rho, \pi_\lambda^G \right\rangle_G + \left\langle \rho^*, \operatorname{Ind}_{\Gamma_1}^G \left(1_{\Gamma_1}\right)\right\rangle_G \left\langle \rho^*, \pi_\lambda^G \right\rangle_G \\ &= \left\langle \rho, \operatorname{Ind}_{\Gamma_2}^G \left(1_{\Gamma_2}\right)\right\rangle_G \left\langle \rho, \pi_\lambda^G \right\rangle_G + \left\langle \rho^*, \operatorname{Ind}_{\Gamma_2}^G \left(1_{\Gamma_2}\right)\right\rangle_G \left\langle \rho^*, \pi_\lambda^G \right\rangle_G. \end{aligned} \tag{3.12}$$

Note that a function f belongs to a subspace of $E(\Delta, \lambda)$ [resp. $L^2\left(G/\Gamma_j, \mathbb{C}\right)$] isomorphic to the representation space of ρ if and only if $\overline{f}$ belongs to a subspace isomorphic to the representation space of ρ^* and one deduces that $\left\langle \rho, \operatorname{Ind}_{\Gamma_j}^G \left(1_{\Gamma_j}\right)\right\rangle_G = \left\langle \rho^*, \operatorname{Ind}_{\Gamma_j}^G \left(1_{\Gamma_j}\right)\right\rangle_G$, $j = 1, 2$, and $\left\langle \rho, \pi_\lambda^G \right\rangle_G = \left\langle \rho^*, \pi_\lambda^G \right\rangle_G$. The last two equalities and (3.12) show that $\left\langle \rho, \operatorname{Ind}_{\Gamma_1}^G \left(1_{\Gamma_1}\right)\right\rangle_G = \left\langle \rho, \operatorname{Ind}_{\Gamma_2}^G \left(1_{\Gamma_2}\right)\right\rangle_G$. Finally, if the real representation $\left(\pi_\lambda^G\right)^{\mathbb{R}}$ is of quaternionic type, then $\pi_\lambda^G = \rho \oplus \rho$ and one also gets $\left\langle \rho, \operatorname{Ind}_{\Gamma_1}^G \left(1_{\Gamma_1}\right)\right\rangle_G = \left\langle \rho, \operatorname{Ind}_{\Gamma_2}^G \left(1_{\Gamma_2}\right)\right\rangle_G$. Consequently, in all cases one has

$$\left\langle \rho, \operatorname{Ind}_{\Gamma_1}^G \left(1_{\Gamma_1}\right)\right\rangle_G = \left\langle \rho, \operatorname{Ind}_{\Gamma_2}^G \left(1_{\Gamma_2}\right)\right\rangle_G$$

for any $\rho \in \hat{G}_K$, what shows that the representations $\operatorname{Ind}_{\Gamma_1}^G \left(1_{\Gamma_1}\right)$ and $\operatorname{Ind}_{\Gamma_2}^G \left(1_{\Gamma_2}\right)$ are K-equivalent.

Q.E.D.

Remark 3.9 In Theorem 3.5 one can relax the requirement that Γ_1 and Γ_2 act freely. In this case, the assumption is that M/Γ_1 and M/Γ_2 are isospectral as orbifolds.

Now let us turn to some examples. It is well-known that if $M = G/K$ is a symmetric space of compact type of rank 1, then the eigenspaces of Δ are irreducible for G. In particular, if $G = O(n+1)$, $K = O(n)$, then $M = G/K$ is the n-sphere S^n. In this special case there are examples of finite subgroups Γ_1 and Γ_2 of G, which are non-conjugate in G and such that the representations $\operatorname{Ind}_{\Gamma_1}^G \left(1_{\Gamma_1}\right)$ and $\operatorname{Ind}_{\Gamma_2}^G \left(1_{\Gamma_2}\right)$ are K-equivalent in the sense of Definition 3.1. Next let us give examples of such subgroups using Theorem 3.5.

Example 3.5 Let us consider the examples of Ikeda [1989], namely lenticular spaces, that is quotients of spheres with respect to cyclic groups (see Chapter 1, Example 2.16). More precisely, let $S^{2n+1} \subset \mathbb{C}^{n+1}$ and Γ_α be the cyclic group generated by the element φ_α of $O(2n+2)$ given by

$$\varphi_\alpha\left(z^1,\ldots,z^{n+1}\right)=\left(\exp\left(2\pi i\alpha^1\right)z^1,\ldots,\exp\left(2\pi i\alpha^{n+1}\right)z^{n+1}\right),$$

where $\alpha=\left(\alpha^1,\ldots,\alpha^{n+1}\right)$. If all the α^j are rational numbers, one obtains a finite group Γ_α which acts freely on S^{2n+1}. Ikeda succeeded to construct pairs of isospectral lenticular spaces which are not isometric. Moreover, for each $p\in\{0,\ldots,n\}$, he has been able to construct a pair of groups $\left(\Gamma_\alpha,\Gamma_\beta\right)$ such that S^{2n+1}/Γ_α and S^{2n+1}/Γ_β are j-isospectral in the sense of Definition 4.1, Chapter 5, for any $j\le p$, but are not $(p+1)$-isospectral. It is obvious that the induced representations $\operatorname{Ind}_{\Gamma_\alpha}^G\left(1_{\Gamma_\alpha}\right)$ and $\operatorname{Ind}_{\Gamma_\beta}^G\left(1_{\Gamma_\beta}\right)$ are K-equivalent but not equivalent. For example, if one takes $n=2$, $\alpha=\left(\frac{3}{11},\frac{4}{11},\frac{5}{11}\right)$ and $\beta=\left(\frac{1}{11},\frac{2}{11},\frac{3}{11}\right)$, then the quotients S^5/Γ_α and S^5/Γ_β are o-isospectral (that is isospectral) but not 1-isospectral.

Example 3.6 Consider the action of $G=O(n+1)$ on $\mathbb{R}^{n+1}$ and let us describe the examples of Urakawa [1982]. These are subgroups Γ_j of $O(n+1)$ generated by reflections $(n\ge 3)$. Consider a root system in the Euclidean space $\mathbb{R}^{n+1}$, and let C be a closed Weyl chamber. Bérard and Besson [1980] (see also Bérard [1980]) determined the spectra of spherical domains of the form $C\cap S^n$ in terms of the Weyl group exponents. Urakawa [1982] noticed that if one chooses two root systems R_1 and R_2 whose sets of Weyl group exponents coincide, then the corresponding closed Weyl chambers C_1,C_2 give rise to isospectral domains $D_1=C_1\cap S^n$, $D_2=C_2\cap S^n$ in the sphere. He gave examples of Coxeter groups Γ_1,Γ_2 associated with the root systems R_1 and R_2 in $\mathbb{R}^{n+1}$, such that the spherical domains $D_1=C_1\cap S^n$ and $D_2=C_2\cap S^n$ are Neumann (and Dirichlet) isospectral, hence isospectral as orbifold quotients of S^n, as in Gordon, Webb, Wolpert [1992], since a Γ_j-invariant eigenfunction on S^n is one which extends by reflection across walls of the Weyl chamber C_j and hence has zero normal derivative on the boundary of D_j. By Remark 3.9, the induced representations $\operatorname{Ind}_{\Gamma_1}^G\left(1_{\Gamma_1}\right)$ and $\operatorname{Ind}_{\Gamma_2}^G\left(1_{\Gamma_2}\right)$ are K-equivalent, where $K=O(n)$. Urakawa's examples of isospectral domains occur in all Euclidean spaces of dimension at least 4. One of his simplest example is given by the Coxeter groups Γ_1,Γ_2 associated with the root systems $R_1=A_3\times A_1$ and $R_2=I_2(3)\times I_2(4)$ in $\mathbb{R}^4$. Then a Weyl chamber for R_1 is given by $C_1=\left\{\sum_{j=1}^4\alpha_j u_j \,\middle|\, \alpha_j\ge 0,\ j=1,2,3,4\right\}$, where $u_1=(0,0,1,0)$, $u_2=(1,-1,1,0)$, $u_3=(1,1,1,0)$, $u_4=(0,0,0,1)$, while a Weyl chamber for

R_2 is given by $C_2 = \left\{ \sum_{j=1}^{4} \alpha_j v_j \middle| \alpha_j \geq 0,\ j = 1,2,3,4 \right\}$, where $v_1 = (1,0,0,0)$, $v_2 = (1,\sqrt{3},0,0)$, $v_3 = (0,0,1,0)$, $v_4 = (0,0,1,1)$.

Now following Gordon and Webb [1994], let us describe a slight modification of Urakawa's construction to obtain a pair of **convex** domains in $\mathbb{R}^n$ for each $n \geq 4$, which are non-isometric yet isospectral for both Dirichlet and Neumann boundary conditions. With the same notations as in Example 3.6, Urakawa [1982] showed that for $0 < \varepsilon < 1$, the truncated cones $\Omega_1(\varepsilon) = \{ tx \in \mathbb{R}^{n+1} \mid x \in D_1,\ \varepsilon < t < 1 \}$ and $\Omega_2(\varepsilon) = \{ tx \in \mathbb{R}^{n+1} \mid x \in D_2,\ \varepsilon < t < 1 \}$ are isospectral domains in $\mathbb{R}^{n+1}$ for both Dirichlet and Neumann boundary conditions. His argument breaks down for $\varepsilon = 0$ due to the presence of the radial variable t in the denominator. In what follows, we shall prove the following.

Theorem 3.6 **(Gordon, Webb** [1994]). The convex domains $\Omega_1 = \Omega_1(0)$ and $\Omega_2 = \Omega_2(0)$ in $\mathbb{R}^{n+1}$ are isospectral for both Dirichlet and Neumann boundary conditions.

Proof. First notice that Ω_1 and Ω_2 are isospectral for Dirichlet boundary conditions. Indeed, using the mini-max characterization of eigenvalues for Dirichlet boundary conditions, the k-th eigenvalue $\mu_k^j(\varepsilon)$ corresponding to $\Omega_j(\varepsilon)$ is given by

$$\mu_k^j(\varepsilon) = \inf_V \sup_{f \in V} \frac{\int |df|^2}{\int |f|^2},$$

where the infimum is taken over all k-dimensional subspaces V of the space $C_0^\infty(\Omega_j(\varepsilon))$ of smooth functions compactly supported in the interior of $\Omega_j(\varepsilon)$. Since $\Omega_j(\varepsilon) \supseteq \Omega_j(\varepsilon')$ for $\varepsilon > \varepsilon'$, and since the domains $\Omega_j(\varepsilon)$ exhaust Ω_j, one gets that $\mu_k^j(0) = \lim_{\varepsilon \downarrow 0} \mu_k^j(\varepsilon)$, $j = 1,2$. Since $\mu_k^1(\varepsilon) = \mu_k^2(\varepsilon)$ for each $\varepsilon > 0$, it follows that $\mu_k^1(0) = \mu_k^2(0)$, that is Ω_1 and Ω_2 are isospectral for Dirichlet boundary conditions.

Now let us prove that the convex domains Ω_1 and Ω_2 are isospectral for Neumann boundary conditions. Let us consider the action of $G = O(n+1)$ on the unit ball B^{n+1} in $\mathbb{R}^{n+1}$. The isotropy group at every point except the origin is $O(n)$, so the principal isotropy group of the action of G on B^{n+1} is $K = O(n)$. By the above, $\mathrm{Ind}_{\Gamma_1}^G(1_{\Gamma_1})$ and $\mathrm{Ind}_{\Gamma_2}^G(1_{\Gamma_2})$ are K-equivalent, so by Remarks 3.5 and 3.6, B^{n+1}/Γ_1 and B^{n+1}/Γ_2 are isospectral orbifolds with boundary: for each domain, the singular locus is the union of the walls of the Weyl chambers, and the boundary is the intersection of the domain with S^n (here Neumann boundary conditions on the orbifold boundary, that is on S^n, are imposed). But the orbifold isospectrality of B^{n+1}/Γ_1 and B^{n+1}/Γ_2 is precisely the Neumann isospectrality of the underlying domains Ω_1 and Ω_2.

Q.E.D.

Remark 3.10 Adapting Sunada's technique to the case of orbifolds by using ideas of Bérard [1989], [1992], Gordon, Webb, Wolpert [1992] have constructed a pair of non-congruent simply connected plane domains which are both Dirichlet and Neumann isospectral. Their method allows a broad collection of other types of isospectral surfaces. Thus they settle in the negative the famous question of Mark Kac [1966] whether the spectrum of a plane domain determines the domain up to congruence. Buser, Conway, Doyle, Semmler [1994] gave a lot of new examples, together with a simple proof of isospectrality which shows how to explicitly transplant eigenfunctions on one domain to eigenfunctions on the other.

Now, let us turn to another special case of a converse to Pesce's theorem.

Let $\left(\mathbb{R}^n, g_{can}\right)$ be the standard Riemannian manifold (see Chapter 1, Example 2.1) and G its isometry group, that is G is the semi-direct product of $\mathbb{R}^n$ and $O(n)$. Recall that the group law on G is given by

$$(x,A)(y,B)=(x+Ay,AB),$$

where $x,y\in\mathbb{R}^n, A,B\in O(n)$. Thus $\mathbb{R}^n$ can be viewed as the homogeneous space G/K, where $K=O(n)$. It is possible to apply Mackey's theory in order to classify the irreducible representations of a semi-direct product of an abelian group with a compact group (see Mackey [1968]). Thus any irreducible representation of G which occur in $\hat{G}_K$ is equivalent to a representation of G of type ρ_α, which we shall define. If α belongs to the dual $\left(\mathbb{R}^n\right)^*$ of $\mathbb{R}^n$, let us denote $K_\alpha=\left\{A\in O(n)\,|\,\alpha A=\alpha\right\}$. Let us define the representation ρ_α of G in $L^2\left(K_\alpha\setminus K,\mathbb{C}\right)$ by

$$\left[\left(\rho_\alpha(x,A)\right)(f)\right](K_\alpha B)=e^{2i\pi\alpha(Bx)}f(K_\alpha BA),$$

where $x\in\mathbb{R}^n, A,B\in O(n)$ and $f\in L^2\left(K_\alpha\setminus K,\mathbb{C}\right)$. Moreover, two such representations ρ_α and ρ_β are equivalent if and only if $\|\alpha\|=\|\beta\|$, where $\|\ \|$ denotes the norm associated to the scalar product on $\left(\mathbb{R}^n\right)^*$ induced by the canonical scalar product on $\mathbb{R}^n$.

Now let us consider a lattice Γ of $\mathbb{R}^n$. Γ can be identified with a discrete subgroup of G via the injection $\Gamma\ni\gamma\mapsto(\gamma,I_n)\in G$, where I_n is the $n\times n$ identity matrix. Γ acts freely and properly discontinuous on $\mathbb{R}^n$ and $\Gamma\setminus G$ is compact. If H is an irreducible subspace of $\left(L^2(\Gamma\setminus G,\mathbb{C})\right)_K$, then $\dim H^K=1$. Moreover, if $f\in H^K$, then f descends in a function on $\mathbb{R}^n/\Gamma=\Gamma\setminus\left(G/K\right)$ which is an eigenfunction of the Laplace-Beltrami operator associated to the flat metric (see Helgason [1984], p. 410). Moreover, if the restriction of $\mathrm{Ind}_\Gamma^G(1_\Gamma)$ to H is equivalent to ρ_α, then f is an eigenfunction corresponding to the eigenvalue $4\pi^2\|\alpha\|^2$. Conversely, if f is an eigenfunction on $\mathbb{R}^n/\Gamma$ corresponding to the eigenvalue λ, then f can be viewed as an element of $\left(L^2(\Gamma\setminus G;\mathbb{C})\right)^K\subset\left(L^2(\Gamma\setminus G;\mathbb{C})\right)_K$ and each non-trivial projection of f on an irreducible subspace is yet an eigenfunction corresponding to the eigenvalue λ. Thus the multiplicity of the eigenvalue λ in the spectrum of $\mathbb{R}^n/\Gamma$ is equal to the number of equivalence classes of representations ρ_α which occur in $\left(L^2(\Gamma \backslash G, \mathbb{C})\right)_K$, when α runs over $\left(\mathbb{R}^n\right)^*$ such that $4\pi^2 \|\alpha\|^2 = \lambda$. But, if $\alpha, \beta \in \left(\mathbb{R}^n\right)^*$ are such that $\|\alpha\|^2 = \|\beta\|^2$, then ρ_α and ρ_β are equivalent. Thus the multiplicity of an eigenvalue λ is equal to the multiplicity of the unique equivalence class which contains an element ρ_α with $4\pi^2\|\alpha\|^2 = \lambda$. Consequently, by Remark 3.6 and the above, if Γ_1, Γ_2 are two lattices of $\mathbb{R}^n$, then the flat tori $\left(\mathbb{R}^n/\Gamma_1, g_{can}/\Gamma_1\right)$ and $\left(\mathbb{R}^n/\Gamma_2, g_{can}/\Gamma_2\right)$ are isospectral if and only if the induced representations $\mathrm{Ind}_{\Gamma_1}^G\left(1_{\Gamma_1}\right)$ and $\mathrm{Ind}_{\Gamma_2}^G\left(1_{\Gamma_2}\right)$ are K-equivalent.

Remark 3.11 If G is a compact group of isometries acting freely on a closed Riemannian manifold (M, g), one can prove the existence of a dense set in the set $\mathcal{M}^G$ of G-invariant Riemannian metrics on M for which the converse of Theorem 3.4 is true. More precisely, let K be the principal isotropy group of the action of G on M and S_K^G the set of Riemannian metrics $g \in \mathcal{M}^G$ which are characterized by the property: If Γ_1 and Γ_2 are two discrete subgroups of G such that the closed Riemannian manifolds $\left(M/\Gamma_1, g_1\right)$ and $\left(M/\Gamma_2, g_2\right)$ are isospectral with respect to the Laplace-Beltrami operator, then the induced representations $\mathrm{Ind}_{\Gamma_1}^G\left(1_{\Gamma_1}\right)$ and $\mathrm{Ind}_{\Gamma_2}^G\left(1_{\Gamma_2}\right)$ are K-equivalent. Using a result of Zelditch [1990], Pesce [1997] has shown that if M and G are such that for any real irreducible representation ρ of G admitting K-invariant vectors one has $\dim(\rho) \leq \dim(M)$, then S_K^G is dense in $\mathcal{M}^G$. For example, let M be a closed simply-connected smooth manifold that admits a symmetric Riemannian metric g of rank 1, that is the sphere or the projective spaces, and G the identity component of the isometry group of (M, g). Then $S_K^G = \mathcal{M}^G$. Indeed, any Riemannian metric of $\mathcal{M}^G$ is a multiple of g and it is well-known that the eigenspaces of the Laplace-Beltrami operator associated to a symmetric Riemannian metric of rank 1 are all irreducible for G.

REFERENCES

BÉRARD, P., *Spectres et groupes cristallographiques I: Domaines euclidiens*, Invent. Math. 58(1980), 179-199.

BÉRARD, P., *Spectral Geometry: Direct and Inverse Problems*, Lecture Notes in Math., Vol. 1207, Springer, Berlin, Heidelberg, New York, 1986.

BÉRARD, P., *Variétés riemanniennes isospectrales non isométriques*, Astérisque 177-178(1989), 127-154.

BÉRARD, P., *Transplantation et isospectralité I*, Math. Ann. 292 (1992)547-559.

BÉRARD, P., *Transplantation et isospectralité II*, The Journal of the London Math. Society (Second Series), 48(1993), 565-576.

BÉRARD, P. et BESSON, G., *Spectres et groupes cristallographiques II: Domaines sphériques*, Ann. Inst. Fourier Grenoble 30(3)(1980), 237-248.

BÉRARD, P. et PESCE, H., *Construction de variétés isospectrales autour du théorème de Sunada*, Progress in Inverse Spectral Geometry (Stig I. Anderson; Michel L. Lapidus, Editors), Trends in Mathematics, Birkhäuser Verlag, Basel, Boston, Berlin, 1997, pp. 63-84.

BERGER, M., GAUDUCHON, P., et MAZET, E., *Le spectre d' une variété riemannienne*, Lecture Notes in Math., Vol. 194, Springer, Berlin, Heidelberg, New York, 1971.

BERGERON, N., *Premier nombre de Betti et spectre du laplacien de certaines variétés hyperboliques*, L' Enseignement Mathématique, t. 46(2000), 109-137.

BREDON, G., *Introduction to Compact Transformation Groups*, Academic Press, New York, London, 1972.

BROOKS, R., *Constructing isospectral manifolds*, Amer. Math. Monthly 95(1988), 823-839.

BROOKS, R., *The Sunada method*, Tel Aviv Topology Conference: Rothenberg Festschrift, June 1-5, 1998, pp. 25-35, Contemp. Math., 231, Amer. Math. Soc., Providence, RI, 1999.

BROOKS, R. and TSE, R., *Isospectral surfaces of small genus*, Nagoya Math. J., 107(1987), 13-24.

BROOKS, R., GORNET, R. and GUSTAFSON, W., *Mutually isospectral Riemann surfaces*, Adv. in Math. 138(1998), 306-322.

BROOKS, R., PERRY, P. and YANG, P., *Isospectral sets of conformally equivalent metrics*, Duke Math. J., 58(1989), 131-150.

BRÜNING, J. and LESCH, M., *On the spectral geometry of algebraic curves*, J. Reine Angew. Math. 474(1996), 25-66.

BUSER, P., *Isospectral Riemann surfaces*, Ann. Inst. Fourier, Grenoble 36(1986), 167-192.

BUSER, P., *Geometry and spectra of compact Riemann surfaces*, Progress in Mathematics, Vol. 106, Birkhäuser, Boston, 1992.

BUSER, P., CONWAY, J.H., DOYLE, P. and SEMMLER, K.-D., *Some planar isospectral domains*, Internat. Math. Res. Notices no. 9(1994), 391-400.

CAHN, S. and WOLF, J., *Zeta functions and their asymptotic expansions for compact symmetric spaces of rank one*, Comment. Math. Helv., 51(1976), 1-21.

CASSELS, J.W.S., *An Introduction to the Geometry of Numbers*, Die Grundlehren der mathematischen Wissenschaften, Bd. 99, Springer-Verlag, Berlin, 1959.

CHAVEL, I., *Eigenvalues in Riemannian Geometry*, Academic Press, Inc., New York, London, Tokyo, 1984.

CHAZARAIN, J., *Formule de Poisson pour les variétés Riemanniennes*, Invent. Math. 24(1974), 65-82.

CHEN, B. – Y. and VANHECKE, L., *The spectrum of the Laplacian of Kähler manifolds*, Proc. Amer. Math. Soc., 79(1980), 82-86.

COLIN DE VERDIÈRE, Y., *Spectre du Laplacian et géodésiques périodiques*, Compositio Math., 27(1973), 159-184.

CONWAY, J.H., *The Sensual Quadratic Form*, The Carus Mathematical Monographs, Vol. 26, The Mathematical Association of America, 1997.

CONWAY, J.H. and SLOANE, N.J.A., *Four-dimensional lattices with the same theta series*, International Mathematics Research Notices, No. 4(1992), 93-96.

CONWAY, J.H. and SLOANE, N.J.A., *Sphere Packings, Lattices and Groups*, Third Edition, Grundlehren der mathematischen Wissenschaften 290, Springer-Verlag, New York, 1999.

CORWIN, L. and GREENLEAF, F.P., *Representations of Nilpotent Lie Groups and Their Applications*, Part I: Basic Theory and Examples, Cambridge University Press, Cambridge, 1990.

CRAIOVEANU, M. si PUTA, M., *Introducere in geometria spectrală*, Editura Academiei Române, Bucuresti, 1988.

DAMEK, E. and RICCI, F., *A class of non-symmetric harmonic manifolds*, Bull. Amer. Math. Soc. 27(1992), 139-142.

DETURCK, D., *Audible and inaudible geometric properties*, Rendiconti Seminari Facoltà Scienze Università Cagliari Supplemento al Vol. 58(1988), 1-26.

DETURCK, D. and GORDON, C.S., *Isospectral deformations I: Riemannian structures on two-step nilspaces*, Comm. Pure Appl. Math., 40(1987), 367-387.

DETURCK, D. and GORDON, C.S., *Isospectral deformations II: Trace formulas, metrics and potentials*, Comm. Pure Appl. Math. 42(1989), 1067-1095.

DETURCK, D., GORDON, C.S., GLUCK, H. and WEBB, D., *You cannot hear the mass of a homology class*, Comm. Math. Helv. 64(1989), 589-617.

DONNELLY, H., *G-spaces, the asymptotic splitting of* $L^2(M)$ *into irreducibles*, Math. Ann. 237(1978), 23-40.

DONNELLY, H., *On the wave equation asymptotics of a compact negatively curved surface*, Inv. Math. 45(1978), 115-137.

DOTTI MIATELLO, I. and MIATELLO, R.J., *Isospectral compact flat manifolds*, Duke Math. J., 68(1992), 489-498.

DUISTERMAAT, J.J. and GUILLEMIN, V., *The spectrum of positive elliptic operators and periodic bi-characteristics*, Inv. Math. 24(1975), 39-80.

EJIRI, N., *A construction of nonflat, compact irreducible Riemannian manifolds which are isospectral but not isometric*, Math. Z. 168(1979), 207-212.

GASSMANN, F., *Bemerkungen zur vorstehenden Arbeit von Hurwitz*, Math. Zeit. 25(1926), 665-675.

GERST, I., *On the theory of n-th power residues and a conjecture of Kronecker*, Acta Arith. 27(1970), 121-139.

GORDON, C.S., *You can't hear the shape of a manifold, New Developments in Lie Theory and Their Applications* (J. Tirao, N. Wallach, eds.), Progress in Mathematics, Vol. 105, Birkhäuser Verlag, Boston, 1992, pp. 129-146.

GORDON, C.S., *Isospectral closed Riemannian manifolds which are not locally isometric*, I: J. Diff. Geometry, 37(1993), 639-649; II: Contemporary Mathematics, Vol. 173(1994), 121-131.

GORDON, C.S., *Survey of isospectral manifolds*, Handbook of Differential Geometry, F.J.E. Dillen and L.C.A. Verstraelen, eds., vol. 1, Elsevier Science B.V., 2000a, pp. 747-778.

GORDON, C.S., *Isospectral deformations of metrics on spheres*, Preprint, May 2000b, to appear in Inventiones Mathematicae, 2001.

GORDON, C.S. and GORNET, R., *Spectral geometry on nilmanifolds*, Progress in Inverse Spectral Geometry (S.I. Anderson, M.L. Lapidus, eds.), Trends in Mathematics, Birkhäuser Verlag, Basel, Boston, Berlin, 1997, pp. 23-50.

GORDON, C.S. and SZABÓ, Z.I., *Isospectral deformations of negatively curved Riemannian manifolds with boundary which are not locally isometric*, Preprint, March 2000.

GORDON, C.S. and WEBB, D.L., *Isospectral convex domains in Euclidean space*, Mathematical Research Letters, 1(1994), 539-545.

GORDON, C.S. and WILSON, E.N., *Continuous families of isospectral Riemannian metrics which are not locally isometric*, J. Differential Geometry, 47(1997), 504-529.

GORDON, C.S., GORNET, R., SCHÜTH, D., WEBB, D. and WILSON, E., *Isospectral deformations of closed Riemannian manifolds with different scalar curvature*, Ann. Inst. Fourier 48(1998), 593-607.

GORDON, C.S., WEBB, D.L. and WOLPERT, S., *Isospectral plane domains and surfaces via Riemannian orbifolds*, Invent. Math., 110(1992), 1-22.

GORNET, R., *A new construction of isospectral Riemannian nilmanifolds with examples*, Michigan Math. J. 43, no. 1(1996), 159-188.

GORNET, R., *Continuous families of Riemannian manifolds, isospectral on functions but not on 1-forms*, The Journal of Geometric Analysis, vol. 10, no. 2(2000), 281-298.

GUILLEMIN, V., *Wave trace invariants and a theorem of Zelditch*, Int. Math. Research Notices 12(1993), 303-308.

GUILLEMIN, V., *Wave trace invariants*, Duke Math. J. 83(1996), 287-352.

HELGASON, S., *Groups and Geometric Analysis: Integral Geometry, Invariant Differential Operators and Spherical Functions*, Academic Press, New York, 1984.

HÖRMANDER, L., *Linear partial differential operators*, Springer, Berlin, Heidelberg, New York, 1964.

IKEDA, A., *On the spectrum of a Riemannian manifold of positive constant curvature*, Osaka J. Math., 17(1980), 75-93.

IKEDA, A., *On spherical space forms which are isospectral but not isometric*, J. Math. Soc. Japan, 35(1983), 437-444.

IKEDA, A., *Isospectral problem for spherical space forms, Spectra of Riemannian Manifolds* (M. Berger, S. Murakami, T. Ochiai, eds.), Kaigai Publications, 1983, pp. 57-63.

IKEDA, A., *Riemannian manifolds p-isospectral but not $(p+1)$-isospectral*, Geometry of Manifolds (Matsumoto), Perspect. Math. Vol. 8, Academic Press, Boston, MA, 1989, pp. 383-417.

IKEDA, A., *On space forms of real Grassmann manifolds which are isospectral but not isometric*, Kodai Math. J. 20(1997), no. 1, 1-7.

JACOBSON, N., *Lie Algebras*, Interscience Publishers, New York, London, Sydney, 1962.

KAC, M., *Can one hear the shape of a drum?*, Amer. Math. Monthly, 73(1966), 1-23.

KITAOKA, Y., *Positive definite quadratic forms with same representation numbers*, Arch. Math., 28(1977), 495-497.

KNESER, M., *Lineare Relationen zwischen Darstellungszahlen quadratischer Formen*, Math. Ann., 168(1967), 31-39.

KOMATSU, K., *On the adele ring of algebraic number fields*, Kodai Math. Sem. Rep. 28(1976), 78-84.

MACKEY, W., *Ergodic transformation groups with a pure point spectrum*, Ill. J. Math. 8(1964), 593-600.

MACKEY, G., *Induced Representations of Groups and Quantum Mechanics*, W.A. Benjamin, New York, 1968.

MIATELLO, R.J. and ROSSETTI, J.P., *Isospectral Hantzsche-Wendt manifolds*, J. Reine Angew. Math. 515(1999), 1-23.

MIATELLO, R.J. and ROSSETTI, J.P., *Hantzsche-Wendt manifolds of dimension 7*, Differential Geometry and Applications, Satellite Conference of ICM in Berlin, August 10-14, 1998, Brno, Masaryk University in Brno (Czech Republic) 1999, 379-390.

MILNOR, J., *Eigenvalues of the Laplace operator on certain manifolds*, Proc. Nat. Acad. Sci. USA, 51(1964), 542.

PESCE, H., *Borne explicite du nombre de tores plats isospectraux à un tore donné*, Manuscripta Mathematica, 75(1992), 211-223.

PESCE, H., *Représentations de groupes et variétés riemanniennes isospectrales*, Contemporary Mathematics: Geometry of the Spectrum (R. Brooks, C.S. Gordon, P. Perry, eds.), Vol. 173, AMS, 1994, pp. 231-240.

PESCE, H., *Variétés hyperboliques et elliptiques fortement isospectrales*, Journal of Functional Analysis, 134(1995), 363-391.

PESCE, H., *Représentations relativement équivalentes et variétés riemanniennes isospectrales*, Comment. Math. Helvetici, 71(1996), 243-268.

PESCE, H., *Compacité de l' ensemble des réseaux isospectraux et conséquences*, Topology Vol. 36, No. 3(1997), 695-710.

PESCE, H., *Une réciproque générique du théorème de Sunada*, Compositio Mathematica, 109(1997), 357-365.

PESCE, H., *Quelques applications de la théorie des représentations en géométrie spectrale*, Rendiconti di Matematica, Serie (VII), Vol. 18, Fasc., 1(1998), 1-63.

RAGHUNATHAN, M.S., *Discrete Subgroups of Lie Groups*, Ergebnisse der Mathematik und ihrer Grenzgebiete, 68, Springer-Verlag, Berlin, Heidelberg and New York, 1972.

REID, A.W., *Isospectrality and commensurability of arithmetic hyperbolic 2-and 3-manifolds*, Duke Math. J. 65(1992), 215-228.

SAKAI, T., *Riemannian Geometry*, Translations of Mathematical Monographs, Vol. 149, American Mathematical Society, Providence, Rhode Island, 1996.

SCHIEMANN, A., *Ein Beispiel positiv definiter quadratischer Formen der Dimension 4 mit gleichen Darstellungszahlen*, Arch. Math., 54(1990), 372-375.

SCHIEMANN, A., *Ternary positive definite quadratic forms are determined by their theta series*, Math. Ann., 308(1997), 507-517.

SCHÜTH, D., *Continuous families of isospectral metrics on simply connected manifolds*, Annals of Math., 149(1999), 287-308.

SCHÜTH, D., *Isospectral manifolds with different local geometries*, Habilitationsschrift, SFB256, Preprint no. 644, Bonn, 2000, J. Reine und Angew. Math. 534(2001), 41-94.

SCHÜTH, D., *Isospectral metrics on five-dimensional spheres*, Preprint, February 2001a, math. DG/0102126.

SPATZIER, R.J., *On isospectral locally symmetric spaces and a theorem of von Neumann*, Duke Math. J. 59(1989), 289-294; Correction: Duke Math. J. 60(1990), 561.

SUNADA, T., *Riemannian coverings and isospectral manifolds*, Ann. of Math., 121(1985), 169-186.

SZABÓ, Z.I., *Locally non-isometric yet super isospectral spaces*, Geom. Funct. Anal., Vol. 9(1999), 185-214.

SZABÓ, Z.I., *Isospectral pairs of metrics constructed even on balls and spheres with different local geometries*, Preprint, 2000a.

SZABÓ, Z.I., *Cornucopia of isospectral pairs of metrics constructed on balls and spheres with different local geometries*, Preprint, 2000b.

TANNO, S., *Eigenvalues of the Laplacian of Riemannian manifolds*, Tôhoku Math. Journal, 25(1973), 391-403.

TANNO, S., *A characterization of the canonical spheres by the spectrum*, Math. Z. 175(1980), 267-274.

TAYLOR, M.E., *Qualitative Studies of Linear Equations*, Applied Mathematical Sciences 116, Springer Verlag, 1996.

TITARENKO, S.A., *The Sunada theorem and the latest advances in the inverse problem of spectral geometry (Russian)*, Algebra i Analiz 8(1996), no. 3, 14-38; translation in St. Petersburg Math. J. 8(1997), no. 3, 381-398.

TRICERRI, F. and VANHECKE, L., *Decomposition of a space of curvature tensors on a quaternionic Kähler manifold and spectrum theory*, Simon Stevin, A Quarterly Journal of Pure and Applied Mathematics, Vol. 53(1979), 163-173.

URAKAWA, H., *Bounded domains which are isospectral but non isometric*, Ann. Sci. Ecole Norm. Sup., 15(1982), 441-456.

VIGNÉRAS, M.F., *Variétés Riemanniennes isospectrales et non isométriques*, Ann. of Math., 112(1980), 21-32.

WARNER, G., *Harmonic Analysis on semi-simple Lie Groups*, I, Springer Verlag, Berlin, Heidelberg, New York, 1972.

WATSON, G.L., *Determination of a binary quadratic form by its values at integer points*, Mathematika 26(1979), 72-75; Acknowledgement 27(1980), 188.

WHITTAKER, E.T. and WATSON, G.N., *A Course of Modern Analysis*, Fourth Edition, Cambridge at the University Press, 1969.

WITT, E., *Eine Identität zwischen Modulformen zweiten Grades*, Abh. Math. Sem. Univ. Hamburg, 14(1941), 323-337.

WOLF, J.A., *Spaces of Constant Curvature*, McGraw-Hill, New York, 1967.

WOLPERT, S., *The eigenvalue spectrum as moduli for flat tori*, Trans. Amer. Math. Soc., 244(1978), 312-321.

YAU, S.T., *Open problems in geometry*, Proc. Symp. Pure Math., Vol. 54, part 1, AMS (1992), 1-28.

ZELDITCH, S., *On the generic spectrum of a Riemannian cover*, Ann. Inst. Fourier, Grenoble, 40(1990), 407-422.

ZELDITCH, S., *Isospectrality in the FIO category*, J. Differential Geometry, 35(1992), 689-710.

ZELDITCH, S., *The inverse spectral problem for surfaces of revolution*, J. Differential Geometry, 49(1998), 207-264.

ZELDITCH, S., *Lectures on wave invariants*, in "Spectral Theory and Geometry" (edited by B. Davies and Y. Safarov), ICMS Instructional Conference, Edinburgh, 30 March – 9 April 1998, pp. 284-328, London Mathematical Society, Lecture Note Series 273, Cambridge University Press, 1999.

Chapter 5
SPECTRAL PROPERTIES OF THE LAPLACIANS FOR THE DE RHAM COMPLEX

1. THE HEAT EQUATION ASSOCIATED TO A HODGE-DE RHAM OPERATOR

In Chapters 3 and 4 we have discussed the Laplace-Beltrami operator on functions. If one considers other natural geometric partial differential operators,namely the Hodge-de Rham operators, then as we shall see in Chapter 6 global topological aspects come into play.

In this section we shall present the extension of the spectral theorem given in Chapter 3, for the Hodge-de Rham operators (or p-Laplacians) as well as some of its consequences. The main point is that these results, also as in the named section, can be derived from the properties of corresponding FSHE (or p-FSHE), which are similar with those of the FSHE for functions. The argument differs essentially only in notational detail from the proofs given there, so that we shall only outline it.

Let (M,g) be an n-dimensional Riemannian manifold, $\xi^{(p)}$ the canonical real smooth vector bundle $\Lambda^p T^*M \otimes \Lambda^p T^*M \to M\times M$, $0\le p\le n$, and $\pi^*\left(\xi^{(p)}\right)$ the induced smooth vector bundle over $M\times M\times(0,+\infty)$, where $\Lambda^p T^*M$ denotes the **p-exterior bundle** over M and $\pi: M\times M\times(0,+\infty)\to M\times M$ the canonical projection. We denote by $C^{k,l}_{(p)}$ the set of all sections of $\pi^*\left(\xi^{(p)}\right)$ whose mixed partial derivatives up to k times with respect to $M\times M$ and up to l times with respect to $(0,+\infty)$ exist and are continuous on $M\times M\times(0,+\infty)$.

Definition 1.1 Let (M,g) be a connected, oriented, smooth Riemannian manifold without boundary. A **fundamental solution of the p-heat equation** (or abbreviated p-FSHE) on (M,g), or the heat p-kernel on (M,g) is a continuous section $E^{(p)}$ of $\pi^*\left(\xi^{(p)}\right)$, so that its value at a point $(x,y,t)\in M\times M\times(0,+\infty)$ is $E^{(p)}(x,y,t)\in \Lambda^p T_x^*M\otimes \Lambda^p T_y^*M$, that is $E^{(p)}$ is a continuous (p,p)-form on $M\times M\times(0,+\infty)$, that satisfies the following conditions:

(i) $E^{(p)}\in C^{2,1}_{(p)}$;

(ii) $\Delta^{(p)}E^{(p)}+\dfrac{\partial E^{(p)}}{\partial t}=0$,

where $\Delta^{(p)}$ is the Hodge-de Rham (or p-Laplacian) with respect to the second variable of $E^{(p)}$ associated to g;

(iii) $\lim\limits_{t\downarrow 0}\int\limits_M E^{(p)}(x,y,t)\wedge\left(S^{(p)}\omega\right)(y)=\omega(x)$

for each $x\in M$, the convergence being locally uniform in x for each continuous p-form ω on M with compact support.

Note that the integrand in (iii) is a continuous double form of bidegree (*p*,*n*), which integrates for fixed *x* and *t* to an element in $\Lambda^p T_x^* M$.

On the other hand, note that, in virtue of the duality $T_y M \cong T_y^* M$ given by the Riemannian metric, we can also think of $E^{(p)}(x,y,t)$ as a linear mapping from $\Lambda^p T_y^* M$ to $\Lambda^p T_x^* M$ and sometimes we shall identify $E^{(p)}(x,y,t)$ with this mapping.

Theorem 1.1 Let (*M*,*g*) be as in Definition 1.1. If *M* is closed, then there exists a *p*-FSHE $E^{(p)}$ on (*M*,*g*) for any *p*, $0 \le p \le n$, that satisfies the following additional conditions:

(i) $E^{(p)} \in C^{\infty,\infty}_{(p)}$;

(ii) $\left(E^{(p)}(x,y,t)\alpha(y),\beta(x)\right)_x = \left(\alpha(x),E^{(p)}(x,y,t)\beta(y)\right)_x$ for all $x,y \in M$, all $t \in (0,+\infty)$ and all $\alpha,\beta \in A^p(M)$.

The proof of Theorem 1.1 can be found in Patodi [1971]. In fact, under the above assumptions a *p*-FSHE need to be unique.

From now until the end of this section, we assume that (*M*,*g*) satisfies the hypotheses of Theorem 1.1 (the hypothesis to be oriented is not essential in all situations).

Definition 1.2 A **solution of the *p*-heat equation on (M,g) with the initial condition** a continuous *p*-form ω on *M* is a continuous section *H* of the real vector bundle $\pi_1^*\left(\Lambda^p T^* M\right)$ over $M \times [0,+\infty)$, where $\pi_1 : M \times [0,+\infty) \to M$ denotes the projection on the first factor [so, the value of *H* at a point $(x,t) \in M \times [0,+\infty)$ is $H(x,t) \in \Lambda^p T_x^* M$], that satisfies the following conditions:

(i) $H \in C^{2,1}\left(M,(0,+\infty);\pi_1^*\left(\Lambda^p T^* M\right)\right)$;

(ii) $\Delta^{(p)} H + \dfrac{\partial H}{\partial t} = 0$,

where $\Delta^{(p)}$ is the Hodge-de Rham operator (or *p*-Laplacian) with respect to the first argument of *H*;

(iii) $H(x,0) = \omega(x)$ for all $x \in M$.

Note that the zero section $0 : M \times [0,+\infty) \to \pi_1^*\left(\Lambda^p T^* M\right)$ is a solution of the *p*-heat equation on (*M*,*g*) with the initial condition the zero *p*-form on *M*.

As for the heat equation associated to the Laplace-Beltrami operator (see Chapter 3, §1), one can prove the following

Proposition 1.1 Let $E^{(p)}$ be a *p*-FSHE on (*M*,*g*) and let ω be as in Definition 1.2. Then the function

$$H : M \times [0,\infty) \to \pi_1^*\left(\Lambda^p T^* M\right)$$

defined by

$$H(x,t) = \begin{cases} \displaystyle\int_M \left[E^{(p)}(x,y,t)\right]\left(\omega(y)\right) d\mu_g(y) & x \in M, t > 0, \\ \omega(x), & x \in M, t = 0, \end{cases}$$

is a solution of the *p*-heat equation on (*M*,*g*) with the initial condition ω. Moreover, this solution is unique.

Corollary 1.1. The hypotheses of Theorem 1.1 concerning (M,g) imply the uniqueness of the p-FSHE on (M,g).

If for every $\omega \in L^2A^p(M)$ and for every $t \in (0,+\infty)$, one defines

$$\left[\mathcal{P}_t^{(p)}(\omega)\right](x) = \int_M \left[E^{(p)}(x,y,t)\right](\omega(y))\,d\mu_g(y),\ x \in M,$$

it follows that

Proposition 1.2.

(i) For any $t \in (0,+\infty)$ and p, $0 \le p \le n$, $\mathcal{P}_t^{(p)} : L^2A^p(M) \to L^2A^p(M)$ is a compact self-adjoint operator;

(ii) For any $s,t \in (0,+\infty)$ and p, $0 \le p \le n$, $\mathcal{P}_s^{(p)} \circ \mathcal{P}_t^{(p)} = \mathcal{P}_{s+t}^{(p)}$. In particular, $\mathcal{P}_t^{(p)}$ is a positive operator for any $t \in (0,+\infty)$ and p, $0 \le p \le n$;

(iii) For any $\omega \in L^2A^p(M), t \in (0,+\infty)$ and p, $0 \le p \le n$, we have $\mathcal{P}_t^{(p)}(\omega) \in A^p(M)$;

(iv) For any $\omega \in A^p(M)$, we have

$$\lim_{t \downarrow 0} \mathcal{P}_t^{(p)}(\omega) = \omega,$$

where the convergence is in the L^2-sense, $0 \le p \le n$.

Note that the semigroup property of $\mathcal{P}_t^{(p)}$ is equivalent to the equality

$$\int_M \left[E^{(p)}(x,y,t) \circ E^{(p)}(y,z,s)\right](\cdot)\,d\mu_g(y) = \left[E^{(p)}(x,z,t+s)\right](\cdot)$$

for all $x,z \in M$ and $s,t \in (0,+\infty)$.

Now the proof of the next statement is the same as for the integral operator $\mathcal{P}_t = \mathcal{P}_t^0$ acting on functions (see Chapter 3, §1).

Proposition 1.3. There exists a complete orthonormal sequence of eigenforms of degree p in the Hilbert space $L^2A^p(M)$ for each $\mathcal{P}_t^{(p)}$, $t \in (0,+\infty)$, $0 \le p \le n$. Moreover, these eigenforms are of class C^∞ and all $\mathcal{P}_t^{(p)}$ have the same finite dimensional eigenspaces.

Finally, one can prove that these eigenspaces are also the eigenspaces of the Hodge-de Rham operator $\Delta^{(p)}$. More precisely, the following **spectral theorem** holds.

Theorem 1.2. Let (M,g) be a closed, connected, oriented, n-dimensional Riemannian manifold. The eigenvalue problem

$$\Delta^{(p)}\omega = \lambda\omega$$

has a complete orthonormal system of smooth eigenforms of degree p $\omega_1^{(p)}, \omega_2^{(p)}, \ldots$ in $L^2A^p(M)$ with corresponding eigenvalues $\lambda_1^{(p)}, \lambda_2^{(p)}, \ldots$. These have the following properties:

(i) $0 \le \lambda_1^{(p)} \le \lambda_2^{(p)} \le \cdots, \lambda_k^{(p)} \to \infty$ as $k \to \infty$;

(ii) $E^{(p)}(x,y,t)=\sum_{k=0}^{\infty} e^{-\lambda_k^{(p)}t}\omega_k^{(p)}(x)\otimes\omega_k^{(p)}(y)$,

where the series converges uniformly on $M\times M$ for each $t\in(0,+\infty)$.

Since the eigenforms of degree p $\omega_1^{(p)},\omega_2^{(p)},\ldots$ of $\Delta^{(p)}$ determine a complete orthonormal system of $L^2A^p(M)$, it follows that for $\omega\in L^2A^p(M)$ we have

$$\omega=\sum_{j=1}^{\infty}\left\langle \omega,\omega_j^{(p)}\right\rangle \omega_j^{(p)} \text{ in } L^2A^p(M)$$

(the Fourier series expansion) and

$$\|\omega\|^2=\sum_{j=1}^{\infty}\left\langle \omega,\omega_j^{(p)}\right\rangle^2$$

(Parseval's identity).

An immediate consequence of the spectral theorem for $\Delta^{(p)}$ (Theorem 1.2) is the following statement.

Corollary 1.2. For each $t\in(0,+\infty)$ and for each p, $0\le p\le n$, the series $\sum_{j=1}^{\infty} e^{-\lambda_j^{(p)}t}$ is convergent and its sum is equal to

$$\int_M trace\left[E^{(p)}(x,x,t)\right]d\mu_g(x).$$

Since $\mathcal{P}_t^{(p)}$ is a compact self-adjoint operator, $\mathcal{P}_t^{(p)}$ is a trace class operator for each $t\in(0,+\infty)$ and p, $0\le p\le n$, so that by the definition of the trace and Corollary 1.2,

$$\begin{aligned} trace\left(\mathcal{P}_t^{(p)}\right)&=\sum_{j=1}^{\infty}\left\langle \mathcal{P}_t^{(p)}\left(\omega_j^{(p)}\right),\omega_j^{(p)}\right\rangle=\sum_{j=1}^{\infty}e^{-\lambda_j^{(p)}t}\\ &=\int_M trace\left[E^{(p)}(x,x,t)\right]d\mu_g(x). \end{aligned}$$

One can show that the positive-definite symmetric operator $\Delta^{(p)}:A^p(M)\to A^p(M)$ has a unique self-adjoint extension $\tilde{\Delta}^{(p)}$ to a dense subspace of $L^2A^p(M)$, namely $\left\{\omega=\sum_{j=1}^{\infty}a_j\omega_j^{(p)}\in L^2A^p(M) \mid \sum_{j=1}^{\infty}\left(\lambda_j^{(p)}a_j\right)^2<\infty\right\}$, which is the largest domain we might consider for the Hodge-de Rham operator $\Delta^{(p)}$ (see also Chapter 3, §1). Moreover, one can prove that $\mathcal{P}_t^{(p)}=e^{-t\tilde{\Delta}^{(p)}}$ for any $t\in(0,+\infty)$ and p, $0\le p\le n$, that is $\tilde{\Delta}^{(p)}$ is the infinitesimal generator of the C_0 semigroup $\mathcal{P}_t^{(p)}$.

Let us denote by

$$Z_{(M,g)}^{(p)}(t)=\sum_{j=1}^{\infty}e^{-\lambda_j^{(p)}t},\quad t\in(0,+\infty),$$

the trace of $\mathcal{P}_t^{(p)}$. For an eigenvalue $\lambda\ge 0$ of $\Delta^{(p)}$, let $m_\lambda^{(p)}\in\mathbb{N}^*$ be the dimension of the corresponding eigenspace. $m_\lambda^{(p)}$ is called **the multiplicity of** λ. Then $Z_{(M,g)}^{(p)}$ can be written as

$$Z_{(M,g)}^{(p)}(t) = \sum_\lambda m_\lambda^{(p)} e^{-\lambda t}$$

for each $t \in (0,+\infty)$, where λ runs over all distinct eigenvalues of $\Delta^{(p)}$. Note that $\lambda = 0$ is an eigenvalue of $\Delta^{(p)}$ if and only if $b_p(M) \neq 0$, where $b_p(M)$ denotes the p-th Betti number of M. Indeed, the smooth p-form ω is an eigenform of $\Delta^{(p)}$ corresponding to the eigenvalue 0 if and only if $\omega \neq 0$ is a harmonic p-form on M. But the space of harmonic p-forms on M is isomorphic to the real cohomology vector space $H^p(M)$.

On the other hand, the above series converges uniformly on $[t_0,+\infty)$ for any $t_0 \in (0,+\infty)$, so that $Z_{(M,g)}^{(p)} : (0,+\infty) \to \mathbb{R}$ is a continuous function. It is decreasing, $\lim\limits_{t\to+\infty} Z_{(M,g)}^{(p)}(t) = b_p(M)$ and $\lim\limits_{t\searrow 0} Z_{(M,g)}^{(p)}(t) = +\infty$. As in the case of the continuous real function $Z_{(M,g)} = Z_{(M,g)}^{(0)}$, the function $Z_{(M,g)}^{(p)} : (0,+\infty) \to \mathbb{R}$ determines the eigenvalues of $\Delta^{(p)}$ and their multiplicities (see Chapter 3, Proposition 1.1).

Example 1.1. Let us describe the eigenvalues of the Hodge-de Rham operator $\Delta^{(p)}$ for the flat n-dimensional torus $\left(T^n = \mathbb{R}^n/\Gamma, g_{can}/\Gamma\right)$. If $(x^1,\dots,x^n)$ are coordinates in $\mathbb{R}^n$ and $\omega \in A^p(T^n)$, $0 \le p \le n$, then ω can be written uniquely as:

$$\omega = \sum_{1\le i_1<\cdots<i_p\le n} \omega_{i_1\cdots i_p} dx^{i_1} \wedge\cdots\wedge dx^{i_p} ,$$

which implies that

$$\Delta^{(p)}(\omega) = \sum_{1\le i_1<\cdots<i_p\le n} \Delta\left(\omega_{i_1\cdots i_p}\right) dx^{i_1} \wedge\cdots\wedge dx^{i_p} ,$$

where $\Delta = \Delta^{(0)}$ from the right hand side denotes the Laplace-Beltrami operator. Therefore

$$\Delta^{(p)}(\omega) = \lambda\omega \qquad ,\lambda \in \mathbb{R},$$

if and only if for each $1 \le i_1 < \cdots < i_p \le n$ one has

$$\Delta\left(\omega_{i_1\cdots i_p}\right) = \lambda\omega_{i_1\cdots i_p} .$$

Consequently, λ is an eigenvalue of $\Delta^{(p)}$ if and only if λ is an eigenvalue of $\Delta = \Delta^{(0)}$. Concerning the multiplicities note that

$$m_\lambda^{(p)} = \binom{n}{p} m_\lambda^{(0)} \quad , \quad 0 \le p \le n .$$

Example 1.2. Let G be a connected n-dimensional Lie group and Γ a cocompact discrete subgroup of G. Suppose that the Riemannian metric on $M = G/\Gamma$ induces a left-invariant metric on G. Elements of the exterior algebra $\Lambda^p(\mathfrak{g}^*)$, where $\mathfrak{g}^*$ is the dual space of the Lie algebra $\mathfrak{g}$ of G, may be viewed first as left-invariants p-forms on G and then as elements of $A^p\left(G/\Gamma\right)$. With this interpretation,

$$A^p\left(G/\Gamma\right)=C^\infty\left(G/\Gamma\right)\otimes\Lambda^p\left(\mathfrak{g}^*\right).$$

If $f\in C^\infty(M)$ and $\omega\in A^p(M)$, then

$$\Delta^{(p)}(f\omega)=(\Delta f)\omega+f\left(\Delta^{(p)}\omega\right)-2\nabla_{\operatorname{grad}(f)}\omega \tag{1.1}$$

(see Chapter 2, Proposition 2.13), where $\Delta^{(p)}$ denotes the Hodge-de Rham operator. Viewing functions on M as left Γ-invariant functions on G, the Laplace-Beltrami operator on $C^\infty(M)$ is given by

$$\Delta f=-\sum_{j=1}^{n}U_j^2 f, \tag{1.2}$$

where $(U_1,\ldots,U_n)$ is any orthonormal basis of $\mathfrak{g}$ with respect to g (see Urakawa [1982]). On the other hand, note that

$$\nabla_{\operatorname{grad}(f)}\omega=\sum_{j=1}^{n}\left(U_j f\right)\nabla_{U_j}\omega.$$

It therefore follows from (1.1) and (1.2) that $K\otimes\Lambda^p\left(\mathfrak{g}^*\right)$ is $\Delta^{(p)}$-invariant whenever K is a subspace of $C^\infty\left(G/\Gamma\right)$ invariant under the right action of G.

Now let $G=H_n$ be the real Heisenberg group and Γ_r the cocompact discrete subgroup of H_n defined in Chapter 1, Example 2.19. Let us denote by $K_{r,1}$ the space of all smooth functions on H_n/Γ_r which vanish on the center of H_n, and by $K_{r,2}$ the complementary subspace of $C^\infty\left(H_n/\Gamma_r\right)$ invariant under the right action of H_n. By the remarks above, if g is any left-invariant Riemannian metric on H_n and $\Delta^{(p)}$ is the Hodge-de Rham operator associated to $\left(H_n/\Gamma_r, g\right)$, then $K_{r,j}\otimes\Lambda^p\left(\mathfrak{g}^*\right)$ is $\Delta^{(p)}$-invariant, $j=1,2$. Let us denote by $\sum_j^p(r,g)$ the collection of eigenvalues, with multiplicities, of $\Delta^{(p)}$ on $K_{r,j}\otimes\Lambda^p\left(\mathfrak{g}^*\right)$.

Let h_n^* be the dual space of h_n. Given

$$g=\begin{bmatrix} h & 0\\ 0 & g_{2n+1}\end{bmatrix}$$

as in Chapter 1, Example 2.19, define

$$\#: h_n^*\to h_n \text{ by } \omega(X)=g(\#\omega,X)$$

and define an inner product $\langle\ ,\ \rangle$ on h_n^* by

$$\langle \omega, \theta \rangle = g(\#\omega, \#\theta).$$

Define

$$\eta : \mathbb{R}^{2n} \to \mathbb{R}^{2n} \text{ by } [X,Y] = h(X, \eta(Y))Z, \text{ where } Z = A(0,0,1),$$

that is

$$h(X, \eta Y) = A(X,Y)$$

for $A : \mathbb{R}^{2n} \times \mathbb{R}^{2n} \to \mathbb{R}$ the non-singular alternating bilinear form defined by

$$A(X,Y)Z = [X,Y] \text{ for } X, Y \in \mathbb{R}^{2n} \text{ and } Z = A(0,0,1)$$

(see the conventions made in Chapter 1, Example 2.19). Since

$$\Sigma = \begin{bmatrix} 0 & I_n \\ -I_n & 0 \end{bmatrix}$$

is the matrix of A relative to the standard basis of $\mathbb{R}^{2n}$ and since we are identifying the inner product h with its matrix in the standard basis, the matrix of η in the standard basis is given by $h^{-1}\Sigma$. But $h^{-1}\Sigma$ is similar to the skew-symmetric matrix $h^{-\frac{1}{2}}\, \Sigma\, h^{-\frac{1}{2}}$, so that it has pure imaginary characteristic roots: $\pm\sqrt{-1}d_k^2, 1 \le k \le n$. Since η is skew-symmetric relative to h with characteristic roots $\pm\sqrt{-1}d_1^2, \ldots, \pm\sqrt{-1}d_n^2$, there exists an orthonormal basis $(X_1', \ldots, X_n', Y_1', \ldots, Y_n')$ of $\mathbb{R}^{2n}$ with respect to h such that

$$\eta X_k' = -d_k^2 Y_k', \eta Y_k' = d_k^2 X_k', \quad 1 \le k \le n.$$

Since

$$[X,Y] = h(X, \eta(Y))Z$$

for all $X, Y \in \mathbb{R}^{2n}$, it follows that

$$[X_k', Y_k'] = d_k^2 Z, \quad 1 \le k \le n,$$

and all other brackets of basis vectors are zero.

Given Γ_r as in Chapter 1, Example 2.19, let

$$\mathfrak{a}_r = \{\omega \in h_n^* \mid \omega(Z) = 0 \text{ and } \omega(\log(\Gamma_r)) \subset \mathbb{Z}\},$$

where $\log : H_n \to h_n$ denotes the inverse of $\exp : h_n \to H_n$. Note that $\log(\Gamma_r)$ is a discrete spanning set of h_n. Using the previous notation, for $\omega \in \mathfrak{a}_r$, let

$$\alpha(\omega) = 4\pi^2 \langle \omega, \omega \rangle, C = \frac{g_{2n+1}}{2} \sum_{j=1}^{n} d_j^4,$$

$$D(\omega) = 4\pi^2 g_{2n+1} \langle (\#^{-1} \circ \eta \circ \#)(\omega), (\#^{-1} \circ \eta \circ \#)(\omega) \rangle,$$

$$\beta_{\pm}(\omega) = \alpha(\omega) + C \pm \sqrt{C^2 + D(\omega)}.$$

Then $\Sigma_1^1(r,g)$ is the collection of numbers λ of the form $\lambda = \alpha(\omega)$ or $\lambda = \beta_{\pm}(\omega)$ for some $\omega \in \mathfrak{a}_r$. λ occurs in $\Sigma_1^1(r,g)$ $2n-1$ times for each $\omega \in \mathfrak{a}_r$ such that $\lambda = \alpha(\omega)$, and once for each $\omega \in \mathfrak{a}_r$ such that $\lambda = \beta_+(\omega)$ or $\lambda = \beta_-(\omega)$.

Indeed, for $\omega \in \mathfrak{a}_r$ define

$$f_\omega : H_n \to \mathbb{C} \text{ by } f_\omega(\exp(X)) = \exp\left[2\pi\sqrt{-1}\omega(X)\right], \; X \in h_n.$$

Then f_ω may be viewed as a function on H_n/Γ_r. On the other hand, $\operatorname{grad}(f_\omega) = 2\pi\sqrt{-1}\#(\omega)$. For $\theta \in h_n^*$, (1.1) yields

$$\Delta^{(1)}(f_\omega \theta) = f_\omega\left(4\pi\langle \omega, \omega \rangle \theta + \Delta^{(1)}\theta - 4\pi\sqrt{-1}\nabla_{\#(\omega)}\theta\right). \tag{1.3}$$

Let $\varpi = (\#)^{-1}(Z) / g_{2n+1}$. Thus $\varpi|_{\mathbb{R}^{2n}} = 0$ and $\varpi(Z) = 1$. Using the definition of $\Delta^{(1)}$, for $\theta \in h_n^*$ one obtains

$$\Delta^{(1)}\theta = g_{2n+1}\left(\sum_{j=1}^{n} d_j^4\right)\theta(Z)\varpi = 2C\theta(Z)\varpi. \tag{1.4}$$

Using the formula

$$2g(\nabla_X Y, U) = g([X,Y],U) - g([X,U],Y) - g([Y,U],X)$$

for $X, Y, U \in h_n$ (see Chapter 1, Example 4.5, where G is replaced by H_n, the additional hypotheses on G, that G is compact and semisimple, are not needed in establishing this formula) together with

$$\nabla_X(\#\theta) = \#\nabla_X(\theta),$$

an easy computation yields

$$\nabla_{\#\omega}\theta = -\frac{1}{2}\theta(Z)\left((\#)^{-1} \circ \eta \circ \#\right)(\omega) + \frac{1}{2}\theta((\eta \circ \#)(\omega)) g_{2n+1}\varpi. \tag{1.5}$$

From (1.3), (1.4) and (1.5), we see that if θ belongs to the $(2n-1)$-dimensional subspace orthogonal to both ϖ and $(\#^{-1} \circ \eta \circ \#)(\omega)$, then $f_\omega\theta$ is an eigenform of $\Delta^{(1)}$ for the eigenvalue $\alpha(\omega) = 4\pi^2\langle \omega, \omega \rangle$. Moreover, on the two-dimensional subspace spanned by $f_\omega\theta$ and $f_\omega\left(\left((\#)^{-1} \circ \eta \circ \#\right)(\omega)\right)$, $\Delta^{(1)} - \alpha(\omega) Id$ is described by the 2×2 matrix

$$\begin{bmatrix} g_{2n+1}\sum_{j=1}^{n} d_j^4 & -2\pi\sqrt{-1}g_{2n+1}\left\|(\eta\circ\#)(\omega)\right\|^2 \\ 2\pi\sqrt{-1} & 0 \end{bmatrix} = \begin{bmatrix} 2C & \frac{-\sqrt{-1}}{2\pi}D(\omega) \\ 2\pi\sqrt{-1} & 0 \end{bmatrix}.$$

Since the characteristic roots of this matrix are $C\pm\sqrt{C^2+D(\omega)}$, the statement on $\Sigma_1^1(r,g)$ now follows from the equality

$$K_{r,1}=\bigoplus_{\omega\in\mathcal{A}_r}\mathbb{R}f_\omega,$$

which is a consequence of the general results on compact nilmanifolds (see Moore, [1973]).

Definition 1.3. Assume that we have ordered all the distinct eigenvalues of $\Delta^{(p)}$ in an increasing sequence:

$$(0\le)\lambda_0<\lambda_1<\lambda_2<\cdots.$$

The ***p*-spectrum** of the closed, n-dimensional Riemannian manifold (M,g), denoted by $\mathbf{Spec}^{(p)}(\boldsymbol{M},\boldsymbol{g})(0\le p\le n)$, is the sequence of real numbers

$$\underbrace{\lambda_0,\ldots,\lambda_0}_{m_{\lambda_0}^{(p)}\text{-times}}<\underbrace{\lambda_1,\ldots,\lambda_1}_{m_{\lambda_1}^{(p)}\text{-times}}<\cdots,$$

where each eigenvalue λ_i of $\Delta^{(p)}$ is repeated according to its multiplicity $m_{\lambda_i}^{(p)}$.

As another nice application of the heat equation method, we prove

Theorem 1.3. Let $\omega(\cdot,\cdot)$ be the solution of the heat equation

$$\frac{\partial\omega}{\partial t}+\Delta^{(p)}\omega=0 \tag{1.6}$$

with the initial condition

$$\lim_{t\searrow 0}\omega(x,t)=\omega_0(x), \tag{1.7}$$

$x\in M$, where $\omega_0\in A^p(M)$.

(i) If ω_0 is closed (resp. exact), then $\mathcal{P}_t^{(p)}(\omega_0)=\omega(\cdot,t)$ is closed (resp. exact) for any $t\in(0,+\infty)$;

(ii) If ω_0 is co-closed (resp. co-exact), then $\mathcal{P}_t^{(p)}(\omega_0)=\omega(\cdot,t)$ is co-closed (resp. co-exact) for any $t\in(0,+\infty)$;

(iii) If ω_0 ω0 is closed, the de Rham cohomology class of $\mathcal{P}_t^{(p)}(\omega_0)=\omega(\cdot,t)$ is the same for all $t\in(0,+\infty)$.

(iv) $\left(S^{(p)}\omega\right)(\cdot,t)=S^{(p)}\left(\mathcal{P}_t^{(p)}(\omega_0)\right)$ is the solution of the heat equation

$$\frac{\partial\tilde{\omega}}{\partial t}+\Delta^{(n-p)}\tilde{\omega}=0$$

with the initial condition $S^{(p)}\omega_0 \in A^{n-p}(M)$, where $S^{(p)}$ denotes the star Hodge operator. In particular,

$$S^{(p)}\mathcal{P}_t^{(p)} = \mathcal{P}_t^{(n-p)}S^{(p)}$$

for any $t \in (0,+\infty)$.

Proof. (i), (ii) As we have pointed out before, the initial value problem (1.6)-(1.7) has a smooth solution

$$\omega(\cdot,t) = e^{-t\tilde{\Delta}^{(p)}}(\omega_0)$$

for any initial continuous p-form ω_0 on M, where $\tilde{\Delta}^{(p)}$ is a self-adjoint unbounded operator on the Hilbert space $L^2A^p(M)$ of real L^2 p-forms on M (with respect to the canonical measure associated with the Riemannian metric), namely the unique self-adjoint extension of $\Delta^{(p)}$. $e^{-t\tilde{\Delta}^{(p)}}$ is a well-defined compact operator on $L^2A^p(M)$. To prove (i) and (ii) it suffices to note that $d^p \circ \Delta^{(p)} = \Delta^{(p+1)} \circ d^p$ and $\delta^p \circ \Delta^{(p)} = \Delta^{(p-1)} \circ \delta^p$ [see Chapter 2, Proposition 2.10 (iv)]. Indeed, these two equalities imply, via the spectral theorem for self-adjoint unbounded operators on Hilbert spaces (see Lang [1985]), that
$d^p \circ e^{-t\tilde{\Delta}^{(p)}}(\omega_0) = e^{-t\tilde{\Delta}^{(p+1)}}(d^p(\omega_0))$ and
$\delta^p \circ e^{-t\tilde{\Delta}^{(p)}}(\omega_0) = e^{-t\tilde{\Delta}^{(p-1)}}(\delta^p(\omega_0))$ for any $\omega_0 \in A^p(M)$ and the desired conclusions follow.

(iii) Assume that $\omega_0 \in A^p(M)$ is closed. Then

$$\Delta^{(p)}\omega(\cdot,t) = d^{p-1}\delta^p\omega(\cdot,t)$$

for any $t \in (0,+\infty)$, i.e. $\Delta^{(p)}\omega(\cdot,t)$ is an exact smooth p-form on M for any $t \in (0,+\infty)$. On the other hand, for a fixed $t_0 \in (0,+\infty)$, we have

$$\omega(\cdot,t) - \omega(\cdot,t_0) = \int_{t_0}^{t}\frac{\partial}{\partial s}\omega(\cdot,s)ds = \int_{t_0}^{t}\Delta^{(p)}\omega(\cdot,s)ds = d^{p-1}\int_{t_0}^{t}\delta^p\omega(\cdot,s)ds$$

for all $t \in (0,+\infty)$. This implies that the de Rham cohomology class $[\omega(\cdot,t)]$ of $\omega(\cdot,t)$ coincides with the de Rham cohomology class $[\omega(\cdot,t_0)]$ of $\omega(\cdot,t_0)$ for any $t \in (0,+\infty)$, i.e. the statement (iii) is valid. In the same time, this shows that

$$[\omega(\cdot,t)] = \left[e^{-t\tilde{\Delta}^{(p)}}\omega_0\right] = [\omega_0] \tag{1.8}$$

for any $t \in (0,+\infty)$.

(iv) The first statement is an immediate consequence of Proposition 2.10 (v) of Chapter 2. The second one follows immediately from the uniqueness of the solution of $(n-p)$-heat equation on (M,g) with initial condition (see Proposition 1.1).

Q.E.D.

Remark 1.1. One can also give another proof for the "closed" and "co-closed" assertions in the previous Theorem [(i) and (ii)]. Indeed, firstly let us note that if

$\omega_0 \in A^p(M)$ is closed then

$$\lim_{t\searrow 0} d^p \circ e^{-t\tilde{\Delta}^{(p)}}(\omega_0) = d^p\left[\lim_{t\searrow 0} e^{-t\tilde{\Delta}^{(p)}}(\omega_0)\right] = d^p\omega_0 = 0 .$$

On the other hand, $\left(d^p \circ e^{-t\tilde{\Delta}^{(p)}}\right)(\omega_0)$ is a solution of the heat equation (1.6):

$$\frac{\partial}{\partial t}\left[d^p \circ e^{-t\tilde{\Delta}^{(p)}}(\omega_0)\right] = d^p\left[\frac{\partial}{\partial t} e^{-t\tilde{\Delta}^{(p)}}(\omega_0)\right] = d^p\left(-\Delta^{(p)} e^{-t\tilde{\Delta}^{(p)}}(\omega_0)\right) = -\Delta^{(p+1)}\left[\left(d^p \circ e^{-t\tilde{\Delta}^{(p)}}\right)(\omega_0)\right]$$

By the uniqueness of the solution of the initial value problem (1.6)-(1.7) [with the initial value the zero $(p+1)$-form], it follows that

$$\left(d^p \circ e^{-t\tilde{\Delta}^{(p)}}\right)(\omega_0) = 0$$

for all $t \in (0,+\infty)$. The proof is similar in the case when $\delta^p\omega_0 = 0$.

The vector spaces $A^p(M), 0 \le p \le n$, are infinite-dimensional. On the other hand, the de Rham cohomology of *M*, that is the cohomology of the de Rham complex

$$A^0(M) \xrightarrow{d} A^1(M) \xrightarrow{d^1} \cdots \xrightarrow{d^{n-1}} A^n(M)$$

in which $d^p : A^p(M) \to A^{p+1}(M)$ is the exterior derivative, is isomorphic to the ordinary cohomology of *M*, and is therefore finite dimensional. When formulated in terms of de Rham cohomology one can look for a proof of it in terms of analysis since this is a result about the solution spaces of the heat equations associated to the Hodge-de Rham operators.

Corollary 1.3. The de Rham cohomology spaces of a closed smooth manifold are finite dimensional.

Proof. Let *g* be any smooth Riemannian metric on *M*. For fixed $t \in (0,+\infty)$, the operator $e^{-t\tilde{\Delta}^{(p)}}$ acts on the *p*-th cohomology space $H^p(M)$ of *M* by

$$e^{-t\tilde{\Delta}^{(p)}}([\omega]) = \left[e^{-t\tilde{\Delta}^{(p)}}\omega\right],\ [\omega] \in H^p(M),$$

and (1.8) shows that in fact it acts as the identity on this space, $0 \le p \le \dim M$. Thus, the desired result is a consequence of the compactness of $e^{-t\tilde{\Delta}^{(p)}} : L^2A^p(M) \to L^2A^p(M)$, which implies that the space $H^p(M)$ is locally compact, hence finite dimensional for $0 \le p \le \dim M$.

Q.E.D.

There are relative and absolute boundary conditions for $\Delta^{(p)}$, which are natural generalizations of Dirichlet and Neumann boundary conditions for $\Delta = \Delta^{(0)}$. To describe these boundary conditions let fix some notations. Let (M,g) be a compact, oriented, smooth Riemannian manifold with smooth boundary ∂M. Absolute and relative boundary conditions are given by decomposing uniquely a form $\omega \in A^p(M)$ into tangential and normal components at any point $x \in \partial M$, $\omega = \omega_{\text{tan}} + \omega_{\text{norm}}$, where

$$\omega_{\tan} \in \Lambda^p T_x^*(\partial M), \omega_{\text{norm}} = \theta \wedge \eta, \theta \in \Lambda^{p-1} T_x^*(\partial M),$$

η being the unit inward pointing normal in the cotangent space at $x \in \partial M$, and requiring that

$$\omega_{\text{norm}} = 0, \ \left(d^p \omega\right)_{\text{norm}} = 0$$

for absolute boundary conditions, and

$$\omega_{\tan} = 0, \ \left(\delta^p \omega\right)_{\tan} = 0$$

for relative boundary conditions. Dirichlet and Neumann boundary conditions for functions are generalized by relative and absolute boundary conditions respectively (see Chavel [1984], Conner [1956], Park [2000], Ray and Singer [1971]). Note that the star Hodge operator interchanges absolute and relative boundary conditions. Either of these boundary conditions gives rise to an elliptic, self-adjoint boundary value problem. The boundary value problems can be reformulated in an L^2 framework. Formally, the theory is very similar to the boundaryless case. In particular, for an oriented, n-dimensional Riemannian manifold (M,g) with boundary ∂M one can define $\mathbf{Spec}_D^{(p)}(\boldsymbol{M},\boldsymbol{g})$ and $\mathbf{Spec}_N^{(p)}(\boldsymbol{M},\boldsymbol{g})$, $0 \le p \le n$. One should mention that heat equation methods have been used successfully to study these boundary value problems (see Ray and Singer [1971], Cheeger [1979]).

2. CHARACTERIZING EIGENVALUES OF $\Delta^{(p)}$

Let (M,g) be a closed, oriented, n-dimensional smooth Riemannian manifold and $E\left(\Delta^{(p)},\lambda\right)$ the space of eigenforms of degree p of $\Delta^{(p)}$ with eigenvalue λ.

The real vector space $E\left(\Delta^{(p)},\lambda\right)$ is finite dimensional and $E\left(\Delta^{(p)},\lambda\right)$ is orthogonal to $E\left(\Delta^{(p)},\mu\right)$ for $\lambda \neq \mu$ (see §1).

Since

$$\left\langle \Delta^{(p)}\omega, \omega \right\rangle = \left\| d^p \omega \right\|^2 + \left\| \delta^p \omega \right\|^2$$

for any $\omega \in A^p(M)$, it follows that

$$\lambda = \frac{\left\| d^p \omega \right\|^2 + \left\| \delta^p \omega \right\|^2}{\left\| \omega \right\|^2} \ge 0$$

for any $\omega \in E\left(\Delta^{(p)},\lambda\right) \setminus \{0\}$, i.e. λ coincides with the **Rayleigh-Ritz quotient**

$$R(\omega) = \frac{\left\| d^p \omega \right\|^2 + \left\| \delta^p \omega \right\|^2}{\left\| \omega \right\|^2}$$

of the eigenform $\omega \neq 0$.

First, let us prove the following minimum characterization of eigenvalues of $\Delta^{(p)}$.

Theorem 2.1. Let

$$\lambda_1^{(p)} \le \lambda_2^{(p)} \le \cdots$$

be the sequence of eigenvalues of $\Delta^{(p)}$, each repeated according to its multiplicity, and

$$\omega_1, \omega_2, \ldots$$

the corresponding complete orthonormal sequence of eigenforms of degree p. Then

$$\lambda_{k+1}^{(p)} = \inf_{\omega} \frac{\left\|d^p\omega\right\|^2 + \left\|\delta^p\omega\right\|^2}{\left\|\omega\right\|^2}, \tag{2.1}$$

where ω ranges over the set A_k^p of all $\omega \in A^p(M) \setminus \{0\}$ such that ω is orthogonal to $\omega_1, \ldots, \omega_k$, $k \in \mathbb{N}^*$. Moreover, the equality

$$\lambda_{k+1}^{(p)} \left\|\omega\right\|^2 = \left\|d^p\omega\right\|^2 + \left\|\delta^p\omega\right\|^2$$

holds if and only if $\omega \in E\left(\Delta^{(p)}, \lambda_{k+1}^{(p)}\right)$.

Proof. Set $\alpha_j = \left\langle \omega, \omega_j \right\rangle$, $j \in \mathbb{N}^*$ and fix $l \in \mathbb{N}^*$. Then one gets successively

$$0 \le \left\| d^p\omega - \sum_{j=1}^{l} \alpha_j d^p\omega_j \right\|^2 = \left\|d^p\omega\right\|^2 - 2\sum_{j=1}^{l} \alpha_j \left\langle d^p\omega, d^p\omega_j \right\rangle + \sum_{j=1}^{l} \alpha_j^2 \left\|d^p\omega_j\right\|^2$$

$$= \left\|d^p\omega\right\|^2 - 2\sum_{j=1}^{l} \alpha_j \left\langle \omega, \delta^{p+1}d^p\omega_j \right\rangle + \sum_{j=1}^{l} \alpha_j^2 \left\|d^p\omega_j\right\|^2,$$

and similarly

$$0 \le \left\| \delta^p\omega - \sum_{j=1}^{l} \alpha_j \delta^p\omega_j \right\|^2 = \left\|\delta^p\omega\right\|^2 - 2\sum_{j=1}^{l} \alpha_j \left\langle \omega, d^{p-1}\delta^p\omega_j \right\rangle + \sum_{j=1}^{l} a_j^2 \left\|\delta^p\omega_j\right\|^2,$$

which implies that

$$0 \le \left\|d^p\omega\right\|^2 + \left\|\delta^p\omega\right\|^2 - 2\sum_{j=1}^{l} \alpha_j \left\langle \omega, \Delta^{(p)}\omega_j \right\rangle + \sum_{j=1}^{l} \alpha_j^2 \left\langle \omega_j, \Delta^{(p)}\omega_j \right\rangle$$

$$= \left\|d^p\omega\right\|^2 + \left\|\delta^p\omega\right\|^2 - \sum_{j=1}^{l} \lambda_j^{(p)} \alpha_j^2,$$

that is

$$\left\|d^p\omega\right\|^2 + \left\|\delta^p\omega\right\|^2 \ge \sum_{j=1}^{\infty} \lambda_j^{(p)} \alpha_j^2 . \tag{2.2}$$

If $\omega \in A_k^p$, then $\alpha_j = 0$ for $j = 1, \ldots, k$, so that

$$\sum_{j=1}^{\infty} \lambda_j^{(p)} \alpha_j^2 \ge \lambda_{k+1}^{(p)} \left(\sum_{j=1}^{\infty} \alpha_j^2 \right) = \lambda_{k+1}^{(p)} \left\|\omega\right\|^2 . \tag{2.3}$$

Combining (2.2) and (2.3), one gets

$$\left\|d^p\omega\right\|^2+\left\|\delta^p\omega\right\|^2\geq\lambda_{k+1}^{(p)}\left\|\omega\right\|^2 \text{ for any } \omega\in A_k^p.$$

Notice that the last inequality becomes equality for any $\omega\in E\left(\Delta^{(p)},\lambda_{k+1}^{(p)}\right)$:

$$\left\|d^p\omega\right\|^2+\left\|\delta^p\omega\right\|^2=\left\langle\Delta^{(p)}\omega,\omega\right\rangle=\lambda_{k+1}^{(p)}\left\|\omega\right\|^2.$$

Conversely, if

$$\left\|d^p\omega\right\|^2+\left\|\delta^p\omega\right\|^2=\lambda_{k+1}^{(p)}\left\|\omega\right\|^2,$$

then $\omega\in E\left(\Delta^{(p)},\lambda_{k+1}^{(p)}\right)$. Indeed, (2.3) is a strict inequality if at least one α_j is $\neq 0$, where the index j is such that $\omega_j\notin E\left(\Delta^{(p)},\lambda_{k+1}^{(p)}\right)$. Consequently, if we have equality, it follows necessarily that

$$\omega=\sum_{j,\omega_j\in E\left(\Delta^{(p)},\lambda_{k+1}^{(p)}\right)}\alpha_j\omega_j,$$

which implies that $\omega\in E\left(\Delta^{(p)},\lambda_{k+1}^{(p)}\right)$.

Q.E.D.

Remark 2.1. The equality (2.1) is true also in the case when ω ranges over the set of all L^2 forms of degree p such that $d^p\omega$ and $\delta^p\omega$ are in L^2, and ω is orthogonal to $\omega_1,\ldots,\omega_k$, where $d^p\omega$ and $\delta^p\omega$ are taken in the sense of de Rham currents.

From now on, let us assume that the p-th Betti number $b_p(M)$ of M is zero, $p\geq 1$, the case when $b_p(M)\neq 0$ implying only some minor changes. Next, we shall prove the following mini-max characterization of the k-th eigenvalue of $\Delta^{(p)}$. If $p\geq 1$, note that in this characterization are involved derivatives of the metric g.

Theorem 2.2. Let

$$0<\lambda_1^{(p)}\leq\lambda_2^{(p)}\leq\cdots$$

be the sequence of eigenvalues of $\Delta^{(p)}$, each repeated according to its multiplicity. Then

$$\lambda_k^{(p)}=\inf_{V_k}\sup_{\omega\in V_k\setminus\{0\}}\left\{\frac{\left\|d^p\omega\right\|^2+\left\|\delta^p\omega\right\|^2}{\left\|\omega\right\|^2}\right\},\quad k\in\mathbb{N}^*,$$

where V_k ranges over all k-dimensional linear subspaces of $A^p(M)$.

Proof. Let $\{\omega_j\}_{j\in\mathbb{N}^*}$ be the corresponding complete orthonormal sequence for the completion $L^2A^p(M)$ of $A^p(M)$ consisting of eigenforms of degree p of $\Delta^{(p)}$. Let us denote

$$s^p\left(V_k\right)=\sup_{\omega\in V_k\setminus\{0\}}\left\{\frac{\left\|d^p\omega\right\|^2+\left\|\delta^p\omega\right\|^2}{\left\|\omega\right\|^2}\right\},$$

where V_k is a k-dimensional real linear subspace of $A^p(M)$. Let V_k^0 be the k-dimensional linear subspace of $A^p(M)$ generated by $\omega_1,\ldots,\omega_k$. Since $\omega_k \in V_k^0$ is an eigenform of $\Delta^{(p)}$ corresponding to the eigenvalue $\lambda_k^{(p)}$, it follows that

$$\lambda_k^{(p)} \|\omega_k\|^2 = \|d^p\omega_k\|^2 + \|\delta^p\omega_k\|^2$$

(see Theorem 2.1), which implies that

$$s^{(p)}(V_k^0) \geq \lambda_k^{(p)}.$$

On the other hand, in virtue of Theorem 2.1 and of the fact that

$$0 < \lambda_1^{(p)} \leq \cdots \leq \lambda_k^{(p)}$$

for any $\omega = \sum_{j=1}^{k} \alpha_j \omega_j \in V_k^0$, one gets

$$\begin{aligned}\|d^p\omega\|^2 + \|\delta^p\omega\|^2 &= \sum_{j=1}^{k} \alpha_j^2 \left(\|d^p\omega_j\|^2 + \|\delta^p\omega_j\|^2\right) \\ &= \sum_{j=1}^{k} \lambda_j^{(p)} \alpha_j^2 \|\omega_j\|^2 \leq \lambda_k^{(p)} \|\omega\|^2,\end{aligned}$$

so that

$$s^{(p)}(V_k^0) \leq \lambda_k^{(p)}.$$

Consequently,

$$s^{(p)}(V_k^0) = \lambda_k^{(p)},$$

that is

$$\lambda_k^{(p)} \geq \inf_{V_k} \sup_{\omega \in V_k \setminus \{0\}} \left\{ \frac{\|d^p\omega\|^2 + \|\delta^p\omega\|^2}{\|\omega\|^2} \right\}.$$

Now, let us assume that

$$\lambda_k^{(p)} > \inf_{V_k} \sup_{\omega \in V_k \setminus \{0\}} \left\{ \frac{\|d^p\omega\|^2 + \|\delta^p\omega\|^2}{\|\omega\|^2} \right\}.$$

Then there exists a k-dimensional linear subspace V_k' of $A^p(M)$ such that

$$\lambda_k^{(p)} > s^{(p)}(V_k').$$

But each $\omega = \sum_{j=1}^{\infty} \beta_j \omega_j \in V_k'$ satisfies the inequality

$$\|d^p\omega\|^2 + \|\delta^p\omega\|^2 \leq s^{(p)}(V_k') \|\omega\|^2$$

and since

$$\left\|d^p\omega\right\|^2+\left\|\delta^p\omega\right\|^2\geq\sum_{j=1}^{\infty}\lambda_j^{(p)}\beta_j^2$$

[see inequality (2.2)] one obtains

$$\sum_{j=1}^{\infty}\lambda_j^{(p)}\beta_j^2\leq s^{(p)}\left(V_k'\right)\sum_{j=1}^{\infty}\beta_j^2 . \tag{2.4}$$

This inequality shows that there exists $j_0\geq 1$ such that

$$\lambda_{j_0}^{(p)}\leq s^{(p)}\left(V_k'\right).$$

Let us rewrite (2.4) under the form

$$\sum_{j,s^{(p)}(V_k')<\lambda_j^{(p)}}\left[\lambda_j^{(p)}-s^{(p)}\left(V_k'\right)\right]\beta_j^2\leq\sum_{j,s^{(p)}(V_k')\geq\lambda_j^{(p)}}\left[s^{(p)}\left(V_k'\right)-\lambda_j^{(p)}\right]\beta_j^2 . \tag{2.5}$$

Let $m=\max\left\{j\,\middle|\,\lambda_j^{(p)}\leq s^{(p)}\left(V_k'\right)\right\}$. If one defines the linear mapping

$$F:V_k'\to A^p\left(M\right)$$

by

$$F\left(\omega\right)=\sum_{j=1}^{m}\left\langle\omega,\omega_j\right\rangle\omega_j,\quad\omega\in V_k',$$

then $rank\left(F\right)<k$. Indeed,

$$\lambda_j^{(p)}\leq s^{(p)}\left(V_k'\right)<\lambda_k^{(p)}\text{ for each } j=1,\dots,m,$$

which implies that

$$rank\left(F\right)\leq m<k.$$

Consequently,

$$kr\left(F\right)\left[=\text{the kernel rank (nullity) of } F\right]\geq 1$$

so that there exists $\omega_0\in Ker\left(F\right)\setminus\{0\}$, that is $\left\langle\omega_0,\omega_j\right\rangle=0$ for all j with the property

$$\lambda_j^{(p)}<s^{(p)}\left(V_k'\right).$$

On the other hand, applying (2.5) for $\omega_0\in V_k'$ it follows that each term of the left-hand side should be equal to zero. Hence $\beta_j=\left\langle\omega_0,\omega_j\right\rangle=0$ for each j satisfying $\lambda_j^{(p)}\geq s^{(p)}\left(V_k'\right)$ that is

$$\omega_0=\sum_{j=1}^{\infty}\left\langle\omega_0,\omega_j\right\rangle\omega_j=0,$$

which contradicts $\omega_0\neq 0$.

Q.E.D.

Let $E_d^p(\lambda)$ [resp. $E_\delta^p(\lambda)$] the space of exact (resp. co-exact) eigenforms of $\Delta^{(p)}$ corresponding to the eigenvalue $\lambda \neq 0$, that is

$$E_d^p(\lambda) = E\left(\Delta^{(p)}, \lambda\right) \cap d^{p-1} A^{p-1}(M),\ E_\delta^p(\lambda) = E\left(\Delta^{(p)}, \lambda\right) \cap \delta^{p+1} A^{p+1}(M).$$

Further we will use the following result.

Theorem 2.3. With the above notations, the following equalities are valid:

(i) $E\left(\Delta^{(p)}, \lambda\right) = E_d^p(\lambda) \overset{\perp}{\oplus} E_\delta^p(\lambda)$,

where $\overset{\perp}{\oplus}$ denotes orthogonal direct sum;

(ii) $E_d^p(\lambda) = d^{p-1} E_\delta^{p-1}(\lambda)$;

(iii) $E_\delta^p(\lambda) = \delta^{p+1} E_d^{p+1}(\lambda)$.

Proof. (i) For $\omega \in E\left(\Delta^{(p)}, \lambda\right)$, $\lambda \neq 0$, in virtue of the Hodge-de Rham decomposition theorem one gets

$$\omega = H^{(p)}(\omega) + d^{p-1}\omega_1 + \delta^{p+1}\omega_2,$$

where $H^{(p)}(\omega) \in Ker\left(\Delta^{(p)}\right)$ denotes the harmonic component of the *p*-form ω, $\omega_1 \in A^{p-1}(M)$, $\omega_2 \in A^{p+1}(M)$. On the other hand, the orthogonal direct decomposition

$$A^p(M) = Ker\left(\Delta^{(p)}\right) \overset{\perp}{\oplus} \Delta^{(p)}\left(A^p(M)\right)$$

implies

$$\omega = H^{(p)}(\omega) + \Delta^{(p)}(\omega),$$

so that the equality

$$\omega = \Delta^{(p)}\left(\frac{1}{\lambda}\omega\right) = 0 + \Delta^{(p)}\left(\frac{1}{\lambda}\omega\right)$$

leads to

$$H^{(p)}(\omega) = 0,$$

that is

$$\omega = d^{p-1}\omega_1 + \delta^{p+1}\omega_2. \tag{2.6}$$

Since

$$\Delta^{(p)}\left(d^{p-1} A^{p-1}(M)\right) \subseteq d^{p-1} A^{p-1}(M),\ \Delta^{(p)}\left(\delta^{p+1} A^{p+1}(M)\right) \subseteq \delta^{p+1} A^{p+1}(M)$$

and since (2.6) implies

$$\lambda d^{p-1}\omega_1 + \lambda \delta^{p+1}\omega_2 = \lambda\omega = \Delta^{(p)}(\omega) = \Delta^{(p)}\left(d^{p-1}(\omega_1)\right) + \Delta^{(p)}\left(\delta^{p+1}(\omega_2)\right),$$

the uniqueness of the Hodge-de Rham decomposition shows that

$$d^{p-1}(\omega_1) \in E_d^p(\lambda), \delta^{p+1}(\omega_2) \in E_\delta^p(\lambda),$$

which together with (2.6) proves the desired equality.

(ii) Let $\omega \in E_d^p(\lambda)$. Then

$$\Delta^{(p)}(\omega) = \lambda\omega \text{ and } \omega = d^{p-1}\eta \text{ with } \eta \in A^{p-1}(M).$$

On the other hand, in virtue of the Hodge-de Rham decomposition theorem, one gets

$$\eta = H^{(p-1)}(\eta) + d^{p-2}\alpha + \delta^p\beta,$$

where $\alpha \in A^{p-2}(M)$, $\beta \in A^p(M)$, so that

$$\omega = d^{p-1}\delta^p\beta. \tag{2.7}$$

But

$$\Delta^{(p-1)}(\delta^p\beta) = \delta^p d^{p-1}\delta^p(\beta) = \delta^p(\omega) \tag{2.8}$$

Since

$$d^{p-1}\delta^p(\omega) = \Delta^{(p)}(\omega) = \lambda\omega = \lambda d^{p-1}\delta^p(\beta),$$

that is

$$d^{p-1}(\delta^p\omega - \lambda\delta^p\beta) = 0,$$

it follows that

$$\begin{aligned}\left\|\delta^p(\omega) - \lambda\delta^p(\beta)\right\|^2 &= \left\langle \delta^p(\omega) - \lambda\delta^p(\beta), \delta^p(\omega) - \lambda\delta^p(\beta)\right\rangle \\ &= \left\langle d^{p-1}\left(\delta^p(\omega) - \lambda\delta^p(\beta)\right), \omega - \lambda\beta\right\rangle = 0,\end{aligned}$$

i.e. $\delta^p(\omega) = \lambda\delta^p(\beta)$.

Consequently, (2.8) leads to

$$\Delta^{(p-1)}(\delta^p\beta) = \lambda(\delta^p\beta),$$

that is $\delta^p(\beta) \in E_\delta^{p-1}(\lambda)$ and (2.7) gives the desired result.

The last equality (iii) follows similarly.

Q.E.D.

Corollary 2.1. For every $p = 1,\ldots,n$, the linear mapping

$$d^{p-1} : E_\delta^{p-1}(\lambda) \to E_d^p(\lambda)$$

induced by the exterior derivative d^{p-1} is an isomorphism of norm $\left\|d^{p-1}\right\| = \sqrt{\lambda}$.

Proof. Theorem 2.3 (ii) shows that $d^{p-1} : E_\delta^{p-1}(\lambda) \to E_d^p(\lambda)$ is onto. On the other hand, for every $\omega \in E_\delta^{p-1}(\lambda)$,

$$\left\|d^{p-1}\omega\right\|^2 = \left\langle \delta^p d^{p-1}\omega, \omega\right\rangle = \left\langle \Delta^{(p-1)}(\omega), \omega\right\rangle = \lambda\|\omega\|^2,$$

that is $\left\|d^{p-1}\omega\right\| = \sqrt{\lambda}\,\|\omega\|$ with $\lambda > 0$ and we are done.

Q.E.D.

Note that for $\lambda = 0$, $E\left(\Delta^{(p)},\lambda\right) = E_d^p(\lambda) = E_\delta^p(\lambda)$.

Since the star Hodge operator commutes with $\Delta^{(p)}$ [see Chapter 2, Proposition 2.10(v)], it follows that $S^{(p)}$ induces also an isomorphism of $E_d^p(\lambda)$ onto $E_\delta^{n-p}(\lambda)$ and that $\lambda_j^{(p)} = \lambda_j^{(n-p)}$, $j \in \mathbb{N}^*, 0 \le p \le n$, where $0 \le \lambda_1^{(p)} < \lambda_2^{(p)} < \cdots$ denotes here the set of eigenvalues of $\Delta^{(p)}$.

Previous considerations shows that concerning the eigenvalues of the Hodge-de Rham operators $\Delta^{(p)}, 0 \le p \le n$, it will be enough to look to the restriction of $\Delta^{(p)}$ to exact forms of degree $p, 0 \le p \le n$, since the restriction of $\Delta^{(1)}$ to exact forms of degree one will have the same eigenvalues as $\Delta^{(0)} = \Delta$, the restriction of $\Delta^{(1)}$ to co-exact forms of degree one will have the same eigenvalues as the restriction of $\Delta^{(2)}$ to exact forms of degree two and so on. Consequently, one gets

Corollary 2.2. The knowledge of the eigenvalues of $\Delta^{(p)}, 0 \le p \le n$, is equivalent to knowing the eigenvalues of the restriction of $\Delta^{(q)}$ to the space of exact forms of degree $q, 0 < q \le n$.

Example 2.1. Let $\left(x^1, x^2, \ldots, x^{n+1}\right)$ be the standard coordinate system on $\mathbb{R}^{n+1}$. Denote by $A^p\left(\mathbb{R}^{n+1}\right)$ the space of smooth differential *p*-forms on $\mathbb{R}^{n+1}$ and let $A^*\left(\mathbb{R}^{n+1}\right) = \sum_{p=0}^{n+1} A^p\left(\mathbb{R}^{n+1}\right)$. Put $r^2 = \sum_{i=1}^{n+1}\left(x^i\right)^2$, $X_0 = r\frac{d}{dr} = \sum_{i=1}^{n+1} x^i \frac{\partial}{\partial x^i}$ and $\omega_0 = r dr = \sum_{i=1}^{n+1} x^i dx^i$. Denote by $i(X_0)$ the interior product by X_0 on $A^*\left(\mathbb{R}^{n+1}\right)$ and by $\varepsilon(\omega_0) : A^*\left(\mathbb{R}^{n+1}\right) \to A^*\left(\mathbb{R}^{n+1}\right)$ the linear operator defined by $\varepsilon(\omega_0)\omega = \omega_0 \wedge \omega$ (see also Chapter 2, Remark 1.6). Let d_0^p be the exterior differential acting on $A^*\left(\mathbb{R}^{n+1}\right)$ and $\Delta_0^{(p)} = d_0^{p-1}\delta_0^p + \delta_0^{p+1}d_0^p : A^p\left(\mathbb{R}^{n+1}\right) \to A^p\left(\mathbb{R}^{n+1}\right)$ the Hodge-de Rham operator associated to $\left(\mathbb{R}^{n+1}, g_{can}\right)$, where $\delta_0^p : A^p\left(\mathbb{R}^{n+1}\right) \to A^{p-1}\left(\mathbb{R}^{n+1}\right)$ is the codifferential operator. Define $P^{p,k}\ (p \ge 0, k \ge 0)$ be the set of $\omega \in A^p\left(\mathbb{R}^{n+1}\right)$ of the form

$$\omega = \sum_{1 \le i_1 < \cdots < i_p \le n+1} \omega_{i_1 \ldots i_p} dx^{i_1} \wedge \cdots \wedge dx^{i_p},$$

where $\omega_{i_1 \ldots i_p}$ are homogeneous polynomials of degree *k*. Put $P^{p,k} = \{0\}$ if $p < 0$ or $k < 0$. Note that $P^{0,0} = \left\{\text{the constant functions on } \mathbb{R}^{n+1}\right\}$ and $P^{n+1,0} = \left\{a dx^1 \wedge \cdots \wedge dx^{n+1} \,\middle|\, a \in \mathbb{R}\right\}$. Note that

$$\left[d^{p-1} i^p(X_0) + i^{p+1}(X_0) d^p\right](\omega) = L_{X_0}^p(\omega) = (k+p)\omega \tag{2.9}$$

and

$$\delta_0^{p+1}\varepsilon^p(\omega_0)\omega+\varepsilon^{p-1}(\omega_0)\delta_0^p\omega=(-1)^{np+1}\left(S_0^{(n-p)}L_{X_0}^{n-p}S_0^{(p)}\right)(\omega)=-(n+1-p+k) \quad (2.10)$$

for all $\omega\in P^{p,k}$, where L_{X_0} denotes the Lie derivative with respect to the vector field X_0 and $S_0^{(\,)}$ the star Hodge operator on $A^*\left(\mathbb{R}^{n+1}\right)$. (2.9) and (2.10) imply the following direct sum decomposition

$$P^{p,k}=\left(\ker d_0^p\cap P^{p,k}\right)\oplus\left(\ker i^p(X_0)\cap P^{p,k}\right),\ p+k\neq 0, \quad (2.11)$$

$$P^{p,k}=\left(\ker\delta_0^p\cap P^{p,k}\right)\oplus\left(\ker\varepsilon^p(\omega_0)\cap P^{p,k}\right),n+1-p+k\neq 0. \quad (2.12)$$

Further, let us define the subspace $H^{p,k}$ of $P^{p,k}$ by

$$H^{p,k}=\left\{\omega\in P^{p,k}\left|\Delta_0^{(p)}\omega=0,\delta_0^p\omega=0\right.\right\},$$

which is called **the space of harmonic polynomial *p*-forms on** $\mathbb{R}^{n+1}$. Note that for each $\omega\in H^{p,k}$ one has

$$d_0^p\omega\in H^{p+1,k-1}\text{ and }i^p(X_0)\omega\in H^{p-1,k+1}. \quad (2.13)$$

Indeed, the first property follows easily from the definition of $\Delta_0^{(p+1)}$ and the equality $d_0^p\Delta_0^{(p)}=\Delta_0^{(p+1)}d_0^p$. Concerning the second property, one has

$$\left[\delta_0^{p-1}\circ i^p(X_0)\right](\omega)=-i^{p-1}(X_0)\left(\delta_0^p(\omega)\right)=0$$

and by (2.9),

$$\begin{aligned}\Delta_0^{(p-1)}\left[i^p(X_0)\omega\right]&=\left(d_0^{p-2}\delta_0^{p-1}+\delta_0^pd_0^{p-1}\right)\left[i^p(X_0)\omega\right]=\delta_0^pL_{X_0}^p(\omega)-\delta_0^pi^{p+1}(X_0)d_0^p(\omega)\\&=(k+p)\delta_0^p(\omega)+i^p(X_0)\left(\delta_0^{p+1}d_0^p(\omega)\right)\\&=-i^p(X_0)\left(d_0^{p-1}\left(\delta_0^p\omega\right)\right)=0.\end{aligned}$$

Put $H_d^{p,k}=\left\{\omega\in H^{p,k}\left|d_0^p(\omega)=0\right.\right\}$ and $H_\delta^{p,k}=\left\{\omega\in H^{p,k}\left|i^p(X_0)\omega=0\right.\right\}$. Note that $H^{0,0}=H_d^{0,0}=H_\delta^{0,0}$ and $H^{n+1,0}=H_d^{n+1,0}=H_\delta^{n+1,0}$. By (2.9) and (2.13) one gets

$$H^{p,k}=H_d^{p,k}\oplus H_\delta^{p,k},\quad p+k\neq 0.$$

Note also that $H_d^{0,k}=\{0\}(k>0)$ and $H_\delta^{p,0}=\{0\}(0<p<n+1)$, that can be easily seen by (2.9) and (2.10).

Now let $i:S^n\to\mathbb{R}^{n+1}$ be the inclusion. Since $di^*=i^*d_0$ it follows that $i^*\left(H_d^{p,k}\right)$ consists of exact forms of degree p on S^n. Similarly, the space $i^*\left(H_\delta^{p,k}\right)$ consists of co-exact forms of degree p on S^n. On the other hand, $i^*:\ker\delta_0^p\cap P^{p,k}\to A^p\left(S^n\right)$ is injective. Indeed, let $\omega\in\ker\delta_0^p\cap P^{p,k}$ and assume $i^*\omega=0$. Then ω is in $\ker\left(\varepsilon^p(\omega_0)\right)\cap P^{p,k}$. But $\delta_0^p(\omega)=0$, and so from (2.12) one gets $\omega=0$. Consequently, $i^*:H_d^{p,k}\to A^p\left(S^n\right)$ and $i^*:H_\delta^{p,k}\to A^p\left(S^n\right)$ are injective. Moreover, assuming

$p \le \left[\frac{n}{2}\right]$ and $k>0$, one can show that $i^* H_d^{p,k}$ (resp. $i^* H_\delta^{p,k}$) is an eigenspace consisting of exact (resp. co-exact) forms of degree p of the Hodge-de Rham operator $\Delta^{(p)}$ associated to (S^n, g) corresponding to the eigenvalue $(k+p-1)(k+n-p)$ [resp. $(k+p)(k+n-p-1)$] and no other eigenspace with non-zero eigenvalue of $\Delta^{(p)}$ appears in $A^p(S^n)$ (for details, see Ikeda and Taniguchi [1978]).

Remark 2.2. Since $d^p \circ \Delta^{(p)} = \Delta^{(p+1)} \circ d^p$, $\delta^p \circ \Delta^{(p)} = \Delta^{(p-1)} \circ \delta^p$ [see Chapter 2, Proposition 2.10 (iv)], it follows that

$$d^p E\left(\Delta^{(p)}, \lambda\right) \subseteq E\left(\Delta^{(p+1)}, \lambda\right) \text{ and } \delta^p E\left(\Delta^{(p)}, \lambda\right) \subseteq E\left(\Delta^{(p-1)}, \lambda\right).$$

Note that $E\left(\Delta^{(0)}, \lambda\right) = E_\delta^0(\lambda)$, $E\left(\Delta^{(n)}, \lambda\right) = E_d^n(\lambda)$ and that for $\lambda > 0$, for the induced linear mapping

$$d^p : E\left(\Delta^{(p)}, \lambda\right) = E_d^p(\lambda) \overset{\perp}{\oplus} E_\delta^p(\lambda) \to E\left(\Delta^{(p+1)}, \lambda\right) = E_d^{p+1}(\lambda) \overset{\perp}{\oplus} E_\delta^{p+1}(\lambda)$$

one has $Ker\left(d^p\right) = E_d^p(\lambda)$, $\mathrm{Im}\left(d^p\right) = E_d^{p+1}(\lambda)$, $0 \le p \le n-1$.

Moreover, for $\lambda > 0$, the sequence of real vector spaces

$$0 \to E\left(\Delta^{(0)}, \lambda\right) \xrightarrow{d^0} E\left(\Delta^{(1)}, \lambda\right) \xrightarrow{d^1} \cdots \xrightarrow{d^{n-2}} E\left(\Delta^{(n-1)}, \lambda\right) \xrightarrow{d^{n-1}} E\left(\Delta^{(n)}, \lambda\right) \to 0$$

is exact, so that

$$\sum_{p=0}^{n} (-1)^p \dim E\left(\Delta^{(p)}, \lambda\right) = 0. \tag{2.14}$$

Indeed, we have successively

$$\sum_{p=0}^{n} (-1)^p \dim E\left(\Delta^{(p)}, \lambda\right) = \sum_{p=0}^{n} (-1)^p \left[\dim E_d^p(\lambda) + \dim E_\delta^p(\lambda)\right]$$

$$= \dim E_d^0(\lambda) + \sum_{p=0}^{n-1} (-1)^p \left[\dim E_\delta^p(\lambda) - \dim E_d^{p+1}(\lambda)\right] + (-1)^n \dim E_\delta^n(\lambda) = 0.$$

For $\lambda - 0$, it follows that

$$\dim E\left(\Delta^{(p)}, 0\right) = \dim H^p(M) = b_p(M), \quad 0 \le p \le n,$$

(see §1), so that

$$\sum_{p=0}^{n} (-1)^p \dim E\left(\Delta^{(p)}, 0\right) = \sum_{p=0}^{n} (-1)^p b_p(M) = \chi(M).$$

Consequently, the following equality holds:

$$\sum_{p=0}^{n} (-1)^p \dim E\left(\Delta^{(p)}, \lambda\right) = \begin{cases} 0 & \text{for } \lambda \ne 0, \\ \chi(M) & \text{for } \lambda = 0. \end{cases}$$

With all of these in mind one gets

Corollary 2.3. (**McKean and Singer** [1967]) The following equality holds:

$$\sum_{p=0}^{n}(-1)^p Z_{(M,g)}^{(p)}(t)=\chi(M)$$

for all $t\in(0,+\infty)$.

Indeed,

$$\begin{aligned}\sum_{p=0}^{n}(-1)^p Z_{(M,g)}^{(p)}(t)&=\sum_{p=0}^{n}(-1)^p\sum_{\lambda}\dim E\left(\Delta^{(p)},\lambda\right)e^{-\lambda t}\\&=\sum_{\lambda}\left[\sum_{p=0}^{n}(-1)^p\dim E\left(\Delta^{(p)},\lambda\right)\right]e^{-\lambda t}\\&=\sum_{p=0}^{n}(-1)^p\dim E\left(\Delta^{(p)},0\right)+\sum_{\lambda\neq 0}\left[\sum_{p=0}^{n}(-1)^p\dim E\left(\Delta^{(p)},\lambda\right)\right]e^{-\lambda t}=\chi(M).\end{aligned}$$

The formula contained in Corollary 2.3 gives the expression for the Euler characteristic of M in terms of eigenvalues of the $\Delta^{(p)}$, $0\le p\le n$, which are also global invariants of (M,g).

Remark 2.3. If the dimension of M is even, i.e. $n=2m$, then $\dim E\left(\Delta^{(m)},\lambda\right)$ is also even for $\lambda\neq 0$ (see Millman [1980]).

Indeed, using (2.14) we have successively

$$\begin{aligned}0&=\sum_{p\text{ even}}\dim E\left(\Delta^{(p)},\lambda\right)-\sum_{p\text{ odd}}\dim E\left(\Delta^{(p)},\lambda\right)\\&=\sum_{\substack{p\text{ even}\\p<m}}\dim E\left(\Delta^{(p)},\lambda\right)+\sum_{\substack{p\text{ even}\\p>m}}\dim E\left(\Delta^{(p)},\lambda\right)-\sum_{\substack{p\text{ odd}\\p<m}}\dim E\left(\Delta^{(p)},\lambda\right)\\&-\sum_{\substack{p\text{ odd}\\p>m}}\dim E\left(\Delta^{(p)},\lambda\right)+(-1)^m\dim E\left(\Delta^{(m)},\lambda\right).\end{aligned}$$

Since the star Hodge operator maps eigenspaces of $\Delta^{(p)}$ into eigenspaces of $\Delta^{(n-p)}$ and vice-versa, without changing the eigenvalue, we have

$$\mathbf{Spec}^{(p)}(M,g)=\mathbf{Spec}^{(n-p)}(M,g),\quad 0\le p\le n,$$

which implies that

$$\dim E\left(\Delta^{(m)},\lambda\right)=2(-1)^{m+1}\left[\sum_{\substack{p\text{ even}\\p<m}}\dim E\left(\Delta^{(p)},\lambda\right)-\sum_{\substack{p\text{ odd}\\p<m}}\dim E\left(\Delta^{(p)},\lambda\right)\right].$$

The next result is another mini-max principle.

Theorem 2.4. (**Dodziuk** [1982]). The eigenvalues

$$0<\lambda_1^{(p)}\le\lambda_2^{(p)}\le\cdots$$

of the restriction of the Hodge-de Rham operator $\Delta^{(p)}$ to the space of exact forms of degree

$p, 0 < p \le n$, can be computed by

$$\lambda_k^{(p)} = \inf_{V_k} \sup_{\omega \in V_k \setminus \{0\}} \left\{ \frac{\|\omega\|^2}{\|\theta\|^2} \mid d^{p-1}\theta = \omega \right\}, \tag{2.15}$$

where V_k ranges over all k-dimensional linear subspaces of the space of all smooth exact forms of degree p on M.

Proof. First, notice that taking the supremum in (2.15) can be done in two stages. For each ω we choose $\theta \in A^{p-1}(M)\setminus\{0\}$ to maximize the quotient $\frac{\|\omega\|^2}{\|\theta\|^2}$. One takes any θ such that $d^{p-1}\theta = \omega$ with Hodge-de Rham decomposition

$$\theta = H^{(p-1)}(\theta) + d^{p-2}\alpha + \delta^p \beta$$

and set $\theta_0 = \delta^p\beta \in \delta^p A^p(M)$. Thus

$$\inf_{V_k} \sup_{\omega \in V_k \setminus \{0\}} \left\{ \frac{\|\omega\|^2}{\|\theta\|^2} \mid d^{p-1}\theta = \omega \right\} = \inf_{V_k} \sup_{\omega \in V_k \setminus \{0\}} \left\{ \frac{\|\omega\|^2}{\|\theta_0\|^2} \right\} = \inf_{W_k} \sup_{\omega \in W_k \setminus \{0\}} \left\{ \frac{\|d^{p-1}\theta_0\|^2}{\|\theta_0\|^2} \right\}, \tag{2.16}$$

where W_k ranges over all k-dimensional linear subspaces of $\delta^p A^p(M)$. Since $\delta^{p-1}\theta_0 = 0$ and $\omega = d^{p-1}\theta = d^{p-1}\theta_0$, one can write

$$\|\omega\|^2 = \|d^{p-1}\theta_0\|^2 = \|d^{p-1}\theta_0\|^2 + \|\delta^{p-1}\theta_0\|^2 .$$

Thus, the right-hand side of (2.16) gives the mini-max characterization of the k-th eigenvalue of the restriction of $\Delta^{(p-1)}$ to co-exact forms, which is the same as the k-th eigenvalue of the restriction of $\Delta^{(p)}$ to exact forms.

Q.E.D.

Note that there are no derivatives of the metric g in the previous characterization of the eigenvalues of $\Delta^{(p)}$.

The following corollary will be used to establish the continuous dependence on the Riemannian metric of the eigenvalues of the Hodge-de Rham operators.

Corollary 2.4. Let g,g' be two Riemannian metrics on the closed, n-dimensional smooth manifold $M, \lambda_k^{(p)}(g)$ [resp. $\lambda_k^{(p)}(g')$], $k \in \mathbb{N}^*$, be the k-th eigenvalue of the restriction of $\Delta_g^{(p)}$ (resp. $\Delta_{g'}^{(p)}$) to exact forms of degree p, where $\Delta_g^{(p)}$ (resp. $\Delta_{g'}^{(p)}$) is the Hodge-de Rham operator on p-forms corresponding to the Riemannian metric g (resp. g'). If

$$c\frac{\|d^{p-1}\theta\|_{g'}^2}{\|\theta\|_{g'}^2} \le \frac{\|d^{p-1}\theta\|_g^2}{\|\theta\|_g^2} \quad \text{for every } \theta \in A^{p-1}(M) \text{ with } d^{p-1}\theta \ne 0,$$

where $\| \ \|_g$ (resp. $\| \ \|_{g'}$) denotes the canonical L^2-norm on the real vector space of smooth differential forms of a given degree on M induced by g (resp. g') (see Chapter 2, §2, C) and $c > 0$ is a constant, then

$$c\lambda_k^{(p)}(g') \leq \lambda_k^{(p)}(g).$$

Proof. By Theorem 2.4,

$$\lambda_k^{(p)}(g) = \inf_{V_k} \sup_{\omega \in V_k \setminus \{0\}} \left\{ \frac{\|\omega\|_g^2}{\|\theta\|_g^2} \mid d^{p-1}\theta = \omega \right\}.$$

By assumption,

$$c\frac{\|\omega\|_{g'}^2}{\|\theta\|_{g'}^2} \leq \frac{\|\omega\|_g^2}{\|\theta\|_g^2}$$

for every $\omega = d^{p-1}\theta$. Therefore,

$$c \sup\left(\frac{\|\omega\|_{g'}^2}{\|\theta\|_{g'}^2} \right) \leq \sup\left(\frac{\|\omega\|_g^2}{\|\theta\|_g^2} \right),$$

and

$$c\lambda_k^{(p)}(g') = c \inf_{V_k} \sup_{\omega \in V_k \setminus \{0\}} \left\{ \frac{\|\omega\|_{g'}^2}{\|\theta\|_{g'}^2} \right\} \leq \inf_{V_k} \sup_{\omega \in V_k \setminus \{0\}} \left\{ \frac{\|\omega\|_g^2}{\|\theta\|_g^2} \right\} = \lambda_k^{(p)}(g).$$

Q.E.D.

3. A CONTINUITY PROPERTY OF THE EIGENVALUES OF THE HODGE-DE RHAM OPERATORS

In this section we shall prove that some results from Chapter 3, §4, can be extended to the Hodge-de Rham operators (see Definition 2.3 of Chapter 2). More precisely, we shall discuss the continuous dependence on the Riemannian metric of the eigenvalues of these operators and some of its consequences.

One standard way of deriving this continuity property of the eigenvalues is based on Kato's perturbation theory (see also Remark 3.1). Using Corollary 2.3, which is a consequence of the mini-max principle in the framework of smooth differential forms formulated in Theorem 2.4 (see Dodziuk [1982]), we shall give a direct elementary proof of it (see Puta [1993], Craioveanu and Puta [1996]).

Let M be a closed, connected, n-dimensional smooth oriented manifold, $A^p(M)$ the vector space of real smooth differential forms of degree p on M, $0 \leq p \leq n$, and $\mathcal{M}$ the space of smooth Riemannian metrics on M. For each $g \in \mathcal{M}$ and $p \in \mathbb{N}, 0 \leq p \leq n$, let $\Delta_g^{(p)} : A^p(M) \to A^p(M)$ be the Hodge-de Rham operator, that is $\Delta_g^{(p)} = d^{p-1}\delta_g^p + \delta_g^{p+1}d^p$, where d is the exterior derivative and δ_g the formal adjoint of d, whose construction depends essentially on the Riemannian metric g (see Chapter 2, §2). We have noticed in §1 that the eigenvalue problem

$$\Delta_g^{(p)}\omega = \lambda\omega$$

has a complete orthonormal system of smooth eigenforms in the Hilbert space $L^2A^p(M)$ of square integrable forms of degree p and the corresponding spectrum $\left\{\lambda_k^{(p)}(g) \mid k \in \mathbb{N}\right\}$ has the property

$$0 \le \lambda_0^{(p)}(g) \le \lambda_1^{(p)}(g) \le \cdots,$$

where $\lambda_k^{(p)}(g) \to \infty$ as $k \to \infty$.

Next we shall regard each eigenvalue $\lambda_k^{(p)}(g)$ of $\Delta_g^{(p)}$, $k \in \mathbb{N}, 0 \le p \le n$, as a real function of $g \in \mathcal{M}$, i.e.

$$\lambda_k^{(p)} : \mathcal{M} \to \mathbb{R}.$$

Assume $g, g' \in \mathcal{M}$ are such that

$$\sigma g' \le g \le \tau g' \tag{3.1}$$

for some constants $\sigma, \tau > 0$. Corollary 2.4 of the previous Section allows us to estimate the eigenvalues of $\Delta_g^{(p)}$ in terms of those of $\Delta_{g'}^{(p)}$.

Proposition 3.1. Assume $g, g' \in \mathcal{M}$ satisfy (3.1). Let $\left\{\lambda_k^{(p)}(g) \middle| \, k \in \mathbb{N}^*\right\}$ (respectively $\left\{\lambda_k^{(p)}(g') \middle| \, k \in \mathbb{N}^*\right\}$) be the sequence of eigenvalues of the restriction of $\Delta_g^{(p)}$ (respectively $\Delta_{g'}^{(p)}$) to exact forms of degree p. Then

$$\frac{1}{\sigma}\left(\frac{\sigma}{\tau}\right)^{\frac{n}{2}+p} \lambda_k^{(p)}(g') \le \lambda_k^{(p)}(g) \le \frac{1}{\tau}\left(\frac{\tau}{\sigma}\right)^{\frac{n}{2}+p} \lambda_k^{(p)}(g'). \tag{3.2}$$

Proof. Indeed, an easy calculation shows that, if ω has degree q,

$$\tau^{-q}\sigma^{\frac{n}{2}} \|\omega\|_{g'}^2 \le \|\omega\|_g^2 \le \sigma^{-q}\tau^{\frac{n}{2}} \|\omega\|_{g'}^2,$$

where $\| \; \|_g$ (respectively $\| \; \|_{g'}$) is as in Corollary 2.4. As a consequence, if $\theta \in A^{p-1}(M)$ is such that $d^{p-1}\theta \ne 0$, then

$$\frac{1}{\sigma}\left(\frac{\sigma}{\tau}\right)^{\frac{n}{2}+p} \frac{\|d^{p-1}\theta\|_{g'}^2}{\|\theta\|_{g'}^2} \le \frac{\|d^{p-1}\theta\|_g^2}{\|\theta\|_g^2} \le \frac{1}{\tau}\left(\frac{\tau}{\sigma}\right)^{\frac{n}{2}+p} \frac{\|d^{p-1}\theta\|_{g'}^2}{\|\theta\|_{g'}^2}.$$

(3.2) follows from this and Corollary 2.4.

Q.E.D.

Proposition 3.1 tells us that the eigenvalues of $\Delta_g^{(p)}$ vary continuously when the Riemannian metric g varies continuously in the C^0-topology on $\mathcal{M}$, i.e. $\lambda_k^{(p)} : \mathcal{M} \to \mathbb{R}$ is continuous with respect to the C^0-topology on $\mathcal{M}$. Moreover, we have the following result.

Proposition 3.2. For each $p = 1,\ldots,n$ and for each $k \in \mathbb{N}^*$, the eigenvalue function

$$\lambda_k^{(p)} : \mathcal{M} \to \mathbb{R}$$

is continuous with respect to the C^∞-topology on $\mathcal{M}$, where the sequence $\left\{\lambda_k^{(p)} \middle| \, k \in \mathbb{N}^*\right\}$

is as in Proposition 3.1.

Proof. First recall that the C^∞-topology on $\mathcal{M}$ is in fact a complete metric topology, i.e. it is generated by the distance $d = d' + d''$ on $\mathcal{M}$ defined in Chapter 3, §4, such that $(\mathcal{M},d)$ is complete. For any fixed $\varepsilon > 0$ and $g \in \mathcal{M}$, let $V_\varepsilon(g)$ be the set $\{g' \in \mathcal{M} \mid d''(g,g') < \varepsilon\}$. Notice that

$$\{g' \in \mathcal{M} \mid d(g,g') < \varepsilon\} \subset V_\varepsilon(g).$$

Let $(x^1,\ldots,x^n)$ be a local coordinate system on an open subset U of M. Then for any $g' \in V_\varepsilon(g)$ and any $\theta \in A^p(M)$ with $\operatorname{supp}(\theta) \subset U$, one obtains

$$e^{-\left(\frac{n}{2}+p\right)\varepsilon}\|\theta\|_{g'}^2 \le \|\theta\|_g^2 \le e^{\left(\frac{n}{2}+p\right)\varepsilon}\|\theta\|_{g'}^2, \tag{3.3}$$

where the norm $\| \ \|_g$ (respectively $\| \ \|_{g'}$) is as in Corollary 2.3. Indeed, taking into account the definition of d'', for any $g' \in V_\varepsilon(g)$ the covariant components g_{ij} (respectively g'_{ij}) of g (respectively g') satisfy the inequalities

$$e^{-\varepsilon} g'_{ij} \le g_{ij} \le e^{\varepsilon} g'_{ij},$$

so that if

$$\theta = \sum_{i_1<\cdots<i_p} \theta_{i_1\ldots i_p} dx^{i_1} \wedge \ldots \wedge dx^{i_p},$$

$$\begin{aligned}
\|\theta\|_g^2 &= \int_M \theta \wedge S^{(p)}\theta = \int_M \sum_{k_1,\ldots,k_p} \sum_{i_1<\cdots<i_p} g_{i_1k_1}\cdots g_{i_pk_p}\theta^{k_1\cdots k_p}\theta^{i_1\cdots i_p}\sqrt{\det(g_{ij})}\,dx^1 \wedge \ldots \wedge dx^n \\
&\le e^{p\varepsilon}\int_M \sum_{k_1,\ldots,k_p} \sum_{i_1<\cdots<i_p} g'_{i_1k_1}\cdots g'_{i_pk_p}\theta^{k_1\cdots k_p}\theta^{i_1\cdots i_p}\sqrt{\det(g_{ij})}\,dx^1 \wedge \cdots \wedge dx^n \\
&\le e^{\left(\frac{n}{2}+p\right)\varepsilon}\int_M \sum_{k_1,\ldots,k_p} \sum_{i_1<\cdots<i_p} g'_{i_1k_1}\cdots g'_{i_pk_p}\theta^{k_1\cdots k_p}\theta^{i_1\cdots i_p}\sqrt{\det(g'_{ij})}\,dx^1 \wedge \cdots \wedge dx^n \\
&= e^{\left(\frac{n}{2}+p\right)\varepsilon}\|\theta\|_{g'}^2,
\end{aligned}$$

and similarly

$$\|\theta\|_g^2 \ge e^{-\left(\frac{n}{2}+p\right)\varepsilon}\|\theta\|_{g'}^2.$$

But the inequalities (3.3) are valid-via-a smooth partition of unity on M-for any $\theta \in A^p(M)$. Consequently, (3.3) implies that

$$e^{-(n+2p-1)\varepsilon}\frac{\|d^{p-1}\theta\|_{g'}^2}{\|\theta\|_{g'}^2} \le \frac{\|d^{p-1}\theta\|_g^2}{\|\theta\|_g^2} \le e^{(n+2p-1)\varepsilon}\frac{\|d^{p-1}\theta\|_{g'}^2}{\|\theta\|_{g'}^2}$$

for any $\theta \in A^{p-1}(M)$ such that $d^{p-1}\theta \ne 0$, and using Corollary 2.4 one gets

$$e^{-(n+2p-1)\varepsilon}\lambda_k^{(p)}(g')\leq\lambda_k^{(p)}(g)\leq e^{(n+2p-1)\varepsilon}\lambda_k^{(p)}(g') \tag{3.4}$$

for each $p = 1,\dots,n$ and for each $k\in\mathbb{N}^*$. In particular, the inequalities (3.4) imply the continuity of the real functions $\lambda_k^{(p)}, k\in\mathbb{N}^* : g'\in V_\varepsilon(g)$ implies $\left|\lambda_k^{(p)}(g')-\lambda_k^{(p)}(g)\right|\leq\left[e^{(n+2p-1)\varepsilon}-1\right]\lambda_k^{(p)}(g)$ for each $k\in\mathbb{N}^*$.

Q.E.D.

Remark 3.1. Let us consider the classical formulae

$$\sum_{k=1}^{l}\left[\lambda_k^{(p)}(g)\right]^m=\frac{1}{2\pi i}Trace\int_\Gamma \lambda^m R^{(p)}(g,\lambda)d\lambda$$

expressing the symmetric functions $\sum_{k=1}^{l}\left[\lambda_k^{(p)}(g)\right]^m$ by mean of the resolvent

$$R^{(p)}(g,\lambda)=\left(\Delta_g^{(p)}-\lambda I\right)^{-1},$$

where Γ is a loop in $\mathbb{C}$ enclosing exactly the $\lambda_k^{(p)}$'s, where $k=1,\dots,l$. Since any symmetric function is a polynomial expression of power sums, Proposition 3.2 is a consequence of the continuity of $R^{(p)}$: there exists an open neighbourhood $\mathcal{O}$ of (g_0,λ_0) in $\mathcal{M}\times\mathbb{C}$ such that $(g,\lambda)\mapsto R^{(p)}(g,\lambda)=\left(\Delta_g^{(p)}-\lambda I\right)^{-1}$, is a bounded continuous family on $\mathcal{O}$. This can be proved using Kato's perturbation theory (see Kato [1976]). This gives another proof of the continuity of the real functions $\lambda_k^{(p)},\ k\in\mathbb{N}^*$.

It is possible to generalize Theorem 4.1 of Chapter 3 and Proposition 3.2 as follows.

Let M be a closed smooth manifold, $g(t)$ a smooth 1-parameter family of Riemannian metrics on M, E a complex vector bundle over M and $h(t)$ a smooth 1-parameter family of Hermitian metrics on E. Let $D(t):C^\infty(E)\to C^\infty(E)$ be a smooth 1-parameter family of partial differential operators.

Proposition 3.3. Assume that $D(t)$ is self-adjoint with respect to the inner product defined by $g(t)$ and $h(t)$ and that $D(t)$ is non-negative. Let

$$\lambda_1(t)\leq\cdots\leq\lambda_j(t)\leq\cdots$$

denote the eigenvalues of $D(t)$ repeated according to multiplicity. Then $\lambda_j(t)$ is a continuous function of t.

Proof. By conjugating the operators $D(t)$ appropriately, one may assume without loss of generality that $h(t) = h(0)$ is constant. Similarly one can assume that $d\mu_{g(t)}=d\mu_{g(0)}$ is constant. This simply involves reparametrizing the L^2 spaces under consideration. All separable infinite dimensional L^2 spaces are isometric. To be more specific, let $\dfrac{d\mu_{g(t)}}{d\mu_{g(0)}}=\alpha(t)$. Then $\tilde{D}(t)=\alpha(t)^{\frac{1}{2}}D(t)\alpha(t)^{-\frac{1}{2}}$ is self-adjoint with respect to the inner product induced by $d\mu_{g(0)}$:

$$\begin{aligned}\int_M \left(\tilde{D}(t)\sigma_1 | \sigma_2\right) d\mu_{g(0)} &= \int_M \left(D(t)\alpha(t)^{-\frac{1}{2}}\sigma_1 \middle| \alpha(t)^{-\frac{1}{2}}\sigma_2\right)\alpha(t)\, d\mu_{g(0)} \\ &= \int_M \left(D(t)\alpha(t)^{-\frac{1}{2}}\sigma_1 \middle| \alpha(t)^{-\frac{1}{2}}\sigma_2\right) d\mu_{g(t)} \\ &= \int_M \left(\alpha(t)^{-\frac{1}{2}}\sigma_1 \middle| D(t)\alpha(t)^{-\frac{1}{2}}\sigma_2\right)\alpha(t)\, d\mu_{g(0)} \\ &= \int_M \left(\sigma_1 | \tilde{D}(t)\sigma_2\right) d\mu_{g(0)}\end{aligned}$$

for all $\sigma_1, \sigma_2 \in C^\infty(E)$.

Note that the resulting operators $\tilde{D}(t)$ still form a smooth 1-parameter family and that the eigenvalues are unchanged by conjugation.

Similarly one can straighten out $h(t)$ by conjugating by an appropriate matrix.

Let $\| \ \|_s$ be the Sobolev norms. If k is the order of the operator and if one restricts the domain of t to a compact subset, one can find a constant $C > 0$ so that

$$\|\sigma\|_k \le C\left(\|D(t)\sigma\|_0 + \|\sigma\|_0\right)$$

for all $\sigma \in C^\infty(E)$. The constant C is independent of t. This is Gärding's inequality [see also Chapter 3, Remark 1.4 (ii) for the case of the Laplace-Beltrami operator]. Let $\| \ \|_{k,0}$ be the Sobolev operator norm (for details, see Gilkey [1995]). Given $\varepsilon > 0$, there exists $\delta > 0$ so that if $|t - t_0| < \delta$, then $\|D(t) - D(t_0)\|_{k,0} < \varepsilon$. If $\|\sigma\|_0 = 1$ and if $|t - t_0| < \delta$, then

$$\begin{aligned}\langle D(t)\sigma, \sigma\rangle &= \langle (D(t) - D(t_0))\sigma, \sigma\rangle + \langle D(t_0)\sigma, \sigma\rangle \\ &\le \|(D(t) - D(t_0))(\sigma)\|_0 + \langle D(t_0)\sigma, \sigma\rangle \\ &\le \|D(t) - D(t_0)\|_{k,0}\|\sigma\|_k + \langle D(t_0)\sigma, \sigma\rangle \\ &\le \varepsilon C\left(\|D(t_0)\sigma\|_0 + \|\sigma\|_0\right) + \langle D(t_0)\sigma, \sigma\rangle \\ &= \varepsilon C + \varepsilon C\|D(t_0)\sigma\|_0 + \langle D(t_0)\sigma, \sigma\rangle.\end{aligned}$$

If one takes infimums over σ, the left hand side is $\lambda_1(t)$; this is attained when σ is the first eigenfunction of $D(t)$. On the right hand side, a good choice for an infimum occurs when σ is the first eigenfunction of $D(t_0)$. This leads to the estimate

$$\lambda_1(t) \le \varepsilon C + (1 + \varepsilon C)\lambda_1(t_0) \tag{3.5}$$

provided that $|t - t_0| \le \delta$. Consequently we also have

$$\lambda_1(t_0) \le \varepsilon C + (1 + \varepsilon C)\lambda_1(t)$$

so that

$$\lambda_1(t) \geq \frac{\lambda_1(t_0) - \varepsilon C}{1 + \varepsilon C}. \tag{3.6}$$

Combining the estimates (3.5) and (3.6) one gets

$$\frac{-\varepsilon C\left[1+\lambda_1(t_0)\right]}{1+\varepsilon C} \leq \lambda_1(t) - \lambda_1(t_0) \leq \varepsilon C\left[1+\lambda_1(t_0)\right],$$

which shows that $\lambda_1(t)$ is continuous at $t = t_0$.

More generally, let $\sigma_1, \ldots, \sigma_j$ be an orthonormal system of $C^\infty(E)$ and consider $\langle D(t)\sigma_1, \sigma_1\rangle + \cdots + \langle D(t)\sigma_j, \sigma_j\rangle$. Taking infimums this gives $\lambda_1(t) + \cdots + \lambda_j(t)$. The same argument then shows that these vary continuously and hence by subtraction $\lambda_j(t)$ varies continuously.

Q.E.D.

The elegant argument given above was shown to us by Professor P.B. Gilkey (University of Oregon) (see also Park [1999]).

Let us point out some consequences of Proposition 3.2. The proof of Proposition 3.2 shows that for each real number $\varepsilon > 0$ and each $g, g' \in \mathcal{M}$, the inequality $d(g,g') < \varepsilon$ implies that

$$e^{-(n+2p-1)\varepsilon} \leq \frac{\lambda_k^{(p)}(g)}{\lambda_k^{(p)}(g')} \leq e^{(n+2p-1)\varepsilon}$$

for each $k \in \mathbb{N}^*$, that is if the Riemannian metrics g and g' are close to each other with respect to the distance d, then the ratio $\lambda_k^{(p)}(g) \big/ \lambda_k^{(p)}(g')$ is close to 1 uniformly in $k \in \mathbb{N}^*$. This leads immediately to the following corollary.

Corollary 3.1. For each $p = 1,\ldots,n$ and for each $k \in \mathbb{N}^*$, the multiplicity $m_k^{(p)} = \#\left\{ j \,\middle|\, \lambda_j^{(p)}(g) = \lambda_k^{(p)}(g)\right\}$ of the eigenvalue $\lambda_k^{(p)}(g)$ depends upper-semicontinuously on $g \in \mathcal{M}$: for each $g \in \mathcal{M}$ there exists $\varepsilon > 0$ such that $d(g,g') < \varepsilon$ implies

$$m_k^{(p)}(g') \leq m_k^{(p)}(g).$$

For each $p = 1,\ldots,n$ and $k \in \mathbb{N}^*$, let $\mathcal{S}_k^{(p)}$ be the set of all Riemannian metrics $g \in \mathcal{M}$ for which the first k eigenvalues of $\Delta_g^{(p)}$ on exact forms of degree p are simple, that is they have multiplicity one, i.e.

$$\mathcal{S}_k^{(p)} = \left\{ g \in \mathcal{M} \,\middle|\, \lambda_1^{(p)}(g) < \cdots < \lambda_k^{(p)}(g)\right\}.$$

Corollary 3.2. For each $p = 1,\ldots,n$ and $k \in \mathbb{N}^*$, $\mathcal{S}_k^{(p)}$ is open in $(\mathcal{M},d)$.

Proof. It is sufficient to show that for each $\mathcal{S}_k^{(p)}$ there exists $\varepsilon > 0$ such that $V_\varepsilon(g) \subset \mathcal{S}_k^{(p)}$. Let

$$\varepsilon' = \min\left\{\lambda_2^{(p)}(g) - \lambda_1^{(p)}(g), \ldots, \lambda_k^{(p)}(g) - \lambda_{k-1}^{(p)}(g)\right\} > 0$$

and let us choose $\varepsilon > 0$ so small such that

$$\varepsilon'\left[2\lambda_k^{(p)}(g)\right]^{-1} > e^{(n+2p-1)\varepsilon} - 1.$$

By the inequalities (3.4), for any $g' \in V_\varepsilon(g)$ and $j = 1, \ldots, k-1$, one obtains

$$\begin{aligned}\varepsilon' &\leq \lambda_{j+1}^{(p)}(g) - \lambda_j^{(p)}(g) \leq \left|\lambda_{j+1}^{(p)}(g) - \lambda_{j+1}^{(p)}(g')\right| + \left|\lambda_{j+1}^{(p)}(g') - \lambda_j^{(p)}(g')\right| + \left|\lambda_j^{(p)}(g') - \lambda_j^{(p)}(g)\right| \\ &\leq \left[e^{(n+2p-1)\varepsilon} - 1\right]\left[\lambda_{j+1}^{(p)}(g) + \lambda_j^{(p)}(g)\right] + \left|\lambda_{j+1}^{(p)}(g') - \lambda_j^{(p)}(g')\right| \\ &\leq 2\left[e^{(n+2p-1)\varepsilon} - 1\right]\lambda_k^{(p)}(g) + \left|\lambda_{j+1}^{(p)}(g') - \lambda_j^{(p)}(g')\right|,\end{aligned}$$

so that

$$0 < \varepsilon' - 2\left[e^{(n+2p-1)\varepsilon} - 1\right]\lambda_k^{(p)}(g) \leq \left|\lambda_{j+1}^{(p)}(g') - \lambda_j^{(p)}(g')\right|,$$

for $j = 1,\ldots,k-1$, which implies that $g' \in \mathcal{S}_k^{(p)}$ and therefore $V_\varepsilon(g) \subset \mathcal{S}_k^{(p)}$.

Q.E.D.

4. ASYMPTOTIC EXPANSION FOR THE TRACE OF THE HEAT *P*-KERNEL AND SPECTRAL GEOMETRY

Since it is difficult to study $\mathbf{Spec}^{(p)}(M, g)$ directly, instead one introduces certain functions of the eigenvalues of $\Delta^{(p)}$, which can be used to extract more geometric information (see also Chapter 4, §1). Some useful such functions having applications to spectral geometry are the heat *p*-coefficients, which are discussed below.

Let (M,g) be a closed, oriented, *n*-dimensional smooth Riemannian manifold and $E^{(p)}$ the heat *p*-kernel on (M,g), $0 \leq p \leq n$ (see Definition 1.1). For $(x,y,t) \in M \times M \times (0,+\infty)$, $E^{(p)}(x,y,t)$ is a linear mapping from $\Lambda^p T_y^* M$ into $\Lambda^p T_x^* M$. Let $\Delta^{(p)} : A^p(M) \to A^p(M), 0 \leq p \leq n$, be the Hodge-de Rham operators, $(x^1,\ldots,x^n)$ be a system of local coordinates on M and dx^I, $I = \{1 \leq i_1 < \cdots < i_p \leq n\}$ a local frame for $\Lambda T^* M$. Let g_{ij} be the components of the metric tensor and let g^{ij} denote the inverse matrix, that is

$$\left(dx^i \mid dx^j\right) = g^{ij}$$

(see Chapter 2, §2, C). With respect to this frame, $\Delta^{(p)}$ can be expressed as

$$\begin{aligned}\left(\Delta^{(p)}\omega\right)_{i_1\cdots i_p} &= -\sum_{j,k=1}^{n} g^{jk}\left(\nabla_{\partial/\partial x^j}\nabla_{\partial/\partial x^k}\,\omega\right)_{i_1\cdots i_p} + \sum_{a=1}^{p}\sum_{j,k=1}^{n}(-1)^a g^{jk}\left(c_2 R\right)_{j i_a}\omega_{k i_1\cdots\hat{i}_a\cdots i_p} + \\ &+ \sum_{1\leq a<b\leq p}(-1)^{a+b}\sum_{j,k,l,m=1}^{n} R_{j i_a k i_b}\, g^{jl} g^{km}\,\omega_{l m i_1\cdots\hat{i}_a\cdots\hat{i}_b\cdots i_p},\end{aligned}$$

where

$$\left(\nabla_{\partial/\partial x^j}\nabla_{\partial/\partial x^k}\omega\right)_{i_1\cdots i_p} =$$

$$= \frac{\partial^2\omega_{i_1\cdots i_p}}{\partial x^j\partial x^k} - \sum_{h=1}^{n}\left[\left(\Gamma^h_{ji_1}\frac{\partial}{\partial x^k}+\Gamma^h_{ki_1}\frac{\partial}{\partial x^j}\right)\omega_{hi_2\cdots i_p}+\cdots+\left(\Gamma^h_{ji_p}\frac{\partial}{\partial x^k}+\Gamma^h_{ki_p}\frac{\partial}{\partial x^j}\right)\omega_{i_1\cdots i_{p-1}h}\right]-$$

$$-\sum_{h=1}^{n}\left(\frac{\partial\Gamma^h_{ki_1}}{\partial x^j}\omega_{hi_2\cdots i_p}+\frac{\partial\Gamma^h_{ki_2}}{\partial x^j}\omega_{i_1h\cdots i_p}+\cdots+\frac{\partial\Gamma^h_{ki_p}}{\partial x^j}\omega_{i_1\cdots i_{p-1}h}\right)+$$

$$+\sum_{h,l=1}^{n}\Gamma^h_{ji_1}\left(\Gamma^l_{kh}\omega_{hi_2\cdots i_p}+\Gamma^l_{ki_2}\omega_{hl\cdots i_p}+\cdots+\Gamma^l_{ki_p}\omega_{hi_2\cdots l}\right)+\cdots+$$

$$+\sum_{h,l=1}^{n}\Gamma^h_{ji_p}\left(\Gamma^l_{ki_1}\omega_{li_2\cdots h}+\Gamma^l_{ki_2}\omega_{i_1l\cdots h}+\cdots+\Gamma^l_{kh}\omega_{i_1\cdots i_{p-1}l}\right),$$

$(c_2R)_{ij}$ (resp. R_{ijkl}) are the components of the Ricci (resp. Riemann) tensor field (see Chapter 1, Proposition 4.4) and Γ^i_{jk} the second Christoffel symbols (see Chapter 1, Proposition 3.1). Thus the leading symbol of $\Delta^{(p)}$ is given by the metric tensor field. The first order symbol is linear in the 1-jets of the metric with coefficients which depend smoothly on the g_{ij}'s. The 0-th order symbol is the sum of a term which is linear in the 2-jets of the metric and a term which is quadratic in the 1-jets of the metric with coefficients which depend smoothly on the g_{ij}'s.

The heat p-kernel $E^{(p)}$ has the following fundamental property, namely there exist linear endomorphisms

$$e_k\left(x,\Delta^{(p)}\right):\Lambda^pT^*_xM\to\Lambda^pT^*_xM,\quad k\in\mathbb{N},$$

which depend functorially on a finite number of jets of the total symbol of $\Delta^{(p)}$ with the property

$$E^{(p)}(x,x,t)\sim\sum_{k=0}^{\infty}t^{\frac{k-n}{2}}e_k\left(x,\Delta^{(p)}\right) \tag{4.1}$$

as t approaches 0, i.e. given any $m\in\mathbb{N}$ there exists $k(m)$ such that

$$\left\|E^{(p)}(.,.,t)-\sum_{k\le k(m)}t^{\frac{k-n}{2}}e_k\left(.,\Delta^{(p)}\right)\right\|_{\infty,m}<C_mt^m$$

for $0<t<1$, where $C_m>0$ is a constant. $\|\ \|_{\infty,m}$ denotes here the usual sup norm of the sum of the natural norms of the iterated covariant derivatives up to order m. Despite of the fact that the $e_k(x,\Delta^{(p)})$' s are defined by choosing a coordinate system for M near $x\in M$ and a local frame for Λ^pT^*M, they are invariantly defined, independent of the coordinate system and local frame for Λ^pT^*M. This last remark follows from (4.1) and the fact that $E^{(p)}(x,x,t)$ does not depend on the coordinate system chosen. The $e_k(x,\Delta^{(p)})$ vanish if k is odd and for e_0, e_2 and e_4 one can give explicit formulas (for the proof of all these details, see Gilkey [1984]).

Let us define the scalar invariants

$$a_k\left(x,\Delta^{(p)}\right)=trace\ e_k\left(x,\Delta^{(p)}\right), \tag{4.2}$$

$k\in\mathbb{N}$. $a_k\left(x,\Delta^{(p)}\right)$ is homogeneous of order k in the components (relative to some local frame) of the jets of the total symbol of $\Delta^{(p)}$. Corollary 1.2, (4.1) and (4.2) leads immediately to

$$trace\left(\mathcal{P}_t^{(p)}\right)=Z_{(M,g)}^{(p)}(t)=\sum_{j=1}^{\infty}e^{-\lambda_j^{(p)}t}=\int_M trace\left[E^{(p)}(x,x,t)\right]d\mu_g(x)\sim$$

$$\sim\sum_{k=0}^{\infty}t^{\frac{k-n}{2}}\int_M a_k\left(x,\Delta^{(p)}\right)d\mu_g(x)=\sum_{k=0}^{\infty}t^{\frac{k-n}{2}}a_k\left(\Delta^{(p)}\right),$$

where the

$$a_k\left(\Delta^{(p)}\right)=\int_M a_k\left(x,\Delta^{(p)}\right)d\mu_g(x)$$

are the integrated invariants. Consequently, we have

Theorem 4.1. As $t\searrow 0$, there is an asymptotic expansion

$$trace\left(\mathcal{P}_t^{(p)}\right)=Z_{(M,g)}^{(p)}(t)\sim\sum_{k=0}^{\infty}a_k\left(\Delta^{(p)}\right)t^{\frac{k-n}{2}}. \tag{4.3}$$

The $a_k\left(\Delta^{(p)}\right)$ vanish if k is odd. If k is even, there exist local invariants $a_k\left(x,\Delta^{(p)}\right)$ of the jets of the total symbol of $\Delta^{(p)}$ such that

$$a_k\left(\Delta^{(p)}\right)=\int_M a_k\left(x,\Delta^{(p)}\right)d\mu_g(x).$$

More generally, if $f\in C^{\infty}(M)$, then

$$trace\left(f\mathcal{P}_t^{(p)}\right)\sim\sum_{k=0}^{\infty}a_k\left(f,\Delta^{(p)}\right)t^{\frac{k-n}{2}},$$

where

$$a_k\left(f,\Delta^{(p)}\right)=\int_M f(x)a_k\left(x,\Delta^{(p)}\right)d\mu_g(x).$$

Moreover, this specifies the $a_k\left(x,\Delta^{(p)}\right)$'s uniquely.

The numerical invariants $a_k\left(\Delta^{(p)}\right)$ are called **the heat *p*-coefficients.**

Remark 4.1. By Theorem 1.3 (iv) it follows that $a_k\left(x,\Delta^{(p)}\right)=a_k\left(x,\Delta^{(n-p)}\right)$ for all $x\in M$ and all $k,p\in\mathbb{N}$, $0\le p\le n$, which implies that $a_k\left(x,\Delta^{(p)}\right)=a_k\left(x,\Delta^{(n-p)}\right)$ for all $k,p\in\mathbb{N}$, $0\le p\le n$.

Remark 4.2. One can prove that the coefficients $a_k\left(\Delta^{(p)}\right)$, $k\ge 2$, of the asymptotic expansion (4.3) are universal expressions in the curvature tensor field R associated to

(M,g) and its covariant differentials $\overset{\circ}{\nabla} R, \nabla^2 R, \ldots$, that is these expressions are integrals of polynomials in $R, \overset{\circ}{\nabla} R, \nabla^2 R, \ldots$ whose coefficients depend only on p and the dimension n of M (see also Theorem 4.2).

Remark 4.3. In §1 we have noticed that the knowledge of $Z^{(p)}_{(M,g)}(t)$ is equivalent to knowing $\boldsymbol{Spec}^{(p)}(\boldsymbol{M},\boldsymbol{g})$. On the other hand, note that the asymptotic expansion (4.3), that is the giving of the $a_k\left(\Delta^{(p)}\right)$'s, $k \in \mathbb{N}$, is not sufficient for the determination of $\boldsymbol{Spec}^{(p)}(\boldsymbol{M},\boldsymbol{g})$. Thus, for all flat 2-dimensional tori, which are flat spaces (see Chapter 2, Definition 4.3), by Remark 4.2 one has $a_k\left(\Delta^{(p)}\right) = 0$ for $k \in \mathbb{N}^*, 0 \le p \le 2$.

Remark 4.4. If ζ is a smooth vector bundle over M, Theorem 4.1 is still valid for any self-adjoint second order partial differential operator $D : C^\infty(\zeta) \to C^\infty(\zeta)$ of Laplace type, that is the leading symbol of D is given by the metric tensor field. For example, let (M,g) as before and

$$A^{even}(M) = \underset{q}{\oplus} A^{2q}(M),\ A^{odd}(M) = \underset{q}{\oplus} A^{2q+1}(M).$$

It is immediate that

$$\Delta^{even} = \underset{q}{\oplus} \Delta^{(2q)} : A^{even}(M) \to A^{even}(M)$$

and

$$\Delta^{odd} = \underset{q}{\oplus} \Delta^{(2q+1)} : A^{odd}(M) \to A^{odd}(M)$$

are of Laplace type.

Combining the asymptotic expansion (4.3) given by Theorem 4.1 and Corollary 2.3 one obtains

Corollary 4.1

$$\sum_{p=0}^{n} (-1)^p a_k\left(\Delta^{(p)}\right) = \begin{cases} \chi(M) & \text{if } k = n, \\ 0 & \text{if } k \ne n. \end{cases}$$

Proof. Indeed, Corollary 2.3 and Theorem 4.1 imply

$$\chi(M) = \sum_{p=0}^{n} (-1)^p Z^{(p)}_{(M,g)}(t) \sim \sum_{k=0}^{\infty} t^{\frac{k-n}{2}} \left[\sum_{p=0}^{n} (-1)^p a_k\left(\Delta^{(p)}\right)\right]$$

as $t \searrow 0$. Since the left hand side does not depend on the parameter t, one concludes the desired formula.

Q.E.D.

We note as an immediate consequence that $\chi(M) = 0$ if n is odd.

The remark that the Euler characteristic $\chi(M)$ for n even is given by the local formula

$$\chi(M) = \sum_{p=0}^{n} (-1)^p \int_M a_n\left(x, \Delta^{(p)}\right) d\mu_g(x)$$

has interesting consequences. For example, one has:

Corollary 4.2. Let M_1, M_2 be n-dimensional smooth manifolds and assume that n is even. If $\Phi: M_1 \to M_2$ is a finite smooth covering projection with fiber F, then

$$\chi(M_1) = |F|\chi(M_2),$$

that is the Euler characteristic is multiplicative under finite coverings.

Proof. We can choose the Riemannian metric g_1 on M_1 to be the pull-back of a Riemannian metric g_2 on M_2, so that Φ is a Riemannian covering (see Chapter 1, Example 2.8). Then

$$a_n\left(x, \Delta_{g_1}^{(p)}\right) = a_n\left(\Phi(x), \Delta_{g_2}^{(p)}\right)$$

since a_n is locally defined. This implies

$$\chi(M_1) = \sum_{p=0}^{n}(-1)^p \int_{M_1} a_n\left(x, \Delta_{g_1}^{(p)}\right) d\mu_{g_1}(x)$$

$$= |F| \sum_{p=0}^{n}(-1)^p \int_{M_2} a_n\left(\Phi(x), \Delta_{g_2}^{(p)}\right) d\mu_{g_2}(\Phi(x)) = |F|\chi(M_2).$$

Q.E.D.

Remark 4.5. Let $M = M_1 \# M_2$ be the connected sum of the connected n-dimensional smooth manifolds M_1 and M_2 without boundary. M is the manifold formed by gluing together $M_1 \setminus D_1$ and $M_2 \setminus D_2$ by a diffeomorphism $\partial D_1 \cong \partial D_2$, where $D_1 \subset M_1$ and $D_2 \subset M_2$ are n-disks, and $\partial D_1, \partial D_2$ their boundaries. Assume that n is even. Using the additivity of the $e_k(x,)$, i.e.

$$e_k\left(x, \Delta_1^{(p)} \oplus \Delta_2^{(p)}\right) = e_k\left(x, \Delta_1^{(p)}\right) \oplus e_k\left(x, \Delta_2^{(p)}\right)$$

and that $\chi\left(S^n\right) = 2$, one can show that

$$\chi(M) = \chi(M_1) + \chi(M_2) - 2$$

(see Gilkey [1984]).

Corollary 4.3. Let $c \neq 0$ be a real constant and $\Delta_{c^2 g}^{(p)}$ (respectively $\Delta_g^{(p)}$) the Hodge-de Rham operator corresponding to the Riemannian metric $c^2 g$ (resp. g). Then

$$a_k\left(x, \Delta_{c^2 g}^{(p)}\right) = c^{-k} a_k\left(x, \Delta_g^{(p)}\right), \quad k \in \mathbb{N}.$$

Proof. Since

$$\Delta_{c^2 g}^{(p)} = c^{-2} \Delta_g^{(p)}$$

we conclude that

$$e^{-t\tilde{\Delta}_{c^2 g}^{(p)}} = e^{-tc^{-2}\tilde{\Delta}_g^{(p)}}.$$

Let $f \in C^\infty(M)$. We use Theorem 4.1 to obtain:

$$c^n \sum_k t^{\frac{k-n}{2}} \int_M f(x) a_k\left(x, \Delta_{c^2 g}^{(p)}\right) d\mu_g(x) = \sum_k t^{\frac{k-n}{2}} \int_M f(x) a_k\left(x, \Delta_{c^2 g}^{(p)}\right) d\mu_{c^2 g}(x) \sim$$

$$\sim trace\left(fe^{-t\tilde{\Delta}_{c^2 g}^{(p)}} \right) = trace\left(fe^{-tc^{-2}\tilde{\Delta}_g^{(p)}} \right) \sim \sum_k c^{n-k} t^{\frac{k-n}{2}} \int_M f(x) a_k\left(x, \Delta_g^{(p)}\right) d\mu_g(x).$$

We equate coefficients in the asymptotic expansions to derive

$$\int_M f(x) a_k\left(x, \Delta_{c^2 g}^{(p)}\right) d\mu_g(x) = c^{-k} \int_M f(x) a_k\left(x, \Delta_g^{(p)}\right) d\mu_g(x).$$

Since f was arbitrary, the desired equality follows.

Q.E.D.

The following theorem is due to Patodi [1970] (see also Gilkey [1975]). We give it without proof (see also Remark 4.2).

Theorem 4.2. The first three terms in the asymptotic expansion (4.3) are given for general p, $0 \le p \le n$, by:

(i) $a_0\left(\Delta^{(p)}\right) = (4\pi)^{-\frac{n}{2}} \binom{n}{p} \operatorname{Vol}(M, g);$

(ii) $a_2\left(\Delta^{(p)}\right) = -\dfrac{(4\pi)^{-\frac{n}{2}}}{6} \left[\binom{n}{p} - 6\binom{n-2}{p-1} \right] \displaystyle\int_M \tau \, d\mu_g;$

(iii)
$$a_4\left(\Delta^{(p)}\right) = \frac{(4\pi)^{-\frac{n}{2}}}{360} \int_M \left\{ \left[2\binom{n}{p} - 30\binom{n-2}{p-1} + 180\binom{n-4}{p-2} \right] |R|^2 + \right.$$
$$+ \left[-2\binom{n}{p} + 180\binom{n-2}{p-1} - 720\binom{n-4}{p-2} \right] |c_2 R|^2 +$$
$$\left. + \left[5\binom{n}{p} - 60\binom{n-2}{p-1} + 180\binom{n-4}{p-2} \right] \tau^2 \right\} d\mu_g,$$

where

$$\binom{n}{p} = \begin{cases} \dfrac{n!}{p!(n-p)!} & \text{for } 0 \le p \le n, \\ 0 & \text{for } p < 0 \text{ or } p > n. \end{cases}$$

Corollary 4.4. (**Gauss-Bonnet theorem**) Let M be a closed oriented surface and g a smooth Riemannian metric on it. Then

$$\int_M K \, d\mu_g = 2\pi \chi(M),$$

where K is the Gaussian curvature of (M,g).

Proof. First note that $a_2\left(\Delta^{(0)}\right) = a_2\left(\Delta^{(2)}\right)$ (see Remark 4.1). By Corollary 4.1 and Theorem 4.2, one gets

$$\chi(M)=\sum_{p=0}^{2}(-1)^p a_2\left(\Delta^{(p)}\right)=2a_2\left(\Delta^{(0)}\right)-a_2\left(\Delta^{(1)}\right)$$
$$=-\frac{1}{12\pi}\int_M \tau d\mu_g-\frac{1}{6\pi}\int_M \tau d\mu_g=-\frac{1}{4\pi}\int_M \tau d\mu_g,$$

where $\tau : M \to \mathbb{R}$ is the scalar curvature. Since $\tau=-2K$ (see Chapter 1, Example 4.6), one gets the desired equality.

Q.E.D.

The classical H. Weyl's asymptotic formula (see Chapter 4, §1) has an analogue for the eigenvalues

$$0\le\lambda_1^{(p)}\le\lambda_2^{(p)}\le\cdots$$

of $\Delta^{(p)}$, each distinct eigenvalue repeated according to its multiplicity. In fact, the argument given in Chapter 4, §1, namely taking into account

$$\sum_{j=1}^{\infty}e^{-\lambda_j^{(p)}t}\sim\binom{n}{p}\mathrm{Vol}(M,g)\Big/(4\pi t)^{\frac{n}{2}}$$

as $t\searrow 0$, that is the leading term in the asymptotic expansion of the heat p-kernel and the Karamata theorem, leads immediately to

Corollary 4.5. If the Riemannian manifold (M,g) is as before, then

$$\lim_{j\to\infty}\frac{\left[\lambda_j^{(p)}\right]^{\frac{n}{2}}}{j}=\frac{(4\pi)^{\frac{n}{2}}\Gamma\left(\frac{n}{2}+1\right)}{\binom{n}{p}\mathrm{Vol}(M,g)},$$

for each p, $0\le p\le n$.

Let (M,g), (M',g') be two closed, connected, oriented smooth Riemannian manifolds, $\Delta^{(p)}$ (resp. $\Delta'^{(q)}$) the Hodge-de Rham operator corresponding to (M,g) [resp. (M',g')].

Definition 4.1. (M,g) and (M',g') are said to be ***p*-isospectral** if

$$\boldsymbol{Spec}^{(p)}(\boldsymbol{M},\boldsymbol{g})=\boldsymbol{Spec}^{(p)}(\boldsymbol{M}',\boldsymbol{g}').$$

Let $\Phi:(M,g)\to(M',g')$ be a surjective Riemannian submersion and $\Phi_p^*:A^p(M')\to A^p(M)$ the canonical $\mathbb{R}$-linear mapping induced by Φ (see Chapter 2, Proposition 2.11). Under these assumptions, (M,M',Φ) is a fibre bundle with the Lie group G of isometries of a fibre as structural group (see Hermann, [1960]) and Φ_p^* is injective: if $\sigma:M'\to M$ is a smooth section of this bundle, then $\sigma^*\circ\Phi^*=(\Phi\circ\sigma)^*=1_{A^p(M')}$. Let us discuss shortly the problem of relating the spectral properties of the corresponding Hodge-de Rham operators $\Delta^{(p)}$ $(1\le p\le\dim M)$ and

$\Delta'^{(q)}\left(1\le q\le \dim M'\right)$. Let $E\left(\Delta^{(p)},\lambda\right)$ [resp. $E\left(\Delta'^{(q)},\mu\right)$] be the eigenspace of $\Delta^{(p)}$ (resp. $\Delta'^{(q)}$) corresponding to the eigenvalue λ (resp. μ)$\in[0,+\infty)$ and fix p with $1\le p\le \dim M'$. The following two conditions are equivalent:

(i) $\Delta^{(p)}\circ\Phi_p^*=\Phi_p^*\circ\Delta'^{(p)}$ (see Chapter 2, Proposition 2.11 for the existence of such a mapping Φ);

(ii) $\Phi_p^*:A^p\left(M'\right)\to A^p\left(M\right)$ preserves the eigenforms of $\Delta'^{(p)}$, that is for any eigenvalue $\mu\in[0,+\infty)$ of $\Delta'^{(p)}$, there exists an eigenvalue $\lambda(\mu)\in[0,+\infty)$ of $\Delta^{(p)}$, so that $\lambda(\mu)\ge\mu$ and $\Phi_p^*\left(E\left(\Delta'^{(p)},\mu\right)\right)\subseteq E\left(\Delta^{(p)},\lambda(\mu)\right)$.

It is immediate that (i) implies (ii) with $\lambda(\mu)=\mu$, that is if Φ satisfies condition (i), the eigenvalues does not change. In particular, if Φ is a Riemannian covering, then any eigenvalue μ of $\Delta'^{(p)}$ is also an eigenvalue of $\Delta^{(p)}$ and the multiplicity of μ as eigenvalue of $\Delta'^{(p)}$ is $\le$ than the multiplicity of μ as eigenvalue of $\Delta^{(p)}$ for all p, $0\le p\le \dim M=\dim M'$. If Φ is an isometry, $\Delta^{(p)}$ and $\Delta'^{(p)}$ have the same eigenvalues [since $\Phi^{-1}:(M',g')\to(M,g)$ is also an isometry], and $E\left(\Delta^{(p)},\lambda\right)$ is isomorphic to $E\left(\Delta'^{(p)},\lambda\right)$ for every eigenvalue $\lambda\in[0,+\infty)$, that is $\mathbf{Spec}^{(p)}(\boldsymbol{M},\boldsymbol{g})=\mathbf{Spec}^{(p)}(\boldsymbol{M}',\boldsymbol{g}')$ for all p, $0\le p\le\dim M=\dim M'$.

For the proof of the converse statement, namely (ii) implies (i), see Gilkey and Park [1996].

The condition (ii) means that $\Phi_p^*(\omega')$ is an eigenform of $\Delta^{(p)}$, although with a possibly different eigenvalue, for every eigenform ω' of $\Delta'^{(p)}$. If in addition M' is simply connected, then $\mu=\lambda$, so in this case the eigenvalue does not change (see Gilkey, Leahy and Park [1998]). In general, however, if $p\ge 2$ the eigenvalues can change, that is there exist a surjective Riemannian submersion $\Phi:(M,g)\to(M',g')$, eigenvalues $\mu_0\in[0,+\infty)$ of $\Delta'^{(p)}$ and $\lambda_0\in(0,+\infty)$ of $\Delta^{(p)}$ and an eigenform $\omega'\in E\left(\Delta'^{(p)},\mu\right)$ such that $\mu_0<\lambda_0$ and $\Phi_p^*(\omega')\in E\left(\Delta^{(p)},\lambda\right)$ (see Gilkey and Park [1996]). One does not know if eigenvalues can change if $p=1$.

Let us note that the existence of a surjective Riemannian submersion $\Phi:(M,g)\to(M',g')$ satisfying the condition (i) for a fixed p, $1\le p\le\dim M'$, imposes also a topological restriction on M and M', namely $\beta_p(M')\le\beta_p(M)$, where $\beta_p(M)$ [resp. $\beta_p(M')$] denotes the p-th Betti number of M (resp. M'). Indeed, if Φ satisfies (i), then $\Phi^*\left(\operatorname{Ker}\Delta'^{(p)}\right)\subseteq\operatorname{Ker}\Delta^{(p)}$. Thus $\dim\Phi^*\left(\operatorname{Ker}\Delta'^{(p)}\right)\le\dim\operatorname{Ker}\Delta^{(p)}$. Since $\Phi^*:\operatorname{Ker}\Delta'^{(p)}\to\operatorname{Ker}\Delta^{(p)}$ is an injective linear mapping, it follows that $\dim\Phi^*\left(\operatorname{Ker}\Delta'^{(p)}\right)=\dim\operatorname{Ker}\Delta'^{(p)}$. Hence, from the Hodge-de Rham theorem (see Chapter 6, Corollary 1.1), one gets

$$\beta_p(M') = \dim \operatorname{Ker} \Delta'^{(p)} \leq \dim \operatorname{Ker} \Delta^{(p)} = \beta_p(M).$$

For instance, the canonical Riemannian submersion $\pi : (S^{2n+1}, g_{can}) \to (P^n(\mathbb{C}), g_{can})$ (see Chapter 1, Example 2.14) does not have the above commuting property (i) for $p = 2, 4, \ldots, 2n$, since $\beta_{2k}(P^n(\mathbb{C})) = 1$ and $\beta_{2k}(S^{2n+1}) = 0$ for all k, $1 \leq k \leq n$.

Many of the global geometric properties of a Riemannian manifold are reflected by its spectral properties, in particular by the asymptotic distribution behaviour of the eigenvalues of $\Delta^{(p)}$. For instance, if (M,g) and (M', g') are p-isospectral,

$$\dim M = \dim M' \text{ and } \operatorname{Vol}(M, g) = \operatorname{Vol}(M', g'),$$

so these two quantities are p-spectral invariants. This follows immediately by comparing the two leading terms in the asymptotic expansions of the two corresponding heat p-kernels.

A basic question in spectral geometry is to determine what geometric or topological information is contained in $\mathbf{Spec}^{(p)}(\boldsymbol{M}, \boldsymbol{g})$. In particular, can $\mathbf{Spec}(\boldsymbol{M}, \boldsymbol{g}) = \mathbf{Spec}^{(0)}(\boldsymbol{M}, \boldsymbol{g})$ and $\mathbf{Spec}^{(p)}(\boldsymbol{M}, \boldsymbol{g})$ contain different information? We shall discuss this question in what follows as well as in Chapter 6.

Theorem 4.3. (**Patodi** [1970]) Let (M,g) and (M', g') be closed, connected, oriented, n-dimensional smooth Riemannian manifolds $(n \geq 2)$.

(i) Assume $a_k(\Delta^{(0)}) = a_k(\Delta'^{(0)})$ for $k = 0, 2, 4$ and $a_4(\Delta^{(1)}) = a_4(\Delta'^{(1)})$. If (M', g') has constant sectional curvature c, so does (M, g).

(ii) If $a_4(\Delta^{(p)}) = a_4(\Delta'^{(p)})$ for $p = 0, 1$, then (M,g) is flat if and only if (M', g') is flat.

(iii) Assume $n \geq 4$, $a_k(\Delta^{(0)}) = a_k(\Delta'^{(0)})$ for $k = 0, 2$ and $a_4(\Delta^{(p)}) = a_4(\Delta'^{(p)})$ for $p = 0, 1, 2$. If (M', g') has constant scalar curvature c, so does (M, g). If (M', g') is Einsteinian, so is (M, g).

Proof. (i) The four conditions from hypothesis are respectively equivalent to

$$\operatorname{Vol}(M, g) = \operatorname{Vol}(M', g'), \tag{4.4}$$

$$\int_M \tau \, d\mu_g = \int_{M'} \tau' \, d\mu_{g'}, \tag{4.5}$$

$$\int_M \left(2|R|^2 - 2|c_2 R|^2 + 5\tau^2\right) d\mu_g = \int_{M'} \left(2|R'|^2 - 2|c_2 R'|^2 + 5\tau'^2\right) d\mu_{g'}, \tag{4.6}$$

$$\begin{aligned} &\int_M \left[(2n-30)|R|^2 + (-2n+180)|c_2 R|^2 + (5n-60)\tau^2\right] d\mu_g \\ &= \int_{M'} \left[(2n-30)|R'|^2 + (-2n+180)|c_2 R'|^2 + (5n-60)\tau'^2\right] d\mu_{g'}. \end{aligned} \tag{4.7}$$

If one multiplies (4.6) with $\frac{90-n}{30}$, (4.7) with $\frac{1}{30}$ and one sums, then one obtains

$$\int_M \left(5|R|^2 + 13\tau^2\right) d\mu_g = \int_{M'} \left(5|R'|^2 + 13\tau'^2\right) d\mu_{g'} . \tag{4.8}$$

Assume that (M', g') has constant sectional curvature c. Then

$$\tau' = -n(n-1)c \text{ and } |R'|^2 = 2n(n-1)c^2 .$$

We shall prove that the sectional curvature of (M, g) is constant and equal to c.

First, notice that (4.5) and (4.4) imply the equality

$$\int_M \tau d\mu_g = -n(n-1)c \operatorname{Vol}(M, g). \tag{4.9}$$

Since

$$|R|^2 \geq \frac{2|c_2 R|^2}{n-1} \geq \frac{2}{n(n-1)} \tau^2$$

(see Chapter 1, Propositions 4.6 and 4.7), (4.8) and (4.4) leads to

$$\int_M \tau^2 d\mu_g \leq n^2 (n-1)^2 c^2 \operatorname{Vol}(M, g). \tag{4.10}$$

Moreover, for $n \geq 3$, (4.10) becomes an equality if and only if

$$|R|^2 = \frac{2}{n(n-1)} \tau^2,$$

that is

$$|R|^2 = \frac{2|c_2 R|^2}{n-1},$$

what implies that the sectional curvature σ of (M,g) is constant (see Chapter 1, Proposition 4.7). But the equality

$$\int_M \tau^2 d\mu_g = n^2 (n-1)^2 c^2 \operatorname{Vol}(M, g)$$

shows that in this situation the sectional curvature σ of (M,g) is necessarily equal to c.

Finally, using successively (4.9), the Cauchy-Schwarz inequality and (4.10), one deduces

$$\begin{aligned} n^2 (n-1)^2 c^2 [\operatorname{Vol}(M, g)]^2 &= \left(\int_M \tau d\mu_g \right)^2 \leq \left(\int_M \tau^2 d\mu_g \right) \cdot \left(\int_M d\mu_g \right) \\ &\leq n^2 (n-1)^2 c^2 [\operatorname{Vol}(M, g)]^2 , \end{aligned}$$

and therefore

$$\left(\int_M \tau d\mu_g\right)^2 = \left(\int_M \tau^2 d\mu_g\right)\cdot\left(\int_M d\mu_g\right),$$

which implies that τ is constant. Then, in virtue of (4.9) one obtains

$$\tau = -n(n-1)c,$$

that is one has equality in (4.10), what for $n \geq 3$ implies that (M,g) has constant sectional curvature c.

If $n = 2$, one has

$$\tau = -2\sigma, \tau' = -2c, |R|^2 = 2|c_2 R|^2 = \tau^2 = 4\sigma^2,$$
$$|R'|^2 = 2|c_2 R'|^2 = \tau'^2 = 4c^2,$$

so that (4.9) leads in this case to

$$\int_M \sigma d\mu_g = c\,\mathrm{Vol}(M,g), \tag{4.11}$$

and (4.8), (4.4) imply

$$\int_M \sigma^2 d\mu_g = c^2\,\mathrm{Vol}(M,g).$$

Hence in the case when $n = 2$, one gets the equality

$$\left(\int_M \sigma d\mu_g\right)^2 = \left(\int_M \sigma^2 d\mu_g\right)\cdot\left(\int_M d\mu_g\right),$$

which implies that σ is constant and namely, via (4.11), equal to c.

(ii) In the proof of (i) we have established that

$$a_4\left(\Delta^{(0)}\right) = a_4\left(\Delta'^{(0)}\right) \text{ and } a_4\left(\Delta^{(1)}\right) = a_4\left(\Delta'^{(1)}\right)$$

imply (4.8). Now assume that (M',g') is flat, that is $R' = 0$. Then $\tau' = 0$ and from (4.8) it follows that $|R| = 0$, that is $R = 0$, and therefore (M,g) is flat.

(iii) The assumptions imply (4.4), (4.5), (4.6), (4.7) and

$$\int_M \left(\alpha|R|^2 + \beta|c_2 R|^2 + \gamma\tau^2\right) d\mu_g = \int_{M'} \left(\alpha|R'|^2 + \beta|c_2 R'|^2 + \gamma\tau'^2\right) d\mu_{g'} \tag{4.12}$$

where $\alpha = 2\binom{n}{2} - 30\binom{n-2}{1} + 180$, $\beta = -2\binom{n}{2} + 180\binom{n-2}{1} - 720$, $\gamma = 5\binom{n}{2} - 60\binom{n-2}{1} + 180$. Let

$$A = \begin{pmatrix} 2 & -2 & 5 \\ 2n-30 & -2n+180 & 5n-60 \\ \alpha & \beta & \gamma \end{pmatrix}$$

be the matrix formed with the coefficients that occur in (4.6), (4.7) and (4.12) and let us use elementary line operations to transform A to the matrix

$$A_1 = \begin{pmatrix} 2 & -1 & 5 \\ 1 & -3 & 2 \\ 1 & -2 & 1 \end{pmatrix}.$$

Since A_1 is invertible, A is invertible. Therefore, it follows from (4.6), (4.7) and (4.12) that

$$\int_M |R|^2 \, d\mu_g = \int_{M'} |R'|^2 \, d\mu_{g'}, \int_M |c_2 R|^2 \, d\mu_g = \int_{M'} |c_2 R'|^2 \, d\mu_{g'}, \qquad (4.13)$$
$$\int_M \tau^2 d\mu_g = \int_{M'} \tau'^2 d\mu_{g'}.$$

Consequently, if the scalar curvature τ' of (M', g') is constant $(= c)$, then

$$0 = \int_{M'} (\tau' - c)^2 \, d\mu_{g'} = \int_{M'} \tau'^2 d\mu_{g'} - 2c \int_{M'} \tau' d\mu_{g'} + c^2 \operatorname{Vol}(M', g')$$
$$= \int_M \tau^2 d\mu_g - 2c \int_M \tau d\mu_g + c^2 \operatorname{Vol}(M, g) = \int_M (\tau - c)^2 \, d\mu_g,$$

so that $\tau = c$.

If (M', g') is Einsteinian and $n \geq 4$, then $|c_2 R'|^2 - \frac{\tau'^2}{n} = 0$ (see Chapter 1, Proposition 4.6). Hence

$$0 = \int_{M'} \left(|c_2 R'| - \frac{\tau'^2}{n} \right) d\mu_{g'} = \int_{M'} |c_2 R'|^2 \, d\mu_{g'} - \frac{1}{n} \int_{M'} \tau'^2 d\mu_{g'}$$
$$= \int_M |c_2 R|^2 \, d\mu_g - \frac{1}{n} \int_M \tau^2 d\mu_g = \int_M \left(|c_2 R|^2 - \frac{1}{n} \tau^2 \right) d\mu_g,$$

so $|c_2 R|^2 - \frac{\tau^2}{n} = 0$, that is (M, g) is Einsteinian.

Q.E.D.

Remark 4.6. The proof of Theorem 4.3 (iii) shows that the global integrated invariants

$$\int_M |R|^2 \, d\mu_g, \int_M |c_2 R|^2 \, d\mu_g \text{ and } \int_M \tau^2 d\mu_g$$

are determined by $\mathbf{Spec}^{(p)}(\boldsymbol{M}, \boldsymbol{g})$ for $0 \leq p \leq 2 (n \geq 4)$. More generally, using the same argument, one can check that for any three mutually independent natural numbers $p, p', p'' \, (0 \leq p, p', p'' \leq n)$ the corresponding spectra together determine also these integrals (see Zhou [1997]). Here, two natural numbers p and $p' \, (0 \leq p, p' \leq n)$ are called **independent** if $p \neq p'$ and $p + p' \neq n$.

Remark 4.7. Theorem 4.3 (iii) shows that the property of a closed oriented Riemannian manifold to have constant scalar curvature is p-spectrally determined for

$p=0,1,2$. In contrast, Gordon and Szabó [2000] showed recently that the 0-spectrum alone does not determine this information (see also Chapter 4, Remark 3.7).

Corollary 4.6. Let (M,g) be a closed, connected, oriented, n-dimensional $(n\geq 2)$ Riemannian manifold of constant sectional curvature σ. If (M',g') is a closed, connected, oriented Riemannian manifold such that

$$\mathbf{Spec}^{(p)}(M,g)=\mathbf{Spec}^{(p)}(M',g'),\quad p=0,1,$$

then M' has the dimension n and the sectional curvature of (M',g') is constant and equals σ. In particular, the sphere (S^n,g_0) is completely characterized up to an isometry by its spectra on functions and 1-forms.

Let C be the Weyl conformal curvature tensor field associated to (M,g), $\dim M\geq 3$ (see Chapter 1, §4), $E=c_2R-\frac{1}{n}g$ the Einstein tensor field associated to (M,g) and $a_4(\Delta^{(p)})$ the coefficient of the third term in the asymptotic expansion (4.3). Since

$$|E|^2=|c_2R|^2-\frac{1}{n}\tau^2,$$

by the formula (4.15) of Chapter 1 and by Theorem 4.2 (iii) one derives

$$a_4\left(\Delta^{(p)}\right)=$$

$$=\frac{(4\pi)^{-\frac{n}{2}}\binom{n-4}{p-2}}{360p(p-1)(n-p)(n-p-1)}\int_M\left[\alpha|C|^2+\frac{n-3}{n-2}\beta|E|^2+\frac{(n-2)(n-3)}{n(n-1)}\gamma\tau^2\right]d\mu_g, \tag{4.14}$$

where

$$\alpha=2n^4-(30p+12)n^3+(210p^2-30p+22)n^2-(360p^3+30p^2+12)n+180p^4,$$

$$\beta=-2n^4+(180p+18)n^3-(900p^2+120p+40)n^2+(1440p^3+120p^2+240p+24)n-$$
$$-720p^4-240p^2,$$

$$\gamma=5n^4-(60p+12)n^3+(240p^2+60p+13)n^2-(260p^3+60p^2+60p+6)n+$$
$$+180p^4+60p^2.$$

Now let us mention also the following result.

Theorem 4.4. **(Puta and Török**, [1987]). Let (M,g) and (M',g') be closed, oriented, Riemannian manifolds with

$$\mathbf{Spec}^{(p)}(M,g)=\mathbf{Spec}^{(p)}(M',g')$$

for a fixed $p\in\mathbb{N}$, $p\geq 2$ (this implies $\dim M=\dim M'=n$). Then for each $n\in\mathbb{N}\setminus\{0,1,2\}$, $n\geq p$ such that $\alpha>0$, $\beta>0$, $\gamma>0$, it follows that (M,g) is of constant sectional curvature σ if and only if (M',g') is of constant sectional curvature σ' and $\sigma=\sigma'$.

Proof. We shall follow a standard argument due to Tanno [1974]. If (M',g') has constant sectional curvature σ', then C'=0 (see Chapter 1, Corollary 4.2) and E'=0, so that (4.14) becomes

$$a_4\left(\Delta'^{(p)}\right)=\frac{(4\pi)^{-\frac{n}{2}}\binom{n-4}{p-2}(n-2)(n-3)}{360p(p-1)(n-p)(n-p-1)n(n-1)}\int_{M'}\gamma\tau'^2 d\mu_{g'} . \tag{4.15}$$

Since $\mathbf{Spec}^{(p)}(\boldsymbol{M},\boldsymbol{g})=\mathbf{Spec}^{(p)}(\boldsymbol{M}',\boldsymbol{g}')$, in virtue of the formulae (4.14) and (4.15) one obtains

$$\int_M\left[\alpha|C|^2+\frac{n-3}{n-2}\beta|E|^2+\frac{(n-2)(n-3)}{n(n-1)}\gamma\tau^2\right]d\mu_g=\frac{(n-2)(n-3)}{n(n-1)}\int_{M'}\gamma\tau'^2 d\mu_{g'} . \tag{4.16}$$

Now, using again the hypothesis $\mathbf{Spec}^{(p)}(\boldsymbol{M},\boldsymbol{g})=\mathbf{Spec}^{(p)}(\boldsymbol{M}',\boldsymbol{g}')$, one gets

$$\int_M \tau d\mu_g=\int_{M'}\tau' d\mu_{g'} ,$$

which implies

$$\int_M \tau^2 d\mu_g\geq\int_{M'}\tau'^2 d\mu_{g'} \tag{4.17}$$

since τ' is constant. From (4.16), (4.17) and our assumption we conclude that (M,g) is of constant sectional curvature σ and moreover $\sigma=\sigma'$.

Q.E.D.

Remark 4.8. For each fixed arbitrary $p\in\mathbb{N},\ p\geq 2$, one can always find by explicit computations using a computer program, the particular values of $n\in\mathbb{N}\setminus\{0,1,2\}$, $n\geq p$, for which α, β, γ are strictly positive. It is not hard to see that for each $p\in\mathbb{N},\ p\geq 24$, these values of n can be found approximately in the following union of intervals: $[1,02p;1,2p]\cup[1,32p;1,6p]\cup[2,6p;4,4p]\cup[5,84p;79p]$. In the Table below for each $p\in\mathbb{N},\ 2\leq p\leq 10$, one indicates the particular values of $n\in\mathbb{N}\setminus\{0,1,2\},\ n\geq p$, for which Theorem 4.4 is valid. These values were obtained using Bairstow's method for the determination of the roots of a polynomial with integer coefficients, that allows to establish the sign of the polynomial.

p	n
2	3,6,7,14,[17,178]
3	$3,[9,12]\cup[19,263]$
4	$4,6,[11,16]\cup[25,347]$
5	$5,7,\ [14,20]\cup[30,432]$
6	$9,\ [16,25]\cup[35,517]$
7	$10,11,\ [19,29]\cup[40,602]$
8	$11,12,\ [22,33]\cup[45,687]$
9	$[12,14]\cup[24,38]\cup[50,771]$
10	$[14,16]\cup[27,42]\cup[55,856]$

Following the classical paper of Kac [1966] and thinking of the eigenvalues as the frequencies of the normal modes of vibration of an idealized elastic medium, the "drum", we say that a geometric property of (*M*,*g*) can be "heard" if it is determined by the $\mathbf{Spec}^{(p)}(\boldsymbol{M},\boldsymbol{g})$. Thus the above results, in particular Theorems 4.1 and 4.2, Corollary 4.1 show that $\dim M, \mathrm{Vol}(M,g)$, the coefficients of the asymptotic expansion of the function $Z^{(p)}_{(M,g)}$, the integral of curvatures of (*M*,*g*), $\chi(M)$,... can be heard. In fact, $\mathbf{Spec}^{(p)}(\boldsymbol{M},\boldsymbol{g}), 0\le p\le \dim M$, can determine the aspects of dimension one and dim *M* of the geometry of (*M*,*g*), but not aspects of intermediary dimensions (see DeTurck [1988]). On the other hand, DeTurck, Gluck, Gordon and Webb [1989] have exhibited specific geometric invariants on a particular closed 6-dimensional smooth Riemannian manifold that cannot be "heard". Essentially, these invariants are functions defined by using the volumes of higher than one-dimensional minimizing cycles of homology classes on *M*. They allow to show that certain isospectral metrics on *M* are not isometric and in turn help us to answer the question "How can a drum change shape, while sounding the same?"

Now it is natural to ask: Let (*M*,*g*) and (M',g') be two *p*-isospectral closed *n*-dimensional Riemannian manifolds for some $p, 0\le p\le n$. Then are (*M*,*g*) and (M',g') isometric? The answer is no and the first counterexample was given by Milnor [1964]: there exist two 16-dimensional flat tori that are *p*-isospectral for each $p, 0\le p\le 16$, but which are not isometric. Later, Kneser [1967] (resp. Kitaoka [1977], resp. Conway and Sloane [1992]) has constructed *p*-isospectral flat tori in dimension 12 (resp. 8, resp. 4) for any *p*, which are not isometric. Starting from these examples, it is possible to construct non-isometric *p*-isospectral *n*-dimensional flat tori for each $p, 0\le p\le n$, and for any $n\ge 4$. On the other hand, two 2(resp. 3)-dimensional flat tori are *p*-isospectral $(0\le p\le 2,\ \text{resp. } 0\le p\le 3)$ if and only if they are isometric (see Chapter 4, §2, Berger, Gauduchon and Mazet [1971], Craioveanu and Puta [1988], Pesce [1992]). Kneser (unpublished) and Wolpert [1978] have established that there exists only a finite number of isospectral flat tori which are two by two isometric.

Let us note that the isometry classes of Riemannian Heisenberg manifolds (see Chapter 1, Example 2.19) are also completely described (see Gordon [1986]$_a$, Gordon and Wilson [1986]). Pesce [1992] gives in any dimension an explicit bound of the number of isometry classes of flat tori isospectral to a given flat torus. The problem of the existence of a universal bound for this number of isometry classes in dimension $n \geq 4$ is open. In the case when the number of isometry classes of closed Riemannian manifolds p-isospectral to a given closed Riemannian manifold is finite, it would be interesting to find bounds for this number in terms of the geometry of the given manifold and to establish under what assumptions these bounds depend only on the dimension.

Vignéras [1978] has given examples of non-isometric Riemannian surfaces that are p-isospectral for any p. The construction of all these examples is based essentially on the possibility of describing their p-spectra. Sunada [1985] gave a method that allows the generalizations of these examples. This method reduces partially the construction of p-isospectral closed Riemannian manifolds to a well-understood problem of finite groups (see also Chapter 4, §3, Remark 3.2). Indeed, let M be a closed smooth manifold and assume that the finite group G acts without fixed points on M. Let H_1, H_2 be subgroups of G and take the corresponding quotient manifolds $M_1 = M/H_1$ and $M_2 = M/H_2$. Choosing a Riemannian metric on M/G, it induces Riemannian metrics g (resp. g_1, resp. g_2) on M (resp. M_1, resp. M_2) such that the corresponding canonical projections become Riemannian coverings.

Theorem 4.5. (**Sunada** [1985]) If H_1 and H_2 are not conjugate subgroups of G and there exists a bijection $F: H_1 \to H_2$ such that for any $h_1 \in H_1$, there exists $g = g_{h_1} \in G$ such that $F(h_1) = g h_1 g^{-1}$, then the closed Riemannian manifolds (M_1, g_1) and (M_2, g_2) are p-isospectral for any p.

The p-isospectrality of these two closed Riemannian manifolds follows also immediately from a direct computation showing that the operators $e^{-t\Delta_{g_1}^{(p)}}$ and $e^{-t\Delta_{g_2}^{(p)}}$ have the same trace.

For examples of finite groups G and subgroups H_1, H_2 of G that satisfy the hypothesis of Theorem 4.5 see Bérard [1989], Brooks [1988], Buser [1988], Sunada [1985] (see also Chapter 4, §3).

Theorem 4.5 can be stated in representation theoretic terms: If the discrete groups H_1 and H_2 of isometries of M are "representation equivalent" subgroups of a larger group G of isometries, i.e. if the quasi-regular representations of G on $L^2\left(G/H_1\right)$ and $L^2\left(G/H_2\right)$ are equivalent, then M/H_1 and M/H_2 are p-isospectral for any p (see also Chapter 4, §3).

Let us consider another interesting example given by DeTurck and Gordon [1989]. Let G be a simply-connected Lie group, which is nilpotent of rank m and Γ a cocompact discrete subgroup of G. Let (M,g) be a Riemannian manifold without boundary and assume that one has a left G-action on M by isometries, which is faithful and proper. Also assume that Γ acts freely on M and that the (left) quotient manifold M/Γ is closed.

Theorem 4.6. (**Bérard** [1989], **DeTurck and Gordon** [1989]). Let $\varphi: G \to G$ be an automorphism, but not an interior automorphism of G, with the following property: for any $\gamma \in \Gamma$, there exists $g = g_\gamma \in G$ such that $\varphi(\gamma) = g\gamma g^{-1}$. Then $\varphi(\Gamma)$ acts freely on M, the quotient manifold $M/\varphi(\Gamma)$ is closed and the closed Riemannian manifolds

$\left(M/_{\varphi(\Gamma)}, g\right)$ and $\left(M/_{\Gamma}, g\right)$ are *p*-isospectral for any *p*.

For examples of Lie groups *G* and cocompact discrete subgroups Γ of *G* satisfying the above assumptions, see Gordon [1986]$_a$, Gordon and Wilson [1984], DeTurck and Gordon [1989]. In fact, DeTurck and Gordon [1989] have generalized Sunada's construction in such a way that Theorems 4.5 and 4.6 became consequences of a much more general result.

The methods described in Theorems 4.5 and 4.6 always produce *p*-isospectral manifolds for any *p*; in most, but not all, cases, they are non-isometric.

The above examples of closed Riemannian manifolds are *p*-isospectral for any *p*. It is natural to ask if this is always the case. Examples of closed Riemannian manifolds that are 0-isospectral but not *p*-isospectral for all *p* provide us with a means of identifying properties that distinguish these spectra. Such examples are sparse, however, as almost all known examples of isospectral closed Riemannian manifolds can now be explained by Sunada's method (see Chapter 4, §3) or its generalizations (see DeTurck and Gordon [1989], Bérard [1992] and Pesce [1996]). Gordon [1986]$_a$, [1986]$_b$ has given examples of pairs of non-isometric, 0-isospectral Riemannian Heisenberg manifolds which are not 1-isospectral. Recently, Gornet [2000] showed that there exist continuous families of closed Riemannian manifolds that are isospectral on functions but not on 1-forms. On the other hand, for a given *p*, Ikeda [1989] exhibits examples of *k*-isospectral lenticular spaces for $0 \le k \le p$, that are not (*p*+1)-isospectral. Miatello and Rossetti [2001] succeeded to give many examples of pairs of closed flat Riemannian manifolds which are *p*-isospectral for certain values of *p* only. For instance, they constructed pairs of closed flat Riemannian manifolds of dimension $n = 2p$, $p \ge 2$, not homeomorphic to each other, which are *p*-isospectral but not *q*-isospectral for $q \ne p, 0 \le q \le n$. Also, they gave examples of such manifolds *p*-isospectral if and only if *p* is odd, one of them orientable and the other not.

5. LOWER BOUNDS FOR THE SMALLEST POSITIVE EIGENVALUE OF THE HODGE-DE RHAM OPERATOR

Let (*M*,*g*) be an *n*-dimensional closed (i.e. compact and without boundary), connected, orientable and smooth Riemannian manifold. In this section we shall give a lower bound for the first positive eigenvalue of the Hodge-de Rham operator $\Delta^{(p)} : A^p(M) \to A^p(M)$ acting on the space $A^p(M)$ of smooth *p*-forms (see Chapter 2, Definition 2.3). Our argument here follows that of Chanillo and Trèves [1997]$_a$. The bound is related to the parameters of a certain kind of open covering $\{U_i\}$ of the manifold *M* by finitely many geodesically convex sets. The parameters are the number of sets U_i, their geodesic radii and bounds on the differential of the exponential map in each U_i. More precisely, let r_0 be the injectivity radius of (*M*,*g*).

Definition 5.1. Any covering $\{U_1, \ldots, U_N\}$ of *M* by open sets is called **admissible** for (*M*,*g*) if there are two positive numbers $A, r,\ r << \min(1, r_0)$, such that the following properties hold, for some $R > 4^n r$, and every $i = 1, \ldots, N$:

(i) There exists $x_i \in U_i$ such that the exponential map $E_i \overset{not}{=} \exp_{x_i} : T_{x_i}M \to M$ is a diffeomorphism of the ball $B(x_i,R)=\left\{v\in T_{x_i}M \,\middle|\, \left[g_{x_i}(v,v)\right]^{1/2}<R\right\}$ onto $E_i(B(x_i,R))$;

(ii) $\frac{1}{A}\le \|dE_i(v)\|\le A$ for all $v\in B(x_i,R)$;

(iii) $U_i=E_i(B(x_i,r))$.

We shall refer to the positive numbers *N*,*r* and *A* as the parameters of the admissible covering $\{U_1,\dots,U_N\}$.

Let us denote by $L^2A^p(M)$ the Hilbert space of L^2 *p*-forms. The norm (resp. interior product) in $L^2A^p(M)$ will be denoted by $\|\ \|$ (resp. $\langle\ ,\ \rangle$) (see Chapter 2, §1, C).

Let $\delta^p=(-1)^{np+n+1}S^{(n-p+1)}d^{n-p}S^{(p)}$ be the formal transpose of the exterior derivative d^{p-1}, where the linear isometry $S^{(p)}:L^2A^p(M)\to L^2A^{n-p}(M)$ is the extension of the star Hodge operator: for any compactly supported $\omega\in A^p(M)$,

$$\|\omega\|^2=\int_M \omega\wedge S^{(p)}\omega$$

(see Chapter 2, §2, D). All the results in this section follow from

Theorem 5.1. Suppose there exists an open covering $\{U_1,\dots,U_N\}$ of *M* which is admissible for (*M*,*g*) with parameters *N*,*r* and *A*. Then there is a constant $C>0$ depending solely on *n* and *A*, such that the following is true, whatever the integer *p*, $1\le p\le n$: If $\omega\in A^p(M)$ is *d*-exact, that is there exists $\omega_1\in A^{p-1}(M)$ such that $d^{p-1}\omega_1=\omega$, then the equation $d^{p-1}\theta=\omega$ has a solution $\theta\in A^{p-1}(M)$ such that

$$\|\theta\|\le CrN^{4p}\|\omega\|. \tag{5.1}$$

The proof of this theorem is based on solving the equation $d^{p-1}\theta=\omega$ in a bounded convex open subset of the Euclidean space with good bounds on the L^2-norm of the solution, but without any pretense to obtain the best possible bounds (for details, see Chanillo and Trèves [1997]a, §2). Theorem 5.1 does not provide estimates of the L^2-norm of an arbitrary solution of the equation $d^{p-1}\theta=\omega$, since one can add to θ any closed smooth $(p-1)$-form. But it does yield an estimate for the canonical solution, that is, the solution orthogonal in $L^2A^p(M)$ to all closed forms. Indeed this solution minimizes the L^2-norm over the affine subspace of all L^2 solutions.

The following statement is an immediate consequence of Theorem 5.1.

Corollary 5.1. Let *p* be an integer, $0\le p\le n-1$. Under the same assumptions as in Theorem 5.1, if $\omega\in A^p(M)$ is δ–exact, then the equation $\delta^{p+1}\theta=\omega$ has a solution $\theta\in A^{p+1}(M)$ such that

$$\|\theta\|\le CrN^{4(n-p)}\|\omega\|,$$

where *C* is the same positive constant as in (5.1).

Proof. If $\omega \in A^p(M)$ is δ-exact, then $S^{(p)}\omega \in A^{n-p}(M)$ is d-exact. Apply Theorem 5.1 to the equation $d^{n-p-1}\theta_1 = (-1)^{np+1} S^{(p)}\omega$, where $\theta_1 \in A^{n-p-1}(M)$ and set $\theta = S^{(n-p-1)}\theta_1 \in A^{p+1}(M)$. Then $\delta^{p+1}\theta = \omega$ and

$$\|\theta\| = \left\|S^{(n-p-1)}\theta_1\right\| = \|\theta_1\| \le CrN^{4(n-p)}\left\|S^{(p)}\omega\right\| = CrN^{4(n-p)}\|\omega\|.$$

Q.E.D.

Theorem 5.2. Let p be an integer, $0 \le p \le n$, and $\lambda_1^{(p)}$ be the smallest positive eigenvalue of the Hodge-de Rham operator $\Delta^{(p)} : A^p(M) \to A^p(M)$. Under the same hypotheses as in Theorem 5.1, one has

$$\lambda_1^{(p)} \ge C^{-2} r^{-2} N^{-4(n+1)}.$$

Proof. Let $\omega \in A^p(M)$ be orthogonal to the space $\ker \Delta^{(p)}$ of harmonic p-forms on M, and $\theta \in A^p(M)$ be such that $(d^{p-1}\delta^p)(\theta) + (\delta^{p+1}d^p)(\theta) = \omega$ and be also orthogonal to $\ker \Delta^{(p)}$ (see also next chapter, §1). Then one has

$$\|\omega\| \ge \left\|d^{p-1}\delta^p\theta\right\| + \left\|\delta^{p+1}d^p\theta\right\|. \tag{5.2}$$

On the other hand, $\delta^p\theta$ is the solution of the equation $d^{p-1}\tilde{\theta} = d^{p-1}\delta^p\theta$, which is orthogonal to all d-closed forms, and $d^p\theta$ is the solution of the equation $\delta^{p+1}\tilde{\theta} = \delta^{p+1}d^p\theta$ which is orthogonal to all δ-closed forms. It follows from (5.2), Theorem 5.1 and Corollary 5.1 that

$$\|\omega\| \ge C^{-1}r^{-1}\left[N^{-4p}\left\|\delta^p\theta\right\| + N^{-4(n-p)}\left\|d^p\theta\right\|\right]. \tag{5.3}$$

Using the Hodge-de Rham decomposition theorem, one can write

$$\theta = \theta_1 + \theta_2,$$

where $\theta_1 = \delta^{p+1}\omega_1 \in A^p(M)$ [resp. $\theta_2 = d^{p-1}\omega_2 \in A^p(M)$] is orthogonal to the space of d (resp. δ)-closed p-forms, $\omega_1 \in A^{p+1}(M)$, $\omega_2 \in A^{p-1}(M)$. Note that $\langle\theta_1, \theta_2\rangle = 0$, $d^p\theta = d^p\theta_1$ and $\delta^p\theta = \delta^p\theta_2$. Again by Theorem 5.1 and Corollary 5.1 one gets

$$\begin{cases} \left\|d^p\theta\right\| \ge C^{-1}r^{-1}N^{-4(p+1)}\|\theta_1\|, \\ \left\|\delta^p\theta\right\| \ge C^{-1}r^{-1}N^{-4(n-p+1)}\|\theta_2\|. \end{cases} \tag{5.4}$$

Putting (5.4) into (5.3), one obtains

$$\|\omega\| \ge C^{-1}r^{-1}N^{-4(n+1)}\left[\|\theta_1\| + \|\theta_2\|\right] \ge C^{-1}r^{-1}N^{-4(n+1)}\|\theta\|. \tag{5.5}$$

Taking $\omega = \lambda_1^{(p)}\theta$, where $\theta \in E\left(\Delta^{(p)}, \lambda_1^{(p)}\right) \setminus \{0\}$, (5.5) yields the desired inequality in Theorem 5.2.

Q.E.D.

Remark 5.1. Such eigenvalue estimates prove useful in establishing the local exactness in certain differential complexes defined by complex vector fields (for details, see Chanillo-Trèves [1997]b).

Now let us give an application of Theorem 5.2. Let $f : S^n \to \mathbb{R}$ be a real analytic function, $n \geq 2$, where $S^n \subset \mathbb{R}^{n+1}$ is the unit sphere, and K the subset of the closed interval $f(S^n)$ consisting of the noncritical values of f. The set of critical values of f is finite and equal to the boundary ∂K of $K : \partial K = \{y_0, y_1, \ldots, y_k\} \subset \mathbb{R}$ with

$$\min_{S^n} f = y_0 < y_1 < \cdots < y_k = \max_{S^n} f .$$

The preimage $f^{-1}(K)$ is a disjoint union of finitely many open and connected subsets of S^n whose boundaries are pieces of analytic varietes. Let $\mathcal{O}$ be a connected component of $f^{-1}(K)$, i.e. $f(\mathcal{O}) = (y_{j-1}, y_j)$ for some $j, 1 \leq j \leq k$. If $y \in f(\mathcal{O})$, then the level set of f in $\mathcal{O}$, $L_y = f^{-1}(y) \cap \mathcal{O}$ is a compact submanifold without boundary of S^n of class C^ω, and codimension one. Moreover, one has:

Lemma 5.1. Every level set L_y in a connected component $\mathcal{O}$ of $f^{-1}(K)$ is connected.

Proof. Assume the contrary, namely that L_y has at least two distinct connected components Γ and Γ'. Let $\tilde{\Gamma}$ be the set of points of $\mathcal{O}$ which lie on some integral curve of $\operatorname{grad}(f)$ that intersects Γ. Since $\operatorname{grad}(f) \neq 0$ at every point of $\mathcal{O}$, it is clear that $\tilde{\Gamma}$ contains a tubular neighborhood of Γ. On the other hand, since the flow of $\operatorname{grad}(f)$ defines local diffeomorphisms, $\tilde{\Gamma}$ is open and closed in $\mathcal{O}$ endowed with the relative topology. Consequently, $\tilde{\Gamma} = \mathcal{O}$ and each point of Γ' should lie on an integral curve γ of $\operatorname{grad}(f)$ intersecting Γ, which would imply that f takes the value y at two distinct points of γ. This would contradict the general fact that f is a bijection of anyone of its integral curves in $\mathcal{O}$ onto $f(\mathcal{O})$.

Q.E.D.

Clearly $\mathcal{O}$ can be identified to the product $(y_{j-1}, y_j) \times L_u$ for any $u \in (y_{j-1}, y_j)$: a pair (y, z), $y_{j-1} < y < y_j$, $z \in L_u$, is identified to the unique point $x \in L_y$ joined to z by an arc of an integral curve of $\operatorname{grad}(f)$. Let each manifold L_y be equipped with the Riemannian metric induced by the Riemannian metric g_{can} of S^n and denote by $\lambda_1^{(p)}(y)$ the smallest positive eigenvalue of the Hodge-de Rham operator $\Delta^{(p)}$ on L_y acting on p-forms. Then one obtains:

Theorem 5.3. **(Chanillo-Trèves** [1997]a). Let p be an integer, $0 \leq p \leq n-1$. Then there are positive numbers $\alpha, \beta > 0$ such that

$$\lambda_1^{(p)}(y) \geq \alpha \left[dist(y, \partial K) \right]^\beta ,$$

for every $y \in K$.

Proof. The sought estimates will be simpler if one assumes the diameter of $f(S^n) < 1$, which can always be achieved by rescaling the metric. Let

$d(y) = dist(y, \partial K), y \in f(S^n)$, and note that $d(y) \leq d_0 < 1$. Further, let us use the properties of the level sets L_y established in Chanillo [1994]. Lemma 2, loc. cit., combining the implicit function theorem with the Lojasiewicz inequality shows that the injectivity radius of L_y is $\geq d(y)^{\sigma_0}$ for a suitably large constant $\sigma_0 > 0$ independent of $y \in K$. According to Lemmas 2 and 5, ibid., if *N,r* and *A* are the positive numbers in Definition 5.1 for $M = L_y$, then one can take

$$N \leq d(y)^{-\rho}, \ r = d(y)^{\sigma}, \ A = 100$$

with a suitable large number ρ, also independent of *y*, and $\sigma = 100\sigma_0$. If one applies Theorem 5.2 with these values of *N,r,A* one gets rightaway Theorem 5.3.

Q.E.D.

Remark 5.2. Theorem 5.3 states that the first positive eigenvalue of the Hodge-de Rham operator on the regular level sets of a C^{ω}-function $f : S^n \to \mathbb{R}, n \geq 2$, is bounded below by a constant times a power of the distance between the value of *f* and the set of critical values of *f*. This result has been proved for the Laplace-Beltrami operator in full generality by Chanillo [1994].

The lower bound for the first positive eigenvalue of the Hodge-de Rham operator on the regular level sets of a real analytic function on S^n given in Theorem 5.3 is crucial in the proof of the existence of solutions in certain classes of differential complexes. In fact, this result of Chanillo and Trèves represents one stage in the very technical proof given by them, under the additional hypothesis of real analyticity, for a conjecture raised by Trèves in 1983.

Remark 5.3. The situation concerning eigenvalue estimates with best possible bounds for the Hodge-de Rham operators is much worse than in the case of the Laplace-Beltrami operator. For a commentary on the difficulties one encounters, see Dodziuk's appendix in Chavel [1984].

Remark 5.4. Colbois and Courtois [1990] showed that, in contrast to the case of the Laplace-Beltrami operator, Cheeger's isoperimetric constant does not permit one to control the first nonzero eigenvalue of the *p*-spectrum ($p \neq 0$, the dimension of the manifold). They gave some examples of compact Riemannian manifolds of any dimension $n \geq 3$ which have arbitrarily small first nonzero eigenvalue of the *p*-spectrum $(1 \leq p \leq n-1)$, whereas Cheeger's constant stays bounded away from zero. They proved that the property for *M* to have Riemannian metrics with uniformly bounded sectional curvature and diameter and arbitrarily small first nonzero eigenvalue of the *p*-spectrum $(1 \leq p \leq n-1)$ implies that these metrics have arbitrarily small injectivity radius and that this fact has topological implications for *M*.

REFERENCES

BÉRARD, P., *Variétés riemanniennes isospectrales non isométriques*, Séminaire Bourbaki, 41ème année, Vol. 1988-89, no. 705, Astérisque 177-178 (1989), 127-154.

BÉRARD, P., *Transplantation et isospectralité I*, Math. Ann. 292(1992), 547-559.

BÉRARD, P., *Transplantation et isospectralité II*, J. London Math. Soc. (2)48(1993), no. 3, 565-576.

BERGER, M., GAUDUCHON, P. and MAZET, E., *Le spectre d'une variété Riemannienne*, Lecture Notes in Math., Vol. 194, Springer-Verlag, Berlin, 1971.

BROOKS, R., *Constructing isospectral manifolds*, Amer. Math. Monthly 95(1988), 823-839.

BUSER, P., *Cayley graphs and planar isospectral domains*, in: Lecture Notes in Mathematics, 1339, Springer-Verlag, 1988, pp. 64-77.

CHANILLO, S., *The first eigenvalue of analytic level surfaces on spheres*, Math. Research Letters 1(1994), 159-166.

CHANILLO, S. and TRÈVES, F., *On the lowest eigenvalue of the Hodge Laplacian*, J. Differential Geometry 45$(1997)_a$, 273-287.

CHANILLO, S. and TRÈVES, F., *Local exactness in a class of differential complexes*, J. of the American Math. Society, Vol. 10$(1997)_b$, 393-426.

CHAVEL, I., *Eigenvalues in Riemannian Geometry*, Academic Press, Inc., New York, London, 1984.

CHEEGER, J., *Analytic torsion and the heat equation*, Ann. Math. 109(1979), 259-322.

COLBOIS, B. and COURTOIS, G., *A note on the first nonzero eigenvalue of the Laplacian acting on p-forms*, Manuscripta Math. 68(1990), 143-160.

CONNER, P.E., *The Neumann problem for differential forms on Riemannian manifolds*, Mem. Am. Math. Soc. 20(1956).

CONWAY, J.H. and SLOANE, N.J.A., *Four-dimensional lattices with the same theta series*, International Mathematics Research Notices, No. 4(1992), 93-96.

CRAIOVEANU, M. and PUTA, M., *Introducere în Geometria Spectrală*, Editura Academiei Române, Bucuresti, 1988.

CRAIOVEANU, M. and PUTA, M., *On the continuous dependence on the Riemannian metric of the eigenvalues of the Hodge-de Rham-Laplace operators*, Proc. of the 24th National Conference of Geometry and Topology, Timisoara, July 5-9, 1994, Editura Mirton Timisoara (1996), pp. 85-95.

CRAIOVEANU, M., PUTA, M., and RASSIAS, Th.M., *On the geometry of the Hodge-de Rham Laplace operators*, in: Geometry in Partial Differential Equations, eds. A. Pràstaro and Th.M. Rassias, World Scientific Publ. Co. (1994), pp. 51-67.

DETURCK, D.M., *Audible and inaudible geometric properties*, Rendiconti Seminari Facolta Scienze Università Cagliari, Supplemento al Vol. 58(1988), 1-26.

DETURCK, D.M., and GORDON, C.S., *Isospectral deformations I: Riemannian structures on two-step nilspaces*, Comm. Pure Appl. Math. 40(1987), 367-387.

DETURCK, D.M., and GORDON, C.S., *Isospectral deformations II: trace formulas, metrics and potentials*, Comm. Pure Appl. Math. 42(1989), 1067-1095.

DETURCK, D.M., GLUCK, H., GORDON, C.S., and WEBB, D., *You cannot hear the size of a homology class*, Comment. Math. Helv. 64(1989), 589-617.

DODZIUK, J., *Eigenvalues of the Laplacian on forms*, Proc. Amer. Math. Soc., 85, (3)(1982), 437-443.

GILKEY, P.B., *The spectral geometry of a Riemannian manifold*, J. Differential Geometry 10(1975), 601-618.

GILKEY, P.B., *Invariance Theory, the Heat Equation and the Atiyah-Singer Index Theorem*, Publish or Perish, Wilmington, Delaware, 1984; Second Edition, CRC Press, Boca Raton, Ann Arbor, London, Tokyo, 1995.

GILKEY, P.B., LEAHY, J.V. and PARK, J.H., *Eigenvalues of the form valued Laplacian for Riemannian submersions*, Proc. Amer. Math. Soc. 126(1998), 1845-1850.

GILKEY, P.B. and PARK, J.H., *Riemannian submersions which preserve the eigenforms of the Laplacian*, Illinois J. Math. 40(1996), 194-201.

GORDON, C.S., *Riemannian manifolds isospectral on functions but not on 1-forms*, J. Diff. Geometry 124(1986)$_a$, 79-96.

GORDON, C.S., *The Laplace spectrum versus the length spectra of Riemannian manifolds*, Contemporary Math. 51(1986)$_b$, 63-80.

GORDON, C.S., *Isospectral closed Riemannian manifolds which are not locally isometric*, J. Differential Geometry 37(1993), 639-649.

GORDON, C.S. and WILSON, E.N., *Isospectral deformations of compact solvmanifolds*, J. Diff. Geometry 19(1984), 241-256.

GORDON, C.S., and WILSON, E.N., *The spectrum of the Laplacian on Riemannian Heisenberg manifolds*, Michigan Math. Journal 33(1986), 253-271.

GORNET, R., *Continuous families of Riemannian manifolds, isospectral on functions but not on 1-forms*, The Journal of Geometric Analysis, 10(2000), 281-298.

HERMANN, R., *A sufficient condition that a mapping of Riemannian manifolds be a fibre bundle*, Proc. Amer. Math. Soc. 11(1960), 236-242.

IKEDA, A., *Riemannian manifolds p-isospectral but not (p+1)-isospectral*: in the volume "Geometry of Manifolds", edited by K. Shiohama, Perspectives in Mathematics, Vol. 8, Academic Press, Inc., (1989), 383-417.

IKEDA, A., and TANIGUCHI, Y., *Spectra and eigenforms of the Laplacian on* S^n *and* $P^n(\mathbb{C})$, Osaka J. Math. 15 (1978), 515-546.

KAC, M., *Can one hear the shape of a drum?* Amer. Math. Monthly 73(1966), 1-23.

KATO, T., *Perturbation Theory for Linear Operators*, Second Ed., Springer Verlag, Berlin, Heidelberg, New York, 1976.

KITAOKA, Y., *Positive definite quadratic forms with the same representation numbers*, Arch. Math. 28(1977), 495-497.

KNESER, M., *Lineare Relationen zwischen Darstellungszahlen quadratischer Formen*, Math. Ann. 168(1967), 31-39.

LANG, S., $SL_2(\mathbb{R})$, GTM #105, Springer-Verlag, Berlin, Heidelberg, New York, 1985.

McKEAN, H.P. and SINGER, I.M., *Curvature and the eigenvalues of the Laplacian*, J. Diff. Geometry 1(1967), 43-69.

MIATELLO, R.J. and ROSSETTI, J.P., *Isospectral flat manifolds with diagonal holonomy*, Preprint, 2000.

MIATELLO, R.J. and ROSSETTI, J.P., *Flat manifolds isospectral on p-forms*, Jour. Geom. Analysis, 2001 (to appear).

MILLMAN, R.S., *Remarks on the spectrum of the Laplace-Beltrami operator in the middle dimension*, Tensor N.S. 34(1980), 94-96.

MILNOR, J., *Eigenvalues of the Laplace operator on certain manifolds*, Proc. Nat. Acad. Sci. 51(1964), 542.

MOORE, C.C., *Representations of solvable and nilpotent groups and harmonic analysis on nil and solvmanifolds*, in: Harmonic analysis on homogeneous spaces (Williamstown, Mass., 1972), pp. 3-44, Proc. Sympos. Pure Math., 26, Amer. Math. Soc., Providence, R.I., 1973.

PARK, JEONG HYEONG, *Continuous variation of eigenvalues and Garding's inequality*, Differential Geometry and its Applications, Vol. 10(1999) 187-189.

PARK, JEONG HYEONG, *The spectral geometry of Riemannian submersions for manifolds with boundary*, Rocky Mountain Journal of Mathematics 30(2000), 353-369.

PATODI, V.K., *Curvature and the fundamental solution of the heat operator*, J. Indian Math. Soc. 34(1970), 269-285.

PATODI, V.K., *Curvature and the eigenforms of the Laplace operator*, J. Diff. Geometry 5(1971), 233-249.

PESCE, H., *Borne explicite du nombre de tores plats isospectraux à un tore donné*, Manuscripta Math. 75(1992), 211-223.

PESCE, H., *Représentations de groupes et variétés isospectrales*, Contemp. Math. 173(1994), 231-240.

PESCE, H., *Représentations relativement équivalentes et variétés riemanniennes isospectrales*, Comment. Math. Helvetici 71(1996), 243-268.

PUTA, M., *Some continuity properties for the eigenvalues of the Laplacian on p-forms*, An. Univ. Timisoara, XXXI, fasc. 2(1993), 77-86.

PUTA, M. and TÖRÖK, A., *On the spectrum of the Laplacian on p-forms*, An. Univ. Timisoara, XXV(1987), 67-73.

RAY, D.B. and SINGER, I.M., *R-torsion and the Laplacian on Riemannian manifolds*, Adv. Math. 7(1971), 145-210.

SUNADA, T., *Riemannian coverings and isospectral manifolds*, Ann. of Math. 121(1985), 169-186.

TANNO, S., *The spectrum of the Laplacian for 1-forms*, Proc. Amer. Math. Soc. 45(1974), 125-129.

TSAGAS, Gr. and KOCKINOS, C., *The geometry and the Laplace operator on the exterior 2-forms on a compact Riemannian manifold*, Proc. Amer. Math. Soc. 73(1979), 109-116.

URAKAWA, H., *Bounded domains which are isospectral but not congruent*, Ann. Sci. École Norm. Sup. (4)15(1982), 441-456.

VIGNÉRAS, M.F., *Exemples de sous-groupes discrets non conjugués de PSL(2, $\mathbb{R}$) qui ont même fonction zêta de Selberg*, C.R. Acad. Sci. Paris 287(1978), 47-49.

WOLPERT, S., *The eigenvalue spectrum as moduli for flat tori*, Trans. Amer. Math. Soc. 244(1978), 312-321.

ZHOU, Z., *A note on spectral invariants*, J. Central China Normal Univ. (Natur. Sci.) 31(1997), no. 4, 398-400.

Chapter 6
APPLICATIONS TO GEOMETRY AND TOPOLOGY

In the theory of the Hodge-de Rham operators, the interplay among analysis, topology and geometry is very striking. We discuss in this Chapter some aspects of this interplay, making obvious the power of the heat equation method.

1. THE HODGE-DE RHAM DECOMPOSITION THEOREM

Let (M,g) be a closed, oriented, n-dimensional smooth Riemannian manifold. A fundamental result relating the p-Laplacian $\Delta^{(p)}$, associated to (M,g), to the topology of M is the Hodge-de Rham decomposition theorem. In this Section we shall give a proof of this theorem based on the heat equation for $\Delta^{(p)}$, whose idea is to find a best representative within the de Rham cohomology class of a closed form $\omega_0 \in A^p(M)$. This proof was pioneered by Milgram and Rosenbloom [1951]. The form $\omega_0 \in A^p(M)$ is modified to a form

$$\omega_t = \omega_0 + d^{p-1}\alpha_t, \text{ where } \alpha_t \in A^{p-1}(M).$$

The representative ω_t is expected to be optimal, if $\|\omega_t\|$ is minimal. Since

$$\frac{d}{dt}\left(\frac{1}{2}\|\omega_t\|^2\right) = \langle \dot{\omega}_t, \omega_t \rangle = \langle d^{p-1}\dot{\alpha}_t, \omega_t \rangle = \langle \dot{\alpha}_t, \delta^p \omega_t \rangle,$$

an extremal of $\|\omega_t\|^2$ is characterized by $\delta^p \omega_t = 0$, or because $d^p \omega_t = 0$, equivalently $\Delta^{(p)}\omega_t = 0$. Moreover, the minimum of $\|\omega_t\|^2$ is approached by steepest descent for the choice of $\dot{\alpha}_t = -\delta^p \omega_t$, for which

$$\frac{d}{dt}\left(\frac{1}{2}\|\omega_t\|^2\right) = -\|\delta^p \omega_t\|^2$$

is minimal. But then

$$\dot{\omega}_t = d^{p-1}\dot{\alpha}_t = -d^{p-1}\delta^p \omega_t = -\Delta^{(p)}\omega_t,$$

since $d^p \omega_t = 0$. Thus for a given initial form $\omega_0 \in A^p(M)$, we are led to consider the initial value problem on $A^p(M)$:

$$\begin{cases} \left(\dfrac{\partial}{\partial t} + \Delta^{(p)}\right)\omega(x,t) = 0, \\ \lim\limits_{t \searrow 0} \omega(x,t) = \omega_0(x). \end{cases} \tag{1.1}$$

The unique solution of (1.1) is given by

$$\omega(x,t) = \int_M E^{(p)}(x,y,t) \wedge S^{(p)}\omega_0(y),$$

where $E^{(p)}$ is the p-FSHE on (M,g) (see Chapter 5, §1).

For $\omega \in A^p(M)$ and $t>0$, let

$$\left[\mathcal{P}_t^{(p)}(\omega)\right](x)=\int_M \left[E^{(p)}(x,y,t)\wedge S^{(p)}\omega(y)\right]=\sum_{k=1}^{\infty} e^{-\lambda_k^{(p)}t}\langle \omega_k,\omega\rangle \omega_k(x), \tag{1.2}$$

where the notations are as in Theorem 1.2, Chapter 5. Notice that

$$\mathcal{P}_t^{(p)}: A^p(M)\to A^p(M) \text{ for all } t\in(0,\infty)$$

and

$$\lim_{t\searrow 0}\mathcal{P}_t^{(p)}=1_{A^p(M)}$$

[see Chapter 5, Proposition 1.2 (iii), (iv)]. Moreover, $\mathcal{P}_t^{(p)}$ is a self-adjoint operator for all $t\in(0,\infty)$:

$$\left\langle \mathcal{P}_t^{(p)}(\alpha),\beta\right\rangle=\left\langle \alpha,\mathcal{P}_t^{(p)}(\beta)\right\rangle \tag{1.3}$$

for all $\alpha,\beta\in A^p(M)$ [see Chapter 5, Proposition 1.2 (i)]. Indeed, let

$$f(s,t)=\left\langle \mathcal{P}_s^{(p)}(\alpha),\mathcal{P}_t^{(p)}(\beta)\right\rangle,\qquad s,t\in(0,\infty).$$

Then

$$\begin{aligned}\frac{\partial}{\partial s}f(s,t)&=\left\langle \frac{\partial}{\partial s}\mathcal{P}_s^{(p)}(\alpha),\mathcal{P}_t^{(p)}(\beta)\right\rangle=\left\langle -\Delta^{(p)}\mathcal{P}_s^{(p)}(\alpha),\mathcal{P}_t^{(p)}(\beta)\right\rangle\\&=\left\langle \mathcal{P}_s^{(p)}(\alpha),-\Delta^{(p)}\mathcal{P}_t^{(p)}(\beta)\right\rangle=\left\langle \mathcal{P}_s^{(p)}(\alpha),\frac{\partial}{\partial t}\mathcal{P}_t^{(p)}(\beta)\right\rangle\\&=\frac{\partial}{\partial t}f(s,t).\end{aligned}$$

It follows that $f(s,t)$ is a function of $s+t$, so that

$$\lim_{s\searrow 0}f(s,t)=\lim_{s\searrow 0}f(t,s),$$

which implies (1.3).

Since the solution to the initial value problem (1.1) is unique, one can derive the semi-group property of $\mathcal{P}_t^{(p)}$:

$$\mathcal{P}_s^{(p)}\circ\mathcal{P}_t^{(p)}=\mathcal{P}_{s+t}^{(p)} \tag{1.4}$$

for all $s,t\in(0,\infty)$ [see also the proof of Proposition 1.3 (ii) from Chapter 3 as well as Proposition 1.2 (ii), Chapter 5].

We examine next $\mathcal{P}_t^{(p)}(\omega)$ for $\omega\in A^p(M)$ in its dependence of t.

Lemma 1.1. If $\omega\in A^p(M)$, then $\left\|\mathcal{P}_t^{(p)}(\omega)\right\|^2$ is a non-increasing real function of t for $t>0$.

Proof. We have

$$\frac{d}{dt}\left\|\mathcal{P}_t^{(p)}(\omega)\right\|^2=\frac{d}{dt}\left\langle\mathcal{P}_t^{(p)}(\omega),\mathcal{P}_t^{(p)}(\omega)\right\rangle=2\left\langle\frac{d}{dt}\mathcal{P}_t^{(p)}(\omega),\mathcal{P}_t^{(p)}(\omega)\right\rangle$$
$$=-2\left\langle\Delta^{(p)}\mathcal{P}_t^{(p)}(\omega),\mathcal{P}_t(\omega)\right\rangle=-2\left(\left\|d^p\mathcal{P}_t^{(p)}(\omega)\right\|^2+\left\|\delta^p\mathcal{P}_t^{(p)}(\omega)\right\|^2\right)\le 0.$$

Q.E.D.

Now we turn to the discussion of the behaviour of $\mathcal{P}_t^{(p)}(\omega)$ for $t\to\infty$. The result is as follows.

Theorem 1.1. Let $\omega\in A^p(M)$.

(i) $\mathcal{P}_t^{(p)}(\omega)$ converges uniformly for $t\to\infty$;

(ii) $H^{(p)}(\omega)=\lim\limits_{t\to\infty}\mathcal{P}_t^{(p)}(\omega)\in A^p(M)$ is $\Delta^{(p)}$-harmonic.

Proof. (i) By the previous Lemma, there exists

$$\lim_{t\to\infty}\left\|\mathcal{P}_t^{(p)}(\omega)\right\|^2=l.$$

On the other hand, using (1.4) and (1.3) one gets

$$\begin{aligned}\left\|\mathcal{P}_{t+2s}^{(p)}(\omega)-\mathcal{P}_t^{(p)}(\omega)\right\|^2&=\left\|\mathcal{P}_{t+2s}^{(p)}(\omega)\right\|^2+\left\|\mathcal{P}_t^{(p)}(\omega)\right\|^2-2\left\langle\mathcal{P}_{t+2s}^{(p)}(\omega),\mathcal{P}_t^{(p)}(\omega)\right\rangle\\&=\left\|\mathcal{P}_{t+2s}^{(p)}(\omega)\right\|^2+\left\|\mathcal{P}_t^{(p)}(\omega)\right\|^2-2\left\langle\mathcal{P}_s^{(p)}\circ\mathcal{P}_{t+s}^{(p)}(\omega),\mathcal{P}_t^{(p)}(\omega)\right\rangle\\&=\left\|\mathcal{P}_{t+2s}^{(p)}(\omega)\right\|^2+\left\|\mathcal{P}_t^{(p)}(\omega)\right\|^2-2\left\langle\mathcal{P}_{t+s}^{(p)}(\omega),\mathcal{P}_s^{(p)}\circ\mathcal{P}_t^{(p)}(\omega)\right\rangle\\&=\left\|\mathcal{P}_{t+2s}^{(p)}(\omega)\right\|^2+\left\|\mathcal{P}_t^{(p)}(\omega)\right\|^2-2\left\|\mathcal{P}_{t+s}^{(p)}(\omega)\right\|^2,\end{aligned}$$

which tends to $l+l-2l=0$ for $t\to\infty$. Thus $\left\{\mathcal{P}_t^{(p)}(\omega)\right\}_{t>0}$ has the Cauchy property in $L^2A^p(M)$, so that converges in $L^2A^p(M)$ for $t\to\infty$.

To prove uniform convergence, let us fix $s\in(0,+\infty)$. Then, using the Cauchy-Schwarz inequality in $L^2A^p(M)$ and the continuity of $E^{(p)}$, we obtain for all $t\in(0,+\infty)$ and $x\in M$:

$$\begin{aligned}\left|\mathcal{P}_{t+2s}^{(p)}(\omega)-\mathcal{P}_{t+s}^{(p)}(\omega)\right|(x)&=\left|\mathcal{P}_s^{(p)}\circ\left(\mathcal{P}_{t+s}^{(p)}(\omega)-\mathcal{P}_t^{(p)}(\omega)\right)\right|(x)\\&=\left|\int_M E^{(p)}(x,y,s)\wedge S^{(p)}\left(\mathcal{P}_{t+s}^{(p)}(\omega)-\mathcal{P}_t^{(p)}(\omega)\right)(y)\right|\\&=\left|\left\langle E^{(p)}(x,\cdot,s),\mathcal{P}_{t+s}^{(p)}(\omega)-\mathcal{P}_t^{(p)}(\omega)\right\rangle\right|\\&\le\left\|E^{(p)}(x,\cdot,s)\right\|\left\|\mathcal{P}_{t+s}^{(p)}(\omega)-\mathcal{P}_t^{(p)}(\omega)\right\|\\&\le\sup_{z\in M}\left\|E^{(p)}(z,\cdot,s)\right\|\left\|\mathcal{P}_{t+s}^{(p)}(\omega)-\mathcal{P}_t^{(p)}(\omega)\right\|,\end{aligned}$$

which converges to 0 for $t\to\infty$ by the above remark. It follows that $\mathcal{P}_t^{(p)}(\omega)$ converges uniformly on M, as $t\to\infty$, to a continuous p-form which we denote by $H^p(\omega)$.

(ii) Similarly, using again the Cauchy-Schwarz inequality, we notice first that for fixed $s\in(0,+\infty)$ and all $x\in M$,

$$\begin{aligned}\left|\mathcal{P}_{s+t}^{(p)}(\omega)-\mathcal{P}_s^{(p)}\left(H^{(p)}(\omega)\right)\right|(x)&=\left|\mathcal{P}_s^{(p)}\circ\left(\mathcal{P}_t^{(p)}(\omega)-H^{(p)}(\omega)\right)\right|(x)\\&=\left|\int_M E^{(p)}(x,y,s)\wedge S^{(p)}\left(\mathcal{P}_t^{(p)}(\omega)-H^{(p)}(\omega)\right)(y)\right|\\&=\left|\left\langle E^{(p)}(x,\cdot,s),\mathcal{P}_t^{(p)}(\omega)-H^{(p)}(\omega)\right\rangle\right|\\&\le\sup_{z\in M}\left\|E^{(p)}(z,\cdot,s)\right\|\cdot\left\|\mathcal{P}_t^{(p)}(\omega)-H^{(p)}(\omega)\right\|,\end{aligned}$$

which implies for $t\to\infty$:

$$0=\lim_{t\to\infty}\left[\mathcal{P}_{s+t}^{(p)}(\omega)-\mathcal{P}_s^{(p)}\left(H^{(p)}(\omega)\right)\right]=H^p(\omega)-\mathcal{P}_s^{(p)}\left(H^p(\omega)\right),$$

that is

$$\mathcal{P}_s^{(p)}\left(H^{(p)}(\omega)\right)=H^p(\omega), \tag{1.5}$$

for all $s\in(0,+\infty)$ [see Chapter 5, Proposition 1.2 (iii)], from which one has that $H^{(p)}(\omega)\in A^p(M)$.

On the other hand, it follows from (1.5) that

$$\Delta^{(p)}\left(H^{(p)}(\omega)\right)=\Delta^{(p)}\left(\mathcal{P}_s^{(p)}\left(H^{(p)}(\omega)\right)\right)=-\frac{\partial}{\partial s}\mathcal{P}_s^{(p)}\left(H^{(p)}(\omega)\right)=-\frac{\partial}{\partial s}H^{(p)}(\omega)=0,$$

i.e. $H^{(p)}(\omega)$ is $\Delta^{(p)}$-harmonic.

Since M is closed, note that for $p=0$, the limit function $H^{(0)}(f)$, $f\in C^\infty(M)$, is constant.

For $\omega\in A^p(M)$, let us define now

$$\left[G^{(p)}(\omega)\right](x)=\int_0^\infty\left[\mathcal{P}_t^{(p)}(\omega)-H^{(p)}(\omega)\right](x)\,dt, \tag{1.6}$$

$x\in M$.

Lemma 1.2. $G^{(p)}(\omega)$ is well-defined and $G^{(p)}(\omega)\in A^p(M)$.

Proof. It follows from (1.2) that

$$H^{(p)}(\omega)=\lim_{t\to\infty}\mathcal{P}_t^{(p)}(\omega)=\sum_{j,\lambda_j^{(p)}=0}\langle\omega,\omega_j\rangle\omega_j$$

since the contribution of the strictly-positive eigenvalues $\lambda_j^{(p)}$ disappears for $t\to\infty$. Consequently,

$$\mathcal{P}_t^{(p)}(\omega)-H^{(p)}(\omega)=\sum_{j,\lambda_j^{(p)}\neq0}e^{-\lambda_j^{(p)}t}\langle\omega,\omega_j\rangle\omega_j\,. \tag{1.7}$$

For fixed $s \in (0,+\infty)$ and each $x \in M$ and $t \in (0,+\infty)$, using (1.5) and (1.7) one gets

$$\begin{aligned}
\left|\mathcal{P}_{t+s}^{(p)}(\omega) - H^{(p)}(\omega)\right|(x) &= \left|\mathcal{P}_s^{(p)} \circ \left[\mathcal{P}_t^{(p)}(\omega) - H^{(p)}(\omega)\right]\right|(x) \\
&= \left|\int_M E^{(p)}(x,y,s) \wedge S^{(p)}\left[\mathcal{P}_t^{(p)}(\omega) - H^{(p)}(\omega)\right](y)\right| \\
&\leq \sup_{z\in M}\left\|E^{(p)}(z,\cdot,s)\right\| \cdot \left\|\mathcal{P}_t^{(p)}(\omega) - H^{(p)}(\omega)\right\| \\
&= c\left\|\sum_{j,\lambda_j^{(p)}\neq 0} e^{-\lambda_j^{(p)}t}\langle\omega,\omega_j\rangle\omega_j\right\| \\
&\leq ce^{-\lambda_k^{(p)}t}\left\|\sum_j\langle\omega,\omega_j\rangle\omega_j\right\| = c\|\omega\|e^{-\lambda_k^{(p)}t},
\end{aligned}$$

where $\lambda_k^{(p)}$ denotes the first non-zero eigenvalue of $\Delta^{(p)}$.

It follows that

$$\begin{aligned}
\left|\int_0^{t+s}\left[\mathcal{P}_{t+s}^{(p)}(\omega) - H^{(p)}(\omega)\right](x)dt\right| &\leq \int_0^{t+s}\left|\mathcal{P}_{t+s}^{(p)}(\omega) - H^{(p)}(\omega)\right|(x)dt \\
&\leq c\|\omega\|\int_0^{t+s} e^{-\lambda_k^{(p)}t}dt = \frac{c\|\omega\|}{\lambda_k^{(p)}}\left[1 - e^{-\lambda_k^{(p)}(t+s)}\right].
\end{aligned}$$

Since the last parenthesis converges as $t \to \infty$, it follows that the right hand side of (1.7) exists as the limit for $t \to \infty$ of the integral $\int_0^t$, uniformly in x. Since the integrand is a smooth p-form for each $t \in (0,+\infty)$, one deduces that $G^{(p)}(\omega) \in A^p(M)$.

Q.E.D.

Now we can prove the **Hodge-de Rham decomposition theorem** (see de Rham [1960]).

Theorem 1.2. (**Hodge-de Rham**). For each $p, 0 \leq p \leq n$, the following orthogonal decomposition of $A^p(M)$ is true:

$$A^p(M) = \operatorname{Im} H^{(p)} \overset{\perp}{\oplus} \operatorname{Im} d^{p-1} \overset{\perp}{\oplus} \operatorname{Im} \delta^{p+1}.$$

Proof. Indeed, for each $\omega \in A^p(M), H^{(p)}(\omega)$ is $\Delta^{(p)}$-harmonic [see Theorem 1.1 (ii)], so that

$$\begin{aligned}\Delta^{(p)}\left[G^{(p)}(\omega)\right] &= \Delta^{(p)}\int_0^\infty\left[\mathcal{P}_t^{(p)}(\omega)-H^{(p)}(\omega)\right]dt\\ &= \int_0^\infty \Delta^{(p)}\left[\mathcal{P}_t^{(p)}(\omega)-H^{(p)}(\omega)\right]dt = \int_0^\infty \Delta^{(p)}\mathcal{P}_t^{(p)}(\omega)\,dt\\ &= \int_0^\infty -\frac{\partial}{\partial t}\left(\mathcal{P}_t^{(p)}(\omega)\right)dt = \lim_{t\downarrow 0}\mathcal{P}_t^{(p)}(\omega)-\lim_{t\to\infty}\mathcal{P}_t^{(p)}(\omega)\\ &= \omega - H^{(p)}(\omega),\end{aligned}$$

i.e.

$$\omega = H^{(p)}(\omega)+\Delta^{(p)}\left(G^{(p)}(\omega)\right) = H^{(p)}(\omega)+d^{p-1}\left[\delta^p G^{(p)}(\omega)\right]+\delta^{p+1}\left[d^p G^{(p)}(\omega)\right].$$

The other conditions of the orthogonal decomposition are easily verified.

Q.E.D.

An immediate application of Theorem 1.2 is the unique representability of de Rham cohomology classes by harmonic forms.

Corollary 1.1. (**Hodge-de Rham**) Let the hypotheses be as before. Then $H^p(M)\cong Ker\Delta^{(p)}, 0\le p\le n$, where $H^p(M)$ denotes the real p-th cohomology space of M.

Proof. The mapping

$$\Phi : Ker\left(\Delta^{(p)}\right)\to H^p(M), \Phi(\omega)=[\omega]= \text{ the de Rham cohomology class of } \omega,$$

is a well-defined linear morphism, since $\Delta^{(p)}\omega=0$ if and only if $d^p\omega=0$ and $\delta^p\omega=0$ [see Chapter 2, Proposition 2.10 (iii)].

Φ is injective. Indeed, if $\Phi(\omega)=0$, then $\omega=d^{p-1}\alpha$, where $\alpha\in A^{p-1}(M)$. But $\omega\in\ker\left(\Delta^{(p)}\right)$, so that $\delta^p\omega=0$, i.e. $\delta^p\left(d^{p-1}\alpha\right)=0$. Therefore,

$$\left\|d^{p-1}\alpha\right\|^2=\left\langle d^{p-1}\alpha, d^{p-1}\alpha\right\rangle=\left\langle\alpha,\delta^p\left(d^{p-1}\alpha\right)\right\rangle=0,$$

which implies that $d^{p-1}\alpha=0$, or equivalently $\omega=0$.

Φ is surjective. Let $[\omega]\in H^p(M)$, where ω is a closed form of degree p. Consider the Hodge-de Rham decomposition

$$\omega=H^{(p)}(\omega)+d^{p-1}\alpha+\delta^{p+1}\beta,$$

with $H^{(p)}: A^p(M)\to Ker\left(\Delta^{(p)}\right)$ the orthogonal projection to harmonic forms of degree p. Then

$$0=d^p\omega=d^p\delta^{p+1}\beta$$

implies

$$0=\left\langle d^p\delta^{p+1}\beta,\beta\right\rangle=\left\langle\delta^{p+1}\beta,\delta^{p+1}\beta\right\rangle=\left\|\delta^{p+1}\beta\right\|^2,$$

hence $\delta^{p+1}\beta = 0$. It follows that

$$\omega = H^{(p)}(\omega) + d^{p-1}\alpha ,$$

which shows that ω is cohomologous to its harmonic representative $H^{(p)}(\omega)$, i.e. $[\omega] = \left[H^{(p)}(\omega)\right]$. Consequently,

$$\Phi\left(H^{(p)}(\omega)\right) = [\omega].$$

Q.E.D.

2. VANISHING THEOREMS FOR THE REAL COHOMOLOGY OF CLOSED RIEMANNIAN MANIFOLDS

In this Section we shall establish vanishing theorems for the real cohomology of closed Riemannian manifolds based on the positivity of certain curvature expressions, that is theorems which assert that under certain assumptions on curvature some cohomology groups are trivial. More precisely, using the heat equation, we prove that if the curvature term in the pointwise Weitzenböck formula is strictly positive, then any closed form is exact. As illustrations, we present the case of strictly positive Ricci curvature as well as the case of a positive curvature operator (see Gallot and Meyer [1975]), exhibiting very nicely the interaction between spectral properties (the heat kernel), topology (cohomology) and geometry (curvature).

Let (M,g) be a closed, oriented, n-dimensional smooth Riemannian manifold $(n \geq 2)$. The curvature tensor field R of the Levi-Civita connection on M determines at each point $x \in M$ an operator

$$\tilde{R}_x : \Lambda^2 T_x M \to \Lambda^2 T_x M ,$$

called the **curvature operator** at x associated to R: if $(e_1,\ldots,e_n)$ is an orthonormal basis for T_xM, then with respect to the orthonormal basis $\left(\left(e_i \wedge e_j\right)_{1\leq i<j\leq n}\right)$ for $\Lambda^2 T_x M, \tilde{R}_x$ is defined by

$$\tilde{R}_x\left(e_l \wedge e_k\right) = \sum_{1\leq i<j\leq n} g_x\left(R_x\left(e_l, e_k\right)e_j, e_i\right)e_i \wedge e_j .$$

Similarly, the Ricci tensor field c_2R on M (see Chapter 1, Definition 4.4) defines at each point $x \in M$, the operator

$$Ric_x : T_x M \to T_x M,\; Ric_x\left(e_i\right) = -\sum_{j=1}^{n}\left(c_2 R\right)_x\left(e_i, e_j\right)e_j ,$$

called the **Ricci curvature operator**. Notice that for each $x \in M$, $\tilde{R}_x$ and Ric_x are symmetric operators and $g_x\left(Ric_x(u), v\right) = -\left(c_2 R\right)_x(u,v)$ for all $u, v \in T_x M$.

The heat equation method yields the following vanishing theorems.

Theorem 2.1. Let (M,g) be as before and assume the Ricci curvature operator Ric_x of (M,g) to be strictly positive for each $x \in M$. Then $H^1(M) = 0$.

Thus a manifold M with the de Rham cohomology group $H^1(M) \neq 0$ admits no

Riemannian metric with strictly negative Ricci curvature.

Theorem 2.2. Let (*M*,*g*) be as before and assume the curvature operator $\tilde{R}_x$ to be positive for each $x \in M$. Then $H^p(M)=0$ for $1 \le p \le n-1$, i.e. *M* has the same real cohomology as the *n*-dimensional sphere S^n, that is $H^*(M) \cong H^*(S^n)$. Moreover, if *M* is simply-connected and $n \ge 4$, *M* is even homeomorphic to S^n.

Now let us prove the above theorems. First, let us recall the following **pointwise Weitzenböck formula**: for $0 < p \le n$ and any $\omega \in A^p(M)$,

$$\left(\Delta^{(p)}\omega \middle| \omega\right) = \frac{1}{2}\Delta\left(|\omega|^2\right) + \left|\overset{\circ}{\nabla}\omega\right|^2 + \left(W^{(p)}(\omega)\middle|\omega\right),$$

where $\overset{\circ}{\nabla}$ denotes the covariant differential (see Chapter 1, Definition 3.3), and $W^{(p)} : A^p(M) \to A^p(M)$ is an algebraic operator involving the curvature tensor field associated to the Levi-Civita connection on (M,g). For $p=1$, the operator $W^{(p)}$ is well known to be given as follows.

Lemma 2.1. **(Bochner's formula)** If $\omega \in A^1(M)$, then

$$\left(W^{(1)}(\omega)\right)(x) = \omega_x \circ Ric_x, \quad x \in M,$$

where $Ric_x : T_xM \to T_xM$ is the Ricci curvature operator.

Indeed, the Bochner-Lichnerowicz formula (see Chapter 2, Proposition 2.4) is actually valid for any $\omega \in A^1(M)$:

$$\left(\Delta^{(1)}(\omega)\middle|\omega\right) = \frac{1}{2}\Delta\left(|\omega|^2\right) + \left|\overset{\circ}{\nabla}(\omega)\right|^2 - (c_2R)\left(\omega^\#, \omega^\#\right),$$

where $\# : A^1(M) \to \mathfrak{X}(M)$ denotes the canonical $C^\infty(M)$-isomorphism induced by the Riemannian metric *g* defined in Chapter 1, §2 (for a proof in the general case, see Lichnerowicz [1958]). On the other hand, let us define $W^{(1)} : A^1(M) \to A^1(M)$ by

$$W^{(1)}(\omega) = \left[Ric\left(\omega^\#\right)\right]^b,$$

$\omega \in A^1(M)$, where $Ric : \mathfrak{X}(M) \to \mathfrak{X}(M)$ is viewed here as the smooth tensor field of type (1,1) given by

$$Ric(X) = -\sum_{j=1}^{n}(c_2R)\left(X, E_j\right)E_j, \quad X \in \mathfrak{X}(M),$$

where $(E_1, \ldots, E_n)$ is a local field of orthonormal frames on (*M*,*g*), and $b : \mathfrak{X}(M) \to A^1(M)$ is the inverse of the isomorphism # (see Chapter 1, §2). Equivalently, $W^{(1)}(\omega) = \omega \circ Ric$: indeed, since *Ric* is a symmetric operator with respect to *g*, one gets

$$\left[W^{(1)}(\omega)\right](X)=g\left(Ric\left(\omega^{\#}\right),X\right)=g\left(\omega^{\#},Ric(X)\right)=\omega\left(Ric(X)\right),$$

for any $X\in\mathfrak{X}(M)$. Finally note that

$$-(c_2R)\left(\omega^{\#},\omega^{\#}\right)=g\left(Ric\left(\omega^{\#}\right),\omega^{\#}\right)=\left(\left[Ric\left(\omega^{\#}\right)\right]^{b}|\omega\right)=\left(W^{(1)}(\omega)|\omega\right)$$

for any $\omega\in A^1(M)$.

For arbitrary p, $0<p<n$, the self-adjoint operator $W^{(p)}$ can be described as follows.

Let $x\in M$. The symmetric curvature operator is positive definite for $1\leq p\leq n-1$, that is

$$\left(W^{(p)}(\omega)\middle|\,\omega\right)\geq c|\omega|^2 \tag{2.1}$$

for all $\omega\in A^p(M)$, where $c>0$.

Proof. Let $\alpha_1>0$ be the smallest eigenvalue. By Lemma 2.2 we have then

$$\begin{aligned}\left(W^{(p)}(\omega)\middle|\,\omega\right)&=-\left(\sum_{k=1}^{m}\alpha_k A_k\cdot A_k\cdot\omega\,\middle|\,\omega\right)\\&=\sum_{k=1}^{m}\alpha_k\left(A_k\cdot\omega\,\middle|\,A_k\cdot\omega\right)\geq\alpha_1\sum_{k=1}^{m}|A_k\cdot\omega|^2.\end{aligned}$$

The sum

$$\sum_{k=1}^{m}|A_k\cdot\omega|^2$$

is non-zero for non-zero ω, because the representation of the Lie algebra $o(m)$ on p-forms is irreducible for $p\neq 0,n$. In fact,

$$\sum_{k=1}^{m}|A_k\cdot\omega|^2=p(n-p)|\omega|^2$$

(see Gallot and Meyer [1975], Cor. 2.6). Hence

$$\left(W^{(p)}(\omega)\middle|\,\omega\right)\geq c|\omega|^2,$$

where $c=p(n-p)\alpha_1$.

Q.E.D.

Remark 2.1. In the case of 1-forms, we have by Lemma 2.3 the same inequality (2.1), except that now c is the minimum eigenvalue of the Ricci curvature operator.

Let us consider now the solution $\omega(\cdot,t)=e^{-t\Delta^{(p)}}(\omega_0)$ of the heat equation $\left(\dfrac{\partial}{\partial t}+\Delta^{(p)}\right)\omega=0$ with initial condition $\lim_{t\downarrow 0}\omega(\cdot,t)=\omega_0$, $1\leq p\leq n-1$. By Theorem 1.3 (i), Chapter 5, for a closed form ω_0 of degree p, the forms $\omega(\cdot,t)$ are also closed forms for $t\in(0,+\infty)$. Let us compute the pointwise derivative

$$\frac{1}{2}\frac{\partial}{\partial t}|\omega(\cdot,t)|^2 = -\left(\Delta^{(p)}\omega(\cdot,t)\middle|\,\omega(\cdot,t)\right). \tag{2.2}$$

By the pointwise Weitzenböck formula

$$\left(\Delta^{(p)}\theta\,|\,\theta\right) = \frac{1}{2}\Delta\left(|\theta|^2\right) + \left|\overset{\circ}{\nabla}\theta\right|^2 + \left(W^{(p)}(\theta)\,|\,\theta\right)$$

valid for all $\theta \in A^p(M)$ (see Lichnerowicz [1955]), we have then

$$\left(\Delta^{(p)}\omega(\cdot,t)\middle|\,\omega(\cdot,t)\right) = \frac{1}{2}\Delta^{(p)}\left(|\omega(\cdot,t)|^2\right) + \left|\overset{\circ}{\nabla}\omega(\cdot,t)\right|^2 + \left(W^{(p)}\omega(\cdot,t)\middle|\,\omega(\cdot,t)\right),$$

which by (2.2) implies that

$$\frac{1}{2}\left(\frac{\partial}{\partial t} + \Delta^{(p)}\right)|\omega(\cdot,t)|^2 = -\left|\overset{\circ}{\nabla}\omega(\cdot,t)\right|^2 - \left(W^{(p)}\omega(\cdot,t)\middle|\,\omega(\cdot,t)\right) \le -c|\omega(\cdot,t)|^2$$

with $c > 0$.

The maximum principle now implies that

$$e^{ct}\max|\omega(\cdot,t)|^2$$

is a nonincreasing function of t, hence

$$\max|\omega(\cdot,t)|^2 \le e^{-ct}\max|\omega_0|^2$$

for all $t \in [0,+\infty)$. This proves that

$$\lim_{t\to\infty}\omega(\cdot,t) = 0$$

in the C^0-norm. By the smoothing properties of the heat semigroup $e^{-t\Delta^{(p)}}$, it follows that

$$\lim_{t\to\infty}\omega(\cdot,t) = 0$$

in the C^∞-norm. Since by Theorem 1.3 (iii), Chapter 5, the cohomology class of $\omega(\cdot,t)$ is the same for all $t \in [0,+\infty)$, the proof of Theorems 2.1 and 2.2 is complete, except the last statement in Theorem 2.2, which is a direct consequence of a result of Moore [1986].

Remark 2.2. If the Ricci curvature ρ of (M,g) is non-positive and if $\dim H^1(M) = n$, then (M,g) is isometric to an n-dimensional flat torus.

Indeed, by integrating the pointwise Weitzenböck formula $(p=1)$ on M, one gets

$$\left\langle \Delta^{(1)}(\omega),\omega\right\rangle = \left\|\overset{\circ}{\nabla}(\omega)\right\|^2 + \int_M -(c_2R)\left(\omega^\#,\omega^\#\right)d\mu_g .$$

Consequently, by Remark 2.7 of Chapter 2, it follows that $\Delta^{(1)}\omega = 0$ if and only if $\overset{\circ}{\nabla}\omega = 0$. Hence

$$\dim\left(\ker\overset{\circ}{\nabla}\right)=\dim H^1(M)=n.$$

But a form whose covariant differential vanishes is invariant by parallel transport. Therefore it is determined by its value at a point, so that there exist n linearly independent parallel 1-forms, or equivalently, n linearly independent parallel vector fields $X_1,\dots,X_n$. Since $\nabla_X X_j=0$ for any $j=1,\dots,n$ and $X\in\mathfrak{X}(M)$, it follows that $R(X_j,X_k)X_l=0$ for any $j,k,l=1,\dots,n$, that is (M,g) is a flat space. A theorem of Bieberbach says that if (M,g) is a flat closed Riemannian manifold, $\pi_1(M)$ its fundamental group, and $\Gamma=\pi_1(M)\cap\mathbb{R}^n$ the subgroup of translations in $\pi_1(M)$, then Γ is a free abelian normal subgroup of rank n of $\pi_1(M)$ and $\mathbb{R}^n/\Gamma$ is a torus which covers M with the group of deck transformations Γ, and the claimed statement follows from this theorem.

3. LEFSCHETZ FIXED POINT THEOREM

Following P.S. Gilkey (see Gilkey [1984], [1993]) we illustrate with a heat equation method for the proof of the Lefschetz fixed point theorem for the de Rham complex.

Let (M,g) be a closed, n-dimensional smooth Riemannian manifold and $\Phi: M\to M$ a smooth mapping. Let Φ^* denote the natural action by pull-back on $A^p(M)$, $0\le p\le n$. Since

$$d^p\circ\Phi^*=\Phi^*\circ d^p, \tag{3.1}$$

where $d^p: A^p(M)\to A^{p+1}(M)$ denotes the exterior derivative, $0\le p\le n-1$, Φ^* induces a natural action on the de Rham real cohomology groups $H^p(M)=\ker(d^p)/\operatorname{Im}(d^{p-1})$, $0\le p\le n$, denoted also by $\Phi^*: H^p(M)\to H^p(M)$, $\Phi^*([\omega])=[\Phi^*(\omega)]$ for any closed p-form $\omega\in A^p(M)$. **The Lefschetz number** $L(\Phi)$ of Φ is defined by

$$L(\Phi)=\sum_{p=0}^{n}(-1)^p\, Trace\left(\Phi^* \text{ on } H^p(M)\right).$$

This is always an integer. Note that if Φ is the identity map 1_M, then $Trace\left(\Phi^* \text{ on } H^p(M)\right)=\dim H^p(M)$, so that $L(1_M)=\mathfrak{X}(M)$ (= the Euler characteristic of M). More generally, if the smooth mapping $\Phi: M\to M$ is such that $\Phi^*=1_{H^p(M)}: H^p(M)\to H^p(M)$ for any p, $0\le p\le n$, then $Trace\left(\Phi^* \text{ on } H^p(M)\right)$ is just $\dim H^p(M)$ (= the p-th Betti number of M) and therefore $L(\Phi)=\mathfrak{X}(M)$. For example, this is the case when Φ is homotopic with 1_M. This last remark is a particular case of a more general result, namely if $\Phi_1,\Phi_2: M\to M$ are homotopic maps, then $L(\Phi_1)=L(\Phi_2)$. Let us consider another example. If $M=\mathbb{R}^2/\mathbb{Z}^2$ is the two dimensional standard torus, the mapping

$$\varphi:\mathbb{R}^2\to\mathbb{R}^2,\ \varphi(x,y)=(n_1x+n_2y, n_3x+n_4y),\ (x,y)\in\mathbb{R}^2$$

where n_1,n_2,n_3,n_4 are integers, preserves the integer lattice $\mathbb{Z}^2$, so that it defines the smooth mapping $\Phi : M \to M$. Since

$$\Phi^*(1)=1,\ \Phi^*(dx)=n_1dx+n_2dy,\ \Phi^*(dy)=n_3dx+n_4dy,$$
$$\Phi^*(dx\wedge dy)=\Phi^*(dx)\wedge\Phi^*(dy)=(n_1n_4-n_2n_3)dx\wedge dy,$$

it follows that

$$L(\Phi)=1-(n_1+n_4)+(n_1n_4-n_2n_3).$$

If M is the n-sphere S^n, it is easy to see that $L(\Phi)=1+(-1)^n\deg(\Phi)$ for any smooth mapping $\Phi : S^n \to S^n$, where deg(Φ) denotes the topological degree of Φ.

If $\Delta^{(p)} : A^p(M)\to A^p(M)$ denotes the Hodge-de Rham operator (p-Laplacian) associated to (M,g) (see Chapter 2, §2), the operator is of trace class on $L^2A^p(M)$ for $t>0$ (see Chapter 5, §1). Φ^* defines a bounded operator on $L^2A^p(M)$, so that $Trace\left(\Phi^* e^{-t\Delta^{(p)}}\right)$ is well-defined for $t>0$. More explicitly, if $\left\{\omega_k^{i_k}\mid k\in\mathbb{N}, i_k=1,\ldots,m_k\right\}$ is a complete orthonormal system for $L^2A^p(M)$, where $\left\{\omega_k^{i_k}\mid i_k=1,\ldots,m_k\right\}$ is a basis for the eigenspace $E\left(\Delta^{(p)},\lambda_k\right)\subset A^p(M)$ of $\Delta^{(p)}$ corresponding to the eigenvalue λ_k, $m_k=\dim E\left(\Delta^{(p)},\lambda_k\right)$, then

$$Trace\left(\Phi^* e^{-t\Delta^{(p)}}\right)=\sum_{k\in\mathbb{N}}\sum_{i_k=1}^{m_k} e^{-\lambda_k t}\left\langle\Phi^*\left(\omega_k^{i_k}\right),\omega_k^{i_k}\right\rangle,\ t>0. \tag{3.2}$$

Moreover, Corollary 2.3, Chapter 5, can be generalized in order to compute $L(\Phi)$ in terms of these traces.

Proposition 3.1. If $\Phi : M\to M$ is a smooth mapping, then

$$\sum_{p=0}^{n}(-1)^p\, Trace\left(\Phi^* e^{-t\Delta^{(p)}}\right)=L(\Phi).$$

Proof. Let $\pi(\lambda_k,p)$ be the orthogonal projection from $L^2A^p(M)$ to $E(\Delta^{(p)},\lambda_k)$. Note that the mapping $T : A^p(M)\to E\left(\Delta^{(p)},\lambda_k\right), T(\omega)=\sum_{i_k=1}^{m_k}\left\langle\Phi^*(\omega),\omega_k^{i_k}\right\rangle\omega_k^{i_k}$, $\omega\in A^p(M)$, is just the composition $\pi(\lambda_k,p)\circ\Phi^*$; consequently,

$$Trace\left(\pi(\lambda_k,p)\circ\Phi^*\Big|_{E\left(\Delta^{(p)},\lambda_k\right)}\right)=\sum_{i_k=1}^{m_k}\left\langle\Phi^*\left(\omega_k^{i_k}\right),\omega_k^{i_k}\right\rangle,$$

so that (3.2) implies

$$Trace\left(\Phi^* e^{-t\Delta^{(p)}}\right)=\sum_{k\in\mathbb{N}} e^{-\lambda_k t}\, Trace\left(\pi(\lambda_k,p)\circ\Phi^*\Big|_{E\left(\Delta^{(p)},\lambda_k\right)}\right). \tag{3.3}$$

(3.3) leads to

$$\sum_{p=0}^{n}(-1)^p \, Trace\left(\Phi^* e^{-t\Delta^{(p)}}\right)=\sum_{k\in\mathbf{N}} e^{-\lambda_k t}\sum_{p=0}^{n}(-1)^p \, Trace\left(\pi(\lambda_k,p)\circ\Phi^*\Big|_{E\left(\Delta^{(p)},\lambda_k\right)}\right). \tag{3.4}$$

Taking into account (3.1) and the equalities

$$d^p\circ\pi(\lambda_k,p)=\pi(\lambda_k,p+1)\circ d^p,\quad 0\le p\le n-1,$$

for $\lambda_k\neq 0$ one gets a chain map between the following exact sequences

$$\begin{array}{ccccccccc}
0\to E\left(\Delta^{(0)},\lambda_k\right) & \xrightarrow{d^0}\cdots\xrightarrow{d^{p-1}} & E\left(\Delta^{(p)},\lambda_k\right) & \xrightarrow{d^p} & E\left(\Delta^{(p+1)},\lambda_k\right) & \xrightarrow{d^{p+1}}\cdots\xrightarrow{d^{n-1}} & E\left(\Delta^{(n)},\lambda_k\right)\to 0 \\
\Phi^*\downarrow & & \Phi^*\downarrow & & \Phi^*\downarrow & & \Phi^*\downarrow \\
0\to A^0(M) & \xrightarrow{d^0}\cdots\xrightarrow{d^{p-1}} & A^p(M) & \xrightarrow{d^p} & A^{p+1}(M) & \xrightarrow{d^{p+1}}\cdots\xrightarrow{d^{n-1}} & A^n(M)\to 0 \\
\pi(\lambda_k,0)\downarrow & & \pi(\lambda_k,p)\downarrow & & \pi(\lambda_k,p+1)\downarrow & & \pi(\lambda_k,n)\downarrow \\
0\to E\left(\Delta^{(0)},\lambda_k\right) & \xrightarrow{d^0}\cdots\xrightarrow{d^{p-1}} & E\left(\Delta^{(p)},\lambda_k\right) & \xrightarrow{d^p} & E\left(\Delta^{(p+1)},\lambda_k\right) & \xrightarrow{d^{p+1}}\cdots\xrightarrow{d^{n-1}} & E\left(\Delta^{(n)},\lambda_k\right)\to 0.
\end{array}$$

Since this diagram commutes and since the first and third row are long exact sequences of finite dimensional real vector spaces (see Chapter 5, Remark 2.2), a standard result in homological algebra implies

$$\sum_{p=0}^{n}(-1)^p \, Trace\left(\pi(\lambda_k,p)\circ\Phi^*\Big|_{E\left(\Delta^{(p)},\lambda_k\right)}\right)=0 \text{ for } \lambda_k\neq 0$$

[choose convenient bases in three consecutive vector spaces $E\left(\Delta^{(p-1)},\lambda_k\right), E\left(\Delta^{(p)},\lambda_k\right), E\left(\Delta^{(p+1)},\lambda_k\right)$ and compute $Trace\left(\pi(\lambda_k,p)\circ\Phi^*\Big|_{E\left(\Delta^{(p)},\lambda_k\right)}\right)$, $Trace\left(\pi(\lambda_k,p+1)\circ\Phi^*\Big|_{E\left(\Delta^{(p+1)},\lambda_k\right)}\right)$ using these bases].

On the other hand, let us use Corollary 1.1 to identify $E(\Delta^{(p)},0) = H^p(M)$. Under this identification, the mapping

$$\pi(0,p)\circ\Phi^*\Big|_{E\left(\Delta^{(p)},0\right)}: E\left(\Delta^{(p)},0\right)\to E\left(\Delta^{(p)},0\right)$$

corresponds to the action of Φ^* on $H^p(M)$:

$$A^p(M)$$
$$\Downarrow$$
$$\Phi^*(\omega)$$
$$\nearrow \qquad \searrow$$
$$E\left(\Delta^{(p)},0\right) \ni \omega \mapsto \pi(0,p)\circ\Phi^*(\omega) \in E\left(\Delta^{(p)},0\right)$$
$$\downarrow \qquad\qquad\qquad \downarrow$$
$$H^p(M) \ni [\omega] \mapsto \left[\pi(0,p)\circ\Phi^*(\omega)\right] \in H^p(M).$$

Note that any $\omega \in E\left(\Delta^{(p)},0\right)$ is a closed p-form. Hence the equality of the cohomology classes $\left[\pi(0,p)\circ\Phi^*(\omega)\right]$ and $\left[\Phi^*(\omega)\right]$ is an immediate consequence of the decomposition

$$\Phi^*(\omega) = \pi(0,p)\circ\Phi^*(\omega) + d^{p-1}\delta^p\Phi^*(\omega)$$

(see Theorem 1.2). Consequently,

$$\sum_{p=0}^{n}(-1)^p \, Trace\left(\pi(0,p)\circ\Phi^*\Big|_{E\left(\Delta^{(p)},0\right)}\right) = \sum_{p=0}^{n}(-1)^p \, Trace\left(\Phi^* \text{ on } H^p(M)\right) = L(\Phi)$$

by the definition of the Lefschetz number of Φ. Hence

$$\sum_{p=0}^{n}(-1)^p \, Trace\left(\pi(\lambda_k,p)\circ\Phi^*\Big|_{E\left(\Delta^{(p)},\lambda_k\right)}\right) = \begin{cases} L(\Phi) & \text{if } \lambda_k = 0, \\ 0 & \text{if } \lambda_k > 0, \end{cases}$$

and our assertion will then follow, via (3.4), from this identity.

Q.E.D.

Proposition 3.2. If the smooth mapping $\Phi : M \to M$ has no fixed points, then

$$\left|Trace\left(\Phi^* e^{-t\Delta^{(p)}}\right)\right| \le Ct$$

for any $t \in (0,1)$.

Proof. The proof uses some analytical facts concerning the kernel of the heat equation and we refer to Gilkey [1984] for more details.

Let $\Phi^* = \Lambda^p(d\Phi) : \Lambda^p T^*M \to \Lambda^p T^*M$ be the pull-back operation. It induces a mapping $\Phi^*(x)$ from the fiber over $\Phi(x)$ to the fiber over x. If $E^{(p)}$ denotes the fundamental solution of the p-heat equation (see Chapter 5, Definition 1.1), then

$$\left(e^{-t\Delta^{(p)}}\omega\right)(x) = \int_M \left[E^{(p)}(x,y,t)\right](\omega(y))\, d\mu_g(y).$$

We pull-back $E^{(p)}$ to define

$$\Phi^*\left(E^{(p)}\right)(x,y,t) = \Phi^*(x)\, E^{(p)}(\Phi(x),y,t).$$

Then

$$\left(\Phi^* e^{-t\Delta^{(p)}}\omega\right)(x)=\int_M \Phi^*(x)E^{(p)}(\Phi(x),y,t)\omega(y)d\mu_g(y)$$

and from this last equality it follows that

$$Trace\left(\Phi^* e^{-t\Delta^{(p)}}\right)=\int_M Trace\left(\Phi^*\left(E^{(p)}\right)\right)(\Phi(x),x,t)d\mu_g(x).$$

Since Φ has no fixed points, $\Phi(x)\neq x$ for any $x\in M$, one can bound this integral by Ct for any $t\in(0,1)$ if one bounds the distance between x and $\Phi(x)$ away from zero. It is only on the diagonal of $M\times M$ that there is an asymptotic series.

Q.E.D.

Theorem 3.1. (**Lefschetz**). If $\Phi: M\to M$ has no fixed points, then $L(\Phi)=0$.

Proof. Let $t>0$. Then by Proposition 3.1, it follows that

$$L(\Phi)=\sum_{p=0}^{n}(-1)^p\, Trace\left(\Phi^* e^{-t\Delta^{(p)}}\right).$$

Since Φ is fixed point free, in virtue of Proposition 3.2 one gets

$$|L(\Phi)|\leq Ct \text{ for any } t\in(0,1).$$

Taking the limit as $t\downarrow 0$ one obtains $L(\Phi)=0$.

Q.E.D.

Let us give some applications illustrating the sorts of consequences which follow from this theorem.

Corollary 3.1. If the smooth mapping $\Phi: M\to M$ has the property $L(\Phi)\neq 0$, then Φ has at least a fixed point.

Note that despite of the fact that the identity map $1_{S^1}: S^1\to S^1$ has fixed points, $L(1_{S^1})=\chi(S^1)=0$.

Example 3.1. Let $P^n(\mathbb{R})$ be the n-dimensional real projective space and assume that n is even. Then any smooth mapping $\Phi: P^n(\mathbb{R})\to P^n(\mathbb{R})$ has at least a fixed point. Indeed, the de Rham cohomology of $P^n(\mathbb{R})$ (n even) has the form

$$H^p\left(P^n(\mathbb{R})\right)=\begin{cases}\mathbb{R} & \text{if } p=0,\\ 0 & \text{if } p\neq 0,\end{cases}$$

so that $L(\Phi)=1$, and our statement is an immediate consequence of Corollary 3.1.

Example 3.2. Let $P^n(\mathbb{C})$ be the n-dimensional complex projective space. $P^n(\mathbb{C})$ is a connected, closed, $2n$-dimensional smooth manifold. If n is even and $\Phi: P^n(\mathbb{C})\to P^n(\mathbb{C})$ a smooth mapping, then Φ has at least a fixed point. Indeed, the de Rham cohomology ring $H^*\left(P^n(\mathbb{C})\right)$ of $P^n(\mathbb{C})$ is isomorphic with the quotient ring $\mathbb{R}[x]/(x^{n+1})$, where $\mathbb{R}[x]$ is the ring of polynomials with real coefficients in an unknown x and $(x^{n+1})\subset\mathbb{R}[x]$ the ideal generated by the polynomial x^{n+1}, i.e. $H^*\left(P^n(\mathbb{C})\right)$ is the truncated polynomial ring on a generator $x \in H^2(P^n(\mathbb{C}))$. The elements of the de Rham cohomology ring

$$H^*(P^n(\mathbb{C})) = H^0(P^n(\mathbb{C})) \oplus H^2(P^n(\mathbb{C})) \oplus H^4(P^n(\mathbb{C})) \oplus \cdots \oplus H^{2n}(P^n(\mathbb{C}))$$
$$\cong \mathbb{R} \oplus \mathbb{R}x \oplus \mathbb{R}x^2 \oplus \cdots \oplus \mathbb{R}x^n,$$

can be written uniquely in the form $\lambda_0 + \lambda_1 x + \cdots + \lambda_n x^n$, where $\lambda_j \in \mathbb{R}$, $j = \overline{0,n}$. Let

$$\Phi^* : H^2(P^n(\mathbb{C})) \to H^2(P^n(\mathbb{C})),$$

where $\Phi^*(x) = \lambda x$ with $\lambda \in \mathbb{R}$. Since Φ^* is a ring homomorphism, $\Phi^*(x^n) = (\Phi^*(x))^n$, i.e. all the homomorphisms

$$\Phi^* : H^p(P^n(\mathbb{C})) \to H^p(P^n(\mathbb{C})), \quad p \geq 4 \text{ even},$$

are already determined by $\Phi^* : H^2(P^n(\mathbb{C})) \to H^2(P^n(\mathbb{C}))$.

Consequently,

$$L(\Phi) = 1 + (-1)^2 \lambda + (-1)^4 \lambda^2 + \cdots + (-1)^{2n} \lambda^n = 1 + \lambda + \lambda^2 + \cdots + \lambda^n. \quad (3.5)$$

If n is even, there are no real roots for the polynomial (3.5): the roots are in fact given by

$$\lambda_k = \cos\frac{2k\pi}{n+1} + i\sin\frac{2k\pi}{n+1}, \quad k = 1, \ldots, n,$$

which all are complex [in order to see this fact, write $L(\Phi)$ as the quotient $\lambda^{n+1} - 1 / \lambda - 1$, $\lambda \neq 1$]. Hence $L(\Phi) \neq 0$ if n is even, and by Corollary 3.1 Φ must have a fixed point.

Example 3.3. If $\chi(M) \neq 0$, then any smooth mapping $\Phi : M \to M$ homotopic with 1_M has at least a fixed point. Indeed, in this case $\Phi^*\big|_{H^p(M)} = 1_{H^p(M)}$, $0 \leq p \leq n$, so that $L(\Phi) = \chi(M) \neq 0$. For example, if G is a connected, compact Lie group of dimension $n \geq 1$, then $\chi(G) = 0$. For, let $e \in G$ be the unity element of G, $g_0 \in G \setminus \{e\}$ a fixed element, $\Phi_0 : G \to G$, $\Phi_0(g) = g_0 \cdot g$, $g \in G$, the left translation defined by g_0 and $\alpha : [0,1] \to G$ a continuous path such that $\alpha(0) = g_0$, $\alpha(1) = e$. Φ_0 is fixed point free and hence, by Theorem 3.1, $L(\Phi_0) = 0$. On the other hand, Φ_0 and the identity mapping 1_G are homotopic, a homotopy between these mappings being given by $H : G \times [0,1] \to G$, $H(g,t) = \alpha(t) \cdot g$, $(g,t) \in G \times [0,1]$. Hence $\chi(G) = L(\Phi_0) = 0$. Consequently, the spheres of even dimension and the closed oriented surfaces of genus $g \neq 1$ cannot be Lie groups.

Example 3.4. If M is a connected, closed, $2n$-dimensional smooth manifold, n is even (resp. odd) and $F \to P^n(\mathbb{C}) \to M$ is a finite covering, then $|F| = 1$ (resp. $|F| \leq 2$).

Indeed, let $P^n(\mathbb{C}) \to M$ be a covering projection having as fiber the finite set F. Since $P^n(\mathbb{C})$ is simply connected, $F \cong \pi_1(M)$ acts on $P^n(\mathbb{C})$ as deck transformations, so that if $\varphi \in \pi_1(M), \varphi \neq 1_{P^n(\mathbb{C})}$, φ is fixed point free. Example 3.2 shows that this is impossible for n even, so that in this case $F = \{1_{P^n(\mathbb{C})}\}$. Therefore the cardinal of F is one if n is even. If n is odd, let $\varphi^*(x) = \lambda(\varphi)x$, where $\lambda(\varphi) \in \mathbb{R}$. Since $\varphi : P^n(\mathbb{C}) \to P^n(\mathbb{C})$ is fixed point free, $0 = L(\varphi) = 1 + \lambda(\varphi) + \cdots + (\lambda(\varphi))^n$ and this polynomial has only one real root, namely $\lambda(\varphi) = -1$. Moreover, for $F \neq \{1_{P^n(\mathbb{C})}\}$, the mapping $F \to \{-1,1\}, \varphi \mapsto \lambda(\varphi)$, is an injective group homomorphism, what implies $|F| = 2$. Consequently, $|F| \leq 2$ if n is odd.

4. CHERN-GAUSS-BONNET THEOREM

In this last section one deals with the Chern-Gauss-Bonnet theorem. We will present Gilkey's proof of it, which is based on the heat equation approach. We refer the reader to Gilkey [1984], [1993] for more details.

First let us introduce the formal variables

$$g_{ij/\alpha} = \partial_x^\alpha g_{ij}$$

for the partial derivatives of the Riemannian metric tensor field. We will also use the notation $g_{ij/kl\ldots}$ for multiple partial derivatives. These variables depend upon the choice of a coordinate system, that is they are not tensorial. One defines the order in the jets of the metric by

$$ord(g_{ij/\alpha}) = |\alpha|.$$

Let g be a Riemannian metric for a smooth manifold of dimension n and let $x_0 \in M$. One says that a system of coordinates ξ on M is normalized with respect to (g,x_0) if x_0 has the coordinates $(0,\ldots,0)$ and if

$$g_{ij}(\xi,g)(x_0) = \delta_{ij} \text{ and } g_{ij/k}(\xi,g)(x_0) = 0.$$

Such coordinate systems always exist; for example we could take geodesic polar coordinates. We restrict to coordinate systems which are normalized with respect to (g,x_0) and consequently we may omit the variables $\{g_{ij}, g_{ij/k}\}$ from consideration. For example, in such coordinates, the value of the Riemannian curvature tensor field at x_0 for the coordinate frame $\partial/\partial x_i$ is given by

$$R_{ijkl} = \frac{1}{2}(g_{ik/jl} + g_{jl/ik} - g_{jk/il} - g_{il/jk})(\xi,g)(x_0).$$

Let

$$\mathcal{U}_n^g = \mathbb{R}[g_{ij/\alpha}] \text{ for } |\alpha| \geq 2$$

be the polynomial algebra in these variables. Let $\mathcal{U}_{n,m}^g$ be the linear subspace of polynomials which are homogeneous of order m in the jets of the metric.

If g is a metric, ξ a coordinate system and P a polynomial in these variables, one defines $P(\xi, g)(x_0)$ by evaluation. One says that $P \in \mathcal{U}_n^g$ is **invariant** if

$$P(\xi, g)(x_0) = P(\eta, g)(x_0)$$

for any two normalized [with respect to (g, x_0)] coordinate systems ξ and η. One denotes the common value by $P(g)(x_0)$. Let $I_n^g \subset \mathcal{U}_n^g$ be the ring of all invariant polynomials in the derivatives of the metric which are defined for a manifold of dimension n and $I_{n,m}^g$ the linear subspace of invariant polynomials which are homogeneous of order m.

Let $P \in I_n^g$ and A a monomial. We let $c(A,P)$ be the coefficient of A in P:

$$P = \sum_A c(A, P) A .$$

Then $c(A, \cdot)$ defines a linear functional on I_n^g. One says that A is a **monomial** of P if $c(A, P) \neq 0$.

The definition of $I_{n,m}^g$ is purely algebraic. There is a coordinate free characterization that is useful.

Lemma 4.1. Let $P \in I_n^g$. Then $P \in I_{n,m}^g$ if and only if for all (g, x_0) and $c \neq 0$,

$$P(c^2 g)(x_0) = c^{-m} P(g)(x_0).$$

Proof. Let $P \in I_n^g$. Fix $c \neq 0$ and let ξ be normalized with respect to (g, x_0). Let $\eta = c\xi$ be a new coordinate system. Since

$$\partial_i^y = c^{-1} \partial_i^x, \partial_y^\alpha = c^{-|\alpha|} \partial_x^\alpha ,$$

one gets

$$g(\partial_i^y, \partial_j^y) = c^{-2} g(\partial_i^x, \partial_j^x),$$
$$g_{ij/\alpha}(\eta, g) = c^{-2-|\alpha|} g_{ij/\alpha}(\xi, g)$$

and

$$g_{ij/\alpha}(\eta, c^2 g) = c^{-|\alpha|} g_{ij/\alpha}(\xi, g).$$

If A is any monomial of P, then A is homogeneous of order m in the jets of the metric, so that

$$A(\eta, c^2 g)(x_0) = c^{-m} A(\xi, g)(x_0). \tag{4.1}$$

Since η (resp. ξ) is normalized with respect to $(c^2 g, x_0)$ [resp. (g, x_0)], one can use (4.1) to establish one implication of our Lemma:

$$P(c^2 g)(x_0) = \sum_A c(A,P) A(\eta, c^2 g)(x_0) = \sum_A c^{-m} c(A,P) A(\xi, g)(x_0) = c^{-m} P(g)(x_0).$$

The other implication of the statement follows similarly.

Q.E.D.

Remark 4.1. If $P \in I_n^g$, one can decompose

$$P = P^{(0)} + \cdots + P^{(m)}$$

into homogeneous polynomials. One can use Lemma 4.1 to see that the $P^{(k)}$ are all invariant separately. Consequently, I_n^g has a direct sum decomposition

$$I_n^g = I_{n,0}^g \oplus I_{n,1}^g \oplus \cdots \oplus I_{n,k}^g \oplus \cdots$$

and has the structure of a graded algebra.

Remark 4.2. Assume that $P \in I_n^g$ is non-zero as a polynomial. In virtue of Remark 4.1, without loss of generality one can assume that $P \in I_{n,m}^g$. Choose real constants $c_{ij/\alpha}$ for $|\alpha| \geq 2$ such that $P(c_{ij/\alpha}) \neq 0$. Define the germ of a metric g on $\mathbb{R}^n$ by

$$g_{ij}(x) = \delta_{ij} + \sum_{2 \leq |\alpha| \leq m} c_{ij/\alpha} \frac{x^\alpha}{\alpha!}.$$

Using a partition of unity, one can extend g to a closed manifold. Then the standard coordinate system on $\mathbb{R}^n$ is normalized with respect to g and

$$P(\xi, g)(0) = P(c_{ij/\alpha}) \neq 0,$$

that is P is non-zero as a formula. It is for this reason that Gilkey [1984], [1993] works with the algebra of jets: it is a pure polynomial algebra and there are no relations. Consequently one can identify the polynomial and the formula which it defines henceforth. If one works instead with the algebra of covariant derivatives of the curvature tensor, one must introduce the additional relations which correspond to the Bianchi identities and the covariant derivatives of the Bianchi identities. The resulting algebra would no longer be a pure polynomial algebra (see Berger, Gauduchon and Mazet [1971] or Craioveanu and Puta [1988] for more details).

Remark 4.3. Taking $c = -1$ in Lemma 4.1, it follows that $I_{n,m}^g$ is zero if m is odd.

Example 4.1. The scalar curvature τ belongs to $I_{n,2}^g$. Indeed, if the coordinate system ξ is normalized with respect to (g, x_0), then one has

$$\tau(g)(x_0) = 2\sum_{i,j} \left(g_{ij/ij} - g_{ii/jj}\right)(\xi, g)(x_0).$$

The variable $g_{11/22}$ (resp. $g_{11/11}$) is (resp. is not) a monomial of τ.

Example 4.2. $a_m\left(x, \Delta^{(p)}\right)$ [see Chapter 5, (4.2) and Theorem 4.1] defines an element of $I_{n,m}^g$.

Indeed, the total symbol of $\Delta^{(p)}$ is functorial so that $a_m\left(x, \Delta^{(p)}\right)$ is given by a polynomial in the jets of the metric. Since $a_m\left(x, \Delta^{(p)}\right)$ is invariantly defined (see Chapter 5, Theorem 4.1), this polynomial is invariant. We use Lemma 4.1 and Corollary 4.3, Chapter 5, to see that $a_m\left(x, \Delta^{(p)}\right)$ is homogeneous of order m in the jets of the metric.

Now let us recall H. Weyl's theorem on the invariants of the orthogonal group. Let V be a finite dimensional real vector space with a fixed positive definite inner product. Let $O(V)$ be the group of linear bijections of V onto V which preserve this inner product.

Let

$$\otimes^k V = V \otimes \cdots \otimes V$$

denote the k-th tensor product of V. If $T \in O(V)$, one extends T to act orthogonally on $\otimes^k V$. Let $u \mapsto T \cdot u$ denote this action. We let $O(V)$ act trivially on $\mathbb{R}$. One says that the multi-linear map $f : \otimes^k V \to \mathbb{R}$ is $O(n)$-invariant if

$$f(T \cdot u) = f(u) \text{ for every } T \in O(V).$$

By letting $T = -1_V$, one can see that there are no $O(V)$-invariant multi-linear maps if k is odd. For k even, one can construct such invariant mappings as follows. If $k = 2j$ is even, let us define

$$f_0 : V \otimes V \to \mathbb{R}$$

using the inner product and let

$$f_0^{(j)} = f_0 \otimes \cdots \otimes f_0 \ (j\text{-times}) : \otimes^{2j} V = (V \otimes V) \otimes \cdots \otimes (V \otimes V) \to \mathbb{R}.$$

Let $\Sigma(k)$ be the group of all permutations of $\{1,\ldots,k\}$. If $\sigma \in \Sigma(k)$, let $F^\sigma : \otimes^k V \to \otimes^k V$ be the mapping permuting the factors. Then the multi-linear mapping

$$f_{2j}^\sigma = f_0^j \circ F^\sigma : \otimes^{2j} V \to \mathbb{R}$$

is $O(V)$-invariant for any permutation $\sigma \in \Sigma(2j)$.

Theorem 4.1. **(H. Weyl)** The set of mappings $\{f_{2j}^\sigma\}_{\sigma \in \Sigma(2j)}$ constructed above span the space of $O(V)$-invariant multi-linear maps from $\otimes^{2j} V$ into $\mathbb{R}$.

A classical result of E. Cartan says that in geodesic polar coordinates all the jets of the metric of order ≥ 2 can be expressed as universal polynomials in terms of the covariant derivatives of the Riemann curvature tensor field. Thus if one introduces formal variables $R_{i_1 i_2 i_3 i_4 / i_5 i_6 \cdots i_k}$ for the multiple covariant derivatives of the curvature tensor field, one can regard P as a polynomial in these variables.

Further, let us use the metric tensor field to raise and lower indices and identify $TM = T^*M$. The curvature tensor R defines an element of the fourth tensor product $\otimes^4 V$, where V is an n-dimensional real vector space equipped with a positive definite inner product. There are, however, relations among the curvature variables:

$$R_{ijkl} = -R_{jikl}, R_{ijkl} = R_{klij} \text{ and } R_{ijkl} + R_{iklj} + R_{iljk} = 0.$$

Similarly, the multiple covariant differential $\overset{\circ}{\nabla}{}^k R$ defines an element of a linear subspace of $\otimes^{4+k} V$, which is defined by the higher order Bianchi identities. Consequently P defines an $O(n)$-invariant map

$$P : \mathcal{R} \to \mathbb{R}$$

where $\mathcal{R}$ is a suitable $O(n)$-invariant subspace of the direct sum of tensor powers of V. Extend P to be 0 on the orthogonal complement of $\mathcal{R}$. Then P is an $O(n)$-invariant map from the direct sum of tensor powers of V to $\mathbb{R}$. We use Theorem 4.1 for each summand to give a spanning set for the spaces $I_{n,m}^g$, where m is even:

Lemma 4.2. Consider monomial expressions in the multiple covariant derivatives of the Riemann curvature tensor field such that the order should be m and contract indices in pairs relative to a local orthonormal frame. Then all possible such expressions span $I^g_{n,m}$. In particular,

$$\{1\} \text{ spans } I^g_{n,0};$$
$$\{R_{ijij}\} \text{ spans } I^g_{n,2};$$
$$\{R_{ijkl}R_{ijkl}, R_{ijik}R_{ljlk}, R_{ijij}R_{klkl}, R_{ijij/kk}\} \text{ spans } I^g_{n,4}.$$

There is a natural restriction map

$$r : I^g_{n,m} \to I^g_{n-1,m}$$

which is defined algebraically as follows. Let

$$\deg_k(g_{ij/\alpha}) = \delta_{i,k} + \delta_{j,k} + \alpha(k)$$

be the number of times the index k appears in the variable $g_{ij/\alpha}$. Let

$$r(g_{ij/\alpha}) = \begin{cases} g_{ij/\alpha} & \text{if } \deg_n(g_{ij/\alpha}) = 0, \\ 0 & \text{if } \deg_n(g_{ij/\alpha}) > 0. \end{cases}$$

Since $r(g_{ij/\alpha})$ does not involve the last index, $r(g_{ij/\alpha}) \in \mathcal{U}^g_{n-1}$. We extend r to an algebra homomorphism from $\mathcal{U}^g_n$ to $\mathcal{U}^g_{n-1}$ which preserves the grading defined by the order in the jets of the metric:

$$r(\mathcal{U}^g_{n,m}) \subset \mathcal{U}^g_{n-1,m}.$$

The map r is the dual of a natural extension map. Let g' be a Riemannian metric on a smooth manifold M' of dimension $n-1$ and define

$$i(g') = g' + d\theta^2$$

on the product manifold $M = M' \times S^1$, where S^1 is the unit circle with the natural parameter θ. If ξ' is a local coordinate system on M' which is normalized with respect to (g', x_0'), then $i(\xi') = (\xi', \theta)$ is a local coordinate system on M normalized with respect to $(i(g'), (x_0', \theta_0))$, where θ_0 is the base point of S^1. It follows from the definition that

$$r(P)(\xi', g')(x_0') = P(i(\xi'), i(g'))(x_0', \theta_0).$$

This shows that the restriction r is the dual of this natural extension i. Consequently, if P is invariant, $r(P)$ is invariant. Therefore r defines an algebra homomorphism

$$r : I^g_n \to I^g_{n-1}$$

which restricts to linear maps

$$r : I^g_{n,m} \to I^g_{n-1,m}.$$

There is another description of r which will be useful in the proof of Theorem 4.2 (i). In discussing a spanning set for $I^g_{n,m}$ (see Lemma 4.2), the indices range from 1 through n. The restriction r is defined by letting the indices range from 1 through $n-1$. Thus

$$r\left(R_{ijij}\right)=R_{ijij}$$

is its own restriction in a formal sense. Note that $r\left(R_{ijij}\right)=0$ if $n=2$ since there are no non-trivial local invariants over S^1.

Theorem 4.2.

(i) $r: I^g_{n,m} \to I^g_{n-1,m}$ is surjective;

(ii) $r: I^g_{n,m} \to I^g_{n-1,m}$ is bijective if $m<n$;

(iii) If n is even, $r: I^g_{n,n} \to I^g_{n-1,n}$ has 1-dimensional kernel. If n is odd, $I^g_{n,n} \to I^g_{n-1,n}=\{0\}$.

Proof.

(i) If we take a spanning set for $I^g_{n-1,m}$ (see Lemma 4.2), we extend these elements to define invariant polynomials in $I^g_{n,m}$ by extending the range of summation.

(ii) Let $\mathcal{K}_{n,m}$ be the kernel of $r: I^g_{n,m} \to I^g_{n-1,m}$, $0 \neq P \in \mathcal{K}_{n,m}$ and $A=g_{i_1 j_1/\alpha_1}\cdots g_{i_l j_l/\alpha_l}$ be a monomial of P. Since $r(P)=0$, $\deg_n(A)\neq 0$. Since P is invariant under the action of the coordinate permutations, $\deg_i(A)\neq 0$ for all indices i. Since P is invariant if we change the sign of a coordinate function, $\deg_i(A)$ is even. This shows

$$\deg_i(A)\geq 2 \text{ for } 1\leq i\leq n.$$

A total of $2l+m$ indices appear in A. Thus

$$2n\leq \sum_{1\leq i\leq n}\deg_i(A)=2l+m. \tag{4.2}$$

On the other hand, we normalized the coordinate systems so $|\alpha_k|\geq 2$. Thus

$$2l\leq \sum_{1\leq k\leq l}|\alpha_k|\leq m. \tag{4.3}$$

One uses (4.2) and (4.3) to see:

$$2n\leq 2l+m\leq 2m.$$

Consequently, if $m<n$, then $\mathcal{K}_{n,m}=\{0\}$. This proves (ii).

(iii) As the proof is a bit technical, we refer to Gilkey [1984] for details. One should note that at this point it is absolutely crucial that one deals with a free polynomial algebra.

Q.E.D.

Now let us recall the definition of the **Euler integrand**. Let (M,g) be a closed, n-dimensional Riemannian manifold, R_{ijkl} the components of the curvature tensor field of the Levi-Civita connection relative to a local orthonormal frame for the tangent bundle of M. With our definition, the scalar curvature τ is given by

$$\tau = -\sum_{i,j} R_{ijij}$$

(see Chapter 1, §4). Let $E_n = 0$ if n is odd. If $n = 2m$ is even, define $E_n(g) \in I^g_{n,n}$ by

$$\begin{aligned} E_n(g) &= c_n \sum_{\alpha,\beta} sign(\alpha) sign(\beta) R_{\alpha(1)\alpha(2)\beta(1)\beta(2)} \cdots R_{\alpha(n-1)\alpha(n)\beta(n-1)\beta(n)}, \\ c_n &= \frac{1}{(-8\pi)^m m!}, \end{aligned} \tag{4.4}$$

where the sum ranges over all permutations α,β of the group $\Sigma(n)$. For example,

$$E_2(g) = -\frac{1}{2\pi} R_{1212} = \frac{1}{4\pi}\tau$$

and

$$\begin{aligned} E_4(g) &= \frac{1}{32\pi^2} \sum_{i,j,k,l} \left(R_{ijkl}R_{ijkl} - 4R_{ijik}R_{ljlk} + R_{ijij}R_{klkl} \right) \\ &= \frac{1}{32\pi^2} \left(|R|^2 - 4|c_2R|^2 + \tau^2 \right). \end{aligned}$$

Using Lemma 4.2 one sees that

$$I^g_{n,2} = \text{Span}\{\tau\}$$

and

$$I^g_{n,4} = \text{Span}\left\{ |R|^2, |c_2R|^2, \tau^2, \Delta\tau \right\}. \tag{4.5}$$

Since the metric is flat in the case when $n = 1$, clearly $\tau = 0$. If $n \geq 2$, the scalar curvature τ is non-zero and forms a basis for $I^g_{n,2}$. If $n \geq 4$, the scalar invariants of (4.5) are linearly independent and form a basis for $I^g_{n,4}$. If $n = 3$, the invariants of (4.5) are linearly dependent: there is a single additional relation, namely $|R|^2 - 4|c_2R|^2 + \tau^2 = 0$ (see Chapter 1, Example 4.7).

Example 4.3. Let (S^{2n}, g_{can}) be the Riemannian manifold considered in Chapter 1, Example 2.5, where S^{2n} is the unit sphere in $\mathbb{R}^{2n+1}$. Since the curvature tensor field of this Riemannian manifold is given by

$$R_{ijkl} = \delta_{il}\delta_{jk} - \delta_{ik}\delta_{jl},$$

using (4.4) it follows that

$$E_{2n}(g_{can}) = c_{2n}(-1)^n 2^n (2n)! = \frac{(2n)!}{(4\pi)^n n!}.$$

Consequently,

$$\int_{S^{2n}} E_{2n}(g_{can}) d\mu_{g_{can}} = \frac{(2n)!}{(4\pi)^n n!} \text{Vol}(S^{2n}, g_{can}) = 2 = \chi(S^{2n}).$$

(see Chapter 2, Example 1.1).

Remark 4.4. The Euler integrand $E_n(g)\in\mathcal{K}_{n,n}$, where $\mathcal{K}_{n,n}$ denotes the kernel of $r:I^g_{n,n}\to I^g_{n-1,n}$.

Indeed, this is immediate if n is odd since $E_n(g)=0$. Now assume that n is even. Since $sign(\alpha)sign(\beta)=0$ if an index (value) is repeated in either α or in β, both α and β being permutations of the set $\{1,\ldots,n\}$. Thus in particular, some $\alpha(j)=n$ if $sign(\alpha)sign(\beta)\neq 0$. To define r, we took a product with S^1 and hence $R_{ijkl}=0$ if the index n appears. This proves our remark.

Since $0\neq E_n(g)\in\mathcal{K}_{n,n}$ and $\dim(\mathcal{K}_{n,n})=1$ for n even [see Theorem 4.2 (iii)], it follows that

Corollary 4.1. If n is even, $\operatorname{Span}\{E_n(g)\}=\mathcal{K}_{n,n}$.

We can now prove the **Chern-Gauss-Bonnet theorem**. Let

$$a_m(x,d+\delta)=\sum_{p=0}^{n}(-1)^p\,a_m\left(x,\Delta^{(p)}\right).$$

Example 4.2 shows that $a_m\left(x,\Delta^{(p)}\right)\in I^g_{n,m}$. Consequently, $a_m(x,d+\delta)\in I^g_{n,m}$, that is $a_m(x,d+\delta)$ is an invariant of the de Rham complex.

We use Corollary 4.1, Chapter 5, to see

$$\int_M a_m(x,d+\delta)\,d\mu_g=\begin{cases}0 & \text{if } m\neq n,\\ \chi(M) & \text{if } m=n,\end{cases}\tag{4.6}$$

where $\chi(M)$ is the Euler-Poincaré characteristic of M.

Lemma 4.3.

(i) $a_m(x,d+\delta)=0$ if either n is odd or if $m<n$.

(ii) If n is even, then

$$a_n(x,d+\delta)=E_n(g)\ (= \text{the Euler integrand}).$$

Proof.

(i) We assume first that n is odd. Locally, one can always choose an orientation for the tangent bundle TM. If $S^{(p)}$ denotes the star Hodge operator, then $\Delta^{(p)}S^{(n-p)}=S^{(n-p)}\Delta^{(n-p)}$ locally (see Chapter 2, Proposition 2.9). Since these two operators are locally isomorphic, their local invariants agree so

$$a_m\left(x,\Delta^{(p)}\right)=a_m\left(x,\Delta^{(n-p)}\right).\tag{4.7}$$

Since the argument is local, (4.7) holds even if M does not admit a global orientation. Consequently,

$$a_m\left(x,d+\delta\right)=\sum_{p=0}^{n}\left(-1\right)^p a_m\left(x,\Delta^{(p)}\right)=\sum_{p=0}^{n}\left(-1\right)^p a_m\left(x,\Delta^{(n-p)}\right)$$
$$=\left(-1\right)^n\sum_{p=0}^{n}\left(-1\right)^{n-p} a_m\left(x,\Delta^{(n-p)}\right)=\left(-1\right)^n a_m\left(x,d+\delta\right).$$

This proves $a_m\left(x,d+\delta\right)=0$ for n odd. Next we assume n is even. Let $M=M'\times S^1$ with the product Riemannian metric. One decomposes any $\omega\in A(M)$ uniquely in the form

$$\omega=\omega_1+\omega_2\wedge d\theta \text{ for } \omega_k\in A(M'), k=1,2.$$

Let $F\in End\left(A(M)\right)$ defined by

$$F(\omega)=\omega_1\wedge d\theta+\omega_2.$$

Since the metric is flat in the S^1 direction, $F\circ\Delta=\Delta\circ F$. If one decomposes

$$A(M)=A^{even}(M)\oplus A^{odd}(M),$$
$$A^{even}(M)=\bigoplus_q A^{2q}(M),\ A^{odd}(M)=\bigoplus_q A^{2q+1}(M),$$

into the forms of even and odd degree, then F interchanges these two terms. Therefore

$$a_m\left(x,\Delta^{even}\right)=a_m\left(x,\Delta^{odd}\right),$$

where

$$\Delta^{even}=\bigoplus_q \Delta^{(2q)},\ \Delta^{odd}=\bigoplus_q \Delta^{(2q+1)}$$

(see also Chapter 5, Remark 4.4), so that

$$a_m\left(x,d+\delta\right)=a_m\left(x,\Delta^{even}\right)-a_m\left(x,\Delta^{odd}\right)=0$$

for such a product Riemannian metric. This implies

$$r\left(a_m\left(\cdot,d+\delta\right)\right)=0$$

and consequently

$$a_m\left(\cdot,d+\delta\right)=0$$

for $m<n$ [see Theorem 4.2 (ii)], which completes the proof of (i).

(ii) If n is even, Corollary 4.1 implies that there is some universal constant c_n' so that

$$a_n\left(\cdot,d+\delta\right)=c_n' E_n(g).$$

We show $c_n'=1$ by integrating over the manifold $M=S^n$ and using (4.6) and Example 4.3:

$$2=\chi\left(S^n\right)=\int_{S^n} a_n\left(x,d+\delta\right)d\mu_{g_{can}}(x)=c_n'\int_{S^n} E_n\left(g_{can}\right)d\mu_{g_{can}}=2c_n'.$$

Q.E.D.

The equality (4.6) and Lemma 4.3 (ii) completes the proof of the Chern-Gauss-Bonnet theorem:

Theorem 4.3 (Chern-Gauss-Bonnet) If n is even, then

$$\int_M E_n(g)\,d\mu_g = \chi(M).$$

Remark 4.5. This result was first established if $n = 2$ by McKean and Singer [1967]. The general case was first proved by Patodi [1971] who used an argument very different from Gilkey's argument. There also exist other proofs of the Chern-Gauss-Bonnet theorem (see Cycon, Froese, Kirsch and Simon [1987], Elworthy [1988], Parker [1985] or Rosenberg [1997]). For a purely analytic proof of this theorem based on Corollary 2.3 in Chapter 5, see Iwasaki [1995]. The technique of this last proof is to use pseudodifferential operators so that the symbols of asymptotics are given by functions with geometrical interpretations. For a more geometrical proof, see Chern [1944].

Theorem 4.3 and (4.4) show that a closed oriented smooth manifold M has the following topological obstruction in order to admit a Riemannian metric g such that (M,g) is a flat space (see Chapter 1, Definition 4.3).

Corollary 4.2. If M is a closed oriented smooth manifold admitting a Riemannian metric g such that (M,g) is a flat space, then $\chi(M)=0$. In particular, the even dimensional spheres, the complex and quaternionic projective spaces and the closed oriented surfaces of genus >1 do not admit a Riemannian metric g such that the corresponding Riemannian manifolds are flat spaces.

Remark 4.6. The Chern-Gauss-Bonnet theorem raises the question whether it can be refined to show that each Betti number solely is given by the integral of a universal local expression in the Riemannian metric. This is not in general the case. Indeed, assume that $\beta_k(M)=\int_M [u_k(g)](x)\,d\mu_g(x)$, where $u_k(g)$ is such a universal expression in g. If $\Phi:(M',g')\to(M,g)$ is an $|F|$-fold Riemannian covering with fiber F, then as in the proof of Corollary 4.2 of Chapter 5,

$$\beta_k(M')=\int_{M'}[u_k(g')](\Phi(x))\,d\mu_{g'}(\Phi(x))=|F|\int_M [u_k(g)](x)\,d\mu_g(x)=|F|\,\beta_k(M).$$

But the Betti numbers of such a covering space M' of M do not satisfy this relation in general.

REFERENCES

BERGER, M., GAUDUCHON, P. and MAZET, E., *Le Spectre d'une Variété Riemannienne*, Lecture Notes in Math., Vol. 194, Springer-Verlag, Berlin, 1971.

CHERN, S.S., *A simple intrinsic proof of the Gauss-Bonnet formula for closed Riemannian manifolds*, Ann. of Math. 45(1944), 741-752.

CRAIOVEANU, M. and PUTA, M., *Introducere în Geometria Spectrală*, Editura Academiei Române, Bucuresti, 1988.

CRAIOVEANU, M., PUTA, M., and RASSIAS, Th.M., *On the geometry of the Hodge-de Rham Laplace operators*, in: Geometry in Partial Differential Equations, (eds. A. Pràstaro and Th.M. Rassias), World Scientific Publ. Co. (1994), pp. 51-67.

CYCON, H., FROESE, R., KIRSCH, W., and SIMON, B., *Schrödinger Operators with Applications to Quantum Mechanics and Global Geometry*, Springer-Verlag, Berlin, 1987.

ELWORTHY, K.D., *Geometric aspects of diffusions on manifolds*, École d' Été de Probabilités de Saint-Flour XV-XVII, 1985-87 (P.L. Henneguin, ed.), Lecture Notes in Mathematics, Vol. 1362, Springer-Verlag, 1988, pp. 277-426.

GALLOT, S. and MEYER, D., *Opérateur de courbure et Laplacien des formes différentielles d' une variété riemanniennes*, J. Math. Pures et Appl., 54(1975), 259-284.

GILKEY, P.B., *Invariance Theory, the Heat Equation and the Atiyah-Singer Index Theorem*, Publish or Perish, Wilmington, Delaware, 1984.

GILKEY, P.B., *Applications of Spectral Geometry to Geometry and Topology*, Lecture Notes Series, Nr. 12, Research Institute of Mathematics, Global Analysis Research Center, Seoul National University, 1993.

IWASAKI, C., *A proof of the Gauss-Bonnet-Chern theorem by the symbolic calculus of pseudo-differential operators*, Japan. J. Math. (N.S.) 21(1995), no. 2, 235-285.

LICHNEROWICZ, A., *Théorie Globale des Connexions et des Groupes d' Holonomie*, Cremonese, Rome et Dunod, Paris, 1955.

LICHNEROWICZ, A., *Géométrie des Groupes de Transformations*, Dunod, Paris, 1958.

McKEAN, H.P. and SINGER, I.M., *Curvature and the eigenvalues of the Laplacian*, J. Diff. Geometry 1(1967), 43-69.

MILGRAM, A.N. and ROSENBLOOM, P.C., *Harmonic forms and heat conduction*, I, Proc. Nat. Acad. Sci. USA 37(1951), 180-184.

MOORE, J.D., *Compact Riemannian manifolds with positive curvature operators*, Bull. Amer. Math. Soc. 14(1986), 279-282.

PARKER, T., *Supersymmetry and the Gauss-Bonnet theorem*, preprint, 1985.

PATODI, V.K., *Curvature and the eigenforms of the Laplace operator*, J. Diff. Geometry 5(1971), 233-249.

DE RHAM, G., *Variétés Différentiables*, Hermann, Paris, 1960.

ROSENBERG, S., *The Laplacian on a Riemannian Manifold. An Introduction to Analysis on Manifolds*, London Mathematical Society Student Texts 31, Cambridge University Press, 1997.

YANO, K. and BOCHNER, S., *Curvature and Betti Numbers*, Princeton Univ. Press, Princeton, New Jersey, 1953.

YU, Y.L., *Local index theorem for Dirac operator*, Acta Math. Sinica (N.S.) 3(1987), no. 2, 152-169.

Chapter 7
AN INTRODUCTION TO WITTEN-HELFFER-SJÖSTRAND THEORY

1. INTRODUCTION

Witten-Helffer-Sjöstrand theory (abbreviated WHS-theory) is a substantial addition to the Hodge-de Rham theory and a powerful tool for comparing numerical invariants associated to compact manifolds analytically (i.e. by using a Riemannian metric) and combinatorially (i.e. by using a triangulation), see Bismut and Zhang [1992], [1994], Burghelea, Friedlander, Kappeler and McDonald [1996], Burghelea, Friedlander and Kappeler [1996], [2001]. It states in a precise way the relationship between the de Rham complex of a closed manifold M and the cochain complex provided by a certain smooth triangulation on M defined by a pair (f, g), where $f: M \to \mathbb{R}$ is a Morse function and g a Riemannian metric on M. WHS-theory provides a connection between these two complexes with scalar products and allows to relate some of the spectral properties of the Hodge-de Rham operators associated to the metric g (see Chapter 2, §2, D) and the combinatorial Laplacians associated to the triangulation.

The intuition behind the WHS-theory is provided by physics and consists in regarding a closed smooth manifold endowed with a Riemannian metric and a Morse function as an interacting system of harmonic oscillators. This intuition was first noticed and exploited by Witten [1982], in order to provide a short "physicist' proof" of Morse inequalities, a very useful result in topology.

The idea of Witten's proof is to introduce a perturbed elliptic complex for a given Morse function $f: M \to \mathbb{R}$ as follows:

$$\begin{array}{ccccccc}
\cdots \to A^{p-1}(M) & \xrightarrow{d^{p-1}} & A^{p}(M) & \xrightarrow{d^{p}} & A^{p+1}(M) & \to \cdots \\
e^{-tf}\downarrow & & e^{-tf}\downarrow & & e^{-tf}\downarrow & \\
\cdots \to A^{p-1}(M) & \xrightarrow{d_t^{p-1}} & A^{p}(M) & \xrightarrow{d_t^{p}} & A^{p+1}(M) & \to \cdots
\end{array}$$

with

$$d_t^p = e^{-tf} d^p e^{tf}, \quad 0 \le p \le n = \dim M$$

(see also the notations in Chapter 2,§2,D) and to consider the perturbed p-th Hodge-de Rham operator

$$\begin{aligned}
\Delta_t^{(p)} &= \left(d_t^p\right)^* d_t^p + d_t^{p-1}\left(d_t^{p-1}\right)^* \\
&= \Delta^{(p)} + t\left(L_{-\mathrm{grad}_g(f)} + L^*_{-\mathrm{grad}_g(f)}\right) + t^2 \left\|\mathrm{grad}_g(f)\right\|^2 1_{A^p(M)}.
\end{aligned}$$

The considerations of Chapter 5 imply that there are β_p eigenvalues of $\Delta_t^{(p)}$ equal to 0, where β_p is the p-th Betti number of M, $0 \le p \le n$. There exists a gap in the spectrum of $\Delta_t^{(p)}$, namely the open interval $\left(c_1 e^{-c_2 t}, c_3 t\right)$, where c_1, c_2, c_3 are positive constants (see §6). This gap is detected by elementary mini-max characterization of the spectrum of self-adjoint positive operators and simple estimates involving the equations of the harmonic oscillator. The $\Delta_t^{(p)}$'s in the neighborhoods of critical points of f in "admissible coordinates" (see §4) are given by such equations. Moreover, the number of the eigenvalues of $\Delta_t^{(p)}$ in the interval $\left[0, c_1 e^{-c_2 t}\right]$ counted with their multiplicities is equal to the number of critical points of f having index p, so that the simple version of Morse inequalities

$$\beta_p \leq m_p(f) \overset{def}{=} \#\left\{x_j \in M \middle| ind(f, x_j) = p\right\}$$

is an immediate consequence of this fact. Then a revised elliptic complex can be used to prove the final version of the Morse inequalities.

Helffer and Sjöstrand [1985] have completed Witten's picture with their results on Schrödinger operators and have considerably strengthened Witten's mathematical statements. The work of Helffer and Sjöstrand on the Witten theory, based on a large amount of estimates and preliminary results about Schrödinger operators, can be presented in a self-contained manner and in a way appropriate to topological applications, see Bismut and Zhang [1994], Burghelea, Friedlander, Kappeler and McDonald [1996], Burghelea [1997].

The material of this chapter is based on Witten [1982], Cycon, Froese, Kirsch and Simon [1987], Burghelea, Friedlander, Kappeler and McDonald [1996] and Burghelea [1997].

2. ANALYTIC PRELIMINARIES

First, let us review some facts on partial differential operators on manifolds.

Let $\mathbb{N} = \{0, 1, \ldots\}$ be the set of natural numbers, $\alpha = (\alpha_1, \ldots, \alpha_n)$ be a multi-index, where $\alpha_i \in \mathbb{N}$ and

$$|\alpha| = \alpha_1 + \cdots + \alpha_n,$$

$$d_x^\alpha = \left(\partial/\partial x_1\right)^{\alpha_1} \cdots \left(\partial/\partial x_n\right)^{\alpha_n},$$

$$D_x^\alpha = (-i)^{|\alpha|} d_x^\alpha.$$

For the motivation of the appearing of the factor $(-i)^{|\alpha|}$ in the definition of D_x^α, see Chapter 3, §1.

Let U be a non-empty open subset of $\mathbb{R}^n$ and $C_0^\infty(U)$ the space of smooth complex valued functions on U with compact support in U. Let $a_\alpha(\cdot) \in C_0^\infty(U)$ for $|\alpha| \leq k$ and let

$$P = p(x, D) = \sum_{|\alpha| \leq k} a_\alpha(x) D_x^\alpha : C_0^\infty(U) \to C_0^\infty(U)$$

be a partial differential operator (PDO) of order k. Let $P^k(U)$ be the space of all such operators. If $k_1 < k_2$, then $P^{k_1}(U) \subset P^{k_2}(U)$. The leading symbol $\sigma_L(P)$ of P is the highest order part of the polynomial of degree k

$$p(x, \xi) = \sum_{|\alpha| \leq k} a_\alpha(x) \xi^\alpha$$

in the dual variable ξ, that is

$$\sigma_L(P)(x,\xi) = \sum_{|\alpha|=k} a_\alpha(x)\xi^\alpha .$$

Then $\sigma_L(P)(x,\xi)$ is a homogeneous polynomial of degree k in ξ. For example, if $P = -\frac{\partial^2}{\partial x^2} - \frac{\partial^2}{\partial y^2} = D_{(x,y)}^{(2,0)} + D_{(x,y)}^{(0,2)}$ is the Laplacian on $(\mathbb{R}^2, g_{can})$, then $\sigma_L(P)((x,y),(\xi_1,\xi_2)) = \xi_1^2 + \xi_2^2$.

Now let us consider the effect of changes of coordinates on PDOs. Let U_1 and U_2 be non-empty open subsets of $\mathbb{R}^n$ and let $h: U_1 \to U_2$ be a smooth diffeomorphism. If $f_2 \in C_0^\infty(U_2)$, define the pull back $h^*(f_2) \in C_0^\infty(U_1)$ by $(h^*(f_2))(x) = f_2(h(x))$. h^* is an isomorphism from $C^\infty(U_2)$ onto $C_0^\infty(U_1)$. If $P \in P^k(U_2)$, define the pull-back $h^*P = P_1$ on $C_0^\infty(U_1)$ by $h^* \circ P = P_1 \circ h^*$. Then $P_1 \in P^k(U_1)$ and $h^*: P^k(U_2) \to P^k(U_1)$ is an isomorphism. Moreover,

$$\sigma_L(h^*P)(x_1,\xi_1) = \sigma_L(P)\left(h(x_1),\left(dh_{x_1}^T\right)^{-1}(\xi_1)\right).$$

This shows that the PDOs are invariant under changes of coordinates and that the leading symbol is invariantly defined on the cotangent space $T^*\mathbb{R}^n$. Let us illustrate this for the Laplacian on $\mathbb{R}^2$. Let

$$\Delta = -d_{(x,y)}^{(2,0)} - d_{(x,y)}^{(0,2)} = D_{(x,y)}^{(2,0)} + D_{(x,y)}^{(0,2)}$$

be the Laplacian in rectangular coordinates (x,y) on $\mathbb{R}^2$ and take polar coordinates (r,θ) so that $x = r\cos\theta$, $y = r\sin\theta$. Then

$$\frac{\partial}{\partial x} = \cos\theta\frac{\partial}{\partial r} - \frac{1}{r}\sin\theta\frac{\partial}{\partial\theta},$$

$$\frac{\partial}{\partial y} = \sin\theta\frac{\partial}{\partial r} + \frac{1}{r}\cos\theta\frac{\partial}{\partial\theta},$$

$$\frac{\partial^2}{\partial x^2} = \cos^2\theta\frac{\partial^2}{\partial r^2} - \frac{\sin 2\theta}{r}\frac{\partial^2}{\partial r\partial\theta} + \frac{\sin^2\theta}{r^2}\frac{\partial^2}{\partial\theta^2} + \frac{\sin^2\theta}{r}\frac{\partial}{\partial r} + \frac{\sin 2\theta}{r^2}\frac{\partial}{\partial\theta},$$

$$\frac{\partial^2}{\partial y^2} = \sin^2\theta\frac{\partial^2}{\partial r^2} + \frac{\sin 2\theta}{r}\frac{\partial^2}{\partial r\partial\theta} + \frac{\cos^2\theta}{r^2}\frac{\partial^2}{\partial\theta^2} + \frac{\cos^2\theta}{r}\frac{\partial}{\partial r} - \frac{\sin 2\theta}{r^2}\frac{\partial}{\partial\theta},$$

and consequently the Laplacian in polar coordinates becomes

$$\Delta' = -\frac{\partial^2}{\partial r^2} - \frac{1}{r^2}\frac{\partial^2}{\partial\theta^2} - \frac{1}{r}\frac{\partial}{\partial r} = -d_{(r,\theta)}^{(2,0)} - \frac{1}{r^2}d_{(r,\theta)}^{(0,2)} - \frac{1}{r}d_{(r,\theta)}^{(1,0)}.$$

Let (ξ_1,ξ_2) [resp. (ζ_1,ζ_2)] be the dual variables for the (x,y) [resp. (r,θ)] coordinates. Then

$$\sigma_L(\Delta)((x,y),(\xi_1,\xi_2)) = \xi_1^2 + \xi_2^2$$

and

$$\sigma_L(\Delta')((r,\theta),(\zeta_1,\zeta_2)) = \zeta_1^2 + \frac{1}{r^2}\zeta_2^2.$$

Let $\omega = \xi_1 dx + \xi_2 dy = \zeta_1 dr + \zeta_2 d\theta$. Since

$$\begin{aligned} dx &= \frac{\partial x}{\partial r} dr + \frac{\partial x}{\partial \theta} d\theta = \cos\theta\, dr - r\sin\theta\, d\theta, \\ dy &= \frac{\partial y}{\partial r} dr + \frac{\partial y}{\partial \theta} d\theta = \sin\theta\, dr + r\cos\theta\, d\theta, \end{aligned}$$

if follows that

$$\begin{aligned} \zeta_1 &= \cos\theta\, \xi_1 + \sin\theta\, \xi_2, \\ \zeta_2 &= -r\sin\theta\, \xi_1 + r\cos\theta\, \xi_2. \end{aligned}$$

Thus

$$\zeta_1^2 + \frac{1}{r_2}\zeta_2^2 = \xi_1^2 + \xi_2^2,$$

that is $\sigma_L(\Delta)$ is invariantly defined on $T^*\mathbb{R}^n$.

One says that $P \in P^k(U)$ is **elliptic** if the leading symbol

$$\sigma_L(P)(x,\xi) \neq 0 \text{ for all } x \in U \text{ and all } \xi \neq 0 \text{ in } \mathbb{R}^n.$$

The previous considerations can be extended immediately for PDO's defined on smooth vector valued functions.

We refer to Craioveanu-Ratiu [1976] for more details concerning the local theory for PDO's.

We now extend the above definitions to manifolds. Let (M,g) be a smooth closed Riemannian manifold of dimension n and E, F be finite dimensional complex vector bundles over M. Let $C^\infty(E)$ [resp. $C^\infty(F)$] be the space of smooth sections of E (resp. F] and let $P: C^\infty(E) \to C^\infty(F)$ be a $\mathbb{C}$-linear operator. We say that P is a **partial differential operator (PDO) of order *k*** and write $P \in P^k(E,F)$ if for every open chart (U,φ) on M and for every trivializations $\psi_1 : E|_U \cong U \times \mathbb{C}^m$ and $\psi_2 : F|_U \cong U \times \mathbb{C}^l$, there exist a collection of smooth maps

$$(a_\alpha)_{|\alpha| \le k} : \varphi(U) \to L_{\mathbb{C}}(\mathbb{C}^m, \mathbb{C}^l)$$

indexed by all multi-indices α with $|\alpha| \le k$ such that

$$\psi_2 \circ (P(s)) \circ \varphi^{-1} = \sum_{|\alpha| \le k} a_\alpha D^\alpha_{\varphi(\cdot)}(\psi_1 \circ s \circ \varphi^{-1})$$

for $s \in C^\infty(E)$, that is the principal part of $\psi_2 \circ (P(s)) \circ \varphi^{-1}$ may be regarded as a matrix of PDO's of order k for any $((U,\varphi),\psi_1,\psi_2)$. If $P \in P^k(E,F)$ is as before, given $x \in U$, let us identify $T_x^* M$ (resp. E_x, resp. F_x) with $\mathbb{R}^n$ (resp. $\mathbb{C}^m$, resp. $\mathbb{C}^l$) via $(d\varphi_2)^*$ (resp. ψ_1, resp. ψ_2). By definition, the leading symbol $\sigma_L(P)$ of P is given by the $\mathbb{C}$-linear map

$$\sigma_L(P)(x,\xi):\mathbb{C}^m = E_x \to \mathbb{C}^l = F_x,$$
$$\sigma_L(P)(x,\xi) = \sum_{|\alpha|=k} a_\alpha(\varphi(x))\xi^\alpha,\ \xi \in T_x^*M = \mathbb{R}^n.$$

Thus the leading symbol of P is invariantly defined on T^*M. It is possible to give for the leading symbol $\sigma_L(P)$ of a PDO of order k $P \in P^k(E,F)$ an equivalent definition, which is independent of charts and trivializations, namely $\sigma_L(P)$ assign to each $(x,\xi) \in T^*M \setminus \{0\}$ the linear map

$$\sigma_L(P)(x,\xi): E_x \to F_x$$

defined by

$$\sigma_L(P)(x,\xi)(u) = \frac{1}{(-i)^k} P\left[\frac{1}{k!}(f - f(x))^k s\right](x),$$

where $f \in C^\infty(M)$ satisfies $df_x = \xi$ and $s \in C^\infty(E)$ satisfies $s(x) = u$. One can show that this definition does not depend on the choosing of f and s such that $df_x = \xi$ and $s(x) = u$ [see R.S. Palais [1965] for details].

The local situation shows that if E, F, G are complex vector bundles over M and $P \in P^k(E,F), Q \in P^l(F,G)$, then $Q \circ P \in P^{k+l}(E,G)$ and $\sigma_L(Q \circ P) = \sigma_L(Q) \circ \sigma_L(P)$.

If E and F have the same dimension, $P \in P^k(E,F)$ is called an **elliptic PDO of order *k*** if and only if $\sigma_L(P)(x,\xi): E_x \to F_x$ is a linear isomorphism for all $(x,\xi) \in T^*M \setminus \{0\}$.

Let E (resp. F) be an hermitian bundle over M, that is E (resp. F) is a complex vector bundle with a specific choice $(\ ,\ )_E$ [resp. $(\ ,\ )_F$] of hermitian structure. For $s_1, s_2 \in C^\infty(E)$ one can define a smooth complex valued function $(s_1|s_2)_E$ on M by

$$(s_1|s_2)_E(x) = (s_1(x), s_2(x))_E, \quad x \in M,$$

and similarly for any two smooth sections of F. We define $L^2(E)$ using the global integrated inner product

$$\langle s_1, s_2 \rangle_E = \int_M (s_1|s_2)_E \, d\mu_g.$$

We let $L^2(E)$ be the completion of $C^\infty(E)$ in the corresponding norm. Let $P: C^\infty(E) \to C^\infty(F)$. We let P^* be defined by

$$\langle P(s), t \rangle_F = \langle s, P^*(t) \rangle_E$$

whenever $s \in C^\infty(E), t \in C^\infty(F)$, if such a P^* exists. In particular, if E and F are hermitian bundles over M, then for each $P \in P^k(E,F)$ there exists a unique such P^* and $P^* \in P^k(F,E)$. Moreover $\sigma_L(P^*) = \sigma_L(P)^*$.

Proposition 2.1. Let $E=\wedge\left(T^{*}M\right)\otimes\mathbb{C}=\bigoplus_{k=0}^{n}E^{k}$, where $E^{k}=\wedge^{k}\left(T^{*}M\right)\otimes\mathbb{C}$ is the bundle of complex valued differential forms and let $d^{k}:C^{\infty}\left(E^{k}\right)\to C^{\infty}\left(E^{k+1}\right)$ [resp. $\delta^{k}:C^{\infty}\left(E^{k}\right)\to C^{\infty}\left(E^{k-1}\right)$, resp. $\Delta^{(k)}:C^{\infty}\left(E^{k}\right)\to C^{\infty}\left(E^{k}\right)$] be the exterior derivative (resp. codifferential, resp. Hodge-de Rham operator) viewed as PDO of order 1 (resp. 1, resp. 2). Then

(i) $\sigma_{L}\left(d^{k}\right)(x,\xi)(u)=i(\xi\wedge u)$, for all $u\in\wedge^{k}\left(T_{x}^{*}M\right)$,

(ii) $\sigma_{L}\left(\delta^{k}\right)(x,\xi)=-i\left(i_{\xi}\right)$,

where $i_{\xi}:\wedge^{k}\left(T_{x}^{*}M\right)\to\wedge^{k-1}\left(T_{x}^{*}M\right)$ is defined in the proof below, and

(iii) $\sigma_{L}\left(\Delta^{(k)}\right)(x,\xi)(u)=|\xi|^{2}u$, for all $u\in\wedge^{k}\left(T_{x}^{*}M\right)$ where $(x,\xi)\in T^{*}M$.

Proof. (i) If $f\in C^{\infty}(M)$ [resp. $\omega\in C^{\infty}\left(E^{k}\right)$] is such that $f(x)=0$ and $df_{x}=\xi\in T_{x}^{*}M$ [resp. $\omega_{x}=u\in\wedge^{k}\left(T_{x}^{*}M\right)$], then

$$\sigma_{L}\left(d^{k}\right)(x,\xi)(u)=-\frac{1}{i}\left[d^{k}(f\omega)\right]_{x}=i\xi\wedge u$$

since

$$d^{k}(f\omega)=df\wedge\omega+fd^{k}\omega .$$

(ii) It is enough to notice that $\delta^{k+1}=\left(d^{k}\right)^{*}$ (see Chapter 2, §2) so that $\sigma_{L}\left(\delta^{k+1}\right)=\left(\sigma_{L}\left(d^{k}\right)\right)^{*}$, and to check that the adjoint of the exterior multiplication linear map

$$\varepsilon(\xi):\wedge^{k-1}\left(T_{x}^{*}M\right)\to\wedge^{k}\left(T_{x}^{*}M\right),\left[\varepsilon(\xi)\right](u)=\xi\wedge u,$$

is the operator

$$i_{\xi}:\wedge^{k}\left(T_{x}^{*}M\right)\to\wedge^{k-1}\left(T_{x}^{*}M\right)$$

defined by

$$i_{\xi}\left(u_{1}\wedge\ldots\wedge u_{k}\right)=\sum_{j=1}^{k}(-1)^{j+1}\left(\xi|u_{j}\right)_{x}u_{1}\wedge\ldots\wedge\hat{u}_{j}\wedge\ldots\wedge u_{k}$$

$u_{j}\in T_{x}^{*}M$, $j=1,\ldots,k$, where $(\,|\,)_{x}$ denotes the inner product on $T_{x}^{*}M$ induced by g.

(iii) Since $\Delta^{(k)}=d^{k-1}\delta^{k}+\delta^{k+1}d^{k}$, it follows that

$$\begin{aligned}\sigma_{L}\left(\Delta^{(k)}\right)(x,\xi)&=\sigma_{L}\left(d^{k-1}\right)(x,\xi)\circ\sigma_{L}\left(\delta^{k}\right)(x,\xi)+\sigma_{L}\left(\delta^{k+1}\right)(x,\xi)\circ\sigma_{L}\left(d^{k}\right)(x,\xi)\\&=\left[\xi\wedge i_{\xi}+i_{\xi}\circ(\xi\wedge\ )\right].\end{aligned}$$

A direct application of definitions shows that $\xi\wedge i_{\xi}(u)+i_{\xi}(\xi\wedge u)=|\xi|^{2}u$ [it suffices to choose for ξ an element of an orthonormal basis $(e_1,\ldots,e_n)$ for $T_{x}^{*}M$], that is the linear mapping $\sigma_L\left(\Delta^{(k)}\right)(x,\xi)$ is scalar multiplication in $\wedge^k T_x^* M$ by $|\xi|^2$.

Q.E.D.

Definition 2.1. Let $E=\left(E_j\right)_{0\le j\le n}$ be a graded vector bundle over *M*, that is a finite collection of finite dimensional vector bundles $E_j, 0\le j\le n$, over *M*. Let $P=\left(P_j\right)_{0\le j\le n-1}$ be a graded PDO of order *k*, that is a collection of *k*-th order PDO's

$$P_j : C^\infty\left(E_j\right)\to C^\infty\left(E_{j+1}\right),\ j=0,1,\ldots,n-1.$$

We say that (*E*,*P*) is an **elliptic complex** if

(i) $P_{j+1}\circ P_j=0,\ j=0,\ldots,n-2;$

(ii) For each $x\in M$ and for each $\xi\in T_x^*M\setminus\{0\}$,

$$N\left(\sigma_L\left(P_{j+1}\right)(x,\xi)\right)=R\left(\sigma_L\left(P_j\right)(x,\xi)\right),\ j=0,\ldots,n-2,$$
$$N\left(\sigma_L\left(P_0\right)(x,\xi)\right)=0,\ R\left(\sigma_L\left(P_{n-1}\right)(x,\xi)\right)=0,$$

where *N* (resp. *R*) stands for the null space (resp. range), or equivalently, the complex is exact on the leading symbol level.

Example 2.1 (The **de Rham complex**). Let us define (*E*,*P*) as follows:

$$E^k=\left(\wedge^k\left(T^*(M)\right)\right)\otimes\mathbb{C},\ 0\le k\le n=\dim M,\ P=\left(d^k\right)_{0\le k\le n-1},$$

where $d^k : C^\infty\left(E^k\right)\to C^\infty\left(E^{k+1}\right)$ is the exterior derivative. This is an elliptic complex. Indeed, the only thing to check is the exactness of the leading symbol level. If $(x,\xi)\in T^*M$, choose a basis $\{e_1,\ldots,e_n\}$ for T_x^*M such that $\xi=e_1$. Then $\left\{e_{i_1}\wedge\ldots\wedge e_{i_k}\mid 1\le i_1<i_2<\cdots<i_k\le n\right\}$ forms a basis for $\wedge^k\left(T_x^*M\right)$ and

$$\xi\wedge\left(e_{i_1}\wedge\ldots\wedge e_{i_k}\right)=\begin{cases}0 & \text{if } i_1=1,\\ e_1\wedge e_{i_1}\wedge\ldots\wedge e_{i_k} & \text{if } i_1>1.\end{cases}$$

Therefore,

$$\begin{aligned}N\left(\sigma_L\left(d^k\right)(x,\xi)\right)&=\operatorname{Span}\left\{e_{i_1}\wedge\ldots\wedge e_{i_k}\mid 1=i_1<i_2<\cdots<i_k\le n\right\}\\&=R(\xi\wedge\ )\\&=R\left(\sigma_L\left(d^{k-1}\right)(x,\xi)\right),\ k=1,\ldots,n-1,\end{aligned}$$
$$N\left(\sigma_L\left(d^0\right)(x,\xi)\right)=0,\ R\left(\sigma_L\left(d^{n-1}\right)\right)(x,\xi)=0.$$

Let (*E*,*P*) be an elliptic complex and fix a Hermitian structure on E_j, $j=0,\ldots,n$. We use the inner product together with the Riemannian metric *g* on *M* as before to define $L^2(E_j)$ as well as the adjoint of P_j. Consequently, one can construct the associated self-adjoint, positive operators

$$D_j : C^\infty\left(E_j\right)\to C^\infty\left(E_j\right),\ D_j=P_{j-1}\circ P_{j-1}^*+P_j^*\circ P_j,$$

$j=1,\ldots,n-1$, $D_0 = P_0^* \circ P_0$, $D_n = P_{n-1} \circ P_{n-1}^*$. Moreover, one can also express the condition of ellipticity of (*E*,*P*) in terms of D_j, $j=0,\ldots,n$.

Proposition 2.2. Let (*E*,*P*) be a graded vector bundle together with a graded PDO of order *k* such that

$$P_{j+1} \circ P_j = 0,\ \ j=0,1,\ldots,n-2,$$
$$\sigma_L(P_0)(x,\xi)\ \text{ is injective}$$

and

$$\sigma_L(P_{n-1})(x,\xi)\ \text{ is surjective}$$

for any $(x,\xi) \in T^*M \setminus \{0\}$. Then (*E*,*P*) is an elliptic complex if and only if D_j is an elliptic PDO of order 2*k* for *j* = 0,1,…,n.

For the proof, see Gilkey [1984], pp. 37-38.

In the sequel we shall use the same notations P_j (and P_j^*) representing the differential operators with domains $C^\infty(E_j)$ as well as their closed extensions to $L^2(E_j)$, $j=0,1,\ldots,n$.

Theorem 2.1 (**Hodge theorem**). Let (*E*,*P*) be an elliptic complex and let

$$D_j = P_{j-1} \circ P_{j-1}^* + P_j^* \circ P_j,\ \ j=0,1,\ldots,n-1.$$

Then we have

(i) $L^2(E_j) = N(D_j) \oplus R(D_{j-1}) \oplus R(P_j^*)$;

(ii) $N(P_j) = R(P_{j-1}) \oplus N(D_j)$;

(iii) $N(P_{j-1}^*) = R(P_j^*) \oplus N(D_j)$;

(iv) If $H^j(E,P) = N(P_j)/R(P_{j-1})$ denotes the *j*-th cohomology group of the elliptic complex (*E*,*P*), then for each *j* = 0,1,…,*n*, the following isomorphism holds

$$H^j(E,P) \cong N(D_j),$$

where we denote as before $N(D) = \ker(D), R(D) = \operatorname{Im}(D)$ for each linear operator *D*.

Proof. D_j has a self-adjoint extension, which is denoted by the same symbol. We have

$$L^2(E_j) = N(D_j) \oplus R(D_j)$$

(since D_j is elliptic, D_j has closed range).

By definition,

$$R(D_j) \subset R(P_j^*) + R(P_{j-1})$$

and

$$P_j \circ P_{j-1} = 0$$

so that

$$\langle P_j^* \omega, P_{j-1}\theta \rangle = 0$$

for each $\omega \in C^\infty(E_{j+1})$ and each $\theta \in C^\infty(E_{j-1})$.

Then

$$R(P_j^*) \perp R(P_{j-1})$$

and this implies

$$R(D_j) \subseteq R(P_j^*) \oplus R(P_{j-1}).$$

On the other hand

$$R(P_j^*) \subset N(P_j)^\perp$$

and

$$R(P_{j-1}) \subset N(P_{j-1}^*)^\perp.$$

Hence

$$R(P_j^*) \oplus R(P_{j-1}) \subset N(P_j)^\perp + N(P_{j-1}^*)^\perp \subset N(D_j)^\perp = R(D_j).$$

The last inclusion follows from

$$N(D_j) \subset N(P_j) \cap N(P_{j-1}^*).$$

Therefore we obtain the first conclusion

$$R(D_j) = R(P_j^*) \oplus R(P_{j-1})$$

and

$$L^2(E_j) = N(D_j) \oplus R(P_j^*) \oplus R(P_{j-1}).$$

For (ii), since

$$N(P_j) \subset R(P_j^*)^\perp$$

we have

$$N(P_j) \subset N(D_j) \oplus R(P_{j-1}).$$

Conversely, the inclusion

$$N(D_j) \subset N(P_j)$$

is known and

$$R(P_{j-1}) \subset N(P_j)$$

follows from

$$P_j \circ P_{j-1} = 0.$$

Hence

$$N(P_j)=R(P_{j-1})\oplus N(D_j).$$

(iii) is obtained in a similar manner and (iv) is a direct consequence of (ii).

Q.E.D.

As a consequence of the Hodge theorem we have:

Corollary 2.1. For the de Rham complex $\left(\wedge\left(T^*M\right)\otimes\mathbb{C},d\right)$, the j-th cohomology group

$$H^j(M)=N(d^j)/R(d^{j-1})$$

of M is isomorphic to $N(\Delta^{(j)})$ for each $j = 0,1,\ldots,n$ (see also Chapter 6, §1, Corollary 1.1). It follows that the Betti numbers β_j of M are given by:

$$\beta_j=\dim H^j(M)=\dim N\left(\Delta^{(j)}\right)$$

for each $j = 0,1,\ldots,n$.

Remark 2.1. Theorem 2.1 and Corollary 2.1 are also valid in the real case.

3. MORSE INEQUALITIES

The Morse inequalities describe a relationship between the Betti numbers of a smooth compact manifold M and the critical point behaviour of any one of a large class of smooth functions on M called **Morse functions**. To illustrate informally the nature of these inequalities, consider the class of closed oriented surfaces which look like deformed n-hole doughnuts. The Betti numbers of such a surface M are given by:

$$\beta_0=\beta_2=1,$$
$$\beta_1=2\times\ (\text{number of holes in } M).$$

Now imagine these surfaces embedded in three-dimensional space with some choice of coordinate axes. Then every point $x\in M$ has the representation $x=\left(x^1,x^2,x^3\right)$ and we can define the height function $f:M\to\mathbb{R}$ given by

$$f(x)=x^3.$$

Suppose that matters are arranged in such a way that there are only finitely many points at which $df = 0$, and that each of these critical points is either a local maximum, a local minimum, or a saddle point. In this context the Morse inequalities are:

$$m_0\overset{def}{=}\text{number of local minima of } f\geq\beta_0,$$
$$m_1\overset{def}{=}\text{number of saddle points of } f\geq\beta_1,$$
$$m_2\overset{def}{=}\text{number of local maxima of } f\geq\beta_2.$$

The first and the third of these inequalities are clearly true since $\beta_0=\beta_2=1$ and f must have a global maximum and a global minimum. It is also intuitively clear that each hole in M should produce at least two saddle points for f, at the top and bottom of the hole.

This implies the second inequality (see Fig. 1).

To give a statement of the Morse inequalities for a general closed manifold M we recall some definitions concerning critical points and Morse functions.

Definition 3.1. Let us suppose that $f: M \to \mathbb{R}$ is a smooth function. Then a point $x \in M$ is called a **critical point** for f if $df(x) = 0$. This means in any local coordinate system $x^1, \ldots, x^n$ around x, we have

$$\frac{\partial f}{\partial x^1}(x) = 0, \ldots, \frac{\partial f}{\partial x^n}(x) = 0.$$

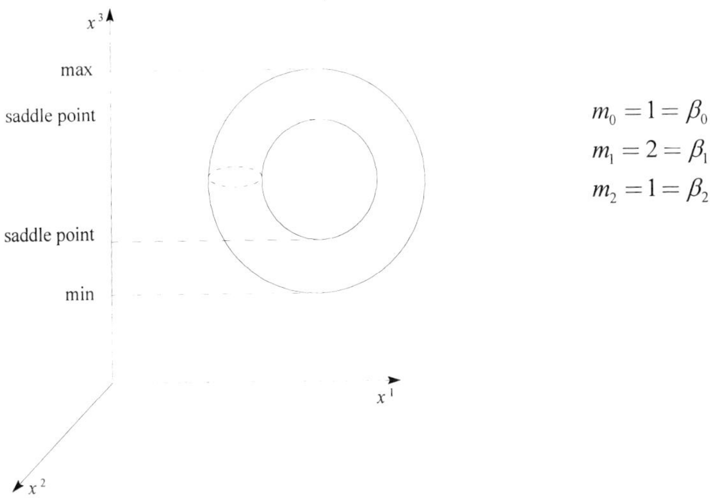

Figure 1: Examples of critical points of the height function

A critical point $x \in M$ is called **nondegenerate** if the Hessian matrix

$$\left[\frac{\partial^2 f}{\partial x^i \partial x^j}(x)\right]$$

is nondegenerate. The **index** $ind(f, x)$ of a critical point $x \in M$ of f is defined to be the number of negative eigenvalues of this Hessian matrix.

Then local nondegenerate maxima and minima have index $n = \dim M$ and 0, respectively while saddle points have some intermediate index.

It is easy to see that the nondegeneracy and the index of a critical point do not depend on the coordinate system used to define the Hessian matrix.

Definition 3.2. A smooth function $f: M \to \mathbb{R}$ is called a **Morse function** if it has finitely many critical points and each critical point is nondegenerate. For a Morse function f and $0 \le p \le n = \dim M$, $m_p(f)$ is defined to be the number of critical points of f with index p.

Theorem 3.1 (Morse). Let $f: M \to \mathbb{R}$ be a smooth function with x_0 a nondegenerate critical point. Then there is a local chart around x_0 such that x_0 is mapped to zero and the local representation of f satisfies

$$f(x) = f(0) + \frac{1}{2} D^2 f(0)(x, x).$$

In particular, nondegenerate critical points of f are isolated.

Proof. We can assume that we are in $\mathbb{R}^n$ and $x_0 = 0, f(x_0) = 0$. Let $\omega_1 = df$ and define the one-form ω_2 by

$$\omega_2(x)\cdot h = D^2 f(0)(h,x).$$

Let

$$\omega_t = t\omega_1 + (1-t)\omega_2.$$

Write

$$\omega_2 = d\varphi,$$

where

$$\varphi(x) = \frac{1}{2} D^2 f(0)(x,x)$$

and define the vector field Z_t by

$$\begin{cases} i(Z_t)\omega_t + (f-\varphi) = 0, \\ Z_t(0) = 0. \end{cases}$$

It is easy to see that Z_t exists near 0 by the nondegeneracy hypothesis. Let ϕ_t be the flow of Z_t. Then we have successively:

$$\frac{d}{dt}(\phi_t^*\omega_t) = \phi_t^*\left(di(Z_t)\omega_t + i(Z_t)d\omega_t + \omega_t - \omega_2\right) = \phi_t^* d\left(i(Z_t)\omega_t + f - \varphi\right) = 0.$$

Thus

$$d\varphi = \omega_2 = \phi_0^*(\omega_2) = \phi_0^*(\omega_0) = \phi_1^*(\omega_1) = \phi_1^*(df) = d(f\circ\phi_1),$$

so ϕ_1 gives (near 0) the coordinate change required.

Q.E.D.

Now the Morse inequalities can be formulated as follows:

Theorem 3.2 (Weak Morse inequalities). For each Morse function $f: M \to \mathbb{R}$ and for each $p = 0,1,\ldots,n$, we have:

$$m_p(f) \geq \beta_p.$$

Theorem 3.3 (Strong Morse inequalities). Let M be a closed orientable manifold and suppose that $f: M \to \mathbb{R}$ is a Morse function on M. Then

$$\sum_{p=0}^{k}(-1)^{k-p} m_p(f) \geq \sum_{p=0}^{k}(-1)^{k-p}\beta_p,\ 0 \leq k < n$$

and

$$\sum_{p=0}^{n}(-1)^p m_p(f) = \sum_{p=0}^{n}(-1)^p \beta_p = \chi(M).$$

For an account of the standard proof of these inequalities, see Milnor [1963]. Witten's proof proceeds via Hodge theory, which we shall examine in the following sections.

4. GENERALIZED TRIANGULATIONS

Let M be a closed (i.e. compact and without boundary) smooth n-dimensional manifold.

Definition 4.1 A pair (f,g), where $f:M\to\mathbb{R}$ is a Morse function (see Definition 3.2) and g a Riemannian metric on M, is called a **generalized triangulation** if the following three conditions C1-C3 are satisfied:

C1. For any (nondegenerate) critical point x_0 of f of index k, there exists a local chart (U,φ) around x_0, where U is an open neighborhood of x_0 in M and $\varphi:U\to D_\varepsilon^n=\left\{(x^1,\dots,x^n)\in\mathbb{R}^n \,\middle|\, \left[(x^1)^2+\cdots+(x^n)^2\right]^{\frac{1}{2}}<\varepsilon\right\}$ a smooth diffeomorphism with $\varphi(x_0)=(0,\dots,0)$, such that

(i) $f\circ\varphi^{-1}(x^1,\dots,x^n)=f(x_0)-\frac{1}{2}\sum_{j=1}^{k}(x^j)^2+\frac{1}{2}\sum_{j=k+1}^{n}(x^j)^2$ [$ind(f,x_0)=k$ is independent of the choice of a coordinate system with respect to which f has the form (i)];

(ii) $(\varphi^{-1})^*(g|_U)$ is the canonical Riemannian metric on the open disc D_ε^n, that is it is given by $g_{ij}(x^1,\dots,x^n)=\delta_{ij},\ i,j=1,\dots,n$.

Local coordinates such that the conditions (i) and (ii) hold are called **admissible**.

For any $x\in M$, let $\gamma_x:\mathbb{R}\to M$ be the integral curve of the vector field $-\mathrm{grad}_g(f)$, which satisfies the initial $\gamma_x(0)=x$. For a critical point x_0 of f, let $W_{x_0}^-$ (resp. $W_{x_0}^+$) be the set $W_{x_0}^-=\left\{x\in M \,\middle|\, \lim_{t\to-\infty}\gamma_x(t)=x_0\right\}$ (resp. $W_{x_0}^+=\left\{x\in M \,\middle|\, \lim_{t\to+\infty}\gamma_x(t)=x_0\right\}$), i.e. $W_{x_0}^-$ (resp. $W_{x_0}^+$) is the unstable (resp. stable) manifold with respect to $-\mathrm{grad}_g(f)$.

In view of (i), it follows that $\varphi_*\left(-\mathrm{grad}_g(f)|_U\right):D_\varepsilon^n\to\mathbb{R}^n$ is given by the mapping $(x^1,\dots,x^n)\mapsto(x^1,\dots,x^k,-x^{k+1},\dots,-x^n)$, and consequently $\varphi(\gamma_x(t))=(x^1e^t,\dots,x^ke^t,x^{k+1}e^{-t},\dots,x^ne^{-t})$, where $D_\varepsilon^n\ni(x^1,\dots,x^n)=\varphi(x),\ x\in U$. Hence $x\in W_{x_0}^-\cap U$ (resp. $W_{x_0}^+\cap U$) if and only if $\lim_{t\to-\infty}\varphi(\gamma_x(t))-(0,\dots,0)$ [resp. $\lim_{t\to+\infty}\varphi(\gamma_x(t))=(0,\dots,0)$], that is $x^{k+1}=\cdots=x^n=0$ (resp. $x^1=\cdots=x^k=0$). Consequently,

$$\varphi\left(W_{x_0}^-\cap U\right)=\left\{(x^1,\dots,x^n)\in D_\varepsilon^n \,\middle|\, x^{k+1}=\cdots=x^n=0\right\}$$

and

$$\varphi\left(W_{x_0}^+\cap U\right)=\left\{(x^1,\dots,x^n)\in D_\varepsilon^n \,\middle|\, x^1=\cdots=x^k=0\right\},$$

so that $W_{x_0}^-$ (resp. $W_{x_0}^+$) is a smooth submanifold of M diffeomorphic to $\mathbb{R}^k$ (resp. $\mathbb{R}^{n-k}$), where $k=ind(f,x_0)$, i.e. $W_{x_0}^-$ (resp. $W_{x_0}^+$) is an open $k\left[\text{resp.}(n-k)\right]$-cell. Note that

$W^-_{x_0} \cap W^+_{x_0} = \{x_0\}$. Since M is closed, one gets $M = \bigcup_{x_0} W^-_{x_0}$, where x_0 runs over the set of all critical points of f. Hence the unstable manifolds $W^-_{x_0}$ provide a partition of M into open cells, where $W^-_{x_0}$ is an open cell of dimension equal to $ind(f, x_0)$.

C2. The Riemannian metric g is so that the vector field $-\mathrm{grad}_g(f)$ satisfies the **Morse-Smale condition**, that is for any two critical points x_0, y_0 of f, $W^-_{x_0}$ and $W^+_{y_0}$ are transversal.

C3. The Morse function f is **self-indexing**, i.e. $f(x_0) = ind(f, x_0)$ for any critical point x_0 of f (in fact, this condition is merely a convenience).

Remark 4.1. Given a smooth simplicial triangulation τ_{sim}, one can construct a generalized triangulation (f, g) such that the unstable manifolds $W^-_{x_0}$ corresponding to $-\mathrm{grad}_g(f)$, with x_0 a critical point of f, are the open simplexes of τ_{sim} (see Poźniak [1990]). This fact justifies the name of generalized triangulation for the pair (f, g) satisfying C1-C3.

Remark 4.2. Given a self-indexing Morse function f and a Riemannian metric g, one can perform arbitrary small C^0-perturbations to g such that the pair consisting of f and the perturbed metric is a generalized triangulation (see Smale [1961]).

C2 implies that $M(x_0, y_0) \overset{def}{=} W^-_{x_0} \cap W^+_{y_0}$ is a smooth manifold of dimension equal to $ind(f, x_0) - ind(f, y_0)$. If $ind(f, x_0) \le ind(f, y_0)$ and $x_0 \ne y_0$, then $M(x_0, y_0) = \varnothing$.

The additive group $(\mathbb{R}, +)$ defines a natural action $\mu : \mathbb{R} \times M(x_0, y_0) \to M(x_0, y_0)$ on $M(x_0, y_0)$ given by $\mu(t, z) = \gamma_z(t)$, $(t, z) \in \mathbb{R} \times M(x_0, y_0)$. If $x_0 \ne y_0$ and $M(x_0, y_0) \ne \varnothing$, this action is free. The corresponding orbit space $M(x_0, y_0)/\mathbb{R}$, denoted by $\tilde{M}(x_0, y_0)$, is a smooth manifold of dimension $ind(f, x_0) - ind(f, y_0) - 1$, diffeomorphic to the submanifold $M(x_0, y_0) \cap f^{-1}(c)$ for any real number c in the open interval $(ind(f, y_0), ind(f, x_0))$. The elements of $\tilde{M}(x_0, y_0)$ are the trajectories from "x_0 to y_0" and such an element will be denoted by γ.

Let us recall that an n-dimensional manifold P with corners is a paracompact, Hausdorff topological space endowed with a maximal smooth atlas with charts (U, φ), where $\varphi : U \to \varphi(U) \subseteq \{(x^1, \ldots, x^n) \in \mathbb{R}^n \mid x^j \ge 0\}$. The set of points of P which correspond (by some and then by any chart) to points in $\mathbb{R}^n$ with exactly $k \ge 1$ coordinates equal to zero is a well-defined subset of P and it will be denoted by P_k. It has the structure of an $(n-k)$-dimensional smooth manifold. The closed subset $\partial P = P_1 \cup P_2 \cup \cdots \cup P_n$ of P is a topological manifold and P is a topological manifold with boundary ∂P. A compact smooth manifold with corners P, whose interior is diffeomorphic to the Euclidean space, is called a **compact smooth cell**.

For any string of critical points $x_0, x_1, \ldots, x_k$ of f with

$$ind(f, x_0) > ind(f, x_1) > \cdots > ind(f, x_k),$$

let us consider the smooth manifold $\tilde{M}(x_0,x_1)\times\tilde{M}(x_1,x_2)\times\cdots\times\tilde{M}(x_{k-1},x_k)\times W_{x_k}^-$ of dimension $ind(f,x_0)-k$, and the smooth map

$$i_{x_0,x_1,\ldots,x_k}:\tilde{M}(x_0,x_1)\times\cdots\times\tilde{M}(x_{k-1},x_k)\times W_{x_k}^-\to M,$$

defined by

$$i_{x_0,x_1,\ldots,x_k}(\gamma_1,\ldots,\gamma_k,y)=i_{x_k}(y),$$

for any $\gamma_j\in\tilde{M}(x_{j-1},x_j)$ and $y\in W_{x_k}^-$, where

$$i_x:W_x^-\hookrightarrow M$$

denotes the inclusion map of W_x^- in M.

Theorem 4.1. Let (f,g) be a generalized triangulation of the closed smooth manifold *M*. For any critical point x_0 of *f*, the smooth manifold $W_{x_0}^-$ has a canonical compactification to a compact manifold with corners $\widehat{W_{x_0}^-}$ and the inclusion map

$$i_{x_0}:W_{x_0}^-\hookrightarrow M$$

has a smooth extension $\widehat{i_{x_0}}:\widehat{W_{x_0}^-}\to M$ such that

(i) $\left(\widehat{W_{x_0}^-}\right)_k=\bigcup_{(x_0,x_1,\ldots,x_k)}\tilde{M}(x_0,x_1)\times\cdots\times\tilde{M}(x_{k-1},x_k)\times W_{x_k}^-$;

(ii) the restriction of $\widehat{i_{x_0}}$ to $\tilde{M}(x_0,x_1)\times\cdots\times\tilde{M}(x_{k-1},x_k)\times W_{x_k}^-$ is given by $i_{x_0,x_1,\ldots,x_k}$.

For the proof of this theorem, see Austin and Braam [1995].

Consequently, the condition C2 implies that the partition $\{W_x^-\mid x \text{ is a critical point of } f\}$ of *M* into open cells is in fact a smooth cell complex.

Given a generalized triangulation (f,g), and for any critical point *x* of *f* an orientation $\mathcal{O}_x$ of W_x^-, one can associate a cochain complex $(C^*(M,(f,g)),\partial^*)$ of real vector spaces, whose differential ∂^* depends on the choosen orientations $\mathcal{O}_x$. Let $C_k(f)$ be the set of all critical points of *f* with index *k*. In order to define this complex, let us introduce the **incidence numbers**

$$I_k:C_k(f)\times C_{k-1}(f)\to\mathbb{Z}$$

defined as follows. Let $(x,y)\in C_k(f)\times C_{k-1}(f)$. If $M(x,y)=\varnothing$, we put $I_k(x,y)=0$. If $M(x,y)\neq\varnothing$, for any $\gamma\in\tilde{M}(x,y)$, the set $\{\gamma\}\times W_y^-$ appears as an open set of the boundary $\partial\widehat{W_x^-}$ and the orientation $\mathcal{O}_x$ induces an orientation on it. If this orientation is the same as the orientation $\mathcal{O}_y$, one sets $\varepsilon(\gamma)=+1$, otherwise one sets $\varepsilon(\gamma)=-1$. In this case, define $I_k(x,y)$ by

$$I_k(x,y)=\sum_{\gamma\in\hat{M}(x,y)}\varepsilon(\gamma).$$

One can check that $I_k(x,y)$ is the incidence number of the open cells W_x^- and W_y^- in the cell complex structure provided by (f,g).

Denote by $(C^*(M,(f,g)),\partial^*)$ the cochain complex of real vector spaces defined by $C^k(M,(f,g))$ is just the set of all maps of $C_k(f)$ into $\mathbb{R}$;

$\partial^{k-1}: C^{k-1}(M,(f,g)) \to C^k(M,(f,g)), (\partial^{k-1}\alpha)(x) = \sum_{y\in C_{k-1}(f)} I_k(x,y)\alpha(y)$, where $\alpha \in C^{k-1}(M,(f,g)), x\in C_k(f)$.

The vector space $C^k(M,(f,g))$ has a canonical basis provided by the family $\{E_x | x\in C_k(f)\}$, where for a fixed $x\in C_k(f)$ the map $E_x : C_k(f)\to\mathbb{R}$ is defined by

$$E_x(y) = \begin{cases} 1 \text{ if } x = y, \\ 0 \text{ if } x \neq y, \end{cases}$$

for any $y\in C_k(f)$. Consequently, $C^k(M,(f,g))$ carries a natural scalar product which makes $\{E_x | x\in C_k(f)\}$ an orthonormal basis.

Since $(C^*(M,(f,g)),\partial^*)$ is the cochain complex associated to the cell complex structure provided by (f,g), it follows that

$$\partial^k \circ \partial^{k-1} = 0 \tag{4.1}$$

for any $k = 1,\ldots,n$.

Let $(A^*(M),d^*)$ be the de Rham complex of *M*, where $A^p(M)$ is the real vector space of smooth differential forms of degree *p* and $d^p : A^p(M)\to A^{p+1}(M)$ is the exterior differential operator (see Chapter 1, §1 and Chapter 2, §2, D).

Theorem 4.2. **(Stokes)** Let *M* be a closed smooth manifold, *P* a compact *m*-dimensional oriented smooth manifold with corners, $P_1 \subset P$ the $(m-1)$-dimensional smooth oriented manifold defined as above, $F: P\to M$ a smooth map and $F_1 : P_1 \to M$ the restriction of *F* to P_1. If $\omega\in A^{m-1}(M)$, then $\int_{P_1} F_1^*(\omega)$ is convergent and

$$\int_P F^*(d^{m-1}\omega) = \int_{P_1} F_1^*(\omega).$$

Let us consider the linear map

$$\mathrm{int}^{(k)} : A^k(M)\to C^k(M,(f,g))$$

defined by

$$\left[\mathrm{int}^{(k)}(\omega)\right](x) = \int_{\widehat{W_x^-}} \omega, \qquad x\in C_k(f).$$

The integration makes sense since $\widehat{W_x^-}$ is a compact smooth manifold with conical singularities (see also the Appendix by F. Laudenbach in Bismut and Zhang [1992]).

The collection $\left\{\mathrm{int}^{(k)} | 0 \le k \le n\right\}$ defines a morphism

$$\mathrm{int}^* : \left(A^*(M), d^*\right) \to \left(C^*(M, (f, g)), \partial^*\right)$$

of cochain complexes, that is $\partial^p \circ \mathrm{int}^{(p)} = \mathrm{int}^{(p+1)} \circ d^p$ for any $p \le n$. In this setting, de Rham theorem can be formulated as follows:

Theorem 4.3. int^* induces an isomorphism in cohomology.

Finally, let us recall that Hodge-de Rham theory provides a canonical and unique representation of each cohomology class of $\left(C^*(M, (f, g)), \partial^*\right)$ by an harmonic form with respect to *g*.

5. WITTEN'S DEFORMATION

Let (*M*,*g*) be a closed, connected, *n*-dimensional smooth Riemannian manifold.

For a given smooth function $f : M \to \mathbb{R}$, we define a new complex (E, d_t) as follows:

$$E = \left\{E^p = \wedge^p\left(T^*(M)\right) \middle| \; p = 0, 1, \ldots, n\right\}$$

and for each $t \ge 0$, let

$$d_t^p = e^{-tf} d^p e^{tf}, \quad p = 0, 1, \ldots, n-1$$

or equivalently

$$d_t^p(\omega) = d^p(\omega) + t df \wedge \omega, \quad \omega \in A^p(M),$$

and

$$d_t = \left\{d_t^p \,|\, p = 0, 1, \ldots, n-1\right\},$$

i.e.

$$\begin{array}{ccccccccc} 0 \to A^0(M) & \to \cdots \to & A^p(M) & \xrightarrow{d^p} & A^{p+1}(M) & \to \cdots \to 0 \\ e^{-tf}\downarrow & & e^{-tf}\downarrow & & e^{-tf}\downarrow & \\ 0 \to A^0(M) & \to \cdots \to & A^p(M) & \xrightarrow{d_t^p} & A^{p+1}(M) & \to \cdots \to 0 \end{array}$$

where $A^p(M) = C^\infty(E^p)$, $0 \le p \le n$.

Since

$$d_t^p \circ d_t^{p-1} = e^{-tf} d^p d^{p-1} e^{tf} = 0$$

and

$$\sigma_L\left(d_t^p\right) = \sigma_L\left(d^p\right),$$

it follows that (E, d_t) is an elliptic complex.

d_t^* is the unique differential in $A^p(M)$ which makes the multiplication by the smooth function e^{tf} an isomorphism of cochain complexes:

$$e^{tf}:\left(A^*(M),d_t^*\right)\rightarrow\left(A^*(M),d^*\right).$$

We define also the **adjoint** $\left(d_t^p\right)^*$ of d_t^p by

$$\left\langle\left(d_t^p\right)^*(\omega),\theta\right\rangle=\left\langle\omega,d_t^p(\theta)\right\rangle,$$

which holds for each $\omega\in A^{p+1}(M)$ and for each $\theta\in A^p(M)$.

Therefore,

$$\left(d_t^p\right)^*(\omega)=e^{tf}\delta^p e^{-tf}(\omega),$$

where δ^p is the codifferential.

Then define the Witten Laplacian associated to f by

$$\Delta_t^{(p)}=\left(d_t^p\right)^*d_t^p+d_t^{p-1}\left(d_t^{p-1}\right)^*.$$

The operators $\Delta_t^{(p)}$ are elliptic, self-adjoint and positive, hence their spectra $\operatorname{spect}\left(\Delta_t^{(p)}\right)$ lie on $[0,+\infty)$. Note that

$$\ker\left(\Delta_t^{(p)}\right)=\left\{\omega\in A^p(M)\,\middle|\,d_t^p\omega=0,\left(d_t^p\right)^*\omega=0\right\}.$$

By the Hodge theorem for elliptic complexes we have successively:

$$\ker\left(\Delta_t^{(p)}\right)\cong\ker\left(d_t^p\right)\Big/\operatorname{Im}\left(d_t^{p-1}\right)\cong\ker\left(d^p\right)\Big/\operatorname{Im}\left(d^{p-1}\right)\cong\ker\left(\Delta^{(p)}\right)$$

and consequently $\beta_p=\dim\ker\left(\Delta_t^{(p)}\right)$ for all $t\geq 0$. The second isomorphism holds, because the map α given by:

$$\alpha:\omega\mapsto\alpha(\omega)=e^{-tf}\omega$$

satisfies the following properties:

(i) $\alpha\big|_{\ker\left(d^p\right)}:\ker\left(d^p\right)\rightarrow\ker\left(d_t^p\right)$ is an isomorphism;

(ii) $\alpha\left(\operatorname{Im}\left(d^{p-1}\right)\right)=\operatorname{Im}\left(d_t^{p-1}\right)$.

Hence if 0 is an eigenvalue of $\Delta^{(p)}$, that is the p-th Betti number β_p of M is $\neq 0$, then it is an eigenvalue of $\Delta_t^{(p)}$ with the same multiplicity β_p for all $t\geq 0$.

For each $\omega\in A^1(M)$, let us define the exterior and interior multiplication operators respectively by

$$\varepsilon(\omega): A^p(M) \to A^{p+1}(M), \varepsilon(\omega)\theta = \omega \wedge \theta, \theta \in A^p(M),$$
$$i_\omega : A^p(M) \to A^{p-1}(M),$$
$$i_\omega(\theta^1 \wedge \ldots \wedge \theta^p) = \sum_{k=1}^{p} (-1)^{k+1} (\omega | \theta^k) \theta^1 \wedge \ldots \wedge \hat{\theta}^k \wedge \ldots \wedge \theta^p,$$

$\theta^1, \ldots, \theta^p \in A^1(M)$, where the smooth real function $(\omega | \theta^k)$ has been defined in Chapter 2, §1 C, and $\hat{\theta}^k$ means that θ^k is deleted. One has

$$\langle \omega \wedge \theta, \alpha \rangle = \langle \theta, i_\omega(\alpha) \rangle \tag{5.1}$$

for each $\theta \in A^{p-1}(M)$ and $\alpha \in A^p(M)$, that is the adjoint of $\varepsilon(\omega)$ relative to $<,>$ is i_ω,

$$i_{\omega_1}(\omega_2 \wedge \theta) = (\omega_1 | \omega_2)\theta - \omega_2 \wedge i_{\omega_1}(\theta) \tag{5.2}$$

for each $\omega_1, \omega_2 \in A^1(M)$ and each $\theta \in A^p(M)$ (see also Chapter 2, Remark 1.6), and

$$\delta(f\theta) = f\delta\theta - i_{df}(\theta) \tag{5.3}$$

for each $f \in C^\infty(M)$ and $\theta \in A^p(M)$.

Indeed, (5.1) and (5.2) can be easily obtained.

Since $i(\text{grad}(f)) = i_{df}$ for every $f \in C^\infty(M)$, (5.3) is just (2.28) in Chapter 2 (see the proof of Proposition 2.11, Chapter 2).

Proposition 5.1. The following equalities hold for any $\omega \in A^p(M)$:

(i) $d_t^p \omega = d^p \omega + t df \wedge \omega$;

(ii) $(d_t^{p-1})^* \omega = \delta^p \omega + t i_{df} \omega$;

(iii) $\Delta_t^{(p)} \omega = \Delta^{(p)} \omega + t P_{df} \omega + t^2 g(df, df)\omega$,

where

$$P_{df}\omega = i_{df} d\omega + d(i_{df}\omega) + \delta(df \wedge \omega) + df \wedge \delta\omega.$$

Proof. For the sake of simplicity we omit to put the corresponding indices for the operators d, d_t, δ and d_t^*.

(i) We have $d_t \omega = e^{-tf} d(e^{tf}\omega) = e^{-tf}(e^{tf} d\omega + te^{tf} df \wedge \omega) = d\omega + t df \wedge \omega$.

(ii) $d_t^* \omega = e^{tf} \delta(e^{-tf}\omega) = e^{tf}(e^{-tf}\delta\omega - i_{de^{-tf}}\omega) = \delta\omega + e^{tf} i_{-te^{-tf}df}\omega = \delta\omega + t i_{df}\omega$.

(iii)
$$\begin{aligned}\Delta_t \omega &= d_t d_t^* \omega + d_t^* d_t \omega = t df \wedge d_t^* \omega + d(d_t^* \omega) + \delta(d_t \omega) + t i_{df}(d_t \omega) \\ &= t df \wedge (\delta\omega + t i_{df}\omega) + d(\delta\omega + t i_{df}\omega) + \delta(t df \wedge \omega + d\omega) + t i_{df}(t df \wedge \omega + d\omega) \\ &= \Delta\omega + t[df \wedge \delta\omega + d(i_{df}\omega) + \delta(df \wedge \omega) + i_{df} d\omega] + t^2 [df \wedge i_{df}\omega + i_{df}(df \wedge \omega)] \\ &= \Delta\omega + t P_{df}\omega + t^2 g(df, df)\omega.\end{aligned}$$

Q.E.D.

In view of the considerations in Chapter 2, §2, D, one can verify that P_{df} is just the zero-th order differential operator

$$L_{-\mathrm{grad}_g(f)} + L^*_{-\mathrm{grad}_g(f)} : A^p(M) \to A^p(M)$$

where $L_{-\mathrm{grad}_g(f)}$ denotes the Lie derivative operator in the direction $-\mathrm{grad}_g(f)$ (see Chapter 1, §1) and $L^*_{-\mathrm{grad}_g(f)}$ the adjoint of the Lie derivative operator $L_{-\mathrm{grad}_g(f)}$ (see Chapter 2, Corollary 2.3). Note that $\Delta_0^{(p)} = \Delta^{(p)}$ and $\Delta_t^{(p)} - \Delta^{(p)} : A^p(M) \to A^p(M)$ is also a zero-th order differential operator.

Now, let us consider the smooth real function $f_k : \mathbb{R}^n \to \mathbb{R}$ defined by $f_k(x) = k - \frac{1}{2}\sum_{j=1}^{k}(x^j)^2 + \frac{1}{2}\sum_{j=k+1}^{n}(x^j)^2$. Denote by $\Delta^{(p)} : A^p(\mathbb{R}^n) \to A^p(\mathbb{R}^n)$ the Hodge-de Rham operator on smooth p-forms on $\mathbb{R}^n$, that is

$$\Delta^{(p)}\left(\sum_I a_I(x^1,\ldots,x^n)dx^I\right) = -\sum_I\left(\sum_{j=1}^{n}\frac{\partial^2}{\partial(x^j)^2}a_I(x^1,\ldots,x^n)\right)dx^I$$

[here $I = (i_1,\ldots,i_p), 1 \le i_1 < i_2 < \cdots < i_p \le n$ and $dx^I = dx^{i_1} \wedge \ldots \wedge dx^{i_p}$] and by

$$\Delta_{t;k}^{(p)} : A^p(\mathbb{R}^n) \to A^p(\mathbb{R}^n)$$

the Witten Laplacian associated to f_k. A straightforward calculation shows that

$$\Delta_{t;k}^{(p)} = \Delta^{(p)} + tn_k^{(p)}1_{A^p(\mathbb{R}^n)} + t^2\left[(x^1)^2 + \cdots + (x^n)^2\right]1_{A^p(\mathbb{R}^n)}, \tag{5.4}$$

where

$$n_k^{(p)} = \begin{cases} -n+2k-2p & \text{if } k \ge p, \\ -n+2p-2k & \text{if } k < p, \end{cases}$$

and $1_{A^p(\mathbb{R}^n)}$ is the identity operator on $A^p(\mathbb{R}^n)$. Note that $n_p^{(p)} = -n$.

Now assume that M is a closed, connected, n-dimensional smooth manifold. Let $f : M \to \mathbb{R}$ be a Morse function and g a Riemannian metric on M such that (f, g) is a generalized triangulation (see Remark 4.2). Let $x \in M$ be a critical point of f with $ind(f, x) = k$ and choose admissible coordinates in the neighborhood of x. Since with respect to these coordinates f has the form

$$k - \frac{1}{2}\left[(x^1)^2 + \cdots + (x^k)^2\right] + \frac{1}{2}\left[(x^{k+1})^2 + \cdots + (x^n)^2\right]$$

and g is a flat Riemannian metric, our previous considerations leads us to the following explicit formula for the Witten Laplacian in admissible coordinates in the neighborhood of x:

Proposition 5.2. If (f, g) is a generalized triangulation, then with respect to admissible coordinates in the neighborhood of a critical point x of f such that $ind(f, x) = k$

the Witten Laplacian $\Delta_t^{(p)}$ has the form $\Delta_{t;k}^{(p)}$ given by (5.4).

Let $A_{rd}^p(\mathbb{R}^n)$ be the space of smooth p-forms $\omega = \sum_I a_I\left(x^1,\ldots,x^n\right)dx^I$ on $\mathbb{R}^n$ with $a_I\left(x^1,\ldots,x^n\right)$ rapidly decaying functions. The operator $\Delta_{t;k}^{(p)}$ acting on $A_{rd}^p(\mathbb{R}^n)$ is globally elliptic in the sense of Shubin [1980] or Hörmander [1985], self-adjoint and positive. This operator is the harmonic oscillator in n variables acting on p-forms and its properties can be derived from the harmonic oscillator in one variable $-\dfrac{d^2}{d\left(x^1\right)^2}+a+b\left(x^1\right)^2$ acting on functions. In particular the following result holds.

Proposition 5.3. (i) $\Delta_{t;k}^{(p)}$ regarded as an unbounded densely defined operator on the L^2-completion of $A_{rd}^p(\mathbb{R}^n)$, is self-adjoint, positive and its spectrum is contained in the set $2t\mathbb{N} = \{2tj \mid j \in \mathbb{N}\}$;

(ii) $\dim\ker \Delta_{t;k}^{(p)} = \begin{cases} 1 \text{ if } k = p, \\ 0 \text{ if } k \neq p; \end{cases}$

(iii) The p-form $\omega_{t;p} = \left(\dfrac{t}{\pi}\right)^{\frac{n}{4}} e^{-t\sum_{j=1}^{n}\frac{\left(x^j\right)^2}{2}} dx^1 \wedge \cdots \wedge dx^p$ is the generator of $\ker \Delta_{t;p}^{(p)}$ and has the L^2-norm equal to 1.

Let $K = \{x_1,\ldots,x_s\}$ be the set of all critical points of f and $ind\left(f,x_j\right) = k_j$, $j = 1,\ldots,s$. For each critical point x_j of f, let us choose once for all an admissible coordinate chart $\left(U_j,\varphi_j\right)$ such that $U_j \cap U_l = \varnothing$ for $j \neq l$, $j,l = 1,\ldots,s$.

If $T_t^{(p)}$ denotes the self-adjoint operator defined by

$$T_t^{(p)} = \Delta_{t;k_1}^{(p)} \times \cdots \times \Delta_{t;k_s}^{(p)} : \overset{s}{\oplus} L^2 A_{rd}^p(\mathbb{R}^n) \to \overset{s}{\oplus} L^2 A_{rd}^p(\mathbb{R}^n),$$

then

$$\dim\ker\left(T_t^{(p)}\right) = m_p(f),\ \ p = 0,1,\ldots,n.$$

This is an immediate consequence of Proposition 5.3(ii): each x_j with $ind\left(f,x_j\right) = p$ has an one-dimensional contribution to the null-space of $T_t^{(p)}$, while each x_j with $ind\left(f,x_j\right) \neq p$ has no contribution to this null-space.

6. THE MAIN RESULTS OF THE WITTEN-HELFFER-SJÖSTRAND THEORY

Assume that (M,g) is a closed, connected, n-dimensional Riemannian manifold and let $f: M \to \mathbb{R}$ be a Morse function such that (f,g) is a generalized triangulation.

The following result provides the first main result of the WHS-theory.

Theorem 6.1. (Witten) Given a generalized triangulation (f,g) of the closed, connected, n-dimensional smooth manifold M, there exist positive constants c_1,c_2,c_3 and

t_0 depending on (f,g), such that for any $t > t_0$ and $p = 0,1,\ldots,n$, $Spect\left(\Delta_t^{(p)}\right) \subset \left[0, c_1 e^{-c_2 t}\right] \cup \left[c_3 t, +\infty\right)$ and the number of the eigenvalues of $\Delta_t^{(p)}$ in the interval $\left[0, c_1 e^{-c_2 t}\right]$ counted with their multiplicities is equal to the number $m_p(f)$ of all critical points of *f* with index *p*.

Recall that $\dim\ker\left(\Delta_t^{(p)}\right) = \beta_p$ (= the *p*-th Betti number of *M*), $p = 0,1,\ldots,n$ (see §5), so that if $\beta_p \neq 0$, 0 is an eigenvalue of $\Delta_t^{(p)}$ and its multiplicity is equal to β_p, and Theorem 6.1 implies in particular that

$$m_p(f) \geq \beta_p$$

for any $p = 0,1,\ldots,n$, i.e. the weak Morse inequalities are true (see Theorem 3.2).

Theorem 6.1 states the existence of a gap in the spectrum of $\Delta_t^{(p)}$, namely the open interval $\left(c_1 e^{-c_2 t}, c_3 t\right)$, which widens to $(0,+\infty)$ when $t \to \infty$. The constants c_1, c_2, c_3 and t_0 determine a constant $T > 0$, such that for $t > T$, 1 is not an eigenvalue of $\Delta_t^{(p)}$,

$$Spect\left(\Delta_t^{(p)}\right) \cap \left[0, c_1 e^{-c_2 t}\right] = Spect\left(\Delta_t^{(p)}\right) \cap [0,1)$$

and

$$Spect\left(\Delta_t^{(p)}\right) \cap \left[c_3 t, \infty\right) = Spect\left(\Delta_t^{(p)}\right) \cap (1,+\infty).$$

This separation property of the eigenvalues of $\Delta_t^{(p)}$ allows to obtain, for *t* sufficiently large, the small complex $\left(A_{sm}^p(M)(t), d_t^p\right)$ as a smooth family of cochain subcomplexes of the deformed de Rham complex $\left(A^p(M), d_t^p\right)$. Moreover, a scaled version of this family of cochain subcomplexes converges for $t \to \infty$ to the cochain complex $\left(C^*(M,(f,g)), \partial^*\right)$ introduced in §4.

More precisely, for $t > T$ let us denote by $A_{sm}^p(M)(t)$ the finite dimensional vector subspace generated by the eigenforms of $\Delta_t^{(p)}$ corresponding to the eigenvalues of $\Delta_t^{(p)}$ smaller than 1. By Theorem 6.1, the dimension of this vector space is equal to $m_p(f)$ (= the number of critical points of *f* of index *p*). The elliptic theory implies that these eigenforms of degree *p*, a priori elements in the space $L^2A^p(M)$ (see Chapter 2, §1, C), are actually in $A^p(M)$ (see also Chapter 5, §1). Since $d_t^p \circ \Delta_t^{(p)} = \Delta_t^{(p+1)} \circ d_t^p$, it follows that $d_t^p\left(A_{sm}^p(M)(t)\right) \subset A_{sm}^{p+1}(M)(t)$, so that $\left(A_{sm}^*(M)(t), d_t^*\right)$ is a finite dimensional cochain complex and a cochain subcomplex of $\left(A^*(M), d_t^*\right)$. Since $e^{tf} \cdot d_t^p = d^p \cdot e^{tf}$, that is the diagram

$$\begin{array}{ccc} A_{sm}^p(M)(t) & \xrightarrow{d_t^p} & A_{sm}^{p+1}(M)(t) \\ e^{tf} \downarrow & & \downarrow e^{tf} \\ A^p(M) & \xrightarrow{d^p} & A^{p+1}(M) \end{array}$$

is commutative for any $p=0,\ldots,n$, the multiplication by e^{tf}, that is $A_{sm}^{p}(M)(t)\ni\omega\mapsto e^{tf}\omega\in A^{p}(M)$, $p=0,\ldots,n$, defines an injective cochain morphism from the cochain complex $\left(A_{sm}^{*}(M)(t),d_{t}^{*}\right)$ into the de Rham complex $\left(A^{*}(M),d^{*}\right)$. Now, for $t>T$, let us introduce a scaled version of the cochain complex $\left(A_{sm}^{*}(M)(t),d_{t}^{*}\right)$, namely instead of the coboundary operators d_{t}^{p}, $p=0,1,\ldots,n$, let us consider the coboundary operators $\tilde{d}_{t}^{p}=\left(\frac{\pi}{t}\right)^{\frac{1}{2}}e^{t}d_{t}^{p}:A_{sm}^{p}(M)(t)\to A_{sm}^{p+1}(M)(t)$, $p=0,1,\ldots,n$. Note that $\left\{F_{t}^{(p)}:A_{sm}^{p}(M)(t)\to A_{sm}^{p}(M)(t),\ p=0,1,\ldots,n\right\}$, where $F_{t}^{(p)}=\left(\frac{\pi}{t}\right)^{\frac{n-2p}{4}}e^{-tp}1_{A_{sm}^{p}(M)(t)}$, defines a cochain isomorphism from the cochain complex $\left(A_{sm}^{p}(M)(t),\tilde{d}_{t}^{*}\right)$, into the cochain complex $\left(A_{sm}^{p}(M)(t),d_{t}^{*}\right)$, which is called the **re-scaling isomorphism**. Indeed, $F_{t}^{(p+1)}\circ\tilde{d}_{t}^{p}=d_{t}^{p}\circ F_{t}^{(p)}$ for any $p=0,1,\ldots,n-1$, and each $F_{t}^{(p)}:A_{sm}^{p}(M)(t)\to A_{sm}^{p}(M)(t)$ is an isomorphism.

For $t>T$, let us consider the composition I_{t}^{*} of the following cochain morphisms

$$\left(A_{sm}^{*}(M)(t),\tilde{d}_{t}^{*}\right)\xrightarrow{F_{t}^{*}}\left(A_{sm}^{*}(M)(t),d_{t}^{*}\right)\xrightarrow{e^{tf}}\left(A^{*}(M),d^{*}\right)\xrightarrow{\mathrm{int}^{*}}\left(C^{*}(M,(f,g)),\partial^{*}\right),$$

where $\mathrm{int}^{(p)}:A^{p}(M)\to C^{p}(M,(f,g))$ is the integration operator on the p-cells of the generalized triangulation (f,g) considered in §4.

In what follows assume that $A_{sm}^{p}(M)(t)\subset A^{p}(M)$ is endowed with the canonical scalar product $\langle\ ,\ \rangle$ defined in Chapter 2, §1, C, and $C^{p}(M,(f,g))$ with the canonical scalar product defined in §4, $0\le p\le n$.

The next theorem provides the second main result of the WHS-theory.

Theorem 6.2. (Helffer-Sjöstrand [1985]) Given a generalized triangulation (f,g) of the closed, connected, n-dimensional smooth manifold M, there exists a constant $T_{1}>0$, depending on (f,g), such that for $t>T_{1}$, 1 is not an eigenvalue of $\Delta_{t}^{(p)}$, $0\le p\le n$, and I_{t}^{*} is an cochain isomorphism. Moreover, for $t>T_{1}$, there exists a family $J_{t}^{(p)}:C^{p}(M,(f,g))\to A_{sm}^{p}(M)(t)$ of isometries of finite dimensional vector spaces such that $I_{t}^{(p)}\circ J_{t}^{(p)}=1_{C^{p}(M,(f,g))}+0\left(\frac{1}{t}\right)$.

Theorem 6.2 says that the small subcomplex is, up to normalization, asymptotically isometric to the combinatorial complex associated to (f,g). It provides, asymptotically, a canonical and unique representation of the full cochain complex $\left(C^{*}(M,(f,g)),\partial^{*}\right)$ and its base $\{E_{x}\,|\,x\text{ is a critical point of }f\}$ inside the de Rham complex $\left(A^{*}(M),d^{*}\right)$.

Now let us give the proofs of Theorems 6.1 and 6.2 based on arguments of Burghelea, Friedlander, Kappeler, McDonald [1996].

The first step is given in the following construction. For $\eta > 0$, let $\gamma_\eta : \mathbb{R} \to [0,1]$ be a smooth function equal to 1 on the interval $\left(-\infty, \frac{\eta}{2}\right]$ and equal to 0 on the interval $[\eta, +\infty)$. For $\varepsilon > 0$, which we will choose later at our convenience, let us define $\tilde{\omega}^{\varepsilon}_{t;p} \in A^p(\mathbb{R}^n)$ with compact support by

$$\tilde{\omega}^{\varepsilon}_{t;p} = \frac{1}{\beta(t)} \gamma_\varepsilon(\|x\|)\omega_{t;p},$$

where $\|x\| = \sqrt{\sum_{j=1}^{n} (x^j)^2}$, $\beta(t) = \|\gamma_\varepsilon(\|x\|)\omega_{t;p}\|_{L^2} = \left(\frac{t}{\pi}\right)^{\frac{n}{4}} \left(\int_{\mathbb{R}^n} \gamma_\varepsilon^2(\|x\|) e^{-t\|x\|^2} dx\right)^{\frac{1}{2}}$ and $\omega_{t;p}$ is the p-form given in Proposition 5.3 (iii), with $\| \ \|_{L^2}$ denoting the L^2-norm. The smooth form $\tilde{\omega}^{\varepsilon}_{t;p}$ has the support contained in the disc of radius ε, agrees with $\omega_{t;p}$ on the disc of radius $\frac{\varepsilon}{2}$ and satisfies

$$\left\langle \tilde{\omega}^{\varepsilon}_{t;p}, \tilde{\omega}^{\varepsilon}_{t;p} \right\rangle = 1 \tag{6.1}$$

with respect to the scalar product $\langle \, , \rangle$ on $A^p_{rd}(\mathbb{R}^n)$. Properties of the operators $\Delta^{(p)}_{t;k} : A^p(\mathbb{R}^n) \to A^p(\mathbb{R}^n)$ defined by (5.4) are given in the following

Proposition 6.1. For a fixed $r \in \mathbb{N}$, there exist positive constants $c, c', c'', t_0, \varepsilon_0$, so that, for $t \geq t_0$ and $\varepsilon < \varepsilon_0$,

$\left| \frac{\partial^{|\alpha|}}{\partial(x^1)^{\alpha_1} \cdots \partial(x^n)^{\alpha_n}} \left(\Delta^{(p)}_{t;p} \tilde{\omega}^{\varepsilon}_{t;p}(x)\right) \right| \leq c e^{-c't}$ for any $x \in \mathbb{R}^n$ and multi-index $\alpha = (\alpha_1, \ldots, \alpha_n)$ with $|\alpha| \leq r$;

$\left\langle \Delta^{(p)}_{t;k}\left(\tilde{\omega}^{\varepsilon}_{t;p}\right), \tilde{\omega}^{\varepsilon}_{t;p} \right\rangle \geq 2t|p-k|$;

$\left\langle \Delta^{(p)}_{t;p}(\omega), \omega \right\rangle \geq c'' t \|\omega\|^2$ for any $\omega \in A^p(M)$ with compact support and $\left\langle \omega, \tilde{\omega}^{\varepsilon}_{t;p} \right\rangle = 0$.

Proof. In what follows we keep the notations used in §5. First notice that

$$\left(\Delta + t n^{(p)}_k + t^2 \|x\|^2\right)\left(e^{\frac{-t\|x\|^2}{2}}\right) = 2t|p-k| e^{\frac{-t\|x\|^2}{2}} \tag{6.2}$$

and

$$\Delta\left(\gamma_\varepsilon(\|x\|)e^{\frac{-t\|x\|^2}{2}}\right)$$
$$= e^{\frac{-t\|x\|^2}{2}}\Delta(\gamma_\varepsilon(\|x\|)) - 2\sum_{j=1}^{n}\frac{\partial}{\partial x^j}\gamma_\varepsilon(\|x\|)\frac{\partial}{\partial x^j}\left(e^{\frac{-t\|x\|^2}{2}}\right) + \gamma_\varepsilon(\|x\|)\Delta\left(e^{\frac{-t\|x\|^2}{2}}\right), \tag{6.3}$$

[see also the formula (2.11) in Chapter 2], where $\Delta = -\sum_{j=1}^{n}\frac{\partial^2}{\partial(x^j)^2}$ is the Laplace-Beltrami operator associated to $(\mathbb{R}^n, g_{can})$ (see Chapter 2, Example 2.1). Then notice that

$$\frac{\partial}{\partial x^j}\gamma_\varepsilon(\|x\|) = \frac{x^j}{\|x\|}\dot{\gamma}_\varepsilon(\|x\|),$$
$$\frac{\partial^2}{\partial(x^j)^2}\gamma_\varepsilon(\|x\|) = \frac{(x^j)^2}{\|x\|^2}\ddot{\gamma}_\varepsilon(\|x\|) + \frac{\|x\|^2-(x^j)^2}{\|x\|^3}\dot{\gamma}_\varepsilon(\|x\|),$$
$$\frac{\partial}{\partial x^j}\left(e^{\frac{-t\|x\|^2}{2}}\right) = -tx^j e^{\frac{-t\|x\|^2}{2}}$$

and take $k = p$ to conclude from (6.3) and (6.2) that

$$\begin{aligned}\left|\left(\Delta + tn_p^{(p)} + t^2\|x\|^2\right)\left(\gamma_\varepsilon(\|x\|)e^{\frac{-t\|x\|^2}{2}}\right)\right| &\le 2\left|\sum_{j=1}^{n}\frac{\partial}{\partial x^j}(\gamma_\varepsilon(\|x\|))\frac{\partial}{\partial x^j}e^{\frac{-t\|x\|^2}{2}}\right| \\ &\quad + \left|\sum_{j=1}^{n}\frac{\partial^2}{\partial(x^j)^2}(\gamma_\varepsilon(\|x\|))e^{\frac{-t\|x\|^2}{2}}\right| \\ &\le \left[\left(2t\|x\| + \frac{n-1}{\|x\|}\right)|\dot{\gamma}_\varepsilon(\|x\|)| + |\ddot{\gamma}_\varepsilon(\|x\|)|\right]e^{\frac{-t\|x\|^2}{2}},\end{aligned}$$

where $\dot{\gamma}_\varepsilon(t) = \frac{d}{dt}\gamma_\varepsilon(t)$ and $\ddot{\gamma}_\varepsilon(t) = \frac{d^2}{dt^2}\gamma_\varepsilon(t)$.

Since $\operatorname{supp}(\dot{\gamma}_\varepsilon)$ and $\operatorname{supp}(\ddot{\gamma}_\varepsilon)$ are contained in $\left[\frac{\varepsilon}{2}, \varepsilon\right]$ one gets

$$\left|\left(\Delta + tn_p^{(p)} + t^2\|x\|^2\right)(\gamma_\varepsilon(\|x\|))e^{\frac{-t\|x\|^2}{2}}\right| \le \left[\|\dot{\gamma}_\varepsilon\|_{L^\infty}\left(2\varepsilon t + 2\frac{n-1}{\varepsilon}\right) + \|\ddot{\gamma}_\varepsilon\|_{L^\infty}\right]e^{\frac{-t\varepsilon^2}{8}} \tag{6.4}$$

To estimate the normalizing constant $\beta(t)$, first notice that

$$\int_{\mathbb{R}^n}\gamma_\varepsilon^2(\|x\|)e^{-t\|x\|^2}dx \ge \int_{\|x\|\le\frac{\varepsilon}{2}} e^{-t\|x\|^2}dx.$$

A change of variables, namely $x^j = \frac{s}{\sqrt{t}} v^j, j = 1,\dots,n, t > 0$, where $s \in \left(0, \frac{\varepsilon\sqrt{t}}{2}\right], \left(v^1,\dots,v^n\right) \in S^{n-1}$, leads to

$$\int_{\|x\|\le\frac{\varepsilon}{2}} e^{-t\|x\|^2} dx = \int_0^{\frac{\varepsilon\sqrt{t}}{2}} \int_{S^{n-1}} e^{-s^2} \left(\frac{s}{\sqrt{t}}\right)^{n-1} v_{g_{can}}\, d\left(\frac{s}{\sqrt{t}}\right) = \mathrm{Vol}\left(S^{n-1}, g_{can}\right) t^{-\frac{n}{2}} \int_0^{\frac{\varepsilon\sqrt{t}}{2}} e^{-s^2} s^{n-1} ds$$

$$= \frac{2\pi^{\frac{n}{2}}}{\Gamma\left(\frac{n}{2}\right)} t^{-\frac{n}{2}} \int_0^{\frac{\varepsilon\sqrt{t}}{2}} e^{-s^2} s^{n-1} ds,$$

where $v_{g_{can}}$ denotes the volume form for $\left(S^{n-1}, g_{can}\right)$ (see also Chapter 2, Example 1.1).

Setting $t_0 = \left(\frac{2}{\varepsilon}\right)^2$, for $t \ge t_0$, one obtains

$$\int_{\mathbb{R}^n} \gamma_\varepsilon^2 \left(\|x\|\right) e^{-t\|x\|^2} dx \ge c_0 t^{-\frac{n}{2}} \int_0^1 e^{-s^2} s^{n-1} ds, \tag{6.5}$$

where $c_0 > 0$ is a constant. Combining (6.4) and (6.5) one concludes that there exists $c_0' > 0$ such that

$$\left|\Delta_{t;p}^{(p)} \tilde{\omega}_{t;p}^{\varepsilon}(x)\right| \le c_0' \left[\left\|\dot{\gamma}_\varepsilon\right\|_{L^\infty} \left(2\varepsilon t + \frac{n-1}{\varepsilon}\right) e^{\frac{-t\varepsilon^2}{16}} + \left\|\ddot{\gamma}_\varepsilon\right\|_{L^\infty} e^{\frac{-t\varepsilon^2}{16}}\right] e^{\frac{-t\varepsilon^2}{16}}. \tag{6.6}$$

This leads to the estimate (i) in the case when $r = 0$. Note that one gets similar estimates for all the partial derivatives of $\left(\Delta + t n_p^{(p)} + t^2 \|x\|^2\right)\left(\gamma_\varepsilon\left(\|x\|\right) e^{-\frac{t\|x\|^2}{2}}\right)$, that is the estimate (i) is true for any fixed $r \in \mathbb{N}$.

(ii) Using Green's formula (see Chapter 2, Proposition 2.3) and (6.3) one gets successively

$$\int_{\mathbb{R}^n} \gamma_\varepsilon(\|x\|) e^{-t\|x\|^2} \Delta(\gamma_\varepsilon(\|x\|)) dx = \sum_{j=1}^{n} \int_{\mathbb{R}^n} \frac{\partial(\gamma_\varepsilon(\|x\|))}{\partial x^j} \frac{\partial\left(\gamma_\varepsilon(\|x\|) e^{-t\|x\|^2}\right)}{\partial x^j} dx$$

$$= \sum_{j=1}^{n} \int_{\mathbb{R}^n} \left(\frac{\partial \gamma_\varepsilon(\|x\|)}{\partial x^j}\right)^2 e^{-t\|x\|^2} dx + \sum_{j=1}^{n} \int_{\mathbb{R}^n} \gamma_\varepsilon(\|x\|) \frac{\partial(\gamma_\varepsilon(\|x\|))}{\partial x^j} 2 \left(\frac{\partial\left(e^{-t\|x\|^2/2}\right)}{\partial x^j}\right) e^{\frac{-t\|x\|^2}{2}} dx$$

$$= \sum_{j=1}^{n} \int_{\mathbb{R}^n} \left(\frac{\partial \gamma_\varepsilon(\|x\|)}{\partial x^j}\right)^2 e^{-t\|x\|^2} dx +$$

$$+ \int_{\mathbb{R}^n} \gamma_\varepsilon(\|x\|) e^{\frac{-t\|x\|^2}{2}} \left[e^{\frac{-t\|x\|^2}{2}} \Delta(\gamma_\varepsilon(\|x\|)) + \gamma_\varepsilon(\|x\|) \Delta\left(e^{\frac{-t\|x\|^2}{2}}\right) - \Delta\left(\gamma_\varepsilon(\|x\|) e^{\frac{-t\|x\|^2}{2}}\right)\right] dx,$$

what implies that

$$\begin{aligned} &\int_{\mathbb{R}^n} \gamma_\varepsilon(\|x\|) e^{\frac{-t\|x\|^2}{2}} \Delta\left(\gamma_\varepsilon(\|x\|) e^{\frac{-t\|x\|^2}{2}}\right) dx \\ &= \int_{\mathbb{R}^n} \left[\sum_{j=1}^{n} \left(\frac{\partial \gamma_\varepsilon(\|x\|)}{\partial x^j}\right)^2 e^{-t\|x\|^2} + \gamma_\varepsilon^2(\|x\|) e^{\frac{-t\|x\|^2}{2}} \Delta\left(e^{\frac{-t\|x\|^2}{2}}\right)\right] dx. \end{aligned} \tag{6.7}$$

Combining (6.2) and (6.7) one obtains

$$\int_{\mathbb{R}^n} \left(\gamma_\varepsilon(\|x\|) e^{\frac{-t\|x\|^2}{2}}\right) \left(\Delta + t n_k^{(p)} + t^2 \|x\|^2\right) \left(\gamma_\varepsilon(\|x\|) e^{\frac{-t\|x\|^2}{2}}\right) dx$$

$$= \int_{\mathbb{R}^n} \left[\left(\gamma_\varepsilon(\|x\|) e^{\frac{-t\|x\|^2}{2}}\right) \Delta\left(\gamma_\varepsilon(\|x\|) e^{\frac{-t\|x\|^2}{2}}\right) + \left(t n_k^{(p)} + t^2 \|x\|^2\right) \gamma_\varepsilon^2(\|x\|) e^{-t\|x\|^2}\right] dx$$

$$= \int_{\mathbb{R}^n} \left[\sum_{j=1}^{n} \left(\frac{\partial \gamma_\varepsilon(\|x\|)}{\partial x^j}\right)^2 e^{-t\|x\|^2} + \gamma_\varepsilon^2(\|x\|) e^{\frac{-t\|x\|^2}{2}} \Delta\left(e^{\frac{-t\|x\|^2}{2}}\right) + \gamma_\varepsilon^2(\|x\|) e^{\frac{-t\|x\|^2}{2}} \left(t n_k^{(p)} + t^2 \|x\|^2\right) e^{\frac{-t\|x\|^2}{2}}\right] dx$$

$$\geq \int_{\mathbb{R}^n} \gamma_\varepsilon^2(\|x\|) e^{\frac{-t\|x\|^2}{2}} \left(\Delta + t n_k^{(p)} + t^2 \|x\|^2\right) \left(e^{\frac{-t\|x\|^2}{2}}\right) dx \geq 2t|p-k| \int_{\mathbb{R}^n} \gamma_\varepsilon^2(\|x\|) e^{-t\|x\|^2} dx.$$

Taking into account the normalization factor $\beta(t)$ one gets

$$\left\langle \Delta_{t;k}^{(p)} \tilde{\omega}_{t;p}^{\varepsilon}, \tilde{\omega}_{t;p}^{\varepsilon} \right\rangle = \frac{\int_{\mathbb{R}^n} \left(\gamma_\varepsilon(\|x\|) e^{\frac{-t\|x\|^2}{2}}\right) \left(\Delta + t n_p^{(p)} + t^2 \|x\|^2\right) \left(\gamma_\varepsilon(\|x\|) e^{\frac{-t\|x\|^2}{2}}\right) dx}{\int_{\mathbb{R}^n} \gamma_\varepsilon^2(\|x\|) e^{-t\|x\|^2} dx} \geq 2t|p-k|.$$

This proves the estimate (ii).

(iii) It suffices to consider $\omega \in A^p(\mathbb{R}^n)$ of the form $\omega = a dx^{i_1} \wedge \ldots \wedge dx^{i_p}$ with $a \in C^\infty(\mathbb{R}^n)$ with compact support. Consequently it remains to prove that there exists $t_0 = t_0(\varepsilon)$ and $c_0(\varepsilon)$ so that for any $a \in C^\infty(\mathbb{R}^n)$ with compact support satisfying

$$\int_{\mathbb{R}^n} a(x)\gamma_\varepsilon(\|x\|)e^{\frac{-t\|x\|^2}{2}}dx = 0 \tag{6.8}$$

the following estimate holds

$$\int_{\mathbb{R}^n} a(x)\left(\Delta - tn + t^2\|x\|^2\right)a(x)dx \geq c_0(\varepsilon)t\int_{\mathbb{R}^n} a^2(x)dx. \tag{6.9}$$

To show (6.9), let us introduce $a_2 = a - a_1$, where

$$a_1(x) = \frac{\int_{\mathbb{R}^n} a(x)e^{\frac{-t\|x\|^2}{2}}dx}{\int_{\mathbb{R}^n} e^{-t\|x\|^2}dx}e^{\frac{-t\|x\|^2}{2}}$$

is the orthogonal projection of the function a onto the uni-dimensional vector subspace generated by the real function $e^{\frac{-t\|x\|^2}{2}}$. Hence a_2 is orthogonal to $e^{\frac{-t\|x\|^2}{2}}$. On the other hand notice that by (6.2), $\left(\Delta - tn + t^2\|x\|^2\right)a_1 = 0$ and, due to Proposition 5.3 (i), the eigenvalues of $\Delta - tn + t^2\|x\|^2$ are contained in the set $2t\mathrm{N}$. Therefore

$$\begin{aligned}\int_{\mathbb{R}^n} a(x)\cdot\left(\Delta - tn + t^2\|x\|^2\right)a(x)dx &\geq \left|\int_{\mathbb{R}^n} a_2(x)\cdot\left(\Delta - tn + t^2\|x\|^2\right)a_2(x)dx\right| \\ &\geq 2t\int_{\mathbb{R}^n} a_2^2(x)dx = 2t\left[\int_{\mathbb{R}^n} a^2(x)dx - \int_{\mathbb{R}^n} a_1^2(x)dx\right].\end{aligned} \tag{6.10}$$

Finally, let us estimate $\int_{\mathbb{R}^n} a_1^2(x)dx = \left|\int_{\mathbb{R}^n} a(x)e^{\frac{-t\|x\|^2}{2}}dx\right|^2$. Using (6.8) and the Cauchy-Schwarz inequality one gets

$$\begin{aligned}\left|\int_{\mathbb{R}^n} a(x)e^{\frac{-t\|x\|^2}{2}}dx\right|^2 &= \left|\int_{\mathbb{R}^n} a(x)\left[1-\gamma_\varepsilon(\|x\|)\right]e^{\frac{-t\|x\|^2}{2}}dx\right|^2 \\ &\leq \int_{\mathbb{R}^n} a^2(x)dx\cdot\int_{\mathbb{R}^n}\left[1-\gamma_\varepsilon(\|x\|)\right]^2 e^{-t\|x\|^2}dx.\end{aligned} \tag{6.11}$$

Further, for $t_0 = t_0(\varepsilon) \geq \left(\frac{2}{\varepsilon}\right)^2$ and $t \geq t_0$ it follows that

$$\int_{\mathbb{R}^n} \left[1-\gamma_\varepsilon\left(\|x\|\right)\right]^2 e^{-t\|x\|^2} dx \le \int_{\|x\|>\frac{\varepsilon}{2}} e^{-t_0\|x\|^2} dx \le c_1 t_0^{-\frac{n}{2}} \int_1^\infty e^{-s^2} s^{n-1} ds, \tag{6.12}$$

where $c_1 > 0$ is a constant. By (6.12) one can choose $t_0 \ge \left(\frac{2}{\varepsilon}\right)^2$ so large that

$$\int_{\mathbb{R}^n} \left[1-\gamma_\varepsilon\left(\|x\|\right)\right]^2 e^{-t\|x\|^2} dx \le \frac{1}{2}. \tag{6.13}$$

Now combine (6.10), (6.11) and (6.13) to conclude that

$$\int_{\mathbb{R}^n} a(x)\left(\Delta - tn + t^2\|x\|^2\right)a(x)dx \ge t\int_{\mathbb{R}^n} a^2(x)dx.$$

Q.E.D.

Let $C_p(f)$ be the set as in §4, $x \in C_p(f)$ and $\left(U_x, \varphi_x : U \to D^n_{2\varepsilon}\right)$ a local chart around x as in Definition 4.1, C1. Choose $\varepsilon > 0$ so that the open neighborhoods U_x of the critical points $x \in C_p(f)$ are pairwise disjoint.

Choose once for all such admissible local charts for any $x \in C_p(f)$. Let us define the p-form $\bar{\omega}_{t;x} \in A^p(M)$ by

$$\bar{\omega}_{t;x}\Big|_{M\setminus\varphi_x^{-1}\left(D^n_{2\varepsilon}\right)} = 0, \bar{\omega}_{t;x}\Big|_{\varphi_x^{-1}\left(D^n_{2\varepsilon}\right)} = \varphi_x^*\left(\tilde{\omega}^\varepsilon_{t;p}\right).$$

The system $\left\{\bar{\omega}_{t;x} \in A^p(M) \middle| x \in C_p(f)\right\}$ provides an orthonormal basis for the vector subspace which they generate. Indeed, the support of $\bar{\omega}_{t;x}$ is contained in U_x and $\left\langle \bar{\omega}_{t;x}, \bar{\omega}_{t;x}\right\rangle = 1$ by (6.1). On the other hand, if $y, z \in C_p(f), y \ne z, \bar{\omega}_{t;y}$ and $\bar{\omega}_{t;z}$ have disjoint support, hence are orthogonal.

For $t \ge t_0$, with t_0 given by Proposition 6.1, let us define $j_t^{(p)} : C^p(M,(f,g)) \to A^p(M)$ to be the $\mathbb{R}$-linear map given by $j_t^{(p)}(E_x) = \bar{\omega}_{t;x}$, where $\left\{E_x : C_p(f) \to \mathbb{R} \middle| x \in C_p(f)\right\}$ is the orthonormal basis for $C^p(M,(f,g))$ described in §4. Obviously $j_t^{(p)}$ is an isometry and in particular an injective mapping.

The key for the proof of Theorem 6.1 is the following mini-max criterion for detecting a gap in the set of eigenvalues of a positive, self-adjoint operator in a Hilbert space.

Proposition 6.2. Let $(H, \langle\, ,\, \rangle)$ be a real Hilbert space, $A : H \to H$ a densely defined positive, self-adjoint operator and $\alpha, \beta \in \mathbb{R}$ so that $0 < \alpha < \beta < +\infty$. Assume that there exist two closed vector subspaces H' and H'' of H with $H_1 \cap H_2 = \{0\}$ and $H = H_1 + H_2$ such that $\left\langle A(x'), x'\right\rangle \le \alpha\|x'\|^2$ for all $x' \in H'$ and $\left\langle A(x''), x''\right\rangle \ge \beta\|x''\|^2$ for all $x'' \in H''$. Then the set of eigenvalues of A does not intersect the open interval (α, β).

Proof. Assume the contrary, namely that $\lambda \in (\alpha, \beta)$ is an eigenvalue of A. Then there exists a sequence $\{x_k\}_{k\in\mathbf{N}} \subset H$ such that $\|x_k\| = 1$ and $u_k = (A - \lambda 1_H)(x_k) \to 0$ when $k \to \infty$. Decompose x_k into the sum $x_k' + x_k''$, where $x_k' \in H', x_k'' \in H''$. Then

$$\begin{aligned}\langle u_k, x_k'\rangle &= \langle (A-\lambda 1_H)(x_k'+x_k''), x_k'\rangle = \langle (A-\lambda 1_H)(x_k'), x_k'\rangle + \langle (A-\lambda 1_H)(x_k''), x_k'\rangle \\ &\leq (\alpha - \lambda)\|x_k'\|^2 + \langle (A-\lambda 1_H)(x_k''), x_k'\rangle\end{aligned}$$

and

$$\begin{aligned}\langle u_k, x_k''\rangle &= \langle (A-\lambda 1_H)(x_k'+x_k''), x_k''\rangle = \langle (A-\lambda 1_H)(x_k''), x_k''\rangle + \langle (A-\lambda 1_H)(x_k'), x_k''\rangle \\ &\geq (\beta - \lambda)\|x_k''\|^2 + \langle x_k', (A-\lambda 1_H)(x_k'')\rangle.\end{aligned}$$

Substract the second inequality from the first one to obtain

$$\langle u_k, x_k' - x_k''\rangle \leq (\alpha-\lambda)\|x_k'\|^2 + (\lambda-\beta)\|x_k''\|^2 \leq -\gamma\left(\|x_k'\|^2 + \|x_k''\|^2\right), \tag{6.14}$$

where $\gamma = \min\{\lambda - \alpha, \beta - \lambda\}$. Since the norms of x_k' and x_k'' are uniformly bounded, and $u_k \to 0$ when $k \to \infty$, it follows that the left hand side of (6.14) approaches 0 when $k \to \infty$. Therefore $\|x_k'\| \to 0$ and $\|x_k''\| \to 0$ as $k \to \infty$, which contradicts $\|x_k' + x_k''\| = 1$.

Q.E.D.

Now let us sketch the proof of Theorem 6.1. Let $H = L^2A^p(M)$ (see Chapter 2, §1, C), $H' = j_t^{(p)}\left(C^p(M,(f,g))\right)$ and $H'' = (H')^{\perp}$ ($\perp$ is taken with respect to the scalar product $\langle\ ,\ \rangle$ defined in Chapter 2, §1, C). Notice first that one concludes from Proposition 6.1 (i) with c, c' and t_0 as in Proposition 6.1 (i), for $t \geq t_0$ and any $\omega' \in H'$,

$$\langle \Delta_t^{(p)}\omega', \omega'\rangle \leq ce^{-tc'}\|\omega'\|^2.$$

Indeed, any $\omega' \in H'$ can be expressed as

$$\omega' = \sum_{x\in C_p(f)} \alpha_x \bar{\omega}_{t;x},$$

where $\alpha_x \in \mathbb{R}$. On the other hand, $\Delta_t^{(p)}$, when restricted to U_x with $x \in C_p(f)$ and expressed in admissible local coordinates in §4, coincides with $\Delta_{t;p}^{(p)}$. Therefore,

$$\langle \Delta_t^{(p)}\omega', \omega'\rangle = \sum_{x\in C_p(f)} \langle \alpha_x \Delta_t^{(p)}(\bar{\omega}_{t;x}), \alpha_x\bar{\omega}_{t;x}\rangle = \sum_{x\in C_p(f)} \alpha_x^2 \langle \Delta_t^{(p)}(\bar{\omega}_{t;x}), \bar{\omega}_{t;x}\rangle \leq ce^{-tc'}\|\omega'\|^2.$$

Let c, c', c'', t_0, be the positive constants given by Proposition 6.1 and define the constants c_1 and c_2 by

$$c_1 = \inf_{x\in M'} \sqrt{g_x\left((\operatorname{grad}_g(f))_x, (\operatorname{grad}_g(f))_x\right)},$$

$$c_2 = \sup_{x\in M} \left|\left(L_{-\operatorname{grad}_g(f)} + L^*_{-\operatorname{grad}_g(f)}\right)_x\right|,$$

where $M' = M \setminus \bigcup_{x \in C_p(f)} \varphi_x^{-1}(D_\varepsilon^n)$ and $\left|\left(L_{-\mathrm{grad}_g(f)} + L^*_{-\mathrm{grad}_g(f)}\right)_x\right|$ denotes the norm of the linear endomorphism $\left(L_{-\mathrm{grad}_g(f)} + L^*_{-\mathrm{grad}_g(f)}\right)_x : \Lambda^p(T_xM) \to \Lambda^p(T_xM)$ corresponding to the scalar product on $\Lambda^p(T_xM)$ induced by g_x (see Chapter 2, §1, C). Using the constants $c, c', c'', c_1, c_2, t_0$ one can show the existence of the positive constants c_3 and ε_1 such that for $t > t_0, \varepsilon < \varepsilon_1$ and any $\omega'' \in H''$ the following estimate holds:

$$\left\langle \Delta_t^{(p)} \omega'', \omega'' \right\rangle \geq c_3 t \left\| \omega'' \right\|^2$$

(see Burghelea, Friedlander, Kappeler, McDonald [1996], pp. 808-810).

Now one can apply Proposition 6.2, whose hypotheses are satisfied for $\alpha = ce^{-c't}, \beta = c_3 t$ and $t > t_0$. This concludes the first part of Theorem 6.1.

Set $Q_t^{(p)}, t > t_0$, be the orthogonal projection of H onto the finite dimensional vector space $A_{sm}^p(M)(t)$ considered above. The following proposition provides an additional important estimate.

Proposition 6.3. For $r \in \mathbb{N}$ one can find $\varepsilon_0 > 0$, and constants $c_4, c_5 > 0$, so that for $t \geq t_0$ as constructed above, any $\varepsilon < \varepsilon_0$ and $v \in C^p(M, (f, g))$ one has

$$\sup_{x \in M} \left| \left(\left(Q_t^{(p)} \circ j_t^{(p)} \right)(v) - j_t^{(p)}(v) \right)_x \right| \leq c_4 e^{-c_5 t} \|v\|.$$

Similar estimates hold for the C^k-norm with $k \leq r$.

For the proof of this Proposition, see Bismut, Zhang [1992], p. 128 and Burghelea, Friedlander, Kappeler, McDonald [1996], pp. 811-812.

Proposition 6.3 implies that for, sufficiently small $\varepsilon > 0$, there exists t_0 so that for $t \geq t_0, i_t^{(p)} = Q_t^{(p)} \circ j_t^{(p)}$ is an isomorphism of $C^p(M, (f, g))$ onto the closed subspace $K^p(t)$ of $A_{sm}^p(M)(t)$. Moreover, for t sufficiently large, $K^p(t) = A_{sm}^p(M)(t)$. To verify this we will show that $\omega \in A_{sm}^p(M)(t)$ and ω orthogonal to $K^p(t)$ imply $\omega = 0$. Indeed, since $Q_t^{(p)}$ is a self-adjoint projector and $Q_t^{(p)}(\omega) = 0$, one has

$$\left\langle \bar{\omega}_{t;x}, \omega \right\rangle = \left\langle Q_t^{(p)}(\bar{\omega}_{t;x}), \omega \right\rangle = 0,$$

where $x \in C_p(f)$. Then, by Proposition 6.1 (iii),

$$\left\langle \Delta_t^{(p)} \omega, \omega \right\rangle \geq c'' t \|\omega\|^2.$$

On the other hand,

$$\left\langle \Delta_t^{(p)} \omega, \omega \right\rangle = \lambda \|\omega\|^2 < \|\omega\|^2$$

because $\omega \in A_{sm}^p(M)(t)$ and hence $\omega = 0$. This finishes the proof of Theorem 6.1.

Finally, let us give an outline of the proof of Theorem 6.2. For $t \geq t_0$, as stated in

Proposition 6.3, let $J_t^{(p)}$ be the isometry defined by

$$J_t^{(p)} = i_t^{(p)} \circ \left(\left(i_t^{(p)} \right)^* \circ i_t^{(p)} \right)^{-\frac{1}{2}} \tag{6.15}$$

and denote by $\Omega_{t;x} = J_t^{(p)}(E_x) \in A^p(M), x \in C_p(f)$. The considerations above show that there exists a constant t_1, such that for $t \geq t_1$ the elements $\Omega_{t;x} \in A_{sm}^p(M)(t), x \in C_p(f)$, provide an orthonormal basis for $A_{sm}^p(M)(t)$. Proposition 6.3 implies that there exist $\varepsilon > 0, t_0 > 0, c_6 > 0$ and $c_7 > 0$ such that for $t \geq t_0$ and $x \in C_p(f)$ one has:

$$\sup_{y \in M \setminus \varphi_x^{-1}(D_\varepsilon^n)} \left| \left(\Omega_{t;x} \right)_y \right| \leq c_6 e^{-\varepsilon t}, \tag{6.16}$$

$$\left| \left(\Omega_{t;x} \right)_y - \left(\bar{\omega}_{t;x} \right)_y \right| \leq c_7 \frac{1}{t} \tag{6.17}$$

for $y \in W_x^- \cap \varphi_x^{-1}(D_\varepsilon^n)$. To prove the estimate $I_t^{(p)}(\Omega_{t;x}) = E_x + 0\left(\frac{1}{t}\right)$, that is Theorem 6.2, it suffices to show that for any cell $W_{x'}^-$,

$$\left| \int_{W_{x'}^-} \left(\Omega_{t;x} \right)_y e^{tf(y)} - \left(\frac{t}{\pi} \right)^{\frac{n-2p}{4}} e^{tp} \delta_{xx'} \right| \leq c_8 \frac{1}{t}$$

for some constant $c_8 > 0$ and any $x, x' \in C_p(f)$. Note that due to (6.16) and (6.17) it suffices to consider the case when $x = x'$. But Proposition 6.3 implies that $\left(i_t^{(p)} \right)^* \circ i_t^{(p)} = 1_{C^p(M,(f,g))} + 0\left(e^{-c_5 t}\right)$, so that the p-forms $\Omega_{t;x}$ expressed in local admissible coordinates on $W_x^- \cap \varphi_x^{-1}(D_\varepsilon^n)$ satisfy the estimate

$$\left(\Omega_{t;x} \right)_y = \left(\frac{t}{\pi} \right)^{\frac{n}{4}} e^{\frac{-t\|y\|^2}{2}} \left[dy^1 \wedge \ldots \wedge dy^p + 0\left(\frac{1}{t} \right) \right]$$

in the case when ε is small enough. Consequently by (6.17) we conclude that

$$\int_{W_x^-} \left(\Omega_{t;x} \right)_y e^{tf(y)} dy =$$

$$= \left(\frac{t}{\pi} \right)^{\frac{n}{4}} e^{pt} \int_{W_x^- \cap \varphi_x^{-1}(D_\varepsilon^n)} e^{\frac{-t\|y\|^2}{2}} \left[dy^1 \wedge \ldots \wedge dy^p + 0\left(\frac{1}{t} \right) \right] + e^{pt} \int_{W_x^{-1} \cap \varphi_x^{-1}(D_\varepsilon^n)} e^{(f(y)-p)t} \left(\Omega_{t;x} \right)_y dy =$$

$$= e^{pt} \left(\frac{t}{\pi} \right)^{\frac{n}{4}} \left(\frac{t}{\pi} \right)^{-\frac{p}{2}} \left[1 + 0\left(\frac{1}{t} \right) \right]$$

for ε small enough. The second integral on the right-side of the above equality decays exponentially when $t \to \infty$ because of (6.16) and of the fact that $f(y) - p \leq 0$ on W_x^-.

Finally let us remark that the work of Witten-Helffer-Sjöstrand has been extended to the de Rham complex of M with coefficients in a Hilbert module of finite type by Burghelea, Friedlander, Kappeler and McDonald [1996]. In this more general situation the spectrum $\Delta_t^{(p)}$ is typically not discrete.

7. STRONG MORSE INEQUALITIES

In this last section we shall prove the **strong Morse inequalities**, i.e. Theorem 3.3.

Assume that M satisfies the assumptions of Theorem 3.3, and let $n = \dim M$. Let g be a Riemannian metric on M such that (f, g) is a generalized triangulation. Fixing t large enough, let us consider the finite dimensional vector space $A_{sm}^p(M)(t)$ defined as in §6 and let us denote by X^p this vector space, $0 \le p \le n$. The next statement is an immediate consequence of Theorem 6.1.

Proposition 7.1. For each $p = 0,1,\dots,n$, we have

$$\dim X^p = m_p(f).$$

Proposition 7.2. For each $p = 0,1,\dots,n$, we have:

(i) $d_t^p(X^p) \subseteq X^{p+1}$;

(ii) $(d_t^{p-1})^*(X^p) \subseteq X^{p-1}$.

Proof. For each $\omega \in X^p$ we have

$$\Delta_t^{(p)}\omega = \lambda^p(t)\omega$$

with

$$\lambda^p(t) < 1.$$

Therefore

$$\begin{aligned}\Delta_t^{(p+1)}(d_t^p\omega) &= \left[(d_t^{p+1})^* d_t^{p+1} + d_t^p (d_t^p)^*\right](d_t^p\omega) = d_t^p d_t^{p*} d_t^p\omega \\ &= d_t^p\left[(d_t^p)^* d_t^p + d_t^{p-1}(d_t^{p-1})^*\right](\omega) = d_t^p\Delta_t^p(\omega) = \lambda^p(t) d_t^p\omega.\end{aligned}$$

This implies $d_t^p\omega \in X^{p+1}$.

Property (ii) can be established in a similar manner.

Q.E.D.

Hence we obtain the following cochain complex:

$$0 \to X^0 \xrightarrow{d_t^0} X^1 \xrightarrow{d_t^1} \cdots \xrightarrow{d_t^{n-1}} X^n \to 0.$$

Proposition 7.3. For each $p = 0,1,2,\dots,n$, we have

$$\dim N(d_t^p)/R(d_t^{p-1}) = \beta_p.$$

Proof. It is easy to see that:

(i) $N(\Delta_t^{(p)}) \subset X^p \cap N(d_t^p)$.

(ii) For each $\omega \in X^p \cap N(d_t^p) \cap N(\Delta_t^{(p)})^{\perp}$ we have

$$\Delta_t^{(p)}\omega = \lambda^p(t)\omega,$$

where

$$\lambda^p(t) \neq 0.$$

For each $\omega \in A^p(M)$ we have

$$\Delta_t^{(p)}\omega = \left[(d_t^p)^* d_t^p + d_t^{p-1}(d_t^{p-1})^*\right](\omega) = d_t^{p-1} d_t^{p-1^*}\omega.$$

Since $(d_t^{p-1})^*\omega \in X^{p-1}$, we see that

$$\omega = \frac{d_t^{p-1} d_t^{p-1^*}\omega}{\lambda^p(t)} \in R(d_t^{p-1}),$$

i.e. those p-forms in $X^p \cap N(d_t^p)$ which have contributions in $N(d_t^p)/R(d_t^{p-1})$, are just $\Delta_t^{(p)}$-harmonic forms. Therefore,

$$N(d_t^p)/R(d_t^{p-1}) \cong N(\Delta_t^{(p)})$$

in the small cochain complex.

Q.E.D.

Now we can pass to the proof of the strong Morse inequalities.

Proof of Theorem 3.3. We start with the following cochain complex:

$$0 \to X^0 \xrightarrow{d_t^0} X^1 \xrightarrow{d_t^1} \cdots \xrightarrow{d_t^{n-1}} X^n \to 0$$

for large t. We have shown that:

(i) $\dim X^p = m_p(f)$,

(ii) $\dim N(d_t^p)/R(d_t^{p-1}) = \beta_p$.

On the other hand, since

$$\dim X^p = \dim N(d_t^p) + \dim R(d_t^p)$$

and

$$\dim N(d_t^p) = \dim R(d_t^{p-1}) + \beta_p,$$

one obtains

$$m_p(f) = \beta_p + \dim R(d_t^p) + \dim R(d_t^{p-1}),$$

where $p = 1,2,\ldots,n$ and we have assumed that

$$d_t^n = 0.$$

It follows that

$$\sum_{p=0}^{m}(-1)^{m-p}\left[m_p(f)-\beta_p\right]=\dim R\left(d_t^m\right)\geq 0$$

for $m=0,1,2,\ldots,n$. Note that the last one is an equality:

$$\sum_{p=0}^{n}(-1)^{n-p}m_p(f)=\sum_{p=0}^{n}(-1)^{n-p}\beta_p=\chi(M).$$

Q.E.D.

Bismut [1986] proved the Morse inequalities by considering the associated heat operator $\exp\left(-st^{-1}\Delta_t^{(p)}\right)$, and also extended this method to prove Bott's inequalities in the degenerate case. To calculate the necessary asymptotics for the heat kernel, Bismut used Brownian motion. However, in the proof of the Morse inequalities of Bott, the Mayer-Vietoris argument is used. This question has received a complete analytical solution by Helffer and Sjöstrand [1985].

WHS-theory was also used to provide a new proof of the equality of analytic and Reidemeister torsion. For a closed Riemannian manifold (M,g), Burghelea, Friedlander, Kappeler, McDonald [1996] have extended the definition of analytic and Reidemeister torsion associated to a unitary representation of $\pi_1(M)$ on a finite dimensional vector space to a unitary representation on a A-Hilbert module W of finite type, where A is a finite von Neumann algebra. If (M,W) is of determinant class and M is of odd dimension they prove generalizing the Cheeger-Müller theorem, that the analytic and Reidemeister torsion are both finite, real numbers and equal. In particular, this proves the conjecture that for closed Riemannian manifolds with positive Novikov-Shubin invariants, the L^2-analytic and L^2-Reidemeister torsions are equal. The interest of the conjecture comes, among other issues, from the geometric significance of the L^2-analytic torsion and the fact that sometimes the L^2-Reidemeister torsion can be computed numerically in an efficient way. The L^2-version of WHS-theory turned out to be so far an unavoidable ingredient in the proof of the equality of these two torsions and of generalizations of this result.

REFERENCES

AUSTIN, D.M., BRAAM,P.M., *Morse-Bott theory and equivariant cohomology*, The Floer memorial volume, Progress in Math., Vol. 133, Birkhäuser Verlag, 1995.

BISMUT, J.-M., *The Witten complex and the degenerate Morse inequalities*, J. Differential Geometry 23(1986), no. 3, 207-240.

BISMUT, J.P., ZHANG, W., *An extension of a theorem by Cheeger and Müller*, Astérisque 205(1992), 1-209.

BISMUT, J.P., ZHANG, W., *Milnor and Ray-Singer metrics on the equivariant determinant of flat vector bundle*, Geometric and Functional Analysis, Vol. 4(1994), 136-212.

BURGHELEA, D., *Lectures on Witten-Helffer-Sjöstrand theory*, Proceedings of the Third International Workshop on Differential Geometry and its Applications and the First German-Romanian Seminar on Geometry, Sibiu, Romania, September 18-23, 1997, General Mathematics, Vol. 5(1997), 85-99.

BURGHELEA, D., FRIEDLANDER, L., KAPPELER, T., *Asymptotic expansion of the Witten deformation of the analytic torsion*, J. of Funct. Anal. 137(1996), 320-363.

BURGHELEA, D., FRIEDLANDER L., KAPPELER, T., *Relative torsion for representations in finite type Hilbert modules*, Communications in Contemporary Math., to appear in issue 1, 2001.

BURGHELEA, D., FRIEDLANDER, L., KAPPELER, T. and McDONALD, P., *Analytic and Reidemeister torsion for representations in finite type Hilbert modules*, Geometric and Functional Analysis, Vol. 6, No. 5(1996), 751-859.

CHANG, KUN-CHING, *Infinite Dimensional Morse Theory and Multiple Solution Problems*, Birkhäuser, Boston, 1993.

CRAIOVEANU, M., RATIU, T.S., *Elemente de Analiză Locală*, Monografii Matematice No.6,7, Tipografia Universitătii din Timisoara, 1976.

CYCON, H.L., FROESE, R.G., KIRSCH, W., SIMON, B., *Schrödinger Operators with Application to Quantum Mechanics and Global Geometry*, Texts and Monographs in Physics, Springer Verlag, Berlin, Heidelberg, New York, London, Paris, Tokyo, 1987.

DAVIES, E.B., *The twisting trick for double well Hamiltonians*, Comm. Math. Phys. 85(1982), 471-479.

GILKEY, P.S., *Invariance Theory, the Heat Equation, and the Atiyah-Singer Index Theorem*, Mathematics Lecture Series 11, Publish or Perish, Inc., Wilmington, Delaware, 1984.

HELFFER, B., *Étude du Laplacien de Witten associé à une fonction de Morse dégénérée*, (1985), preprint.

HELFFER, B., SJÖSTRAND, J., *Multiple wells in the semi-classical limit I*, Comm. Partial Differ. Equations 9(1984), 337-408.

HELFFER, B., SJÖSTRAND, J., *Puits multiples en mécanique semi-classique. IV: Étude du complexe de Witten*, Commun. Partial Differ. Equations 10(1985), 245-340.

HENNIART, G., *Les inégalités de Morse*, Séminaire Bourbaki 1983/1984, 617, 43-60.

HÖRMANDER, L., *The analysis of linear partial differential operators*, Vol. III, Springer Verlag, New York, 1985.

LAUDENBACH, F., *On the Thom-Smale complex (Appendix in Bismut, Zhang [1992])*, Astérisque, 205(1992), 209-233.

MILNOR, J., *Morse Theory*, Ann. Math. Stud. No. 51, Princeton, 1963.

PALAIS, R.S., *Seminar on the Atiyah-Singer Index Theorem*, Ann. Math. Study 57, Princeton Univ. Press, 1965.

POŹNIAK, M., *Triangulation of smooth compact manifolds and Morse theory*, Warwick preprint 11(1990).

SIMON, B., *Semiclassical analysis of low eigenvalues. Nondegenerate minima: Asymptotic expansions*, Ann. Inst. Henri Poincaré 38(1983), 295-307.

SHUBIN, M., *Pseudodifferential Operators and Spectral Theory*, Springer Verlag, New York, 1980.

SMALE, S., *On the gradient dynamical systems*, Annals of Math. 74(1961), 199-206.

WITTEN, E., *Supersymmetry and Morse theory*, J. Diff. Geometry 17(1982), 661-692.

Chapter 8
OPEN PROBLEMS AND COMMENTS

Problem 8.1. Which symmetric tensor fields R_{ij} are locally the Ricci tensor field of a metric?

Comments. DeTurck [1981] proved that R_{ij} is locally a Ricci tensor field if R_{ij} is invertible, but found some obstructions if R_{ij} is not invertible. The natural guess is that R_{ij} is locally a Ricci tensor field if there is some metric g so that the second Bianchi identity is satisfied locally. This is an open problem even in the real analytic case. A similar problem for the scalar curvature was solved by Kazdan and Warner [1975a,b], see also Kazdan [1985].

Problem 8.2 (Yau). Gao and Yau [1986] have proved that every compact 3-dimensional manifold admits a metric with negative Ricci curvature. Does there exist a similar result for a compact manifold with dimension greater than 3?

Problem 8.3 (Yau). Can one say anything about the homotopy type of the space of metrics with negative Ricci curvature?

Problem 8.4 (Yau). Compact manifolds with negative Ricci curvature do not admit a continuous group of isometries. If a finite group acts smoothly on a compact manifold, does the manifold support an equivariant metric with negative Ricci curvature?

Problem 8.5 (Berger). What type of topological information can we extract from the o-spectrum alone?

Comments. For closed surfaces, the Gauss-Bonnet theorem implies that the o-spectrum determines the Euler characteristic [see also Chapter 4, Corollary 1.2 (ii)]. Many examples reveal various global invariants which are not o-spectrally determined such as the fundamental group or Betti numbers. Vignéras [1980] constructed examples of isospectral hyperbolic 3-manifolds, which were the first known examples of isospectral Riemannian manifolds with different fundamental groups. Moreover, Miatello and Rossetti [1999] produced many pairs of non-homeomorphic, o-isospectral Hantzsche-Wendt manifolds in any dimension $n \geq 7$, n odd. Also, Miatello and Rossetti [2000] constructed pairs (M,g), (M',g') of dimension $n \geq 6$, which are o-isospectral and such that for the corresponding Betti numbers one has $\beta_p(M) < \beta_p(M')$ for $0 < p < n$.

Problem 8.6 Under what additional assumptions does the 0-spectrum of a closed, connected Riemannian manifold determine its geometry? What type of geometrical information can we extract from the o-spectrum alone? Identify specific geometric invariants and properties which are not o-spectrally determined (in particular, see also Problem 8.8).

Comments. The problem in general is still unsolved, but there are interesting partial solutions for spaces of constant curvature. For instance Wolpert [1979] has proven the following result:

Theorem (Wolpert). Let $\mathscr{R}_2(G)$ be the set of all compact Riemannian surfaces (M,g) of fixed genus G, $G \geq 2$. Then in $\mathscr{R}_2(G)$ the spectrum generically determines the geometry of (M,g).

Let us mention also Ikeda's result (see Ikeda [1980a]) namely:

Theorem (Ikeda). Let $(M,g),(M',g')$ be isospectral spherical space forms. Then we have:

(i) If $\dim M = \dim M' = 3$, both manifolds are isometric.

(ii) If M is homogeneous, both manifolds are isometric.

On the other hand, the only way to identify specific geometric properties which are not o-spectrally determined is through explicit constructions of o-isospectral closed Riemannian manifolds. Thus, Szabó [1999] constructed a pair of isospectral closed Riemannian manifolds, one of which is homogeneous and the other not. Miatello and Rossetti [2001] gave an example of a pair of o-isospectral flat manifolds, one of them is Kähler, and the other not admitting any Kähler structure. Gordon and Szabó [2000] constructed pairs of isospectral closed Riemannian manifolds one of which has constant scalar curvature and the other variable scalar curvature. It seems that one may not be able to tell from the o-spectrum alone whether a closed Riemannian manifold is Einstein (see also Problem 8.8 and subsequent Comments).

Problem 8.7 Is the sphere S^n with the canonical Riemannian metric, $n \in \mathbb{N}^*$, completely characterized by its spectrum? Do there exist examples of pairs of isospectral closed, n-dimensional Riemannian manifolds, $n \geq 7$, one of which has constant sectional curvature and the other variable sectional curvature?

Comment. If $n \leq 6$ the answer for the first question is yes (see Corollary 1.4 of Chapter 4), but for $n \geq 7$ the problem is open.

Problem 8.8 (Gordon) Give an example of a pair of isospectral closed Riemannian manifolds one which is Einstein and the other is not.

Comments. Patodi [1970] showed that from the p-spectra, $p = 0,1,2$, one can tell whether a closed Riemannian manifold is Einstein, and whether it has constant sectional curvature (see Chapter 5, Corollary 4.4). Recent work of Gordon and Szabó [2000] suggests the possibility that one may not be able to tell from the 0-spectrum whether a closed Riemannian manifold is Einstein.

Problem 8.9 Do there exist isospectral, non-isometric Riemann surfaces of genus two or three?

Comments. Buser [1986] has constructed pairs of isospectral Riemann surfaces in genus 5 and every genus 7 and higher. Brooks and Tse [1987] supplied examples in genus 4 and 6. They also found examples of isospectral surfaces of genus three. However, the metrics on these surfaces are not of constant curvature. When the constant curvature metric is imposed, the resulting isospectral Riemann surfaces are isometric.

Problem 8.10 Do there exist isospectral, non-isometric convex plane domains (resp. with smooth boundary)?

Comments. Using certain pairs of subgroups of finite groups satisfying the condition (3.8) of Chapter 4 and Bérard's transplantation method, Buser, Conway, Doyle and Semmler [1994] have constructed a whole gallery of isospectral domains. Their examples include a pair of special domains (drums) each with a distinguished point such that when they are struck with the same intensity at their distinguished points, they sound identical, i.e. each frequency is excited to the same amplitude on the two drums. All the isospectral plane domains constructed by this transplantation method are non-convex and have corners.

Problem 8.11 (Gordon) Do there exist orientable Riemannian manifolds with boundary which are Neumann but not Dirichlet isospectral (or which are Dirichlet but not Neumann isospectral)?

Comments. Using an orbifold version of Sunada's theorem (see Chapter 4, §3), Bérard and Webb [1995] have constructed a pair of smooth surfaces with boundary which are Neumann isospectral (i.e. the Laplace-Beltrami operators of the two surfaces acting on functions satisfying Neumann boundary conditions are isospectral) but not Dirichlet isospectral. The non-orientability of one of the surfaces is a key feature of the construction. All other known examples of Riemannian manifolds with boundary which are Neumann isospectral are also Dirichlet isospectral and conversely (see also Chapter 4, §3).

Problem 8.12 Let (M,g) be a closed n-dimensional Riemannian manifold. Does there exist an upper bound of λ_1 only in terms of its volume?

Comments. Yang and Yau [1980] proved the following result:

Theorem. Let (M,g) be an orientable Riemannian surface of genus G with area A. Then we have:

$$\lambda_1 \leq 8\pi(G+1)A^{-1}.$$

For manifolds of dimension greater than 2, an analogous upper bound for λ_1 in terms of its volume is false in general. In fact Urakawa [1977] and Tanno [1979] respectively have shown that for odd dimensional spheres S^n and compact Lie groups with non-trivial commutator subgroup, the inequality

$$\lambda_1 \cdot [vol(M,g)]^{2/n} \leq c_n$$

does not hold for a constant c_n independent of the metric.

Thus the problem remains open only for compact nonorientable surfaces.

Problem 8.13 We have seen in Chapter 3, §5, that for each $M \subset \mathbb{R}^m$, where M is a compact Riemannian manifold with smooth boundary, the following inequality holds:

$$\lambda_{n+1} \leq \frac{m+4}{m}\lambda_n .$$

If $k(m)$ denotes the smallest number such that

$$\lambda_{n+1} \leq k(m)\lambda_n$$

it is natural to ask: Does there exist a precise value of $k(m)$?

Comments. In Payne, Pólya and Weinberger [1956] it is conjectured that the value $k(2)$ is achieved for a disk, for which

$$\lambda_{n+1} \leq (2.539)\lambda_n .$$

In Hile and Protter [1980] it is proved that

$$\lambda_{n+1} \leq \frac{m+3+\sqrt{m^2+10m+9}}{2m}\lambda_n$$

but the problem remains open.

Problem 8.14 Let J be the space of all increasing sequences of nonnegative real numbers which tend to infinity. Does there exist for each $(\lambda_n)_{n\in\mathbb{N}} \in J$ a compact Riemannian manifold such that its spectrum should be $(\lambda_n)_{n\in\mathbb{N}}$?

Comment. There are known only some partial answers for domains in $\mathbb{R}^n$ (see for details Hile and Protter [1980]).

Problem 8.15 (Friedlander, Mazzeo). Suppose (M,g) is a smooth compact, Riemannian manifold with boundary and let

$$0 < \mu_1 < \mu_2 \leq \dots$$

and

$$0 = \nu_1 < \nu_2 \le \dots$$

denote the sequences of Dirichlet and Neumann eigenvalues respectively, for its Laplacian. Recently Friedlander [1991] proved that we have

$$\nu_{k+1} \le \mu_k, \quad k = 1, 2, \dots$$

when M is an arbitrary compact set in $\mathbb{R}^n$ with smooth boundary. It is possible to extend this result to an arbitrary compact Riemannian manifold with boundary?

Comment. Recently Mazzeo [1991] has given an affirmative answer to the problem for some particular cases, but the general case remains still open.

Problem 8.16 (Yau). It is well known that Uhlenbeck [1976], [1982] proved that for a generic metric on a closed manifold, every eigenvalue is simple, that is with multiplicity one. Can one give an effective condition on the metric to assure that the metric is generic in this sense? Can one describe the domains where all eigenvalues are simple? Does there exist a domain with only a finite number of multiple eigenvalues?

Comment. Metrics with a special property are unlikely to be generic.

Problem 8.17 (Berger) Let M be a compact manifold and $\mathcal{M}$ the space of all Riemannian metrics on M. We can define a map

$$\mathbf{Spec} : \mathcal{M} \to \mathbb{R}^{\mathbb{N}}$$

assigning to each $g \in \mathcal{M}$ its 0-spectrum $\mathbf{Spec}(M, g)$. Study this map.

Comment. It is now known that the question of the injectivity of this map is false. The problem of describing its image looks too formidable today, even to talk about it.

Problem 8.18 Let us consider the map

$$\mathbf{Spec} : \mathcal{M} \to \mathbb{R}^{\mathbb{N}}$$

as in the previous problem, and on $\mathcal{M}$ the C^k-topology. Then it is natural to ask: Is the map **Spec** a proper one?

Comment. Here the topology on $\mathbb{R}^{\mathbb{N}}$ is crucial. If one puts a relatively weak topology on $\mathbb{R}^{\mathbb{N}}$, e.g. the topology of convergence of each factor (product topology) then the map **Spec** is not a proper map. For instance Chavel [1984] proved that there are examples of Riemannian manifolds (M_i, g_i) whose 0-spectra converge pointwise to the spectrum of a Riemannian manifold (M,g) but M is not diffeomorphic to M_i for each i.

Problem 8.19 Describe the structure of the isospectral set of a closed Riemannian manifold (M, g) [that is the set of all isometry classes of closed Riemannian manifolds which are isospectral to (M, g)] in some natural topology, such as the C^∞-topology.

Comments. A major open question concerns whether isospectral sets of manifolds are necessarily compact relative to the C^∞-topology. This topology is defined as follows: A sequence (M_j, g_j) of Riemannian manifolds is said **to converge** to a Riemannian manifold (M, g) if there exist smooth diffeomorphisms $\phi_j : M \to M_j$ such that the sequence of Riemannian metrics $\{\phi_j^*(g_j)\}$ converges to g in the C^∞ sense. Note that this condition depends only on the isometry classes of manifolds. A set $\mathcal{S}$ of isometry classes of closed Riemannian manifolds is said to be **compact** in the C^∞-topology if every sequence in $\mathscr{S}$ contains a convergent subsequence. In particular, a compact set of closed Riemannian manifolds can contain only finitely many diffeomorphism types.

One of the questions concerning the local structure is the existence of local isospectral deformations of Riemannian metrics. An **isospectral deformation** of g is a family of Riemannian metrics $t \mapsto g_t, t \in [0,1]$, that depends smoothly on t such that $g_0 = g$ and, for all $t \in [0,1]$, the spectra of Δ_{g_t} are mutually identical (counting multiplicities). A closed Riemannian manifold (M,g) is said to be **spectrally rigid** if g does not belong to any continuous family of pairwise isospectral but nonisometric Riemannian metrics. Let us list some results in this direction:

Guillemin and Kazhdan [1980a] have shown that a negatively curved closed surface is spectrally rigid. In other words there are no nontrivial isospectral deformations of negatively curved closed surfaces. Guillemin, Kazhdan [1980b] and Min-Oo [1986] obtained rigidity results in higher dimensions but with stronger curvature hypotheses. Croke and Sharafutdinov [1998] proved that negatively curved manifolds of arbitrary dimension are spectrally rigid. One should mention that the negative curvature assumption cannot be removed; many examples exist of isospectral families of Riemannian metrics of mixed curvature. Very recently, Gordon [2000b] provided the first examples of isospectral deformations of positively curved metrics in marked contrast with the spectral rigidity of negatively curved metrics on closed manifolds. More precisely, she succeeded to prove the existence of smooth families of isospectral, non-isometric Riemannian metrics on the sphere of dimension n for every $n \geq 8$. Moreover, the metrics can be chosen arbitrarily close to the round metric g_{can}, that is given $\varepsilon > 0$, the metrics on the sphere can be chosen so that the sectional curvature σ satisfies $1-\varepsilon < |\sigma| < 1+\varepsilon$.

In contrast to spectral rigidity, compactness theorems give us global information about an isospectral set. Let us list some of these results:

(i) Osgood, Phillips and Sarnak [1988], [1989] proved that isospectral sets of closed surfaces are compact in the C^∞-topology. Also the set of all bounded plane domains with a given spectrum is compact in the C^∞-topology.

The heat invariants alone are not enough to obtain compactness of isospectral sets. Indeed, as pointed out in Chapter 4, §1, all Riemann surfaces of a given volume have the same heat invariants. Osgood, Phillips and Sarnak used the heat invariants along with a global invariant, the determinant of the Laplace-Beltrami operator, to obtain their result. Since the spectrum determines the genus of a closed surface [see Chapter 4, Corollary 1.2(ii)], one knows a priori that isospectral sets of closed surfaces only contain finitely many diffeomorphism types.

The use of the determinant breaks down in higher dimensions and other global spectral invariants are needed. A number of compactness results have been obtained, especially in dimension three, under various hypotheses.

(ii) Chang and Yang [1990] proved that if (M,g) is a closed, oriented, connected, three-dimensional Riemannian manifold, then the set of smooth functions $f \in C^\infty(M)$ such that $(M, e^f g)$ is isospectral to (M,g) is compact in the C^∞-topology on $C^\infty(M)$.

They used the global invariant λ_1, the first non-zero eigenvalue of the Laplace-Beltrami operator. Dropping the condition that the conformal class of metrics be fixed, compactness results have been obtained under a priori restrictions on curvature.

(iii) Brooks, Perry and Petersen [1992] proved that if $\{(M_j, g_j)\}$ is a sequence of isospectral, closed, n-dimensional Riemannian manifolds which either all have negative sectional curvature or have a uniform lower bound on their sectional curvatures, then $\{M_j\}$ contains only finitely many homeomorphism types. Also they proved the compactness of isospectral sets which contain a closed Riemannian manifold (M, g) sufficiently close to a Riemannian manifold of constant curvature. In case when $n = 3$, they proved in [1993] that if $\{(M_j, g_j)\}$ all have negative sectional curvature or have a uniform lower bound on their Ricci curvatures, then $\{M_j\}$ contains only finitely many diffeomorphism types and a subsequence of $\{(M_j, g_j)\}$ converges in the C^∞-topology (see also Anderson [1991] for similar results).

(iv) Zhou [1997] proved that on any given closed manifold, the set of all Riemannian metrics whose sectional curvatures are uniformly bounded in absolute value by some arbitrary constant C and which are isospectral to a given such metric, is compact in the C^∞-topology.

Even in two dimensions, however, very little is known about the structure of isospectral sets of Riemannian metrics beyond the compactness result. For example, even though there are no isospectral deformations of negatively curved surfaces, as we noted before, one cannot conclude that isospectral sets of negatively curved surfaces are always finite, although no counterexamples are known.

Problem 8.20 (Zhou). It is well known that there are only finitely many closed isospectral surfaces with constant curvature -1. Now, we consider a set I of isospectral metrics (on a fixed closed surface) of curvature $K \le -\varepsilon_0 < 0$. In Sarnak [1990] it is proved that if I is contained in any finite dimensional subspace of Riemannian metrics on the surface, then I must be finite. Then it is interesting to give an answer to the following problem: Is I in fact a finite set?

Problem 8.21 (Brooks). A natural question is whether the techniques given in Chapter 4, exhaust the methods of finding isospectral manifolds.

Comment (Brooks). Historical experience suggests that the answer to this question should be "no", but perhaps in sufficiently well-defined cases the answer may be "yes" (see also Chapter 4, Remark 3.4).

Problem 8.22 (Yau). Let (M,g) be a connected, closed Riemannian manifold and denote by f_j, $j = 1, 2, \ldots$, the sequence of eigenfunctions for the Laplace-Beltrami operator associated to (M,g) corresponding to the non zero eigenvalues (see Chapter 3, Corollary 1.4). Show that the number of critical points of f_j is increasing with j.

Problem 8.23 (Miatello and Rossetti). For every given subset S of $\{0, 1, \ldots, n\}$, $n \ge 2$, does there exist pairs of n-dimensional closed Riemannian manifolds which are p-isospectral if and only if $p \in S$?

Comments. Examples of o-isospectral, non-homeomorphic, n-dimensional closed Riemannian manifolds, which are not p-isospectral have been constructed by Ikeda [1989] for lens spaces $(n \ge 5)$, Gordon [1986] and Gornet [2000] in the context of nilmanifolds $(n \ge 5)$ and Schüth [1999]. For each $p \in \mathbb{N}$, Ikeda [1989] has constructed lens spaces which are k-isospectral for $0 \le k \le p$, but not $(p+1)$-isospectral. Miatello and Rossetti

[2001] gave many examples which are p-isospectral for certain values of p only.

Problem 8.24 Does there exist an extension of the Uhlenbeck's results (see Uhlenbeck [1976]) at the level of p-forms?

Comment. There is an obstruction due to Millman's formula (see Millman [1980] or Craioveanu and Puta [1988]), but under this restriction the problem is still open.

Problem 8.25 (Miatello and Rossetti). Does there exist examples of o-isospectral n-dimensional nilmanifolds which are not k-isospectral for $0<k<n$?

Comments. Almost all of the o-isospectral nilmanifolds in the references mentioned in Comments to Problem 8.23 are of the form $\Gamma\setminus M, \Gamma'\setminus M$, where Γ and Γ' are lattices in the same simply connected nilpotent Lie group M, hence, by a theorem of Nomizu, their real cohomology coincides with the Lie algebra cohomology of μ, the Lie algebra of M, hence they both have the same Betti numbers. It seems that examples of o-isospectral nilmanifolds $\Gamma\setminus M, \Gamma'\setminus M'$ with Betti numbers satisfying $\beta_k(\Gamma\setminus M)<\beta_k(\Gamma'\setminus M')$ for $0<k<n$, hence not k-isospectral for $0<k<n$, have not yet be given. Their construction would involve finding an o-isospectral pair $\Gamma\setminus M, \Gamma'\setminus M'$ with $M\neq M'$, where one can compare the dimensions of $H^k(\mu)$ and $H^k(\mu')$ for each value of k.

Problem 8.26 If $\mathbf{Spec}^{(p)}(\boldsymbol{M},\boldsymbol{g})=\mathbf{Spec}^{(p)}(\boldsymbol{S}^n,\boldsymbol{g}_{can})$ for some fixed p, is (M,g) isometric to (S^n, g_{can})?

Comments. The answer is "yes" for $p=0$ and $n\leq 6$, see Chapter 4, Corollary 1.4. Let us mention also some relevant results in this direction:

Theorem (Patodi [1970]). If $\mathbf{Spec}^{(p)}(\boldsymbol{M},\boldsymbol{g})=\mathbf{Spec}^{(p)}(\boldsymbol{S}^n,\boldsymbol{g}_{can})$ for p=0 and p=1, then (M,g) is isometric to (S^n, g_{can}) (see Chapter 5, Corollary 4.5).

Theorem (Goldberg and Gauchman [1987]). Let (M,g) be a compact normal contact Riemannian manifold. If $\mathbf{Spec}^{(2)}(\boldsymbol{M},\boldsymbol{g})=\mathbf{Spec}^{(2)}(\boldsymbol{S}^{2n+1},\boldsymbol{g}_{can})$, then (M,g) is isometric to (S^{2n+1}, g_{can}).

Theorem (Puta and Török [1987], Marchiafava, Puta and Török [1989]). The sphere S^n with the canonical metric is completely characterized in the class of compact, connected, oriented boundaryless Einstein manifolds by the p-spectrum whenever its dimension n and the value of p match the following conditions:

$$(n,p)\notin\{(15,2),(15,13),(16,2);(16,14)\}$$

and

$$n(n-1)-6p(n-p)\neq 0.$$

For the particular cases $p=1, p=2, p=3, p=4, p=5$ we find again the results of Tanno [1973], Tsagas and Kockinos [1979], Strese [1982] and Puta [1985].

The general problem remains still open.

Problem 8.27 (Gilkey, Park). Do there exist a Riemannian submersion $\phi:(M',g')\to(M,g)$, where M' and M are closed, eigenvalues $\lambda\in[0,+\infty)$ of $\Delta^{(1)}$ and $\mu\in(0,+\infty)$ of $\Delta'^{(1)}$, and an eigenform $\omega\in E(\Delta^{(1)},\lambda)$ such that $\lambda<\mu$ and

$\phi^*(\omega) \in E(\Delta'^{(1)}, \mu)$?

Comment. If $p \geq 2$, Gilkey and Park [1996] showed that there exist a Riemannian submersion $\phi : (M', g') \to (M, g)$, eigenvalues $\lambda \in [0, +\infty)$ of $\Delta^{(p)}$ and $\mu \in (0, +\infty)$ of $\Delta'^{(p)}$ and an eigenform $\omega \in E(\Delta^{(p)}, \lambda)$ such that $\lambda < \mu$ and $\phi^*(\omega) \in E(\Delta'^{(p)}, \mu)$.

Problem 8.28 (Chen, Vanhecke). If $\mathbf{Spec}^{(p)}(M, g) = \mathbf{Spec}^{(p)}(P^n(\mathbb{C}), g_{FS})$ for a fixed value of *p*, examine if (*M*,*g*) is holomorphically isometric to $P^n(\mathbb{C})$?

Comments. The answer is "yes" in the following cases:

(i) *p*=0 and $n \leq 6$, see Tanno [1973];

(ii) *p*=2 and $n \geq 2$, see Chen and Vanhecke [1980] and Goldberg [1984].

Let us mention also in this direction the following result:

Theorem (Puta and Török [1991]). The complex projective space $P^n(\mathbb{C})$ with the Fubini-Study metric g_{FS} is completely charecterized in the class of compact Kähler-Einstein manifolds by the *p*-spectrum, whenever its dimension and the value *p* match the conditions:

$$(m, p) \notin \{(8,2);(8,14)\}$$

and

$$m(2m-1) - 3p(2m-p) \neq 0,$$

where $m = \dim_{\mathbb{C}}(M)$.

For the particular cases *p*=1 and *p*=6 we find again the results of Tanno [1974] and Puta, Török [1988].

The general problem remains still open.

Problem 8.29 (Yau). In Riemannian geometry, there is a close relationship between the spectrum of the Laplace-Beltrami operator and the length of closed geodesics. Are there analogues of those formulae for the Hodge-de Rham operators?

Comments. Under certain assumptions, the lengths of the periodic geodesics are determined by the spectrum of the Laplace-Beltrami operator (see Chapter 3, §1). Moreover, this spectrum determines also whether the geodesic flow along the periodic geodesics is of elliptic type or hyperbolic type. Duistermaat and Guillemin [1975] mention that there are analogues of those formulae for the Hodge-de Rham operators, but without proof. Guillemin wrote a follow-up paper (Fourier integral operators for systems) to the Duistermaat-Guillemin paper, but did not publish it. Sandoval [1999] carries this out for general Dirac type operators. She gives very detailed information on the wave trace expansion, not just the principal term. In the case of the Hodge-de Rham operators, the principal symbol is a scalar and the subprincipal symbol is zero, so one can basically reduce to the scalar case and the result could be proved more quickly.

Problem 8.30 (Berger, Tanno). Let (*M*,*g*) be a closed, connected, *n*-dimensional Riemannian manifold and $\lambda_1^{(p)}(g)$ the smallest positive eigenvalue of $\Delta_g^{(p)}$. Does there exist a constant *k*(*M*) such that

$$\lambda_1^{(p)}(g)\ \mathrm{Vol}(M, g)^{\frac{2}{n}} \leq k(M)$$

for any Riemannian metric g?

Comments. In the case $p = 0$, Hersch [1970] has proved that on S^2 we have:

$$\lambda_1(g)\ \text{Area}\left(S^2, g\right) \leq 8\pi$$

for every Riemannian metric g on S^2 (see Chapter 3, Theorem 5.7). Yang and Yau [1980] have proved that the inequality above holds for a compact surface S of genus γ with $k(S) = 8\pi(\gamma+1)$. Bleecker [1983], Urakawa [1979] and others constructed examples of manifolds of dimension $n \geq 3$ for which the above inequality is false. Finally, Xu [1992], Colbois and Dodziuk [1994] showed that the inequality is false for every Riemannian manifold of dimension $n \geq 3$.

In the case $2 \leq p \leq n-2,\ n \geq 4$, Gentile and Pagliara [1995] have proved that the inequality

$$\lambda_1^{(p)}(g)\ \text{Vol}(M, g)^{\frac{2}{n}} \leq k(M)$$

for every Riemannian metric g is also false.

The problem remains still open for 1-forms. Tanno [1983] showed that on S^3 $\lambda_1^{(1)}(g)$ is bounded for the 1-parameter family of metrics in the examples of Bleecker and Urakawa, which gave unbounded first eigenvalue for functions. More generally, the problem how can we estimate $\lambda_1^{(p)}$ in terms of computable geometric quantities remains still open.

Problem 8.31 (Gornet). Do there arise continuous families of nilmanifolds that are o-isospectral and 1-isospectral from the Gordon-Wilson method?

Comments. A nontrivial isospectral deformation is a continuous family of Riemannian metrics such that the corresponding closed Riemannian manifolds are pairwise isospectral but not isometric. Gordon and Wilson [1984] constructed the first nontrivial isospectral deformations, namely continuous families of isospectral, nonisometric Riemannian nilmanifolds (see Chapter 4, Remark 3.7). DeTurck and Gordon [1989] extended the construction and explained it under the Sunada framework (see Chapter 4, §3). Brooks and Gordon [1990] used this generalization to construct conformally equivalent isospectral deformations of closed Riemannian manifolds. Ouyang [1991] and Pesce [1993] independently showed that isospectral deformations of two-step nilmanifolds must arise from the Gordon-Wilson method. Moreover, Schüth [1995] showed that if one uses representation theoretic techniques alone to obtain continuous families of isospectral nilmanifolds, the deformation must arise from the Gordon-Wilson method.

Problem 8.32 (Brooks). Does there exist a Riemann surface of constant curvature -1 such that for all potentials Φ, the spectrum of $\Delta + \Phi$ determines Φ uniquely?

Comments. Using the idea pointed out in Chapter 4, Remark 3.2, Brooks [1987] was able to prove that there exists a Riemann surface and two potentials Φ_1 and Φ_2 such that the corresponding Schrödinger operators $\Delta + \Phi_1$ and $\Delta + \Phi_2$ are isospectral. It was shown by Guillemin and Kazhdan [1980a] that the above property holds for a generic surface (not constant curvature), but their proof fails for all constant curvature surfaces. One expects that a generic Riemann surface has this property.

Problem 8.33 If one of the Hodge-de Rham operators associated to a given C^k-Riemannian metric g on a closed connected n-dimensional smooth manifold M, $n \geq 2$, does have an eigenvalue λ of multiplicity $m \geq 2$, one may ask whether g can be deformed without changing the multiplicity of the eigenvalue. Give sufficient conditions for the set of C^k-Riemannian metrics near g which admit an eigenvalue near λ of multiplicity m to be locally a manifold of finite codimension in the Banach space of all C^k symmetric covariant 2-tensor fields on M.

Comments. For generic Riemannian metrics on M, all eigenvalues of the Laplace-Beltrami operator have multiplicity one (see Chapter 3, §4). If the Laplace-Beltrami operator Δ_g of a given C^k-Riemannian metric g does have an eigenvalue λ of multiplicity $m \geq 2$, Micheletti and Pistoia [1998a] gave a sufficient condition for the set of C^k-Riemannian metrics near g which admit an eigenvalue near λ of multiplicity m to be locally a manifold of codimension $\frac{1}{2}m(m+1)-1$ in the Banach space of all C^k symmetric covariant 2-tensor fields on M (see also Micheletti and Pistoia [1998b] for more general results). Moreover, they have shown that such a condition is fulfilled when $n=2$ and $m=2$.

Problem 8.34 (Th.M. Rassias [1986]). Examine if the first eigenvalue for the Laplace-Beltrami operator on an embedded minimal surface of S^3 is 2?

Remark. Techniques coming from Spectral Geometry should be useful for the understanding and probably the solution of the following problems:

Problem 8.35 (Th.M. Rassias [1986]). Let (M,g) be an n-dimensional Riemannian manifold having nonnegative sectional curvature. Is the kth Betti number of M at most equal to the kth Betti number of the n-torus T^n?

Problem 8.36 (M. Shiffman [1939]). Characterize the Morse type of a minimal surface by properties similar to the number of conjugate points (or the number of negative characteristic roots in the associated problem) in the case of single integral problem in the Calculus of Variations.

Problem 8.37 (Th.M. Rassias). Does there exist a harmonic homeomorphism of the open unit ball B in $\mathbb{R}^n$ onto $\mathbb{R}^n$, where $n=3,4,\ldots$ i.e., do there exist harmonic functions $f_1, f_2, \ldots, f_n$, defined in B such that the mapping $f=(f_1, f_2, \ldots, f_n)$ is a homeomorphism of B onto all of $\mathbb{R}^n$?

Comments. One can prove that there is no harmonic homeomorphism of the open unit disk in the plane onto the whole plane. The problem for $n=3$ was posed in Rassias [1986].

REFERENCES

ANDERSON, M.T., *Remarks on the compactness of isospectral sets in low dimensions*, Duke Math. Journ. 63(3) (1991), 699-713.

BÉRARD, P. and WEBB, D.L., *One cannot hear the orientability of surfaces*, C.R. Acad. Sci. Paris 320 No. 1(1995), 533-536.

BERGER, M., GAUDUCHON, P. and MAZET, E., *Le spectre d'une variété Riemannienne*, Lect. Notes Math., Vol. 194, Springer-Verlag, New York, 1971.

BLEECKER, D., *The spectrum of a Riemannian manifold with a unit Killing vector field*, Trans. Amer. Math. Soc. 275(1983), 409-416.

BROOKS, R., *On manifolds of negative curvature with isospectral potentials*, Topology 26(1987), 63-66.

BROOKS, R., *The Sunada method*, Tel Aviv Topology Conference: Rothenberg Festschrift, June 1-5, 1998, pp. 25-35, Contemp. Math., 231, Amer. Math. Soc., Providence, RI, 1999.

BROOKS, R. and GORDON, C.S., *Isospectral families of conformally equivalent Riemannian metrics*, Bull. Am. Math. Soc. (N.S.) 23(1990), 433-436.

BROOKS, R., GORNET, R. and GUSTAFSON, W., *Mutually isospectral Riemann surfaces*, Adv. in Math. 138(1998), 306-322.

BROOKS, R., PERRY, P., PETERSEN, P., *Compactness and finiteness theorems for isospectral manifolds*, J. Reine Angew. Math. 426(1992), 67-89.

BROOKS, R., PERRY, P. and PETERSEN, P., *Finiteness of diffeomorphism types of isospectral manifolds*, Proc. Symp. Pure Math. 54(1993), 89-94.

BROOKS, R., PERRY, P. and YANG, P., *Isospectral sets of conformally equivalent metrics*, Duke Math. J. 58(1989), 131-150.

BROOKS, R. and TSE, R., *Isospectral surfaces of small genus*, Nagoya Math. J. 107(1987), 13-24.

BUSER, P., *Isospectral Riemann surfaces*, Ann. Inst. Fourier 36(1986), 167-192.

BUSER, P., CONWAY, J., DOYLE, P. and SEMMLER, K.D., *Some planar isospectral domains*, Internat. Math. Res. Notices No. 9(1994), 391-400.

CHANG, S.Y.A. and YANG, P., *Isospectral conformal metrics on 3-manifolds*, Journ. Amer. Math. Soc. 3(1990), 117-145.

CHAVEL, I., *Eigenvalues in Riemannian Geometry*, Academic Press, New York, 1984.

CHAZARAIN, J., *Formule de Poisson pour les variétés Riemanniennes*, Invent. Math. 24(1974), 65-82.

CHEN, B.Y. and VANHECKE, L., *The spectrum of the Laplacian of Kähler manifolds*, Proc. Amer. Math. Soc. 79(1980), 82-86.

COLBOIS, B. and DODZIUK, J., *Riemannian metrics with large* λ_1, Proc. Amer. Math. Soc., 122(1994), 905-906.

COLIN DE VERDIÈRE, Y., *Spectre du Laplacian et longeurs des géodésiques périodiques II*, Compositio Math. 27(1973), 159-184.

CRAIOVEANU, M. and PUTA, M., *Introducere în Geometria Spectrala*, Ed. Academiei Române, Bucuresti, 1988.

CROKE, C. and SHARAFUTDINOV, V., *Spectral rigidity of a compact negatively curved manifold*, Topology 37(1998), 1265-1273.

DETURCK, D., *Existence of metrics with prescribed Ricci curvature: local theory*, Invent. Math. 65(1981), 179-208.

DETURCK, D. and GORDON, C.S., *Isospectral deformations II: trace formulas, metrics and potentials*, Comm. Pure Appl. Math. 42(1989), 1067-1095.

DUISTERMAAT, J.J. and GUILLEMIN, V.W., *The spectrum of positive elliptic operators and periodic bicharacteristics*, Invent. Math. 29(1975), 39-79.

FRIEDLANDER, L., *Some inequalities between Dirichlet and Neumann eigenvalues*, Arch. Ration. Mech. Anal. 116, No. 2(1991), 153-160.

GAO, L.Z. and YAU, S.T., *The existence of negatively Ricci curved metrics on three manifolds*, Invent. Math. 85(1986), 637-652.

GENTILE, O. and PAGLIARA, V., *Riemannian metrics with large first eigenvalue on forms of degree p*, Proc. Amer. Math. Soc., 123, No. 3(1995), 3855-3858.

GILKEY, P.B. and PARK, J.H., *Riemannian submersions which preserve the eigenforms of the Laplacian*, Illinois J. Math. 40(1996), 194-201.

GOLDBERG, S.I., *A characterization of complex projective space*, C.R. Math. Rep. Acad. Sci. Canada, vol. VI, no. 4 (1984), 193-198.

GOLDBERG, S.I. and GAUCHMAN, H., *Characterizing S^m by the spectrum of the Laplacian on 2-forms*, Proc. Amer. Math. Soc. 99(1987), 750-756.

GORDON, C.S., *Riemannian manifolds isospectral on functions but not on 1-forms*, Journ. Diff. Geometry (1) 24(1986), 79-96.

GORDON, C.S., *Survey of isospectral manifolds*, Handbook of Differential Geometry, F.J.E. Dillen and L.C.A. Verstraelen, eds., Vol. 1, Elsevier Science B.V., 2000a, pp. 747-778.

GORDON, C.S., *Isospectral deformations of metrics on spheres*, Preprint, May 2000b.

GORDON, C.S. and SZABÓ, Z.I., *Isospectral deformations of negatively curved Riemannian manifolds with boundary which are not locally isometric*, Preprint, March 2000.

GORDON, C.S. and WILSON, E.N., *Isospectral deformations of compact solvmanifolds*, J. Differential Geometry 19(1984), 241-256.

GORDON, C.S. and WILSON, E.N., *Laplacian spectrum on Riemannian Heisenberg manifolds*, Michigan Math. Journ. 33(1986), 253-273.

GORNET, R., *Continuous families of Riemannian manifolds, isospectral on functions but not on 1-forms*, Journ. Geom. Anal. 10(2)(2000), 281-298.

GUILLEMIN, V.W. and KAZHDAN, D., *Some inverse spectral results for negatively curved 2-manifolds*, Topology 19(1980a), 301-312.

GUILLEMIN, V.W. and KAZHDAN, D., *Some inverse spectral results for negatively curved n-manifolds*, Proc. Symp. Pure Math., Geometry of Laplace Operator, Amer. Math. Soc., 36(1980b), 153-180.

HERSCH, J., *Quatre propriétés isopérimétriques des membranes sphériques homogènes*, C.R. Acad. Sci. Paris, Ser. A 270(1970), 1645-1648.

HILE, G.N. and PROTTER, M.H., *Inequalities for the eigenvalues of the Laplacian*, Indiana Univ. Math. Journ. 29(1980), 523-538.

IKEDA, A., *On the spectrum of a Riemannian manifold of positive constant curvature*, Osaka Journ. Math. 17(1980a), 75-93.

IKEDA, A., *On lens spaces which are isospectral but not isometric*, Ann. Scient. Éc. Norm. Sup. 13(1980b), 303-315.

IKEDA, A., *Riemannian manifolds p-isospectral but not* $(p+1)$*-isospectral*, Geometry of Manifolds (Matsumoto), Perspectives in Math. Vol. 8, Academic Press, Boston, MA, 1989, pp. 383-417.

KAZDAN, J.L., *Prescribing the curvature of a Riemannian manifold*, Regional Conference Series in Math. No. 57(1985), Amer. Math. Soc., Providence, Rhode Island.

KAZDAN, J.L. and WARNER, F.W., *A direct approach to the determination of Gaussian and scalar curvature functions*, Invent. Math. 28(1975a), 227-230.

KAZDAN, J.L. and WARNER, F.W., *Existence and conformal deformation of metrics with prescribed Gaussian and scalar curvatures*, Ann. of Math. (2) 101(1975b), 317-331.

MARCHIAFAVA, S., PUTA, M. and TÖRÖK, A., *On the spectrum of the p-Laplacian*, Proceed. of the XX-th NCGT, Timisoara, October 5-7(1989), 125-129.

MAZZEO, R., *Remarks on a paper of Friedlander concerning inequalities between Neumann and Dirichlet eigenvalues*, Duke Math. Journ. 63. (3) (1991), 41-49.

MIATELLO, R.J. and ROSSETTI, J.P., *Isospectral Hantzsche-Wendt manifolds*, J. Reine Angew. Math. 515(1999), 1-23.

MIATELLO, R.J. and ROSSETTI, J.P., *Flat manifolds isospectral on p-forms*, Journ. Geom. Analysis, 2001 (to appear).

MICHELETTI, A.M. and PISTOIA, A., *Multiple eigenvalues of the Laplace-Beltrami operator and deformation of the Riemannian metric*, Discrete and Continuous Dynamical Systems, Vol. 4(1998a), 709-720.

MICHELETTI, A.M. and PISTOIA, A., *About the deformations of the Riemannian metrics, which preserve the multiplicity of eigenvalues of the Laplace-Beltrami operator*, Atti Sem. Mat. Fis. Univ. Modena, Supplemento al Vol. XLVI (1998b), 875-888.

MILLMAN, R.S., *Remarks on the spectrum of the Laplace-Beltrami operator in the middle dimension*, Tensor 34(1980), 94-96.

MIN-OO, M., *Spectral rigidity for manifolds with negative curvature operator*, Contemp. Math., Nonlinear Problems in Geometry, (51), (1986), 99-103.

OSGOOD, B., PHILLIPS, R. and SARNAK, P., *Compact isospectral sets of surfaces*, Journ. Funct. Anal. 80(1988), 212-234.

OSGOOD, B., PHILLIPS, R. and SARNAK, P., *Moduli space, heights and isospectral sets of plane domains*, Ann. of Math. 129(1989), 293-362.

OUYANG, H., *On isospectral deformations on two-step nilmanifolds*, Ph.D. Dissertation, Washington University, (1991).

PATODI, V.K., *Curvature and the fundamental solution of the heat operator*, Indian Journ. Math. Soc. 34(1970), 269-285.

PAYNE, L.E., POLYA, G. and WEINBERGER, H.F., *On the ratio of consecutive eigenvalues*, Journ. Math. Physics 35(1956), 289-298.

PESCE, H., *Calcul du spectre d' une nilvariété de rang deux et applications*, Trans. Amer. Math. Soc. 339(1993), 433-461.

PUTA, M., *The spectrum of the Laplacian for 5-forms*, Expo. Math. 3(1985), 75-79.

PUTA, M. and TÖRÖK, A., *On the spectrum of the Laplacian on p-forms*, An. Univ. Timisoara, Ser. Stiinte Mat. 25(1987), 67-73.

PUTA, M. and TÖRÖK, A., *The spectrum of the 6-Laplacian on Kähler manifolds*, Casopies pro Pestovani Matematiky no. 3, (1988), 246-251.

PUTA, M. and TÖRÖK, A., *The spectrum of the p-Laplacian on Kähler manifolds*, Rend. Matemat. Ser VII vol. 11(1991), 257-271.

RASSIAS, Th.M., *Foundations of Global Nonlinear Analysis*, Teubner-Texte zur Mathematik, Band 86, Leipzig, 1986.

SANDOVAL, M.R., *Wave-trace asymptotics for operators of Dirac type*, Commun. in Partial Differential Equations 24, no. 9-10, (1999), 1903-1944.

SARNAK, P., *Determinants of Laplacians; Heights and Finiteness. Analysis, etc.*, Academic Press, 1990.

SCHÜTH, D., *Continuous families of quasi-regular representations of solvable Lie groups*, J. Funct. Anal. 134(1995), 247-259.

SCHÜTH, D., *Continuous families of isospectral metrics on simply connected manifolds*, Annals of Math. 149(1999), 287-308.

SHIFFMAN, M., *The Plateau problem for non-relative minima*, Annals of Math., 40(1939), 834-854.

STRESE, H., *On the spectrum of the Laplacian for p-forms*, Math. Nachr. 106(1982), 35-40.

SZABÓ, Z.I., *Locally non-isometric yet super isospectral spaces*, Geom. Funct. Anal. (GAFA) 9(1999 (in preprint 1992)), 185-214.

TANNO, S., *Eigenvalues of the Laplacian of Riemannian manifolds*, Tôhoku Math. Journ. 25(1973), 391-403.

TANNO, S., *The spectrum of the Laplacian for 1-forms*, Proc. Amer. Math. Soc. 45(1974), 125-129.

TANNO, S., *The first eigenvalue of the Laplacian on spheres*, Tôhoku Math. Journ. 31(1979), 179-185.

TANNO, S., *Geometric expressions of eigen-1-forms of the Laplacian on spheres, Spectral Riemannian manifolds*, Kaigai, Kyoto, 1983, 115-128.

TSAGAS, G. and KOCKINOS, C., *The geometry and the Laplace operator on the exterior 2-forms on a compact Riemannian manifold*, Proc. Amer. Math. Soc. 73(1979), 109-116.

UHLENBECK, K., *Generic properties of eigenfunctions*, Amer. Journ. Math., 98(1976), 1059-1078.

UHLENBECK, K., *Removable singularities in Yang-Mills fields*, Comm. Math. Phys. 83(1982), 11-30.

URAKAWA, H., *On the least positive eigenvalue of the Laplacian for Riemannian manifolds*, Proc. Japan Acad. 53(1977), 229-231.

URAKAWA, H., *On the least positive eigenvalue of the Laplacian for compact group manifolds*, J. Math. Soc. Japan 31(1978), 209-226.

VIGNERAS, M.F., *Variétés Riemanniennes isospectrales et non isométriques*, Ann. of Math. 112(1980), 21-32.

XU, Y., *Diverging eigenvalues and collapsing Riemannian metrics*, Institute for Advanced Study, preprint 1992.

YANG, P. and YAU, S.T., *Eigenvalues of the Laplacian of compact Riemann surfaces and minimal submanifolds*, Ann. Scuola Norm. Sup. Pisa cl. Sci. (4) 7(1980), 55-63.

ZHOU, G., *Compactness of isospectral compact manifolds with bounded curvatures*, Pacific J. Math. 181(1997), no. 1, 187-200.

WOLPERT, S., *The length spectra as moduli for compact Riemann surfaces*, Ann. of Math. 109(1979), 323-351.

APPENDIX
SPECTRAL PROPERTIES OF MATRICES

1. REVIEW OF MATRIX ALGEBRA

Let $\mathcal{M}_{m\times n}(\mathbb{K})$ be the set of all $m\times n$ matrices with entries in $\mathbb{K}$, where $\mathbb{K}=\mathbb{R}$ or $\mathbb{K}=\mathbb{C}$. We shall denote each matrix $A\in\mathcal{M}_{m\times n}(\mathbb{K})$ by

$$A=\left[a_{ij}\right] \text{ where } a_{ij}\in\mathbb{K} \text{ for } 1\le i\le m, 1\le j\le n$$

and we shall say that A is an $\boldsymbol{m\times n}$ **$\mathbb{K}$-matrix**, i.e. A is a real or a complex $m\times n$ matrix.

Following the classical convention we shall use the term **scalars** for real or complex numbers. An $m\times n$ matrix is a **rectangular matrix**; when $m=n$ we use the term **square matrix**.

For a square matrix $A=[a_{ij}]$ we refer to a_{ij} when $i\neq j$ as the off-diagonal terms. By a **diagonal matrix** we mean a matrix in which the off-diagonal entries are zero.

The **adjoint** of a matrix A, denoted by A^*, is defined by

$$A^*=\left[\bar{a}_{ji}\right],\ 1\le j\le n,\ 1\le i\le m,$$

where

$$A=\left[a_{ij}\right],\ 1\le i\le m,\ 1\le j\le n,$$

and the superbar denotes complex conjugation. Note that A^* is an $n\times m$ matrix. It is customary to define the **transpose** of A, denoted by A^T, by

$$A^T=\left[a_{ji}\right],\ 1\le j\le n,\ 1\le i\le m.$$

We shall use only the adjoint, and of course $A^T=A^*$ for real matrices.
The determinant det(A) or $|A|$ is a scalar function of a square matrix $A=\left[a_{ij}\right]$, namely

$$\det(A)=\sum_{\sigma}\operatorname{sgn}(\sigma)a_{1\sigma(1)}a_{2\sigma(2)}\cdots a_{m\sigma(m)},$$

where the sum is taken over all permutations σ of the set $\{1,2,\cdots,m\}$ and $\operatorname{sgn}(\sigma)$ denotes the sign of σ.

Let $A=\left[a_{ij}\right]$ be a square $m\times m$ matrix. By the **trace** of A (this notion is defined only for square matrices) we define the sum of its diagonal entries, i.e.

$$Tr(A)=\sum_{i=1}^{m}a_{ii}.$$

For any $A=\left[a_{ij}\right]\in\mathcal{M}_{m\times n}(\mathbb{K})$, we note that both AA^* and A^*A are square matrices. Then it follows that

$$Tr(AA^*)=Tr(A^*A)=\sum_{i=1}^{m}\sum_{j=1}^{n}\left|a_{ij}\right|^2.$$

One can prove the following properties

(i) The map $\langle\cdot,\cdot\rangle$ given by

$$\langle\cdot,\cdot\rangle : \mathcal{M}_{m\times n}(\mathbb{K})\times\mathcal{M}_{m\times n}(\mathbb{K}) \ni (A,B)\mapsto\langle A,B\rangle \stackrel{def}{=} Tr\left(AB^*\right)\in\mathbb{K}$$

is an inner product on $\mathcal{M}_{m\times n}(\mathbb{K})$.

(ii) The map $\|\cdot\|$ given by

$$\|\cdot\| : \mathcal{M}_{m\times n}(\mathbb{K}) \ni A\mapsto\|A\| \stackrel{def}{=} \sqrt{\langle A,A\rangle}\in\mathbb{R}_+$$

is a norm on $\mathcal{M}_{m\times n}(\mathbb{K})$.

As a consequence we deduce that

$$\left\|A^n\right\|\le\|A\|^n$$

for each $A\in\mathcal{M}_{m\times m}(\mathbb{K})$ and for each $n\in\mathbb{N}$.

2. EIGENVECTORS AND EIGENVALUES

A large amount of information about a matrix or a linear operator is carried by its eigenvectors and eigenvalues. In addition, the theory of eigenvectors and eigenvalues has important applications to systems of linear recurrence relations and systems of linear differential equations. We shall describe in this section some of the basic theory.

Definition 2.1. Let A be an $n\times n$ matrix over a field of scalars $\mathbb{K}$, i.e. $\mathbb{K}=\mathbb{R}$ or $\mathbb{K}=\mathbb{C}$. An **eigenvector** of A is a non-zero n-column vector X over $\mathbb{K}$ such that

$$AX=\lambda X$$

for some $\lambda\in\mathbb{K}$. The scalar λ is then called the **eigenvalue** of A associated with the vector X.

From the Definition 2.1 it follows that the eigenvectors associated to λ, together with the zero vector, form the null space of the matrix $A-\lambda I_n$, where I_n is the $n\times n$ identity matrix. This subspace is often referred to as the **eigenspace for the eigenvalue** λ. Now, $(A-\lambda I_n)X=0$ is a linear system of n equations in n unknowns. The condition for the existence of a non-trivial solution of the system is that the coefficient matrix has zero determinant, i.e.

$$\det\left(A-\lambda I_n\right)=0.$$

Conversely, if the scalar $\lambda\in\mathbb{K}$ satisfies this equation, there exists a non-zero solution of the system, and thus λ will be an eigenvalue.

Definition 2.2. The polynomial of degree n in x given by:

$$p_A(x)=\det\left(A-xI_n\right)=\begin{vmatrix} a_{11}-x & a_{12} & \cdots & a_{1n} \\ a_{21} & a_{22}-x & \cdots & a_{2n} \\ \cdots & \cdots & \cdots & \cdots \\ a_{n1} & a_{n2} & \cdots & a_{nn}-x \end{vmatrix}$$

is called the **characteristic polynomial** of A. The equation obtained by setting

$$\det(A - xI_n) = 0 \tag{2.1}$$

is called the **characteristic equation** and its roots are called characteristic roots of A.

Thus the eigenvalues of A are the roots of the characteristic equation (2.1) which lie in the field $\mathbb{K}$.

Now, via the Fundamental Theorem of Algebra we have:

Theorem 2.1. Let A be an $n \times n$ complex matrix.

(i) The eigenvalues of A are precisely the n roots of the characteristic polynomial $p_A(x) = \det(A - xI_n)$;

(ii) The eigenvectors of A associated with an eigenvalue λ are the non-zero vectors in the null space of the matrix $A - \lambda I_n$.

It is not hard to see that the characteristic polynomial $p_A(x)$ of the matrix A can be written in the following form:

$$p_A(x) = (-x)^n + Tr(A)(-x)^{n-1} + \cdots + \det(A). \tag{2.2}$$

The rest of the coefficients in the characteristic polynomial are not so easy to describe, but they are in fact expressible as subdeterminants of det(A). For instance the term of degree n-2 in $p_A(x)$ is just $(-x)^{n-2}$ times the sum of all the 2×2 determinants of the form

$$\begin{vmatrix} a_{ii} & a_{ij} \\ a_{ji} & a_{jj} \end{vmatrix}$$

where $i<j$.

A formula which is useful, but mainly for theoretical rather than computational purposes, is Leverrier's algorithm, which states that the inverse of the characteristic matrix is

$$(A - \lambda I_n)^{-1} = \frac{\lambda^{n-1} I_n + \lambda^{n-2} B_1 + \lambda^{n-3} B_2 + \cdots + B_{n-1}}{p_A(\lambda)},$$

where the k_i and B_i are determined successively by

$$B_1 = A + k_1 I_n,\ B_i = AB_{i-1} + k_i I_n,\ i = 2,3,\cdots,n-1,$$

$$k_1 = -Tr(A),\ k_i = -\frac{1}{i} Tr(AB_{i-1}),\ i = 2,3,\cdots,n.$$

The following is the **Cayley-Hamilton theorem**:

Theorem 2.2. Let $A \in \mathcal{M}_{n\times n}(\mathbb{K})$ be an $n \times n$ matrix over the field $\mathbb{K}$ and $p_A(x) = \det(A - xI_n)$ its characteristic polynomial. Then $p_A(A) = 0$, i.e.

$$(-1)^n A^n + (-1)^{n-1} Tr(A) A^{n-1} + \cdots + \det(A) I_n = 0.$$

Let us introduce now the concept of **similar matrices** which will play a fundamental role in all that follows.

Definition 2.3. Two matrices A and B in $\mathcal{M}_{n\times n}(\mathbb{K})$ are said to be **similar** if there exists an invertible n×n matrix S such that

$$B = SAS^{-1}.$$

Proposition 2.1. Similar matrices have the same characteristic polynomial. Hence they have the same eigenvalues, trace and determinant.

Proof. The characteristic polynomial of $B = SAS^{-1}$ is

$$\det\left(SAS^{-1} - xI_n\right) = \det\left(S\left(A - xI_n\right)S^{-1}\right) = \det S \cdot \det\left(A - xI_n\right)\det\left(S^{-1}\right)$$
$$= \det S \cdot \det\left(A - xI_n\right)\left(\det\left(S\right)\right)^{-1} = \det\left(A - xI_n\right).$$

The statements about trace and determinant now follow from the relation (2.2).

Q.E.D.

Remark 2.1. In general the similar matrices do not have the same eigenvectors. Indeed, the condition for X to be an eigenvector of SAS^{-1} with eigenvalue λ is

$$\left(SAS^{-1}\right)X = \lambda X$$

which is equivalent to

$$AS^{-1}X = \lambda S^{-1}X.$$

Thus X is an eigenvector of SAS^{-1} iff $S^{-1}X$ is an eigenvector of A.

In view of the close relationship between square matrices and linear operators on finite-dimensional vector spaces, it is not hard to see that one can also define eigenvectors and eigenvalues for a linear operator.

Let $T : V \to V$ be a linear operator on a vector space V over the field $\mathbb{K}$. An eigenvector of T is a non-zero vector u of V such that

$$T\left(u\right) = \lambda u$$

for some scalar $\lambda \in \mathbb{K}$; here λ is the eigenvalue of T associated with the eigenvector u.

Suppose now that V is a finite-dimensional vector space over $\mathbb{K}$ with dimension n. Choose an ordered basis for V, say $\hat{e}$. Then with respect to this ordered basis, T is represented by an $n \times n$ matrix over $\mathbb{K}$, say $\left[T\right]_{\hat{e}}$; this means that

$$\left[T\left(u\right)\right]_{\hat{e}} = \left[T\right]_{\hat{e}} \cdot \left[u\right]_{\hat{e}}.$$

The condition $T(u) = \lambda u$ becomes then

$$\left[T\right]_{\hat{e}}\left[u\right]_{\hat{e}} = \lambda\left[u\right]_{\hat{e}},$$

which is just the condition for $\left[u\right]_{\hat{e}}$ to be an eigenvector of the matrix $\left[T\right]_{\hat{e}}$. The eigenvalues of T and $\left[T\right]_{\hat{e}}$ are the same.

If the ordered basis of V is changed, the effect is to replace $\left[T\right]_{\hat{e}}$ by a similar matrix. Of course any such matrix will still have the same eigenvalues as T via Proposition 2.1.

All of these remarks allow us to carry over to linear operators concepts such as characteristic polynomial and trace, which were introduced for matrices.

Now we can prove some density theorems. More precisely we have:

Theorem 2.3. The set $\mathcal{S}_1$ of all matrices $A \in \mathcal{M}_{n \times n}\left(\mathbb{R}\right)$ that have n distinct eigenvalues is dense in $\mathcal{M}_{n \times n}\left(\mathbb{R}\right)$.

Proof. It is known that each $A \in \mathcal{M}_{n \times n}\left(\mathbb{R}\right)$ can be written as the sum of two matrices

$$A = S + N,$$

where

$$S = \begin{bmatrix} \lambda_1 & & & & & \\ & \ddots & & & & \\ & & \lambda_r & & & \\ & & & D_1 & & \\ & & & & \ddots & \\ & & & & & D_s \end{bmatrix},$$

$$D_k = \begin{bmatrix} a_k & -b_k \\ b_k & a_k \end{bmatrix}$$

and

$$N = \begin{bmatrix} 0 & & & & & & \\ 1 & \ddots & & & & & \\ & \ddots & & & & & \\ & & 1 & 0 & & & \\ & & & I_2 & O_2 & & \\ & & & & \ddots & \ddots & \\ & & & & & I_2 & O_2 \end{bmatrix},$$

$$I_2 = \begin{bmatrix} 1 & 0 \\ 0 & 1 \end{bmatrix}, \; O_2 = \begin{bmatrix} 0 & 0 \\ 0 & 0 \end{bmatrix}.$$

The characteristic roots of A are $\lambda_1, \cdots, \lambda_r$ and $a_1 \pm ib_1, \cdots, a_s \pm ib_s$ (see Remark 3.2).

Given $\varepsilon > 0$, let $\lambda_1', \cdots, \lambda_r'$, $a_1', \cdots, a_s'$ be distinct real numbers such that

$$\left|\lambda_j' - \lambda_j\right| < \varepsilon \text{ and } \left|a_k' - a_k\right| < \varepsilon .$$

Set

$$D_k' = \begin{bmatrix} a_k' & -b_k \\ b_k & a_k' \end{bmatrix}, \; S' = \begin{bmatrix} \lambda_1' & & & & & \\ & \ddots & & & & \\ & & \lambda_r' & & & \\ & & & D_1' & & \\ & & & & \ddots & \\ & & & & & D_s' \end{bmatrix}$$

and $A' = S' + N$. Then $\|A' - A\| \le \varepsilon$, and the characteristic roots of A' are the n distinct numbers

$$\lambda_1', \cdots, \lambda_r', \; a_1' \pm ib_1, \cdots, a_s' \pm ib_s .$$

(N is obviously nilpotent). This proves that $\mathcal{S}_1$ is dense.

Q.E.D.

Remark 2.2. Theorem 2.3 has a nice extension at the level of Laplace-Beltrami operators on closed Riemannian manifolds - see for details Chapter 3, §4.

Theorem 2.4. The set

$$\mathcal{S}_2 = \{A \in \mathcal{M}_{n\times n}(\mathbb{R}) | \text{all characteristic roots of } A \text{ have non-zero real part}\}$$

is dense in $\mathcal{M}_{n\times n}(\mathbb{R})$.

Proof. In the proof of density of $\mathcal{S}_1$ in Theorem 2.3, we can take the numbers $\lambda_1', \cdots, \lambda_r'$, $a_1', \cdots, a_s'$ (the real parts of eigenvalues of A') to be non-zero; this proves density.

Q.E.D.

We shall finish this section with some remarks on the continuity of eigenvalues of a complex matrix with respect to its entries.

For each complex number λ and real number $\varepsilon > 0$, we let $B_\varepsilon(\lambda)$ [resp. $S_\varepsilon(\lambda)$] be the open disk [resp. circle] of radius ε centered at λ.

Theorem 2.5. Let $f(z) = z^n + a_1 z^{n-1} + \cdots + a_n$ be a polynomial of degree n and complex coefficients having distinct roots $\lambda_1, \ldots, \lambda_q$ with multiplicities $n_1, \ldots, n_q$, respectively, such that $n_1 + \cdots + n_q = n$. Given any small enough $\varepsilon > 0$ there exists a $\delta > 0$ so that, if

$$g(z) = z^n + b_1 z^{n-1} + \cdots + b_n, \ |a_j - b_j| < \delta, \ j = 1, 2, \ldots, n, \tag{2.3}$$

then g has precisely n_k roots in $B_\varepsilon(\lambda_k)$ for each $k = 1, 2, \ldots, q$.

Proof. We assume without loss of generality that ε is smaller than all differences $\frac{|\lambda_k - \lambda_l|}{2}$ for all $k \neq l$; and denote $S_k \overset{def}{=} S_\varepsilon(\lambda_k)$ for each k. These circles are disjoint and the interiors contain exactly n_k roots of f counted with multiplicities, thus from residue theorem one has

$$n_k = \frac{1}{2\pi i} \int_{S_k} \frac{f'(z)}{f(z)} dz$$

(the integral evaluated counterclockwise for each S_k).

Consider the polynomial g in (2.3) as a function of $b = (b_1, \ldots, b_n)$ and z. As such it is continuous, and the same holds for $\frac{g'(b,z)}{g(b,z)}$ provided that $g(b,z) \neq 0$. By uniform continuity, when z is on the closure of the union of the disks $B_\varepsilon(\lambda_k)$ and b is near $a = (a_1, \ldots, a_n)$, it follows that $g(z) \neq 0$ on every S_k and

$$\left| \frac{f'(z)}{f(z)} - \frac{g'(z)}{g(z)} \right| < \frac{1}{\varepsilon}$$

when g is as in (2.3) and δ is sufficiently small. But then

$$\frac{1}{2\pi i}\int_{S_k}\frac{g'(z)}{g(z)}dz - n_k < 1,$$

which implies that this integral - which is an integer that counts the number of zeros of g in $B_\varepsilon(\lambda_k)$ - is also equal to n_k, and the conclusion of the theorem follows.

A more elementary proof can be obtained following an idea of Uherka and Sergot [1977]. An extensive study of results on the analysis of polynomials and their derivatives, as well as properties of their zeros can be found in G.V. Milovanović, D.S. Mitrinović and Th.M. Rassias [1994]. Consider the following two n-th degree polynomials, one having roots λ_i and the other having roots λ_i^*:

$$p(x) = x^n + a_1x^{n-1} + \cdots + a_n = (x-\lambda_1)\cdots(x-\lambda_n),$$
$$p^*(x) = x^n + a_1^*x^{n-1} + \cdots + a_n^* = (x-\lambda_1^*)\cdots(x-\lambda_n^*).$$

Now, the continuous dependence of the roots of a polynomial on its coefficients is a consequence of the following assertion: Suppose λ^* is a root of p^* of multiplicity m and $\varepsilon > 0$. Then for $|a_i - a_i^*|$ sufficiently small $(i = 1,2,\ldots,n)$ p has exactly m roots within ε of λ^*. For the proof let us remind three facts which are immediate consequences of elementary algebra and limit theorems:

(i) If λ is a root of $p(x)$, then $|\lambda| \le \max\left(1, \sum_{i=1}^{n}|a_i|\right)$.

(ii) If a_i approaches a_i^* $(i = 1,2,\ldots,n)$, then $p(x) \to p^*(x)$ for all real or complex numbers x.

(iii) If λ_i approaches λ_i^* $(i = 1,2,\ldots,n)$, then $p(x) \to p^*(x)$ for all x.

We shall give the proof only for the first property. If $|\lambda| < 1$ then all is clear. Let us suppose that $|\lambda| > 1$. Then we have successively:

$$\lambda^n + a_1\lambda^{n-1} + \cdots + a_n = 0,$$
$$\lambda^n = -a_1\lambda^{n-1} - \cdots - a_n,$$
$$\lambda = -a_1 - \cdots - \frac{a_n}{\lambda^{n-1}},$$
$$|\lambda| = \left|a_1 + \cdots + \frac{a_n}{\lambda^{n-1}}\right| \le |a_1| + \cdots + \left|\frac{a_n}{\lambda^{n-1}}\right| \le |a_1| + \cdots + |a_n|$$

as required.

Now we can pass to the proof of our assertion. Suppose that it is not true. Then there exists a sequence $(p_k)_{k\in\mathbb{N}}$ of polynomials, whose coefficients converge to those coefficients of p^*, such that p_k has fewer than m roots within ε of λ^*. Since the coefficients of (p_k) converge they are bounded. Hence, if

$$p_k(x) = \left(x - \lambda_1^{(k)}\right)\cdots\left(x - \lambda_n^{(k)}\right)$$

then by (i), the root sequence $\left(\lambda_1^{(k)},\ldots,\lambda_n^{(k)}\right)$ is bounded (in the n-dimensional complex space with Euclidean distance) and has a subsequence convergent to some point $(\mu_1,\ldots,\mu_n)$ with most $m-1$ of the μ_i being equal λ^*.

If the corresponding sequence of polynomials is $\left(p_{k_i}\right)$ then by (ii), $p_{k_i}(x)\to p^*(x)$ for all x. But by (iii), $p_{k_i}(x)\to(x-\mu_1)\cdots(x-\mu_n)$ which cannot be $p^*(x)$ since at most $m-1$ of the μ_i equal λ^*. This contradiction finishes the proof of the assertion.

Let us finish with the remark that the real roots of a polynomial may not depend continuously on the coefficients, as is shown by the example:

$$p(x)=\left(x^2+2x+c_1\right)\left(x^2+4x+c_2\right)$$

in a (real or complex) neighborhood of $c_1=1, c_2=4$.

Q.E.D.

We now list some useful consequences:

Corollary 2.1. Let n be any fixed positive integer. For each $a=(a_1,\ldots,a_n)$, let

$$r_1(a)\le r_2(a)\le\cdots\le r_n(a)$$

be the real parts of the zeros of $P_a(z)=z^n+a_1z^{n-1}+\cdots+a_n$ arranged in non-decreasing order. Then, as a function $\mathbb{C}^n\to\mathbb{R}$, each r_i is continuous.

Proof. Fix any a and let $r_i\overset{def}{=}r_i(a)$ for each i. Let $\lambda_1,\ldots,\lambda_q$ be the distinct zeros of P_a with multiplicities $n_1,\ldots,n_q$, respectively, and arranged so that their real parts are non-decreasing. Pick any

$$\varepsilon<\min\left\{\frac{\left|\operatorname{Re}\lambda_i-\operatorname{Re}\lambda_j\right|}{2}\,\middle|\,\operatorname{Re}\lambda_i\ne\operatorname{Re}\lambda_j\right\}$$

and find δ as in Theorem 2.5. We claim that

$$\left|r_i(a)-r_i(b)\right|<\varepsilon$$

for each i whenever all $\left|a_i-b_i\right|<\delta$,which will prove continuity.

Fix any $i=1,\ldots,q$, and observe that

$$r_{n_{i-1}+1}(a)=\cdots=r_{n_i}(a)=\operatorname{Re}\lambda_i \tag{2.4}$$

(denoting $n_0=0$). For b close to a, the polynomial P_b has exactly n_i roots at distance less than ε from λ_i, so the real parts of these n_i zeros are at distance less than ε from (2.4).

Assume that $\operatorname{Re}\lambda_i>\operatorname{Re}\lambda_j$ for $j=1,\ldots,\alpha,\alpha<i$, and that $\operatorname{Re}\lambda_i<\operatorname{Re}\lambda_j$ with $j=\beta,\ldots,q$, for some $\beta>i$. From the choice of ε and the fact that there are at distance less than ε from $r_1(a),\ldots,r_{n_\alpha}(a)$, it follows that

$$r_1(b),\ldots,r_{n_\alpha}(b)$$

are at distance larger than ε from $r_{n_i}(a)$ and the same is true for $r_j(b)$, $j \geq \beta$. Thus the zeros at ε-distance from (2.4) are precisely $r_{n_\alpha+1}(b),\ldots,r_{n_\beta-1}(b)$, and in particular

$$\left|r_j(b)-r_j(a)\right|<\varepsilon$$

for $j=n_{i-1}+1,\ldots,n_i$. This shows that r_j is continuous, for each such j. Since i was arbitrary, the result holds for all r_j's.

Q.E.D.

Corollary 2.2. For each positive integer n, there exist n continuous functions:

$$\lambda_1,\ldots,\lambda_n:\mathbb{C}^{n\times n}\to\mathbb{R}$$

such that $\lambda_1(A)\leq\cdots\leq\lambda_n(A)$ for all A, and if A is Hermitian (see Definition 4.1 of this Appendix), then $\lambda_1(A),\ldots,\lambda_n(A)$ are the eigenvalues of A.

Proof. Just apply the previous corollary to the characteristic polynomial and use the fact that the eigenvalues are all real when the matrix is Hermitian (see Theorem 4.1 of this Appendix).

Q.E.D.

Let us mention also the following result which has a lot of applications in control theory.

Theorem 2.6. Let A and B be two $n\times n$ complex matrices, and consider the linear operator

$$L_{A,B}:X\in\mathcal{M}_{n\times n}(\mathbb{C})\mapsto AX-XB^T\in\mathcal{M}_{n\times n}(\mathbb{C}).$$

Then each eigenvalue of $L_{A,B}$ is of the form $\lambda-\mu$, for some eigenvalues λ and μ of A and B, respectively.

Proof. First consider the case of matrices A and B for which all of the possible n^2-differences

$$\lambda_i-\mu_j$$

are distinct, where $\lambda_1,\ldots,\lambda_n$ and $\mu_1,\ldots,\mu_n$ are, respectively, the eigenvalues of A and B. In particular, each of A and B has distinct eigenvalues. In this case there exist two bases $(u_1,\ldots,u_n)$ and $(w_1,\ldots,w_n)$ of eigenvectors such that

$$Au_i=\lambda_i u_i,\ Bw_i=\mu_i w_i,\ i=1,\ldots,n.$$

Let $X_{ij}=u_i w_j^T$ for each pair i,j. We claim that these are eigenvectors of $L_{A,B}$ with corresponding eigenvalues $\lambda_i-\mu_j$. Indeed,

$$AX_{ij}-X_{ij}B^T=Au_iw_j^T-u_i\left(Bw_j\right)^T=\lambda_iu_iw_j^T-u_i\left(\mu_jw_j\right)^T=\left(\lambda_i-\mu_j\right)X_{ij}.$$

Since there are n^2 such eigenvalue/eigenvector pairs, every eigenvalue has the required form.

We now use continuity of eigenvalues in order to obtain the general result. Given any two matrices A,B, assume that there would be some eigenvalue σ of $L_{A,B}$ that is not of the form $\lambda-\mu$. Let $\varepsilon>0$ be so that

$$\left|\sigma-(\lambda-\mu)\right|>\varepsilon$$

for all eigenvalues λ and μ of A and B respectively. Pick $\delta>0$ so that, if $\left\|\tilde{A}-A\right\|<\delta$ and $\left\|\tilde{B}-B\right\|<\delta$, then all eigenvalues of $\tilde{A}$ [resp. $\tilde{B}$, $L_{\tilde{A},\tilde{B}}$] are at distance less than $\frac{\varepsilon}{3}$ from eigenvalues of A [resp. B, $L_{A,B}$]. Theorem 2.5 is used here; note that any fixed matrix representation of $L_{A,B}$ is continuous on A,B.

It is always possible to find a pair $\left(\tilde{A},\tilde{B}\right)$ so that $\left\|\tilde{A}-A\right\|<\delta$ and $\left\|\tilde{B}-B\right\|<\delta$ and all of the n^2 differences $\tilde{\lambda}_i-\tilde{\lambda}_j$ are distinct (Just perturb the Jordan forms for A and B - see the next section). Using one such pair and applying the conclusion in the special case proved first, there must exist some eigenvalues $\tilde{\lambda}$ and $\tilde{\mu}$ of $\tilde{A}$ and $\tilde{B}$, respectively, so that

$$\left|\sigma-\left(\tilde{\lambda}-\tilde{\mu}\right)\right|<\frac{\varepsilon}{3}$$

while at the same time $\left|\bar{\lambda}-\lambda\right|<\frac{\varepsilon}{3}$ and $\left|\bar{\mu}-\mu\right|<\frac{\varepsilon}{3}$ for some eigenvalues of A and B, respectively. This contradicts (2.5).

Q.E.D.

3. DIAGONALIZABLE MATRICES. TRIANGULARIZABLE MATRICES. JORDAN CANONICAL FORM

Let A be a square matrix over the field $\mathbb{K}$, i.e. the elements of A are complex or real numbers.

Definition 3.1. The matrix A is said to be **diagonalizable** over $\mathbb{K}$ if it is similar to a diagonal matrix D over $\mathbb{K}$, that is, there is an invertible matrix S over $\mathbb{K}$ such that $A=SDS^{-1}$ or equivalently $D=S^{-1}AS$. One also says that S **diagonalizes** A.

Proposition 3.1. Let A be an $n\times n$ matrix over the field $\mathbb{K}$, and let $\lambda_1,\ldots,\lambda_r$ be distinct eigenvalues of A with associated eigenvectors $X_1,\ldots,X_r$. Then $\left\{X_1,\ldots,X_r\right\}$ is a linearly independent subset of $\mathbb{K}^n$.

Proof. Assume that the statement is false; then there is a positive integer i such that $\left\{X_1,\ldots,X_i\right\}$ is linearly independent, but the addition of the vector X_{i+1} produces a linearly dependent set $\left\{X_1,\ldots,X_{i+1}\right\}$. Thus there are scalars $a_1,\ldots,a_{i+1}$, not all of them zero, such that

$$\alpha_1X_1+\alpha_2X_2+\cdots+\alpha_{i+1}X_{i+1}=0.$$

Multiply both sides of this equation by A, and use the relations $AX_j=\lambda_jX_j$ to get

$$\lambda_1\alpha_1X_1+\cdots+\lambda_{i+1}\alpha_{i+1}X_{i+1}=0.$$

On subtracting λ_{i+1} times the first equation from the second, one gets the relation

$$\left(\lambda_1-\lambda_{i+1}\right)\alpha_1X_1+\cdots+\left(\lambda_i-\lambda_{i+1}\right)\alpha_iX_i=0.$$

Since $X_1,\dots,X_i$ are linearly independent, all the coefficients $(\lambda_j-\lambda_{i+1})\alpha_j$ must vanish. But $\lambda_1,\dots,\lambda_{i+1}$ are all different, so we can conclude that $a_j=0$ for $j=1,2,\dots,i$; hence $a_{i+1}X_{i+1}=0$, and so $a_{i+1}=0$, in contradiction to the original assumption. Therefore the statement is correct.

Q.E.D.

Now we can prove the criterion for diagonalizability. More precisely we have:

Theorem 3.1. Let A be an $n\times n$ matrix over the field $\mathbb{K}$. Then A is diagonalizable if and only if A has n linearly independent eigenvectors in $\mathbb{K}^n$.

Proof. Let us suppose that A has n linearly independent eigenvectors in $\mathbb{K}^n$, say $X_1,\dots,X_n$, and that the associated eigenvalues are $\lambda_1,\dots,\lambda_n$. Define S to be the $n\times n$ matrix whose columns are the eigenvectors, i.e.

$$S=[X_1,\dots,X_n].$$

It is clear via Proposition 3.1 that S is invertible (since its columns are linearly independent). Then we have

$$\begin{aligned}AS&=[AX_1,\dots,AX_n]=[\lambda_1X_1,\dots,\lambda_nX_n]\\&=[X_1,\dots,X_n]\cdot\begin{bmatrix}\lambda_1&0&\cdots&0\\0&\lambda_2&\cdots&0\\\vdots&\vdots&\cdots&\\0&0&\cdots&\lambda_n\end{bmatrix}=SD\end{aligned}$$

or equivalently

$$D=S^{-1}AS$$

as required.

Conversely, assume that A is diagonalizable and that $S^{-1}AS=D$ is a diagonal matrix with the diagonal elements $\lambda_1,\dots,\lambda_n$. Then $AS=SD$. Hence, if X_i is the i-th column of S, then AX_i equals the i-th column of SD, which is just λ_iX_i. It follows that $X_1,\dots,X_n$ are eigenvectors of A associated with eigenvalues $\lambda_1,\dots,\lambda_n$. Since $X_1,\dots,X_n$ are columns of the invertible matrix S, they must be linearly independent. Therefore A has n linearly independent eigenvectors.

Q.E.D.

As a consequence we obtain the following

Theorem 3.2. An $n\times n$ complex (real) matrix which has n distinct eigenvalues is diagonalizable.

It is an immediate observation that if A is diagonalizable, i.e. there exists a nonsingular matrix S such that

$$S^{-1}AS=D=\begin{bmatrix}\lambda_1&&0\\&\ddots&\\0&&\lambda_n\end{bmatrix},$$

then $\lambda_1,\dots,\lambda_n$ are eigenvalues of A. Indeed, we have successively

$$\det(A - xI_n) = \det S^{-1}(A - xI_n)S = \det(S^{-1}AS - xI_n)$$
$$= \det(D - xI_n) = (\lambda_1 - x)\cdots(\lambda_n - x).$$

Hence $\lambda_1, \ldots, \lambda_n$ are eigenvalues of A.

Remark 3.1. It is easy to see that there are matrices which are not diagonalizable. For instance the matrix

$$A = \begin{bmatrix} 1 & a \\ 0 & 1 \end{bmatrix}, \ a \in \mathbb{K} \setminus \{0\},$$

is not diagonalizable. Indeed if A were diagonalizable, it would be similar to the identity matrix I_2 since both its eigenvalues are equal with 1, and $S^{-1}AS = I_2$ for some matrix S. But the last equality implies that

$$A = SI_2S^{-1} = I_2,$$

which can not be true.

Q.E.D.

Example 3.1. Let us find a matrix which diagonalizes

$$A = \begin{bmatrix} 1 & 2 \\ 1 & 1 \end{bmatrix}.$$

The matrix A has real and distinct eigenvalues namely

$$\lambda_1 = 1 + \sqrt{2}, \ \lambda_2 = 1 - \sqrt{2}.$$

Then corresponding eigenvectors are

$$X_1 = \begin{bmatrix} \sqrt{2} \\ 1 \end{bmatrix} \text{ and } X_2 = \begin{bmatrix} -\sqrt{2} \\ 1 \end{bmatrix},$$

respectively.
Hence

$$S = \begin{bmatrix} \sqrt{2} & -\sqrt{2} \\ 1 & 1 \end{bmatrix}.$$

Then

$$S^{-1}AS = \begin{bmatrix} 1+\sqrt{2} & 0 \\ 0 & 1-\sqrt{2} \end{bmatrix}.$$

We have seen that not every complex matrix is diagonalizable. For to compensate this failure we shall prove that every complex matrix is triangularizable.

Definition 3.2. A square matrix A over the field $\mathbb{K}$ is said to be **triangularizable** over $\mathbb{K}$ if there exists an invertible matrix S over $\mathbb{K}$ such that $S^{-1}AS = T$ is upper triangular. We shall say also that S **triangularizes** A.

Theorem 3.3. Every complex square matrix is triangularizable.

Proof. Let $A \in \mathcal{M}_{n\times n}(\mathbb{C})$ be an $n \times n$ complex matrix. We shall show by induction on n that A is triangularizable. Of course if $n = 1$, then A is already upper triangular.

Thus, let $n > 1$ and assume that the result is true for square matrices with $n-1$ rows.

We know that A has at least one eigenvalue λ in $\mathbb{C}$, with associated eigenvector X. Since $X \neq 0$, it is possible to add vectors to X to produce a basis of $\mathbb{C}^n$, say $X = X_1, X_2, \ldots, X_n$. Next, recall that left multiplication of the vectors of $\mathbb{C}^n$ by A gives rise to a linear operator T on $\mathbb{C}^n$. With respect to the ordered basis $(X_1, \ldots, X_n)$ the linear operator T will be represented by a matrix with the special form:

$$B_1 = \begin{bmatrix} \lambda & A_2 \\ 0 & A_1 \end{bmatrix},$$

where A_1 and A_2 are certain complex matrices, A_1 having $n-1$ rows and columns. Notice that the matrices A and B_1 are similar since they represent the same linear operator T. Suppose that in fact

$$B_1 = S_1^{-1} A S_1,$$

where S_1 is an invertible $n \times n$ matrix.

Now by the induction hypothesis there is an invertible matrix S_2 with $n-1$ rows and columns such that

$$B_2 = S_2^{-1} A_1 S_2$$

is upper triangular. Write

$$S = S_1 \begin{bmatrix} 1 & 0 \\ 0 & S_2 \end{bmatrix}.$$

This is a product of invertible matrices, thus S is invertible. Now it is easy to see that

$$\begin{aligned} S^{-1}AS &= \begin{bmatrix} 1 & 0 \\ 0 & S_2^{-1} \end{bmatrix} \left(S_1^{-1} A S_1\right) \begin{bmatrix} 1 & 0 \\ 0 & S_2 \end{bmatrix} = \begin{bmatrix} 1 & 0 \\ 0 & S_2^{-1} \end{bmatrix} B_1 \begin{bmatrix} 1 & 0 \\ 0 & S_2 \end{bmatrix} \\ &= \begin{bmatrix} 1 & 0 \\ 0 & S_2^{-1} \end{bmatrix} \begin{bmatrix} c & A_2 \\ 0 & A_1 \end{bmatrix} \begin{bmatrix} 1 & 0 \\ 0 & S_2 \end{bmatrix} = \begin{bmatrix} c & A_2 S_2 \\ 0 & S_2^{-1} A_1 S_2 \end{bmatrix} = \begin{bmatrix} c & A_2 S_2 \\ 0 & B_2 \end{bmatrix}. \end{aligned}$$

This matrix is clearly upper triangular and thus our assertion follows.

Q.E.D.

Example 3.2. Let us triangularize the matrix

$$A = \begin{bmatrix} 2 & 1 \\ -1 & 4 \end{bmatrix}$$

using Theorem 3.3.

The characteristic polynomial of A is $(x-3)^2$, so both eigenvalues of A are equal to 3. Solving the equation $(A - 3I_2)X = 0$, we find that all the eigenvectors of A are scalar multiples of $X_1 = \begin{bmatrix} 1 \\ 1 \end{bmatrix}$. Thus A is not diagonalizable. Let $T : \mathbb{C}^2 \to \mathbb{C}^2$ be the linear operator arising from left multiplication by A. Add a vector X_2 to X_1 to get a basis

$\hat{f} = (X_1, X_2)$ of $\mathbb{C}^2$, say $X_2 = \begin{bmatrix} 0 \\ 1 \end{bmatrix}$. Denote by $\hat{e}$ the standard basis of $\mathbb{C}^2$. Then the change of basis $\hat{e} \to \hat{f}$ is described by the matrix

$$S_1 = \begin{bmatrix} 1 & 0 \\ 1 & 1 \end{bmatrix}.$$

Therefore the matrix B_1 which represents T with respect to the basis $\hat{f}$ is

$$B_1 = S_1^{-1} A S_1 = \begin{bmatrix} 3 & 1 \\ 0 & 3 \end{bmatrix}.$$

Hence

$$S = S_1^{-1} = \begin{bmatrix} 1 & 0 \\ -1 & 1 \end{bmatrix}$$

triangularizes A.

Q.E.D.

Let us mention also the following result which has some very useful applications:

Theorem 3.4. (*QR*-factorization). Let A be a real $m \times n$ matrix $(m \geq n)$ with rank n. Then A can be written as a product QR, where Q is a real $m \times n$ matrix whose columns form an orthonormal set and R is a real $n \times n$ upper triangular matrix with positive entries on its principal diagonal.

Proof. Let V denote the $\mathbb{R}$-vector subspace of the Euclidean inner product space $\mathbb{R}^m$ generated by the columns of the matrix A. Since A has rank n, the n columns $X_1, \ldots, X_n$ of A are linearly independent, and thus these form a basis for V. Hence the Gram-Schmidt procedure can be applied to this basis to produce an orthonormal basis for V, say $\{Y_1, \ldots, Y_n\}$.

Now we see from the way that the Y_i in the Gram-Schmidt procedure are defined that these vectors have the form

$$\begin{aligned} Y_1 &= b_{11} X_1, \\ Y_2 &= b_{12} X_1 + b_{22} X_2, \\ &\cdots\cdots\cdots\cdots \\ Y_n &= b_{1n} X_1 + b_{2n} X_2 + \cdots + b_{nn} X_n, \end{aligned}$$

for cetain real numbers b_{ij} with b_{ii} positive. Solving the equations for $X_1, \ldots, X_n$, by backsubstitution, we get a linear system of the same general form:

$$\begin{aligned} X_1 &= r_{11} Y_1, \\ X_2 &= r_{12} Y_1 + r_{22} Y_2, \\ &\cdots\cdots\cdots\cdots \\ X_n &= r_{1n} Y_1 + r_{2n} Y_2 + \cdots + r_{nn} Y_n, \end{aligned}$$

for certain real numbers r_{ij} with r_{ii} positive again. These equations can be written in matrix form

$$A=[X_1,X_2,\ldots,X_n]=[Y_1,Y_2,\ldots,Y_n]\begin{bmatrix} r_{11} & r_{12} & \cdots & r_{1n} \\ 0 & r_{22} & \cdots & r_{2n} \\ \vdots & \cdots & \cdots & \cdots \\ 0 & 0 & \cdots & r_{nn} \end{bmatrix}.$$

The columns of the $m\times n$ matrix $Q=[Y_1,\ldots,Y_n]$ form an orthonormal set since they constitute an orthonormal basis of V, while the matrix $R=[r_{ij}]$ is plainly upper triangular.

Q.E.D.

The most important case of this theorem is when A is a non-singular square matrix. Then the matrix Q is $n\times n$ and its columns form an orthonormal set; equivalently it has the property

$$Q^T Q = I_n$$

i.e. Q is an orthogonal $n\times n$ matrix. Therefore we have:

Theorem 3.5. Every invertible real matrix A can be written as a product QR where Q is a real orthogonal matrix and R is a real upper triangular matrix with positive entries on its principal diagonal.

Moreover in this last case the decomposition is unique (see for details Francis [1961]).

Example 3.3. Let us write the following matrix in the QR-factorized form:

$$A=\begin{bmatrix} 1 & 1 & 2 \\ 1 & 2 & 3 \\ 1 & 1 & 1 \end{bmatrix}.$$

Let us apply the Gram-Schmidt procedure to the columns X_1,X_2,X_3 of A, which are linearly independent and thus form a basis for the column space of A. This yields to an orthonormal basis $\{Y_1,Y_2,Y_3\}$ where

$$Y_1=\frac{1}{\sqrt{3}}\begin{bmatrix} 1 \\ 1 \\ 1 \end{bmatrix}=\frac{1}{\sqrt{3}}X_1,$$

$$Y_2=\frac{1}{\sqrt{6}}\begin{bmatrix} -1 \\ 2 \\ -1 \end{bmatrix}=-\frac{2\sqrt{6}}{3}X_1+\frac{\sqrt{6}}{2}X_2,$$

$$Y_3=\frac{1}{\sqrt{2}}\begin{bmatrix} 1 \\ 0 \\ -1 \end{bmatrix}=\qquad -\frac{3\sqrt{2}}{2}X_2+\sqrt{2}X_3.$$

Solving for X_1,X_2,X_3, we obtain:

$$X_1=\sqrt{3}\,Y_1,$$

$$X_2=\frac{4\sqrt{3}}{3}Y_1+\frac{\sqrt{6}}{3}Y_2,$$

$$X_3 = 2\sqrt{3}\,Y_1 + \frac{\sqrt{6}}{2}Y_2 + \frac{\sqrt{2}}{2}Y_3 .$$

Therefore $A = QR$, where

$$Q = \begin{bmatrix} \frac{1}{\sqrt{3}} & -\frac{1}{\sqrt{6}} & \frac{1}{\sqrt{2}} \\ \frac{1}{\sqrt{3}} & \frac{2}{\sqrt{6}} & 0 \\ \frac{1}{\sqrt{3}} & -\frac{1}{\sqrt{6}} & -\frac{1}{\sqrt{2}} \end{bmatrix}$$

and

$$R = \begin{bmatrix} \sqrt{3} & \frac{4\sqrt{3}}{3} & 2\sqrt{3} \\ 0 & \frac{\sqrt{6}}{3} & \frac{\sqrt{6}}{2} \\ 0 & 0 & \frac{\sqrt{2}}{2} \end{bmatrix}.$$

Using now Theorem 3.5 we can present a very nice algorithm, the so called QR algorithm, due to Francis [1961], for the computation of the eigenvalues of a real invertible matrix M. Let $M_0 = M = Q_0 R_0$ be the unique factorization of the matrix M_0 into a product of an orthogonal matrix Q_0 and an upper triangular matrix R_0 with positive diagonal. Then set $M_1 = R_0 Q_0 = Q_0^T M_0 Q_0 = Q_0^{-1} M_0 Q_0$, so that M_0 and M_1 have the same spectrum. The matrix M_1 has its own QR factorization $M_1 = Q_1 R_1$ and we derive $M_2 = R_1 Q_1 = Q_1^T M_1 Q_1 = Q_1^{-1} M_1 Q_1$, etc. In this way, we obtain a sequence of isospectral matrices M_k (symmetric if M_0 is symmetric). Furthermore, under appropriate assumptions on M_0, the sequence M_k converges to an upper triangular matrix M_∞ as $k \to \infty$, and the eigenvalues of M_0 must appear along the diagonal of M_∞.

Francis' QR algorithm has the following dynamical interpretation. Let $M_{QR}(t),\ M_{QR}(0) = M_0$, be the flow generated by the Hamiltonian $H_{QR} = Tr(M_0 \ln M_0 - M_0)$ on $\mathbb{R}^{2n}$ with its canonical symplectic structure, i.e.

$$\begin{cases} \dot{M}_{QR} = \begin{bmatrix} 0 & I_n \\ -I_n & 0 \end{bmatrix} \cdot \operatorname{grad}(H), \\ M_{QR}(0) = M_0, \end{cases}$$

where "*grad*(*H*)" is the gradient of H. Then we have

$$M_{QR}(k) = M_k$$

for each $k = 1, 2, \ldots$. For the proof and other details the reader can consult Deift, Nanda and Tomei [1983].

As we have seen before an $n \times n$ matrix with entries from the field $\mathbb{K}$ can be diagonalized when its eigenvector structure is very strong; there must be a basis in $\mathbb{K}^n$ composed of eigenvectors. Now we shall present the proof of one of the most important theorems in linear algebra, namely that a matrix with entries from a field $\mathbb{K}$ has a Jordan form (a weaker property) when its eigenvector structure may not be as strong but is still sufficiently rich. We shall follow Hall [1991].

Let V be a finite-dimensional vector space over the field $\mathbb{K}$, and let $T : V \to V$ be a linear operator on V. The kernel of T, $\ker(T)$, is the (vector) subspace of all vectors of V taken to the 0-vector, 0, by T. The image of T, $\operatorname{im}(T)$, is the (vector) subspace of all vectors of V of the form $T(u)$, for some $u \in V$. Note that

$$\dim(V) = \dim(\ker(T)) + \dim(im(T)).$$

Indeed, this formula follows from the exactness of the sequence of linear maps

$$\{0\} \to \ker(T) \to V \xrightarrow{T} im(T) \to \{0\},$$

$\ker(T)$ and $\operatorname{im}(T)$ being finite-dimensional.

Definition 3.3. A **Jordan chain** for T in V of length $m \in \mathbb{N}^*$ associated with $\lambda \in \mathbb{K}$ is a set of non-zero vectors $u_1, \ldots, u_m$ such that

$$T(u_i) = \lambda u_i + u_{i-1} \quad \text{for } i = 1, \ldots, m \text{ (we set } u_0 = 0\text{)}.$$

Note that λ is an eigenvalue for T with associated eigenvector u_1. In particular, an eigenvector by itself is a Jordan chain of length 1. A Jordan chain for T associated with $\lambda \in \mathbb{K}$ is also a Jordan chain for $T + a1_V$ associated with $\lambda + a$, where 1_V is the identity linear operator on V and $\alpha \in \mathbb{K}$.

Proposition 3.2. Let $T : V \to V$ be a linear operator of the finite-dimensional $\mathbb{K}$-vector space V. Let W be a T-invariant vector subspace of V, i.e. $T(W) \subseteq W$, and let $s : W \to W$ be the restriction and corestriction of T to W. If W contains $\ker(T)$ and $\operatorname{im}(s) = \operatorname{im}(T)$, then $W = V$.

Proof. Indeed, by hypothesis, $\ker(s) = \ker(T)$, $\operatorname{im}(s) = \operatorname{im}(T)$, so that

$$\dim(V) = \dim(\ker(T)) + \dim(im(T)) = \dim(\ker(s)) + \dim(im(s)) = \dim(W).$$

Q.E.D.

Lemma 3.1. Let $T : V \to V$ be a linear operator of the finite-dimensional $\mathbb{K}$-vector space V. Assume that V has a basis which is a disjoint union of Jordan chains for T:

$$u_{1,1}, \ldots, u_{1,m_1},$$
$$u_{2,1}, \ldots, u_{2,m_2},$$
$$\vdots$$
$$u_{k,1}, \ldots, u_{k,m_k},$$

where the i-th chain $x_{i,1}, \ldots, x_{i,m_i}$ has length $m_i \in \mathbb{N}^*$ and is associated with the eigenvalue $\lambda_i \in \mathbb{K}$ of T. Assume further that $\lambda_1 = \cdots = \lambda_d = 0$ and $\lambda_i \neq 0$ for $i > d$. Then

(i) $\{u_{1,1},\ldots,u_{i,1},\ldots,u_{d,1}\}$ is a basis for $\ker(T)(1 \le i \le d)$;

(ii) The disjoint union of Jordan chains

$$\{u_{1,1},\ldots,u_{1,m_1-1}\},$$
$$\vdots$$
$$\{u_{d,1},\ldots,u_{d,m_d-1}\},$$
$$\{u_{d+1,1},\ldots,u_{d+1,m_{d+1}}\},$$
$$\vdots$$
$$\{u_{k,1},\ldots,u_{k,m_k}\}$$

is a basis for $\operatorname{im}(T)$, where in the case when some m_i are equal to $1(1 \le i \le d)$ we accept that the corresponding Jordan chain is empty.

Proof. First note that the given linearly independent sets are contained in the claimed vector subspaces, so that

$$\dim(\ker(T)) \ge d$$

and

$$\dim(im(T)) \ge \dim(V) - d .$$

On the other hand

$$\dim(\ker(T)) + \dim(im(T)) = \dim(V),$$

thus in each of these inequalities we must have equality. Consequently, the given sets are bases.

Q.E.D.

Theorem 3.6. (Existence of the Jordan form for linear operators). Let $T : V \to V$ be a linear operator of the finite-dimensional $\mathbb{K}$-vector space V. Assume that every non-zero T-invariant vector subspace of V contains an eigenvector for T. Then V possesses a basis which is a disjoint union of Jordan chains for T.

Proof. The proof can be given by induction on the dimension of V. If T is a scalar multiple of 1_V [in particular this is the case when $\dim(V) = 1$], any basis is a basis of Jordan chains of length 1. Assume that this is not the case and that the theorem is true in all dimensions less than $n = \dim(V)$.

By hypothesis T has an eigenvector $u \ne 0$ associated with some eigenvalue $\lambda \in \mathbb{K}$. Let $s = T - \lambda \mathbf{1}_V$. Note that s is a non-zero linear operator and $U = \operatorname{im}(s)$ is an s-invariant vector subspace of V. Let $r : U \to U$ be the restriction and corestriction of s to U. Since $u \in \ker(s)$, it follows that U has the dimension m strictly smaller than n. An r-invariant vector subspace of U is invariant under both s and T. Therefore by induction we may conclude that U has a basis $\mathcal{B}$ which is a disjoint union of Jordan chains for r (and thus for s and T as well). Let the chains of $\mathcal{B}$ be $\mathcal{B}_i$ for $1 \le i \le k$. Here the i-th chain $\mathcal{B}_i$, consisting of $u_{i,1},\ldots,u_{i,m_i}$, is associated with the eigenvalue $\lambda_i \in \mathbb{K}$. Assume that $\lambda_1 = \cdots = \lambda_d = 0$ and $\lambda_i \ne 0$ for $i > d$.

For each i with $1 \le i \le d$, u_{i,m_i} is in $U = \mathrm{im}(s)$, so that we may choose a vector u_{i,m_i+1} such that

$$s\left(u_{i,m_i+1}\right) = u_{i,m_i}. \tag{3.1}$$

Next we choose vectors $u_1, \ldots, u_p$ such that

$$u_1, \ldots, u_p, u_{1,1}, \ldots, u_{d,1}$$

is a basis for $\ker(s)$, which has dimension $p + d = n - m$.

Consider now the set $\mathcal{C}$ which is the union of $\mathcal{B}$, of all the u_{i,m_i+1} for $1 \le i \le d$, and of all the u_j for $1 \le j \le p$. This set contains $m + d + p = n$ vectors. We first note that $\mathcal{C}$ is a union of $k + p$ disjoint Jordan chains for s, namely the original chains $\mathcal{B}_i$ (extended by u_{i,m_i+1} when $1 \le i \le d$) and the new chains $\{u_j\}$ of length 1.

It remains only to prove that $\mathcal{C}$ is a basis for V. Let W be the vector subspace of V spanned by $\mathcal{C}$. As W contains $u_1, \ldots, u_p$ as well as $u_{1,1}, \ldots, u_{d,1}$, W contains $\ker(s)$. Since W contains the basis $\mathcal{B}$ of $U = \mathrm{im}(s)$ and also the vectors u_{i,m_i+1} for $1 \le i \le d$, having the property (3.1), it follows that W is an s-invariant vector subspace of V. Moreover, Lemma 3.1 (ii) shows that the linear operator $s : W \to W$, the restriction and corestriction of $s : V \to V$ to W, has U as its image. By Proposition 3.2, W is in fact equal to V. Thus, the set $\mathcal{C}$ of cardinality n spans the vector space V of dimension n and therefore it is a basis of V.

Q.E.D.

Lemma 3.2. Let $T : V \to V$ be a linear operator of the finite-dimensional $\mathbb{K}$-vector space V. Assume that V has a basis $\mathcal{B}$ which is a disjoint union of k Jordan chains $\mathcal{B}_j$ for T, $1 \le j \le k$. Let λ be an eigenvalue of T, and $m \in \mathbb{N}^*$. Then the number of chains $\mathcal{B}_j$, which are associated with λ and of length at least m, equals

$$\dim\left(\ker\left(T - \lambda \mathbf{1}_V\right) \cap im\left(\left(T - \lambda \mathbf{1}_V\right)^{m-1}\right)\right).$$

Proof. Let $s = T - \lambda \mathbf{1}_V$, so that Jordan chains for T associated with λ are Jordan chains for s associated with 0.

For all $i \ge 0$, $\mathrm{im}\left(s^i\right)$ is an s-invariant vector subspace of V. From Lemma 3.1 (i) we note that to get from a set of Jordan chains for s in $\mathrm{im}\left(s^i\right)$ to a set in $\mathrm{im}\left(s^{i+1}\right)$ we need only drop the last member of each chain associated with 0. Thus to get from a set of Jordan chains in $\mathrm{im}\left(s^0\right) = V$ to a set in $\mathrm{im}\left(s^{m-1}\right)$, we drop the last $m-1$ members of each Jordan chain associated with 0. A Jordan chain associated with 0 whose initial length was less than m will disappear entirely, but a Jordan chain of length at least m will survive in part. By Lemma 3.1 (i), the number of surviving Jordan chains for s associated with 0 in $\mathrm{im}\left(s^{m-1}\right)$ is exactly

$$\dim\left(\ker(s) \cap \mathrm{im}\left(s^{m-1}\right)\right).$$

Q.E.D.

Corollary 3.1. (Uniqueness of the Jordan form). Let $T : V \to V$ be a linear operator of the finite-dimensional $\mathbb{K}$-vector space V. Assume that V has bases $\mathcal{B}^{(1)}$ and $\mathcal{B}^{(2)}$, each $\mathcal{B}^{(i)}$ (with $i = 1, 2$) a disjoint union of k_i Jordan chains $\mathcal{B}_j^{(1)}$ for T, $1 \le j \le k_i$. Let $\lambda \in \mathbb{K}$ be an eigenvalue of T and let $m \in \mathbb{N}^*$. Then the number of chains $\mathcal{B}_j^{(1)}$, which are associated with λ and of length equal to m, equals the number of chains $\mathcal{B}_j^{(2)}$ which are associated with λ and of length equal to m.

Proof. Let $s = T - \lambda \mathbf{1}_V$. By Lemma 3.2 both numbers equal

$$\dim\left(\ker(s) \cap \operatorname{im}\left(s^{m-1}\right)\right) - \dim\left(\ker(s) \cap \operatorname{im}\left(s^{m}\right)\right).$$

Q.E.D.

If $\lambda \in \mathbb{K}$ and $m \in \mathbb{N}^*$, let $J_m(\lambda)$ be the $m \times m$ matrix over $\mathbb{K}$ with λ's on the diagonal, 1's on the super-diagonal, and 0's elsewhere, i.e.

$$J_m(\lambda) = \begin{bmatrix} \lambda & 1 & 0 & 0 & \cdots & 0 & 0 \\ 0 & \lambda & 1 & 0 & \cdots & 0 & 0 \\ \cdot & \cdot & \cdot & \cdot & \cdots & \cdot & \cdot \\ 0 & 0 & 0 & 0 & \cdots & \lambda & 1 \\ 0 & 0 & 0 & 0 & \cdots & 0 & \lambda \end{bmatrix}.$$

A proof by induction leads immediately to the following

Proposition 3.3. If $J_m(\lambda)$ is the $m \times m$ matrix described before, then for each $r \in \mathbb{N}^*$ we have:

$$\left[J_m(\lambda)\right]^r = \begin{bmatrix} \lambda^r & \binom{r}{1}\lambda^{r-1} & \cdots & \binom{r}{m-1}\lambda^{r-m+1} \\ 0 & \lambda^r & \cdots & \binom{r}{m-2}\lambda^{r-m+2} \\ \cdot & \cdot & \cdots & \cdot \\ 0 & 0 & \cdots & \lambda^r \end{bmatrix},$$

where $\binom{r}{j}$ denotes the binomial coefficient $\dfrac{r!}{j!(r-j)!}$.

Definition 3.4. A matrix N with entries from $\mathbb{K}$ is said to be in Jordan form if it is a block diagonal matrix, with all diagonal blocks equal to $J_m(\lambda)$ for various $m \in \mathbb{N}^*$ and $\lambda \in \mathbb{K}$, i.e.

$$N = \begin{bmatrix} J_{m_1}(\lambda_1) & 0 & \cdots & 0 \\ 0 & J_{m_2}(\lambda_2) & \cdots & 0 \\ \vdots & \vdots & \ddots & \vdots \\ 0 & 0 & \cdots & J_{m_k}(\lambda_k) \end{bmatrix}.$$

Lemma 3.3. Let A be an $n \times n$ matrix with entries from the field $\mathbb{K}$. Assume that the characteristic polynomial of A factors into first degree factors over $\mathbb{K}$. Let $V = \mathbb{K}^n$ be the n-dimensional $\mathbb{K}$-vector space of column vectors and $T : V \to V$ the linear operator naturally associated with the matrix A. Then any non-zero T-invariant vector subspace W of V contains an eigenvector.

Proof. Let $\{w_1, \ldots, w_m, u_{m+1}, \ldots, u_n\}$ be a basis for V containing the basis $\{w_1, \ldots, w_m\}$ for W. Define Q to be the matrix whose i-th column is w_i when $1 \le i \le m$ and is u_i when $m+1 \le i \le n$. Then $B = Q^{-1}AQ$ is block upper triangular with an $(n-m) \times m$ block of 0's in the lower left. The characteristic polynomial of B is the product of that of the two diagonal blocks and is equal to that of A. In particular, the upper left $m \times m$ block has a characteristic polynomial which splits into first degree factors over $\mathbb{K}$. Therefore B has an eigenvector in V of the form

$$X = (\alpha_1, \ldots, \alpha_m, 0, \ldots, 0)^T,$$

that is there exists $\lambda \in \mathbb{K}$ such that $BX = \lambda X$. Thus,

$$A(QX) = (AQ)X = Q(BX) = \lambda QX,$$

which means that QX is an eigenvector of A corresponding to the eigenvalue λ, or equivalently,

$$w = \alpha_1 w_1 + \cdots + \alpha_m w_m$$

is an eigenvector of T belonging to W.

Q.E.D.

Theorem 3.7. (Existence of the Jordan form for matrices). Let A be an $n \times n$ matrix with entries from the field $\mathbb{K}$. If the characteristic polynomial of A factors into first degree factors over $\mathbb{K}$, then there is an invertible $n \times n$ matrix P with entries from $\mathbb{K}$ such that $P^{-1}AP$ is in Jordan form.

Proof. Let $V = \mathbb{K}^n$ be the n-dimensional $\mathbb{K}$-vector space of column vectors and $T : V \to V$ the linear operator naturally associated with the matrix A, that is with respect to the standard basis for $\mathbb{K}^n$. By Lemma 3.3 and Theorem 3.6, V has a basis $\mathcal{B}$ which is a disjoint union of Jordan chains $u_{i,1}, \ldots, u_{i,m_i}$, $1 \le i \le k$, for T. Let P be the $n \times n$ matrix with columns

$$u_{1,1}, \ldots, u_{1,m_1}, \ldots, u_{k,1}, \ldots, u_{k,m_k}.$$

Then $P^{-1}AP$, the matrix of T with respect to $\mathcal{B}$, is in Jordan form.

Q.E.D.

Corollary 3.2. Every square complex matrix is similar to a matrix in Jordan form.

Remark 3.2. If A is an $n \times n$ real matrix having as characteristic roots the real numbers $\lambda_1, \ldots, \lambda_k$ and the complex numbers $a_1 + ib_1, \ldots, a_p + ib_p$ with the multiplicities $m_1, \ldots, m_k, n_1, \ldots n_p$ respectively, i.e. $m_1 + \cdots + m_k + 2(n_1 + \cdots + n_p) = n$, then it can be proved that A is similar with a matrix of the following type:

$$\begin{bmatrix} \left(J_{m_1}(\lambda_1)\right)^T & & & & & & & & 0 \\ & \ddots & & & & & & & \\ 0 & & \left(J_{m_k}(\lambda_k)\right)^T & & & & & & \\ \cdots & \cdots & \cdots & \cdots & \cdots & \cdots & \cdots & \cdots & \cdots \\ & & & I_2 & D_1 & & & & \\ & & & & \ddots & \ddots & & & 0 \\ & & & & & \ddots & D_1 & & \\ & & 0 & & & & \ddots & \ddots & \\ & & & & & & & \ddots & D_p \\ & & & & & & & \ddots & \ddots \\ & & & & & & & I_2 & D_p \end{bmatrix},$$

where $D_l = \begin{bmatrix} a_l & -b_l \\ b_l & a_l \end{bmatrix}$, $b_l > 0,\ l = 1,2,\ldots,p,\ I_2 = \begin{bmatrix} 1 & 0 \\ 0 & 1 \end{bmatrix}$.

Moreover each block D_l appears in the matrix as many times as the multiplicity of the root $a_l + ib_l$ indicates. Such a matrix is uniquely determined up to the orders of the blocks (see for details Hirsch and Smale [1974]).

4. EIGENVALUES AND EIGENVECTORS OF REAL SYMMETRIC AND HERMITIAN MATRICES

In this section we shall continue the study of the spectral properties of square matrices with special regard to Hermitian matrices and real symmetric matrices.

Definition 4.1. A complex matrix A is called **Hermitian** if

$$A = A^*.$$

Thus Hermitian matrices are the complex analogues of real symmetric matrices.

Theorem 4.1. Let A be a Hermitian or real symmetric matrix. Then

(i) the eigenvalues of A are all real;

(ii) eigenvectors of A associated with distinct eigenvalues are orthogonal.

Proof. (i) Let λ be an eigenvalue of A with associated eigenvector X, so that $AX = \lambda X$. Taking the complex transpose of both sides of this equation we obtain

$$X^* A = \bar{\lambda} X^*,$$

since $A = A^*$. Now multiply both sides of this equation on the right by X to obtain

$$X^* AX = \bar{\lambda} X^* X = \bar{\lambda} \|X\|^2.$$

However,

$$\left(X^* AX\right)^* = X^* A^* X = X^* AX,$$

so that the scalar $X^* AX$ equals its complex conjugate, and thus it is real. It follows that $\bar{\lambda}\|X\|^2$ is real and so $\bar{\lambda}$ hence λ is real, which completes the proof of (i).

(ii) To prove (ii), take two eigenvectors associated with distinct eigenvalues λ and μ. Thus

$$AX = \lambda X \text{ and } AY = \mu Y.$$

Then

$$Y^* AX = Y^* (\lambda X) = \lambda Y^* X$$

and in the same way

$$X^* AY = \mu X^* Y.$$

On the other hand,

$$(X^* AY)^* = Y^* A^* X = Y^* AX$$

and thus

$$(\mu X^* Y)^* = Y^* AX = Y^* \lambda X = \lambda Y^* X$$

or equivalently

$$\mu Y^* X = \lambda Y^* X$$

because $\mu \in \mathbb{R}$ by the first part of the theorem. This means that

$$(\mu - \lambda) Y^* X = 0,$$

from which it follows that $Y^* X = 0$, since $\lambda \neq \mu$. Thus X and Y are orthogonal.

Q.E.D.

Definition 4.2. A complex matrix U is called **unitary** if

$$U^* = U^{-1},$$

where

$$U^* = (\bar{U})^T.$$

Theorem 4.2. (Schur's theorem). Let A be an arbitrary square complex matrix. Then there is a unitary matrix U such that $U^* AU$ is upper triangular. Moreover, if A is a real symmetric matrix, then U can be chosen to be real and orthogonal.

Proof. Let A be an $n \times n$ complex matrix. The proof is by induction on n. Of course if $n = 1$, then A is already upper triangular. Let $n > 1$. There is an eigenvector X_1 of A, with associated eigenvalue λ_1. We can choose X_1 to be a unit vector in $\mathbb{C}^n$. We can now add vectors to X_1 to form a basis for $\mathbb{C}^n$. Then the Gram-Schmidt procedure (in the complex case) may be applied to produce an orthonormal basis $\{X_1, \dots, X_n\}$ for $\mathbb{C}^n$. Note that X_1 will be an element in this basis.

Let U_0 denote the matrix $[X_1, \dots, X_n]$. Then U_0 is unitary since its columns form an orthonormal basis. Now

$$U_0^* AX_1 = U_0^* (\lambda_1 X_1) = \lambda_1 U_0^* X_1.$$

Also

$$X_i^* X_1 = 0$$

if $i > 1$ while

$$X_1^* X_1 = 1.$$

Hence

$$U_0^* A X_1 = \lambda_1 \begin{bmatrix} X_1^* X_1 \\ \vdots \\ X_n^* X_1 \end{bmatrix} = \begin{bmatrix} \lambda_1 \\ 0 \\ \vdots \\ 0 \end{bmatrix}.$$

Since

$$U_0^* A U_0 = U_0^* A [X_1, \ldots, X_n] = [U_0^* A X_1, \ldots, U_0^* A X_n]$$

we deduce that

$$U_0^* A U_0 = \begin{bmatrix} \lambda_1 & B \\ 0 & A_1 \end{bmatrix},$$

where A_1 is a matrix with $n-1$ rows and columns and B is an $(n-1)$-row vector.

Now, we apply the induction hypothesis on n; there is a unitary matrix U_1 such that $U_1^* A_1 U_1 = T_1$ is upper triangular. Write

$$U_2 = \begin{bmatrix} 1 & 0 \\ 0 & U_1 \end{bmatrix},$$

which is obviously a unitary matrix. If we set now

$$U = U_0 U_2$$

this is also a unitary matrix, since

$$U^* U = U_2^* U_0^* U_0 U_2 = U_2^* U_2 = I.$$

Finally,

$$U^* A U = U_2^* \left(U_0^* A U_0 \right) U_2 = U_2^* \begin{bmatrix} \lambda_1 & B \\ 0 & A_1 \end{bmatrix} U_2 = \begin{bmatrix} c_1 & BU_1 \\ 0 & U_1^* A_1 U_1 \end{bmatrix},$$

which shows that

$$U^* A U = \begin{bmatrix} c_1 & BU_1 \\ 0 & T_1 \end{bmatrix}$$

is an upper triangular matrix as required.

Q.E.D.

Now we can prove the following

Theorem 4.3. (Spectral theorem). Let A be a Hermitian matrix. Then there is a unitary matrix U such that $U^* AU$ is diagonal. If A is a real symmetric matrix, then U

may be chosen to be real and orthogonal.

Proof. From Theorem 4.2, there is a unitary matrix U such that

$$U^*AU = T$$

is upper triangular. Then

$$T^* = U^*A^*U = U^*AU = T$$

and therefore T is Hermitian. But T is upper triangular and T^* is lower triangular. Hence they are equal if and only if all the off-diagonal entries of T are zero, that is T is diagonal.

The real case can be deduced in a similar manner.

Q.E.D.

As a consequence we obtain:

Theorem 4.4. If A is an $n\times n$ Hermitian matrix, there is an orthonormal basis for $\mathbb{C}^n$ which consists entirely of eigenvectors of A. If in addition A is real, there is an orthonormal basis for $\mathbb{R}^n$ consisting of eigenvectors of A.

Example 4.1. Let us find a real orthogonal matrix which diagonalizes the matrix

$$A = \begin{bmatrix} 1 & 3 \\ 3 & 1 \end{bmatrix}.$$

The eigenvalues of A are 4 and -2, and corresponding eigenvectors are $\begin{bmatrix} 1 \\ 1 \end{bmatrix}$ and $\begin{bmatrix} -1 \\ 1 \end{bmatrix}$. These vectors are obviously orthogonal. To get an orthonormal basis of $\mathbb{R}^2$, replace them by the unit vectors

$$\frac{1}{\sqrt{2}}\begin{bmatrix} 1 \\ 1 \end{bmatrix} \text{ and } \frac{1}{\sqrt{2}}\begin{bmatrix} -1 \\ 1 \end{bmatrix}, \text{ respectively,}$$

Finally, let

$$S = \frac{1}{\sqrt{2}}\begin{bmatrix} 1 & -1 \\ 1 & 1 \end{bmatrix},$$

which is an orthogonal matrix. An easy computation shows that

$$S^TAS = \begin{bmatrix} 4 & 0 \\ 0 & -2 \end{bmatrix}.$$

Q.E.D.

Example 4.2. Let us find a unitary matrix which diagonalizes the Hermitian matrix

$$A = \begin{bmatrix} \frac{3}{2} & \frac{i}{2} & 0 \\ -\frac{i}{2} & \frac{3}{2} & 0 \\ 0 & 0 & 1 \end{bmatrix}.$$

The eigenvalues of A are 1,2,1 and the vectors

$$\begin{bmatrix} -\frac{i}{\sqrt{2}} \\ \frac{1}{\sqrt{2}} \\ 0 \end{bmatrix}, \begin{bmatrix} \frac{1}{\sqrt{2}} \\ -\frac{i}{\sqrt{2}} \\ 0 \end{bmatrix}, \begin{bmatrix} 0 \\ 0 \\ 1 \end{bmatrix}$$

are associated unit eigenvectors, respectively.
Therefore

$$U^* AU = \begin{bmatrix} 1 & 0 & 0 \\ 0 & 2 & 0 \\ 0 & 0 & 1 \end{bmatrix},$$

where U is the unitary matrix

$$U = \frac{1}{\sqrt{2}} \begin{bmatrix} -i & 1 & 0 \\ 1 & -i & 0 \\ 0 & 0 & \sqrt{2} \end{bmatrix}.$$

Q.E.D.

Let us mention also the following result.

Theorem 4.5. (Sylvester's law of inertia). Let A be a real symmetric $n \times n$ matrix and S an invertible $n \times n$ matrix. Then A and $S^T AS$ have the same numbers of positive, negative and zero eigenvalues.

Proof. Assume first of all that A is invertible; this is the essential case. Recall that by Theorem 3.5 it is possible to express S in the form QR, where Q is real orthogonal and $R = \left[r_{ij}\right]$ is real upper triangular with positive diagonal entries.

Following Robinson [1991] the idea of proof is to obtain a continuous chain of matrices leading from S to the orthogonal matrix Q; the point of this fact is that

$$Q^T AQ = Q^{-1} AQ$$

certainly has the same eigenvalues as A. Define

$$S(t) = tQ + (1-t)S$$

where $0 \leq t \leq 1$. Thus $S(0) = S$ while $S(1) = Q$. Now write

$$U(t) = tI_n + (1-t)R,$$

so that

$$S(t) = Q \cdot U(t).$$

Next $U(t)$ is an upper triangular matrix and its diagonal entries are $t+(1-t)r_{ii}$; these cannot be zero since $r_{ii} > 0$ and $0 \le t \le 1$. Hence $U(t)$ is invertible, while Q is invertible since it is orthogonal. It follows that $S(t)$ is invertible, thus $\det(S(t)) \neq 0$.

Now consider

$$A(t) = S(t)^T A S(t).$$

Since

$$\det(A(t)) = \det(A) \cdot \det(S(t))^2 \neq 0$$

it follows that $A(t)$ cannot have zero eigenvalues. Now as t takes values from 0 to 1, the eigenvalues of $A(0) = S^T AS$ gradually change to those of $A(1) = Q^T AQ$, that is, to those of A. But in the process no eigenvalue can change sign because the eigenvalues that appear are continuous functions of t (see Corollary 2.1), and they are never zero. Consequently the numbers of positive and negative eigenvalues of $S^T AS$ equal those of A.

Finally if A is singular, then we shall consider the real symmetric matrix $A+\varepsilon I_n$, which may be thought as a perturbation of A. Now $A+\varepsilon I_n$ will be invertible provided that ε is sufficiently small and positive; for $\det(A+xI_n)$ is a polynomial of x of degree n, and so it vanishes in at most n values of x. The previous argument shows that the result is true for $A+\varepsilon I_n$ if ε is small and positive; then by taking the limit as $\varepsilon \to 0$ we can deduce the result for A.

Q.E.D.

Let $T : \mathbb{R}^n \to \mathbb{R}^n$ be a linear symmetric operator,

$$\lambda_1 \le \lambda_2 \le \cdots \le \lambda_n$$

its eigenvalues and $u_1,\ldots,u_n$ the corresponding non-zero eigenvectors. From the spectral theorem we can suppose that $\{u_1,\ldots,u_n\}$ is an orthonormal basis of $\mathbb{R}^n$.

Theorem 4.6. (Rayleigh-Ritz). The following equalities hold:

(i) $\lambda_1 = \inf\left\{ \dfrac{\langle T(u),u\rangle}{\langle u,u\rangle} \middle| u \in \mathbb{R}^n \setminus \{0\} \right\}$.

(ii) $\lambda_k = \inf\left\{ \dfrac{\langle T(u),u\rangle}{\langle u,u\rangle} \middle| u \in \mathbb{R}^n \setminus \{0\}, \langle u,u_1\rangle = \cdots = \langle u,u_{k-1}\rangle = 0 \right\}$, where $\langle \cdot,\cdot \rangle$ is the canonical inner product on $\mathbb{R}^n$.

Proof. (i). For each $u = \sum_{k=1}^{n} a_k u_k \in \mathbb{R}^n \setminus \{0\}$ we have:

$$\frac{\langle T(u),u\rangle}{\langle u,u\rangle}=\frac{\langle \lambda_1 a_1 u_1+\cdots+\lambda_n a_n u_n, a_1 u_1+\cdots+a_n u_n\rangle}{\langle a_1 u_1+\cdots+a_n u_n, a_1 u_1+\cdots+a_n u_n\rangle}=\frac{\sum_{k=1}^{n}\lambda_k a_k^2}{\sum_{k=1}^{n} a_k^2}.$$

Using now the hypothesis, we have successively:

$$\frac{\langle T(u),u\rangle}{\langle u,u\rangle}\geq\frac{\lambda_1 a_1^2+\cdots+\lambda_1 a_n^2}{\sum_{k=1}^{n} a_k^2}=\lambda_1\frac{\sum_{k=1}^{n} a_k^2}{\sum_{k=1}^{n} a_k^2}=\lambda_1.$$

Hence

$$\lambda_1\leq\frac{\langle T(u),u\rangle}{\langle u,u\rangle}.$$

On the other hand, if $u\neq 0$ is an element of the eigenspace generated by u_1, then we have

$$\frac{\langle T(u),u\rangle}{\langle u,u\rangle}=\frac{\lambda_1\langle u,u\rangle}{\langle u,u\rangle}=\lambda_1.$$

Therefore

$$\lambda_1=\inf\left\{\frac{\langle T(u),u\rangle}{\langle u,u\rangle}\middle|u\in\mathbb{R}^n\setminus\{0\}\right\},$$

as required.

The second equality can be obtained in a similar manner.

Q.E.D.

Remark 4.1. It is not hard to see that the equalities:

(i) $\lambda_1=\dfrac{\langle T(u),u\rangle}{\langle u,u\rangle},\quad u\neq 0;$

(ii) $\lambda_k=\dfrac{\langle T(u),u\rangle}{\langle u,u\rangle},\quad u\neq 0,\ \langle u,u_1\rangle=\cdots=\langle u,u_{k-1}\rangle=0,\ k=2,\ldots,n,$

hold if and only if u is an eigenvector of T associated to λ_1, respectively to λ_k for k=2,3,...,n (see also the considerations below).

Q.E.D.

Remark 4.2. The quotient $\dfrac{\langle T(u),u\rangle}{\langle u,u\rangle}$, $u\neq 0$, that is the scalar component of $T(u)$ along u, is usually called **Rayleigh-Ritz quotient of** u with respect to linear operator T, and we have refined it in the spectral theory of the Laplace-Beltrami and Hodge-de Rham operators on closed Riemannian manifolds (see Chapters 3 and 5).

Remark 4.3. If $T:\mathbb{R}^n\to\mathbb{R}^n$ is a linear operator, the function

$R:\mathbb{R}^n\setminus\{0\}\to\mathbb{R}$ given by the Rayleigh-Ritz quotient $R(u)=\dfrac{\langle T(u),u\rangle}{\langle u,u\rangle}$, $u\neq 0$, attains maximum and minimum values. Indeed, R is continuous and since $R(\alpha u)=R(u)$ for $\alpha\in\mathbb{R}^*$, R attains all its values on the closed unit ball of $\mathbb{R}^n$, which is a compact set.

Moreover, R is a differentiable function on $\mathbb{R}^n\setminus\{0\}$ and the gradient, grad(R), of R is given by

$$(\text{grad}(R))(x)=\frac{1}{\langle x,x\rangle}\left[T(x)+T^*(x)-2R(x)x\right],\ x\in\mathbb{R}^n\setminus\{0\}, \tag{4.1}$$

where T^* is the adjoint of T, that is the linear operator of $\mathbb{R}^n$ determined uniquely by the requirement that

$$\langle T^*(u),v\rangle=\langle u,T(v)\rangle \text{ for all } u,v\in\mathbb{R}^n.$$

The proof follows by a straightforward calculation. Indeed, if the functions $f_1,f_2:\mathbb{R}^n\setminus\{0\}\to\mathbb{R}$ are differentiable, and $f_2(x)\neq 0$ for all $x\in\mathbb{R}^n\setminus\{0\}$, then

$$\text{grad}\left(\frac{f_1}{f_2}\right)=\frac{1}{f_2^2}\left[f_2\,\text{grad}(f_1)-f_1\,\text{grad}(f_2)\right]. \tag{4.2}$$

On the other hand, in the particular case when

$$f_1(x)=\langle T(x),x\rangle, f_2(x)=\langle x,x\rangle, x\in\mathbb{R}^n\setminus\{0\},$$

one obtains

$$\langle(\text{grad}(f_1))(x),u\rangle=(df_1)_x(u)=\langle T(u),x\rangle+\langle T(x),u\rangle=\langle(T+T^*)(x),u\rangle$$

for all $x\in\mathbb{R}^n\setminus\{0\}, u\in\mathbb{R}^n$, which implies that

$$\text{grad}(f_1)=T+T^*$$

and, by taking $T=\mathbf{1}_{\mathbb{R}^n}$, one has

$$\text{grad}(f_2)=2\cdot\mathbf{1}_{\mathbb{R}^n},$$

so that in this special case, (4.2) leads to (4.1).

The previous considerations show that if the linear operator T is symmetric, i.e. $T^*=T$, the critical points of the function R (be Remark 4.3 such points exist - for example the maxima and minima points of R) are exactly the non-zero eigenvectors of T, and the critical values of R are its eigenvalues. Indeed, in virtue of (4.1), if $x\in\mathbb{R}^n\setminus\{0\}$ is a critical point of R, then $T(x)=R(x)x$. Conversely, if $T(u)=\lambda u$, $\lambda\in\mathbb{R}$, for some $u\in\mathbb{R}^n\setminus\{0\}$, then $R(u)=\lambda$ and by (4.1) u is a critical point of R. Note also that this argument showing that $R(u)=\lambda$ for some $u\in\mathbb{R}^n\setminus\{0\}$ implies u a critical point of R proves the first statement in Remark 4.1.

We wish to finish this section with the following result.

Theorem 4.7. Let $T:\mathbb{R}^n \to \mathbb{R}^n$ be a linear symmetric operator and $\hat{e}=(e_1,\ldots,e_n)$ an orthonormal basis of $\mathbb{R}^n$ with respect to which the matrix of T has a diagonal form, i.e.

$$[T]_{\hat{e}} = \begin{bmatrix} \lambda_1 & & \\ O & \ddots & O \\ & & \lambda_n \end{bmatrix}.$$

Then we have:

$$\|T\| = \max\{|\lambda_1|,\ldots,|\lambda_n|\}.$$

Proof. Let $|\lambda_k| = \max\{|\lambda_1|,\ldots,|\lambda_n|\}$. First recall that, by definition

$$\|T\| = \sup_{\|x\|=1} \|T(x)\|.$$

Then we can write successively:

$$\|T\| = \sup_{\|x\|=1} \|T(x_1 e_1 + \cdots + x_n e_n)\| = \sup_{\|x\|=1} \left\| [\lambda_1 x_1,\ldots,\lambda_n x_n]^T \right\|$$
$$= \sup_{\|x\|=1} \sqrt{\lambda_1^2 x_1^2 + \cdots + \lambda_n^2 x_n^2} \le \sup_{\|x\|=1} |\lambda_k| \|x\| = |\lambda_k|.$$

Hence

$$\|T\| \le |\lambda_k|.$$

On the other hand,

$$\|T(e_k)\| = |\lambda_k e_k| = |\lambda_k|$$

and then our assertion follows.

Q.E.D.

References

DEIFT, P., NANDA, T. and TOMEI, C., *Ordinary differential equations and the symmetric eigenvalue problem*, Siam J. Numer. Anal., 1(1983), 1-22.

FRANCIS, J., *The QR transformation, a unitary analogue to the LR transformation*, I, Comput. J., 4(1961), 265-271.

HALL, J.I., *Another elementary approach to the Jordan form*, Amer. Math. Monthly, 98(4)(1991), 336-340.

HIRSCH, M. and SMALE, S., *Differential Equations, Dynamical Systems and Linear Algebra*, Academic Press, New York, 1974.

HOFFMAN, K. and KUNZE, R., *Linear Algebra*, 2nd ed., Prentice-Hall, Englewood Cliffs, N.J., 1971.

MILOVANOVIĆ, G.V., MITRINOVIĆ, D.S. and RASSIAS Th.M., *Topics in Polynomials: Extremal Problems, Inequalities, Zeros*, World Scientific, Singapore, New Jersey, London, Hong Kong, 1994.

ROBINSON, D.J.S., *A Course in Linear Algebra with Applications*, World Scientific, Singapore, New Jersey, London, Hong Kong, 1991.

RUDOLPH, LEE, *The structure of orthogonal transformations*, Amer. Math. Monthly, 98(4)(1991), 349-352.

UHERKA, D.J. and SERGOTT, Ann M., *On the continuous dependence of the roots of a polynomial on its coefficients*, Amer. Math. Monthly, 84(5) (1977), 368-370.

Subject Index